Energy, Electrical and Power Engineering 2021–2022

Energy, Electrical and Power Engineering 2021–2022

Editors

Cungang Hu
Pinjia Zhang
Wen-Ping Cao
Xiaoyan Huang

MDPI

Basel • Beijing • Wuhan • Barcelona • Belgrade • Novi Sad • Cluj • Manchester

Editors
Cungang Hu
School of Electrical
Engineering and Automation
Anhui University
Hefei, China

Pinjia Zhang
Department of
Electrical Engineering
Tsinghua University
Beijing, China

Wen-Ping Cao
School of Electrical
Engineering and Automation
Anhui University
Hefei, China

Xiaoyan Huang
Department of
Electrical Machine
Zhejiang University
Hangzhou, China

Editorial Office
MDPI
St. Alban-Anlage 66
4052 Basel, Switzerland

This is a reprint of articles from the Special Issue published online in the open access journal *Energies* (ISSN 1996-1073) (available at: https://www.mdpi.com/journal/energies/special_issues/Energy_Electrical_Power).

For citation purposes, cite each article independently as indicated on the article page online and as indicated below:

Lastname, A.A.; Lastname, B.B. Article Title. *Journal Name* **Year**, *Volume Number*, Page Range.

ISBN 978-3-0365-9913-7 (Hbk)
ISBN 978-3-0365-9914-4 (PDF)
doi.org/10.3390/books978-3-0365-9914-4

Contents

energies

MDPI

Article

Power Flow Optimization and Economic Analysis Based on High Voltage Phase Shifting Transformer

Mengze Yu [1], Jiaxin Yuan [2], Zuohong Li [1], Feng Li [1], Xinyi Yang [2], Weizhe Zhang [2,*], Shunkai Xu [2] and Jiajun Mei [2]

[1] Grid Planning & Research Center, Guangdong Power Grid Corporation, CSG, Shuijungang No. 8, Dongfeng East Road, Guangzhou 510080, China; yumengze@gd.csg.cn (M.Y.); lizuohong@gd.csg.cn (Z.L.); lifeng@gd.csg.cn (F.L.)

[2] School of Electrical and Automation, Wuhan University, No. 299, Bayi Street, Wuchang District, Wuhan 430072, China; yuanjiaxin@whu.edu.cn (J.Y.); 2017302540064@whu.edu.cn (X.Y.); flash_xsk191@whu.edu.cn (S.X.); may2323@whu.edu.cn (J.M.)

* Correspondence: 2017302540053@whu.edu.cn; Tel.: +86-180-9683-9722

Abstract: With the development of power systems, the power flow problem of transmission line is becoming more and more prominent. This paper presents a power flow regulation method based on phase shifting transformer (PST). Firstly, the working principle and performance of PST are analyzed. Then, the simulation model of BPA multi node system is established. PST access reduces the line imbalance to less than 8%. On this basis, considering the influence of saturation effect and leakage reactance, a PST suitable for 220 kV power grid is designed. Three different working conditions are simulated by PSCAD software. Under normal working conditions and N-1 conditions, PST can increase the transmission limit of the ring network section by more than 20%. When a short-circuit fault occurs, PST can also suppress the fault current. For 220 kV practical projects, unified power flow controller (UPFC) has faster response speed and stronger performance, but PST is more economical and reliable, and the equipment cost is 49.86% lower than that of UPFC. The power flow regulation method based on PST has good steady-state effect. It can improve the utilization efficiency of power grid assets and optimize the power flow distribution in a short time and at low cost.

Keywords: phase shifting transformer; leakage reactance; transmission line; power flow control

Citation: Yu, M.; Yuan, J.; Li, Z.; Li, F.; Yang, X.; Zhang, W.; Xu, S.; Mei, J. Power Flow Optimization and Economic Analysis Based on High Voltage Phase Shifting Transformer. *Energies* **2022**, *15*, 2363. https://doi.org/10.3390/en15072363

Academic Editor: Pawel Rozga

Received: 23 February 2022
Accepted: 21 March 2022
Published: 24 March 2022

1. Background

With the development and wide application of new energy, power systems are becoming more and more complex. Transmission lines are characterized by long distance, high voltage, and large capacity [1,2]. The grid connection of new energy leads to the decline of power system reliability, and transmission lines may have overload, circulation, and other problems. How to regulate and optimize the power flow distribution and enhance the reliability of power system has become a hot issue. In the past, the above problems were solved by controlling the operation mode of power system, such as adjusting transformer tap, changing the operation mode of generator set, and switching compensation device [3]. These methods have some limitations and cannot flexibly regulate the power flow.

The flexible AC Transmission Technology (FACTS) provides a method for flexible regulation of power flow. FACTS power flow controllers are mainly divided into two types: power electronic type and electromagnetic type [4]. The power electronic flow controller is based on power electronic equipment, and its typical representative is the unified power flow controller (UPFC). The UPFC can independently regulate the active and reactive power and provide voltage support for the system. In addition, with rapid developments of new energy, the demands for energy storage have been intensively increased [5,6], the DC part of UPFC can install energy storage devices to provide active power support [7], which is more suitable for a new energy power system. The regulating speed of UPFC is very fast, which can quickly respond to the voltage drop on the grid side to improve the low voltage ride through (LVRT) ability [8]. The UPFC has many advantages, but the cost is too high to be applied [9]. The phase shifting transformer (PST) is a power flow controller based on a transformer, which has a simple structure, low cost, and high reliability. The application of some new materials and technologies has effectively improved the reliability of transformers [10,11].

References [12–15] summarize the engineering application of PST, and PST on-load voltage regulating switches can be divided into mechanical type and thyristor type. In current engineering applications, the mechanical on-load voltage regulating switch is more widely used, and studies on the thyristor PST are mostly on theory and simulation. References [16–18] describes the research status of PST and analyzes the advantages and disadvantages of different types of PST and their application scenarios: The thyristor PST has a faster response speed and is suitable for situations requiring a fast dynamic response and a small capacity, but its cost is much higher than that of a mechanical PST. The mechanical switch is suitable for the occasions where the response speed is not high, and the capacity is large. Reference [19] improved the structure of the traditional mechanical PST and proposed a new electromagnetic unified power flow controller (EUPFC), which has a larger adjustment range and a faster response speed than the traditional PST. However, this structure is mostly used for power grids below 110 kV, and its insulation cost will increase significantly when it is applied to high-voltage power grids above 220 kV. Reference [20] analyzed the feasibility of double core symmetric PST to improve steady-state power flow in actual power grid and conducted a simulation study on the control characteristics of PST during switching operation.

PST regulates the line power flow by injecting a compensating voltage. When its voltage regulating switch works at different gears, the equivalent impedance value will also change, affecting the size of the phase shift angle. Therefore, this paper firstly analyzes the influence of PST internal impedance on the phase shift angle regarding the principle aspect. Then, based on BPA and PSCAD software, the simulation study of PST power flow regulation on the multi-node system and actual 220 kV ring network is carried out, and the PST power flow regulation effect under different working conditions is analyzed. Then, combined with the practical engineering application, the economic analysis of PST and UPFC is carried out. Finally, the validity of the theory and simulation is verified by prototype experiment, which provides reference for the implementation of PST engineering in a high voltage power grid.

2. Theoretical Research and Selection of PST

2.1. Working Principle of PST

In transmission line, there is voltage drop and phase angle difference at the beginning and end of the line. When there is no control or regulation device, the power flow of parallel transmission line and ring network is distributed according to impedance. As shown in Figure 1, when the PST is installed in the high-voltage transmission line, it can be equivalent to the ideal transformer and impedance in series.

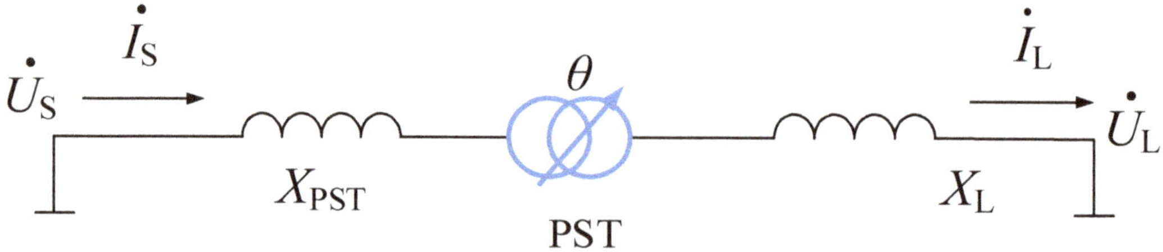

Figure 1. Equivalent model of PST in transmission line.

In Figure 1, U_S and U_L are the voltage amplitude at the head and end of the line. X_L and X_{PST} are the equivalent impedance of line and PST, respectively. δ_S and δ_L are defined as phase angle at the beginning and end of the line, respectively.

Since the equivalent reactance of the transmission line is much greater than its equivalent resistance, its loss can be ignored. After PST is connected, the active power flow of the line is [21,22]:

$$P = \frac{U_S U_L}{X_L + X_{PST}} \sin(\delta_S - \delta_L \pm \theta) \quad (1)$$

It can be concluded from Equation (1) that PST will add a phase angle of $\pm\theta$ to the active power transmitted by the line. When the phase shift angle of PST output is positive, the phase difference between the head and end of the line is reduced, so as to reduce the line active power flow. On the contrary, if the PST output phase shift angle is negative, the line transmission power can be increased.

In this paper, PST is proposed to solve the power flow problem of high-voltage power grid. The steps and work of this method are shown in Figure 2. It is generally divided into the research process of theoretical analysis, simulation calculation, and experimental verification. In topology selection and parameter calculation, the actual power grid demand should be combined, and the influence of leakage reactance and saturation effect should be considered.

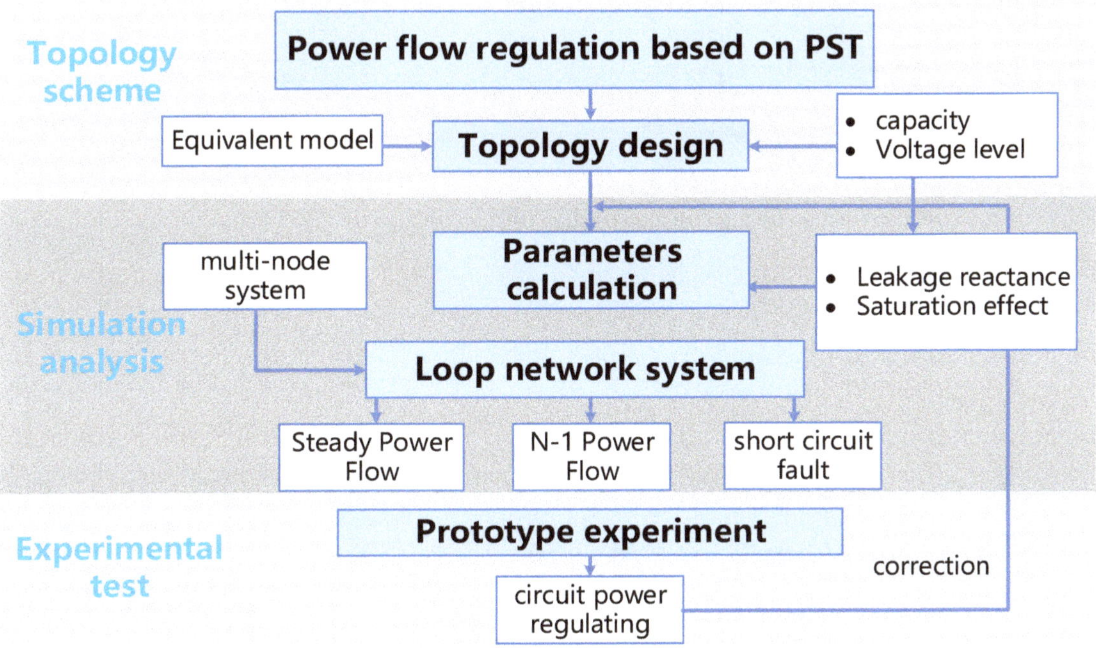

Figure 2. Workflow of the load flow optimization with PST.

2.2. Topology Selection of PST

The uneven distribution of 220 kV transmission lines in Guangdong has led to a serious reduction in the utilization rate of 220 kV transmission lines. The power flow regulation technology based on high voltage PST can transfer the overload line power flow to the light load line and increase the power transmission margin of Huizhou station.

PST can be classified from different angles. According to the output compensation voltage regulation characteristics, it can be sorted into longitudinal regulation, transverse regulation, and oblique regulation [23]. According to the output voltage characteristics of the transformer, it can be sorted into symmetrical type and asymmetric type. According to the structure of transformer body, it can be sorted into single core type and double core type [24]. The topologies of various types of PST are shown in Figure 3.

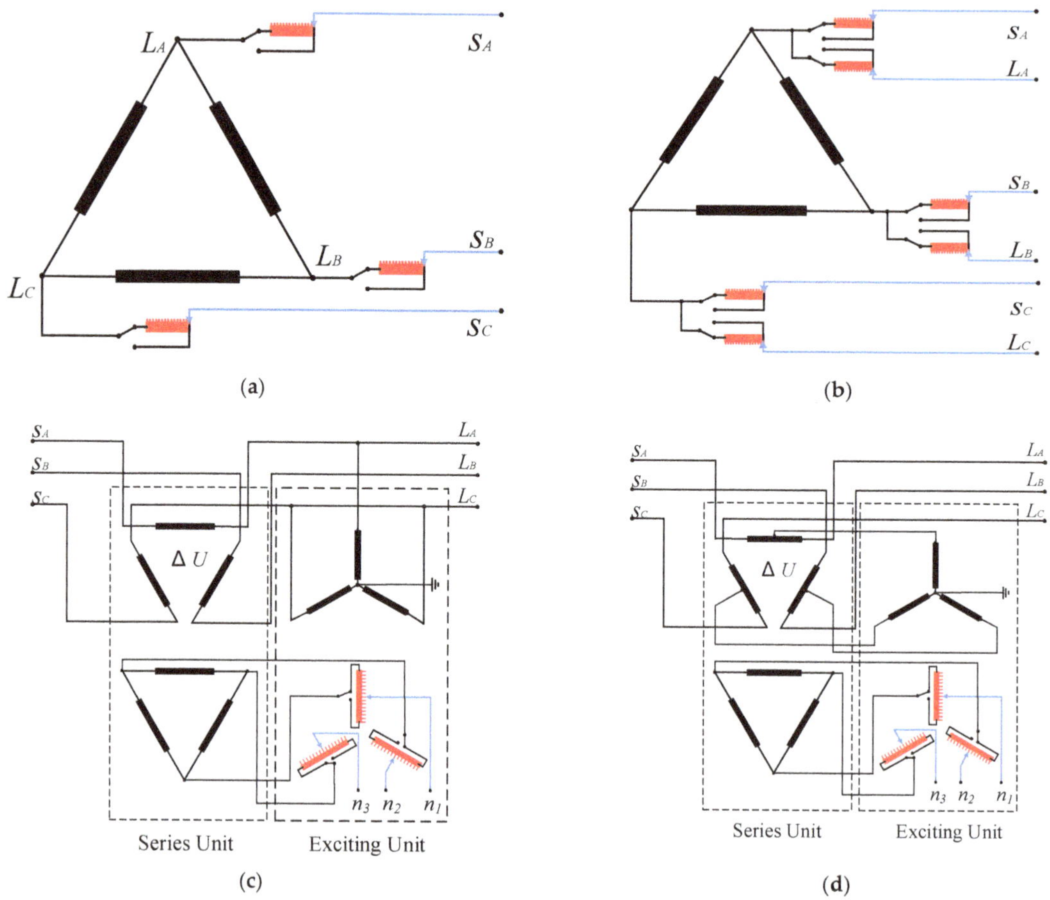

Figure 3. Various types of PST topology: (**a**) Single core asymmetric type, (**b**) Single core symmetrical type, (**c**) Two core asymmetric type, and (**d**) Double core symmetrical type.

2.2.1. Selection of Transformer Structure

The single-core PST is realized by wiring the windings of a transformer. The double core PST is realized by connecting a series transformer and a parallel transformer (excitation transformer) to the system. The performance pairs of single-core and double core PST are shown in Table 1. Single-core PST has a simple structure, and its on-load tap switch is directly connected to the line, so it has high insulation requirements and is mainly applied to 110 kV and below power systems [25]. The double core PST is suitable for high voltage levels. The voltage level of a ring network system in Guangdong is 220 kV, so it is recommended to choose the double core PST.

Table 1. Comparison of single core PST and double core PST.

Performance	Single Core PST	Double Core PST
Structure	Simple	Complex
On-load tap switch	High insulation requirements	Low insulation requirements
Short circuit impedance	<10%	12~19%
Applicable voltage class	Low	High
Cost	Low	High
The phase shift range	Narrow	Wide

2.2.2. Electrical Characteristic Selection

The asymmetric PST changes the voltage amplitude and phase angle simultaneously. Symmetric PST ensures that the voltage amplitude is unchanged before and after compensation and only changes its phase angle [26,27]. The asymmetric and symmetric PST voltage phasor is shown in Figure 4.

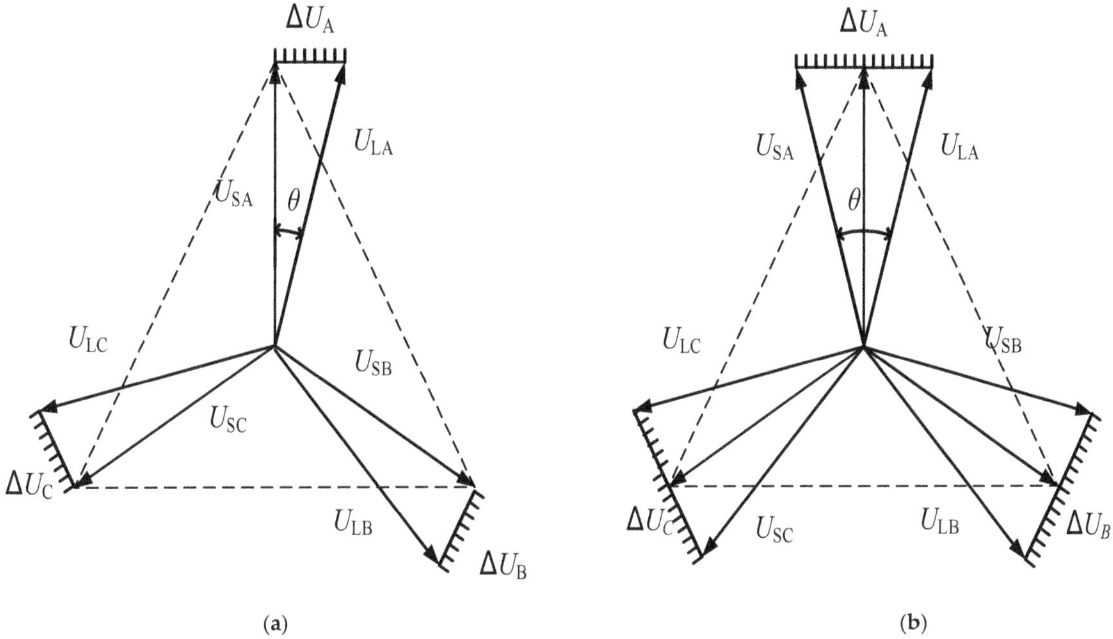

Figure 4. Voltage phasor of PST. (**a**) The asymmetric PST, (**b**) Symmetric PST.

In Figure 4, U_{SA}, U_{SB}, and U_{SC} are the three-phase voltages at the first end of the transmission line, respectively. ΔU_A, ΔU_B, and ΔU_C are the series compensation voltage injected into the line by PST; U_{LA}, U_{LB}, and U_{LC} are the compensated three-phase voltage of the line, respectively.

As can be seen from Figure 3, when the voltages are the same, the output phase shift angle of symmetrical PST is larger. Considering the power flow regulation requirements of the new power system, the phase shift angle required by the 220 kV high-voltage ring network system shall be greater than 20°, and the line voltage amplitude shall be kept within the specified range. Generally, a symmetrical PST is selected. Based on the above analysis, it is recommended that 220 kV ring network PST adopt a dual core symmetrical structure.

2.3. *Equivalent Model of Double Core Symmetric PST*

2.3.1. Double Core Symmetric PST Topology

The double core symmetrical PST includes two independent transformers on the magnetic circuit structure, namely, series transformer BT and excitation transformer ET. The connection mode is shown in Figure 5.

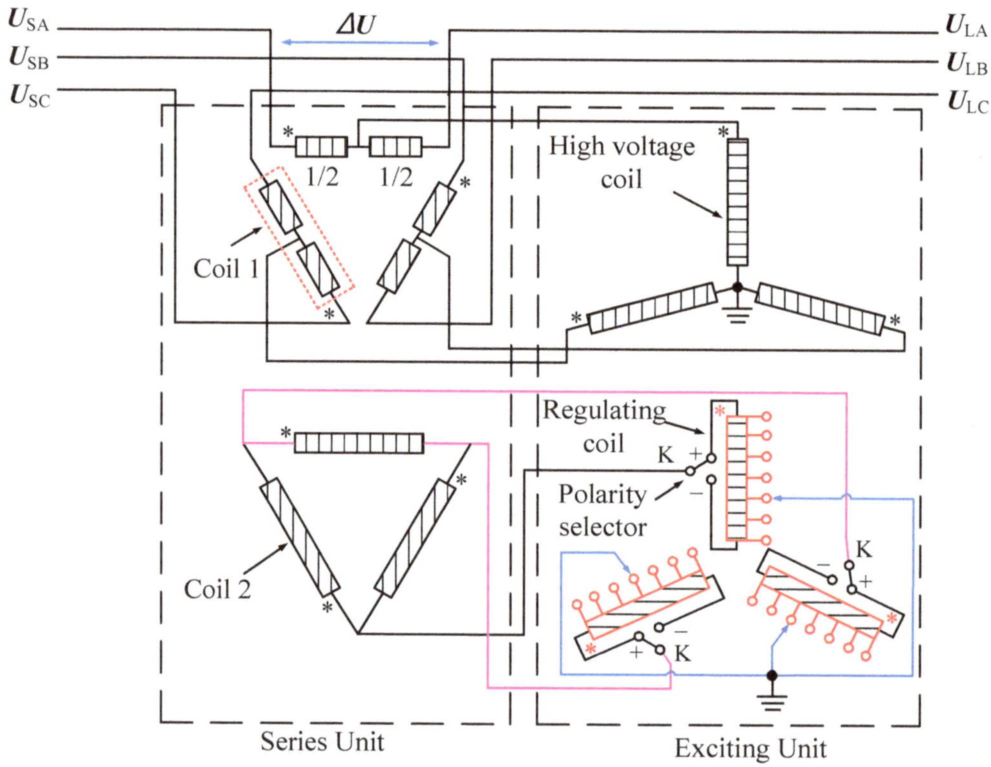

Figure 5. Topological structure of double core symmetrical.

The series transformer includes coil 1 and coil 2. The excitation transformer includes two parts: high voltage coil and voltage regulating coil. The high voltage coil connected in "Y" mode in the excitation transformer provides excitation for the voltage regulating coil. The voltage in the voltage regulating coil can be changed by adjusting the switch gear of the voltage regulating coil (a polarity selector is installed to realize the leading/lagging phase adjustment of the PST after switching the positive and negative poles). Voltage regulating coil and series transformer coil 2 are connected by electrical connection to change the voltage of series transformer coil 2, thus generating phase angle adjustment voltage ΔU in coil 1 to achieve voltage phase angle change between the power supply side and the load side. When rated capacity and voltage grade are low, two transformer bodies can be placed in the same box.

2.3.2. Load Equivalent Model

Under load conditions, the output phase shift angle θ of PST is determined jointly by its no-load phase shift angle α and internal phase shift angle β, which affects not only the structure and winding size of transformer but also the selection of tap switch. The primary and secondary sides of ET are star-shaped wiring, and the midpoint of BT primary side winding leads to the primary side of ET, so the first and secondary sides are divided into two parts with equal equivalent impedance [28]. The voltage in the voltage regulating coil can be changed by adjusting the tap position of the on-load voltage regulating coil on the secondary side of ET so as to generate compensation voltage on the primary side of BT and realize the voltage phase shift between the power side and the load side of the line. Double core symmetrical PST three phase equilibrium establishes its A-phase equivalent circuit, as shown in Figure 6.

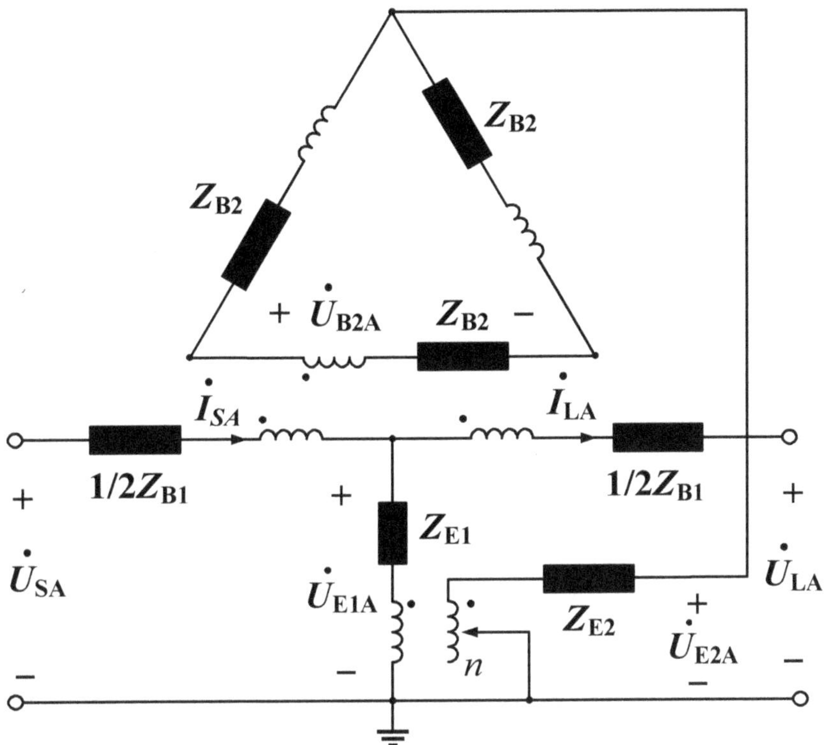

Figure 6. Equivalent circuit of two core symmetrical PST.

In Figure 6, the "T" shaped equivalent circuit of the transformer is used for analysis. Z_{B1} and Z_{B2} are equivalent leakage impedances of the primary side and secondary side of the series transformer; Z_{E1} and Z_{E2} are equivalent leakage impedances of primary side and secondary side of the series transformer. U_{B2A} is the a-phase voltage of the secondary side of the series transformer.

The equivalent impedance of the voltage regulating winding (secondary side) of ET will also change when the adjustable tap position is different. When the maximum adjustable tap position is n_{max} and the selected tap position is n, the impedance of the secondary side is

$$\left.\begin{aligned} Z_{E2n} &= DZ_{E2} \\ D &= \frac{n}{n_{max}} \end{aligned}\right\} \tag{2}$$

At this point, the overall equivalent impedance of the double core symmetric PST can be expressed as [29]:

$$Z_{eq} = Z_{B1} + \frac{1}{3N_B{}^2 + 4(N_E/D)^2}[4(N_E/D)^2 N_B{}^2 Z_{B2} + 12N_B{}^2 Z_{E1} + 12(N_E/D)^2 N_B{}^2 Z_{E2} D] \tag{3}$$

In (3), Z_{eq} is the global equivalent impedance of PST; $N_B = U_{B1A}/U_{B2A}$ is the change transformation ratio of series transformer BT; $N_E = U_{E1A}/U_{E2A}$ is the transformation ratio of the excitation transformer ET.

When the secondary side switching tap of the excitation transformer is n, the output voltage phasor of PST is shown in Figure 7.

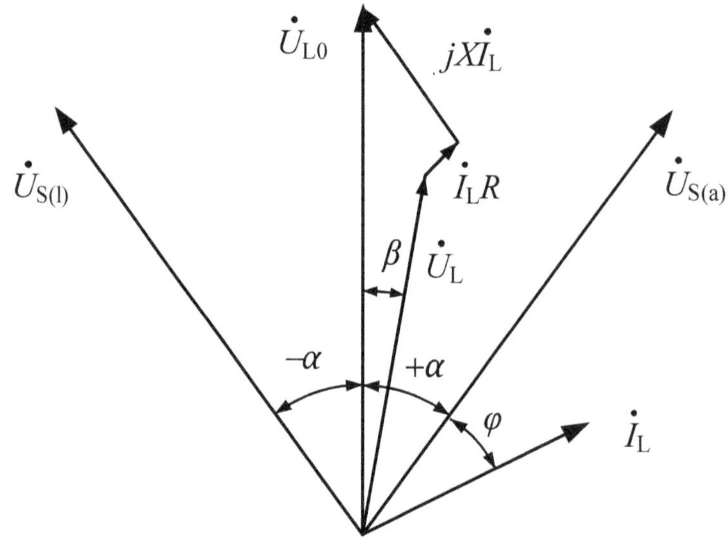

Figure 7. Phasor relation considering internal impedance.

In Figure 7, α is the phase shift angle of PST under no-load condition, and β is the inward phase shift angle. U_S and I_S are the voltage and current on the power supply side of the line. U_{L0} is the load side voltage under no-load condition. $U_{S(L)}$ and $U_{S(a)}$ represent the line power supply side voltage under lagging and advancing conditions, respectively. φ is the power factor angle. X and R are the equivalent reactance and equivalent resistance of PST, respectively, and the equivalent impedance of PST is $Z_{eq} = R + jX$.

At this point, the advancing phase shift angle and the lagging phase shift angle can be expressed as

$$\theta = \beta \pm \alpha \tag{4}$$

The no-load phase shift angle α and internal phase shift angle β of PST can be expressed as [30,31]:

$$e^{j\alpha} = \frac{-2N_E - j\sqrt{3}N_B D}{-2N_E + j\sqrt{3}N_B D} \tag{5}$$

$$\beta = \arctan\frac{|I_L|(X\cos\varphi - R\sin\varphi)}{|U_L| + |I_L|(X\cos\varphi + R\sin\varphi)} \tag{6}$$

3. Multi-Node Network Power Flow Optimization Simulation

In BPA simulation software, IEEE 39-node standard power system is used to simulate and verify the regulation effect of double core symmetric PST on large-scale network power flow. The initial power flow distribution and PST installation position of the system are shown in Figure 8. As can be seen from Figure 8a, area "M" outputs electric energy to the lower area through transmission lines "2–1", "2–3", and "26–27". Combined with Figure 8b, the active power transmitted along the three transmission lines are 119.1, 364.6, and 266.5 MW, respectively. Transmission lines 2–3 and 26–27 are heavy load, while line 2–1 is light load. The three transmission lines have the problem of uneven power flow distribution.

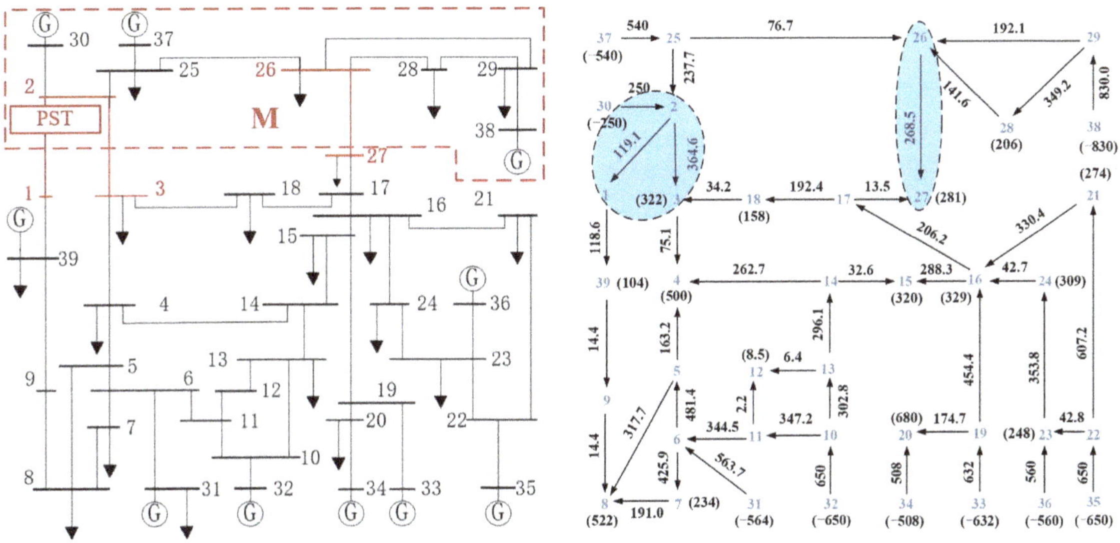

(**a**) (**b**)

Figure 8. Initial power flow of IEEE39 nodes. (**a**) PST location diagram, (**b**) Initial power flow distribution diagram.

In this paper, PST is installed on transmission line "2–1" to transfer heavy load line power flow and improve power flow distribution. After the installation of PST, the active power flow of the three transmission lines "2–1", "2–3", and "26–27" was adjusted to 256.9, 254.3 and 240.3 MW, respectively. At this point, the maximum unbalance of active power on the three transmission lines is less than 8%, which verifies the power flow regulation function of PST, as shown in Figure 9.

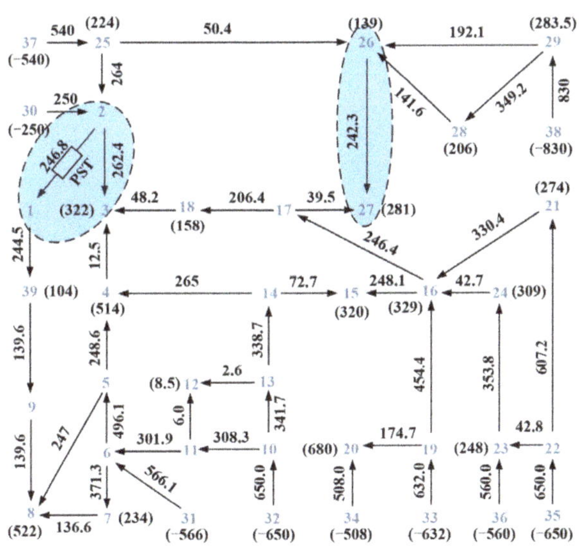

Figure 9. Power flow distribution after installing PST.

4. Loop Network Power Flow Optimization Simulation

4.1. Design of 220 kV Phase Shifting Transformer Parameters

4.1.1. Loop Power Flow

Figure 10 shows the transmission line power flow of the 220 kV loop network of Huizhou–Sandong–Yongyuan of the Guangdong Power Grid. The transmission limit of the Huizhou–Yongyuan transmission line is restricted by the Huizhou–Sandong transmission line. When the load of The Huizhou–Sandong double circuit line is full, the load rate of the Huizhou–Yongyuan line is about 60%, and the transmission limit of the Huizhou–Sandong and Yongyuan section is 1305 MW. According to the research, PST can be installed at the head of Huizhou–Yongyuan line and the head of the Huizhou-Sandong line.

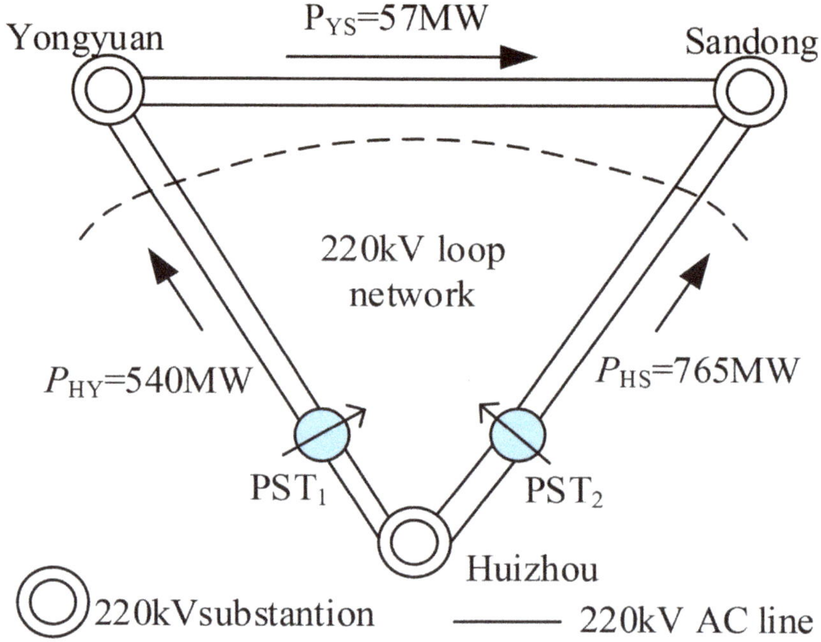

Figure 10. 220 kV transmission lines and PST location.

4.1.2. Parameters Design and Verification

High voltage power systems, transformers can easily generate direct current bias, which causes transformer iron core magnetic saturation. When the iron core was saturated, the magnetic sensing strength B tended to be fixed as the magnetic field strength increased. The inductivity of transformer primary winding is significantly reduced; the number of turns per volt of primary winding is higher than that of primary voltage, but the voltage of secondary winding does not increase, which may cause the transformer to burn out. The shift ratio of the PST saturated with the core is highly altered, and the output voltage and phase angle are greatly biased.

In order to avoid the magnetic saturation effect, the rated working point of the transformer is often designed near the saturation current. When the voltage rises to 1.1~1.2 pu, the transformer core will enter the saturation state. After the shape of the transformer core is determined, the voltage per turn of winding can be calculated, including

$$e_t = 4.44 f B A_t \tag{7}$$

where e_t is the electromotive force per turn of winding; f is the working frequency of the transformer; B is the peak value of magnetic flux density; A_t is the cross-sectional area of the iron core. Generally, in the design of transformer, B takes 1.7 T, which can be expressed as

$$\begin{cases} f = 50\ \text{Hz} \\ e_t = \frac{BA_t}{450} \times 10^5 \end{cases} \tag{8}$$

According to the rated voltage of the transformer, the number of turns of the winding can be determined so as to determine the level voltage of the excitation transformer. The parameters of the excitation transformer and series transformer designed in this paper are shown in Figure 11.

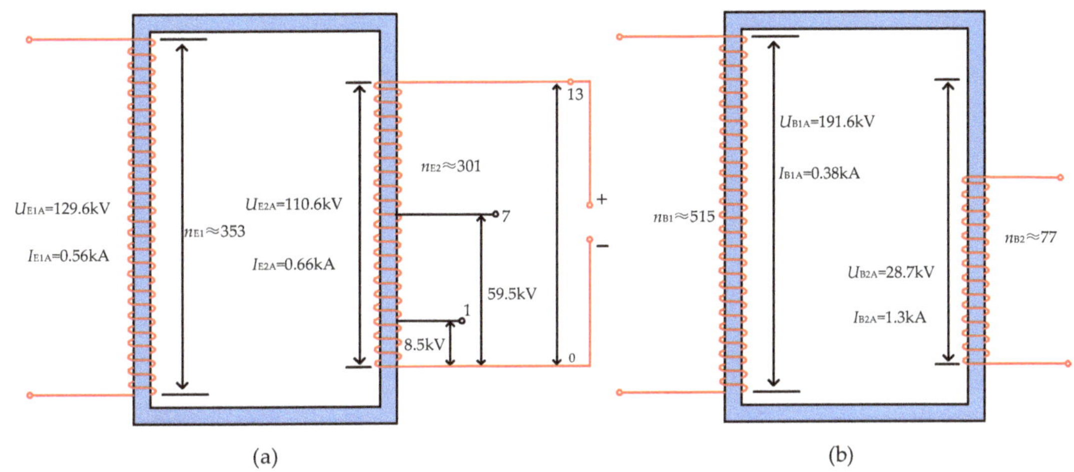

Figure 11. Parameters of the excitation transformer and series transformer. (**a**) Series transformer, (**b**) Excitation transformer.

The equivalent impedance of PST is calculated using the PST model in Section 2.3. Considering leakage reactance, the equivalent impedance of PST varies from 7.15 Ω to 7.84 Ω with the change of phase shift angel. According to the demand of power flow regulation and considering the influence of leakage reactance of transformer, the parameters of 220 kV PST are calculated as shown in Table 2.

Table 2. Technical parameters of 220 kV PST.

Technical Parameters	Value
Rated voltage/kV	230
Rated current/kA	1.3
Rated capacity/MVA	450
Adjustment class	±13
No-load phase shift Angle/°	25
Full-load phase shift Angle/°	20
Short circuit impedance of series transformer/%	7
Short circuit impedance of excitation transformer/%	4.5

Considering the influence of the impedance of PST, simulation software is used to verify parameters of the designed PST. When the voltage at the head of line reaches the rated value, the line current and the output phase shift angle are measured, and the results are shown in Figure 12.

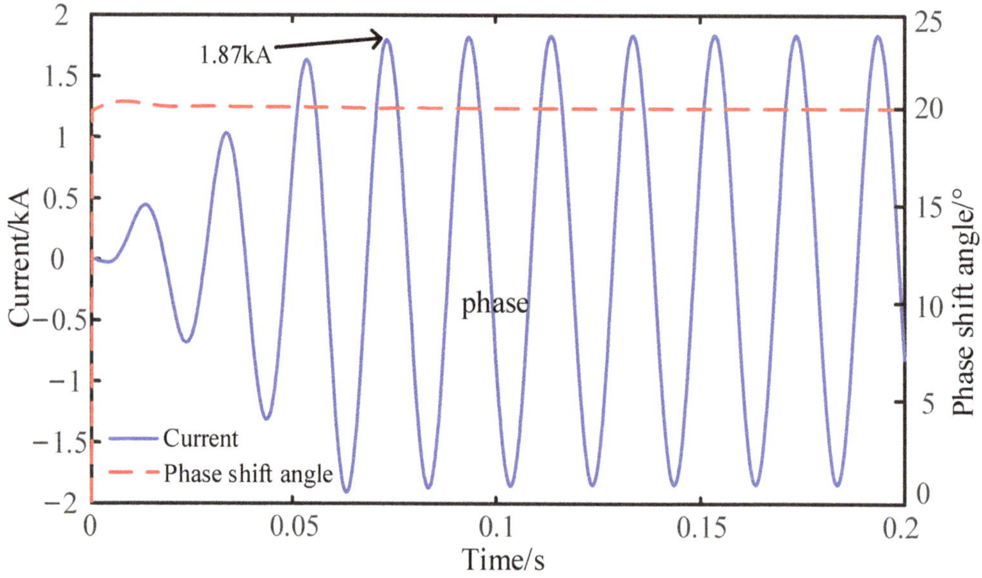

Figure 12. Performance verification of PST.

The simulation verifies that the phase shift angle at full load is 20°, and the internal phase shift angle is about 5°, which meets the performance requirements.

4.2. 220 kV Loop Network Power Flow Optimization Simulation

PSCAD software is used to simulate the actual line power flow, and the power flow regulation effect of PST on Guangdong 220 kV ring network under different working conditions is analyzed.

4.2.1. Steady Power Flow

After PST is installed, the transmission limit P_{max} increases with the increase of phase shift angle θ, as shown in Figure 13.

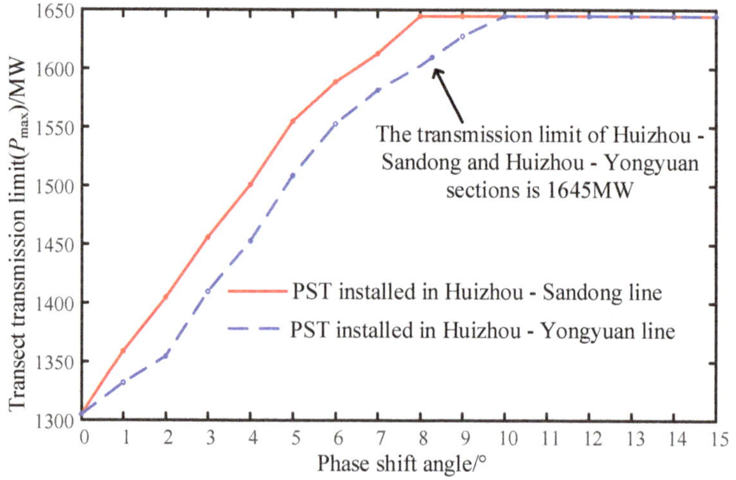

Figure 13. Transmission limit of 220 kV network.

From Figure 13, when PST is installed on the Double-loop line of the Huizhou–Sandong, the transmission limit of Huizhou–Sandong and Huizhou–Yongyuan section reaches 1645 MW when $\theta = 8°$, and the load rates of the four transmission channels (the Huizhou–Sandong double-loop line and the Hui–Yongyuan double-loop line) are close to 100%.

When PST is installed on the Huizhou–Yongyuan double loop line, the transmission limit of the section reaches 1645 MW when the phase shift angle $\theta = 10°$. The transmission and distribution capacity of the existing network topology is significantly improved by PST installation.

4.2.2. N-1 Power Flow

Considering the occurrence of extreme cases, the disconnection of Huizhou–Sandong B Line and Huizhou–Yongyuan B Line is considered respectively, as shown in Figure 14.

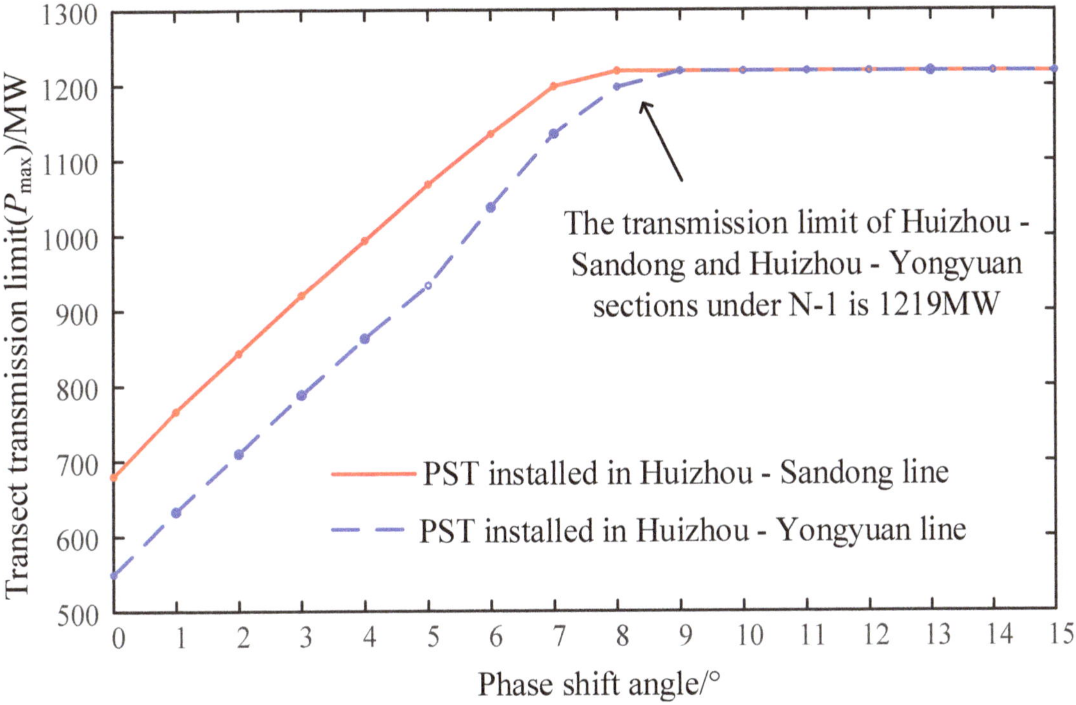

Figure 14. Transmission limit of 220 kV ring network in case of N-1.

From Figure 14, when PST is installed on Hui–Sandong A line, the transmission limit of the Huizhou–Sandongjia and Huizhou–Yongyuan sections reaches 1219 MW when $\theta = 8°$, and the load of the three transmission channels (The Huizhou–Sandong A line and the Huizhou–Yongyuan double loop line) are close to full. When the PST is installed on the Line A of Huizhou–Yongyuan, the transmission limit of the section reaches 1219 MW when $\theta = 9°$.

The transect transmission capacity can be improved by installing phase shifting transformer at the head of Huizhou–Yongyuan line or the head of Huizhou–Sandong line. Considering that the phase shift angle and rated current required to adjust the power flow are smaller when PST is installed on the Huizhou–Sandong line, the capacity of PST can be designed smaller. Therefore, we choose to install PST on the Huizhou–Sandong double-loop line.

4.3. Simulation Analysis under Short Circuit Fault

Based on PSCAD simulation software, a 220 kV voltage level constant load model is built. Explore the influence of PST on the system short-circuit current under the condition of short-circuit fault, and the influence of fault on the output phase shift angle of PST.

Single phase short circuit occurs most frequently in a high voltage system, generally accounting for about 70%. It is set that phase a grounding short circuit fault occurs in the system at 0.3 s, and the fault is removed at 0.4 s. The line short-circuit current with and without PST is shown in Figure 15.

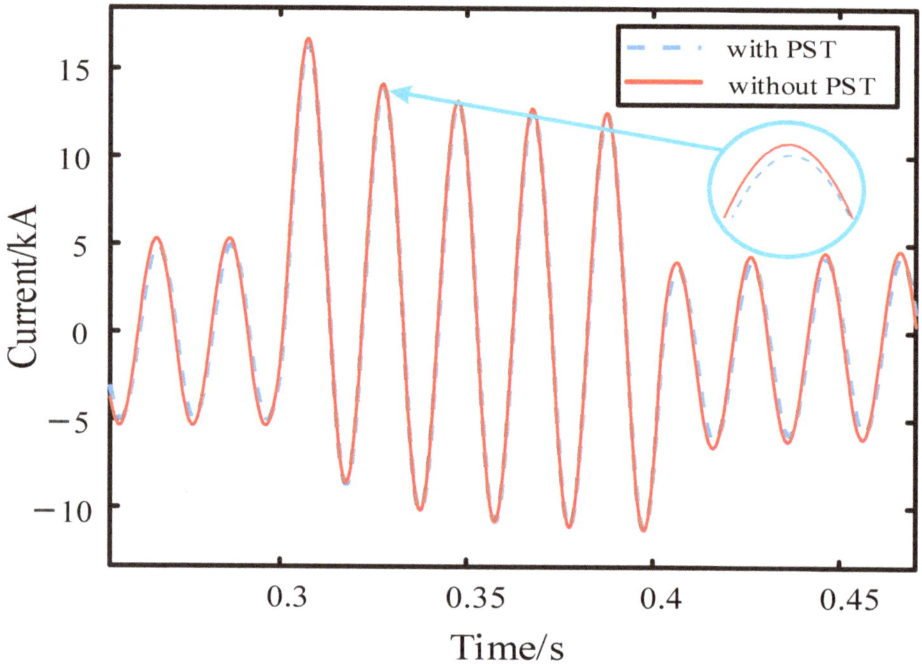

Figure 15. Single phase grounding fault phase current.

It can be seen from Figure 15 that after installing the phase shifting transformer, the short-circuit current value at the head end of the system decreases slightly. When the PST is not installed, the peak value of the short-circuit current of the system is 16.66 kA. After installing the PST, the peak value of the short-circuit current is 16.39 kA. This is because the leakage reactance of the series transformer and excitation transformer windings make the PST show a certain short-circuit impedance when the system fails. To some extent, it limits the size of short-circuit current.

In this case, the phase shift angle of PST output is shown in Figure 16.

As can be seen from Figure 16, in case of phase a grounding fault, the phase shift angle of PST output will increase by an angle. This is because when phase a grounding short circuit occurs, the voltage of phase a decreases, and the voltage of lines B and C increases. This voltage is injected into the excitation transformer, passes through the excitation transformer and series transformer, and finally forms the phase a compensation voltage. Therefore, the phase a compensation voltage increases. The phase shift angle of PST output also increases.

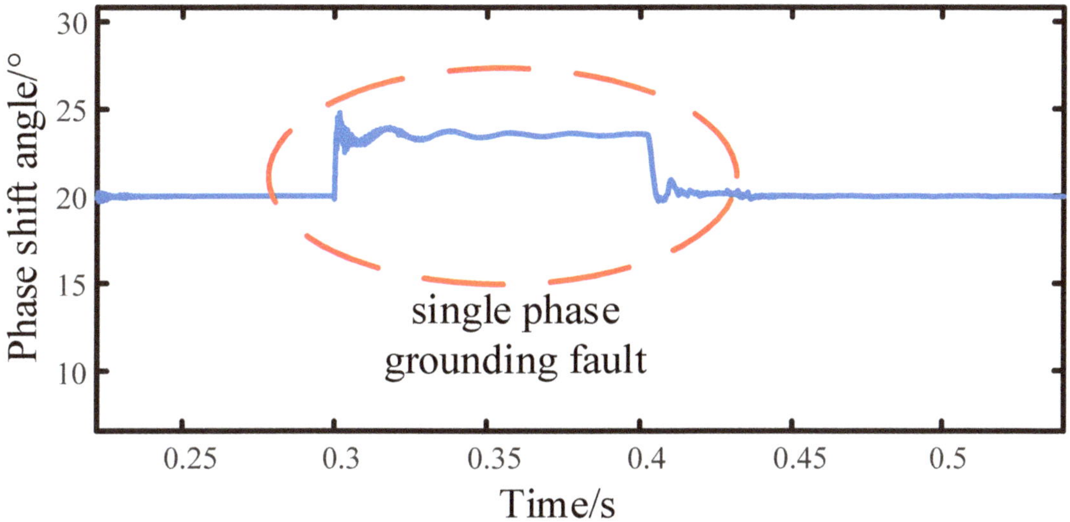

Figure 16. Phase shift angle under single phase grounding.

5. Economic Analysis of Engineering Application

5.1. Cost of PST Analysis

Taking the parameters of 220 kV PST as an example, the total price of PST is defined as C_{PST}, and the unit is ten thousand yuan, while the cost of series transformer is C_{ST}, and the unit is ten thousand yuan. The total capacity of series transformer is V_{ST}, and the unit is MVA. The unit capacity cost of series transformer is d_{ST}, and the unit is ten thousand yuan/MVA. The cost of the excitation transformer is C_{ET}. The total capacity of the excitation transformer is V_{ET}. The unit capacity cost of the excitation transformer is d_{ET}. The cost of the on-load voltage regulator switch is C_R. The total capacity of the on-load voltage regulator switch is V_{RS}. The unit capacity cost of the on-load voltage regulator switch is d_{RS}. The total cost of the PST is shown in (9):

$$C_{PST} = C_{ST} + C_{ET} + C_{RS} \tag{9}$$

The cost of each major component can be expressed as

$$\begin{cases} C_{ST} = V_{ST} \times d_{ST} \\ C_{ET} = V_{ET} \times d_{ET} \\ C_{RS} = V_{RS} \times d_{RS} \end{cases} \tag{10}$$

According to Equations (9) and (10) above, the solution matrices of the total PST cost C_{PST} of component capacity V_{ET}, V_{ST}, and V_{RS} and unit capacity cost d_{ET}, d_{ST}, and d_{RS} can be established, as shown in Equation (11):

$$C_{PST} = [V_{ST}, V_{ET}, V_{RS}] \begin{bmatrix} d_{ST} \\ d_{ET} \\ d_{RS} \end{bmatrix} \tag{11}$$

According to the investigation and inquiry made to the on-load voltage regulating switch plant, the unit cost of the mechanical on-load voltage regulating switch d_{RS} is 13,600 yuan/MVA. According to the contract cost of 350 MVA/230 kV PST provided by Tianwei Protection Transformer Company for Northwest Energy Company of the United States, if the traditional mechanical voltage regulating method is adopted, including switch and construction cost of about 120,000 yuan/MVA, the annual operating cost is about 5000 yuan/MVA. Therefore, the cost of PST can be counted as 125,000 yuan/MVA. Excluding the switch cost, the unit cost of the transformer, d_{ST} and d_{ET}, is about 110,000 yuan/MVA. According to the technical parameters calculated in Table 2, V_{RS} is 219 MVA, V_{ST} is 224 MVA, and V_{ET} is 219 MVA. Therefore, the cost C_{PST} of a single PST in the Huizhou triangle ring network is about 51.7 million yuan, of which C_{RS} is about 2.98 million yuan.

5.2. Comparative Analysis of UPFC and PST Costs

The topology of UPFC (unified power flow controller) is shown in Figure 17. It is mainly composed of three main modules: series transformer (T_{Series}), parallel transformer (T_{Shunt}), and MMC (VSC1 and VSC2). UPFC is the most comprehensive facts device, including all the capabilities of voltage regulation, series compensation, and phase shifting control.

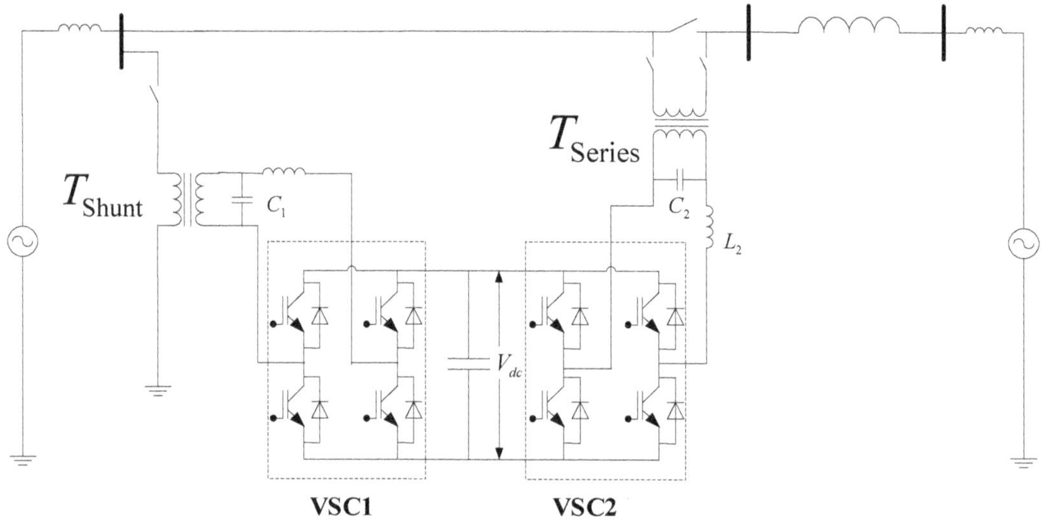

Figure 17. Schematic diagram of UPFC.

UPFC can simultaneously and quickly control the active power and reactive power in the transmission line [32,33]. After power grid failure, UPFC can quickly and independently control the active and reactive power compensation of the wind power system and maintain the voltage stability of grid points by providing reactive power support so as to improve the low-voltage ride through capacity of wind turbines.

The total price of UPFC is defined as C_{UPFC}. The cost of MMC is quite high due to the adoption of a large number of power electronic devices and the consideration of voltage and heat dissipation.

The cost of series transformer is $C_{ST\text{-}UPFC}$, and the unit is ten thousand yuan. The total capacity of series transformer is $V_{ST\text{-}UPFC}$. The unit capacity cost of series transformer is $d_{ST\text{-}UPFC}$. The cost of the parallel transformer is $C_{PT\text{-}UPFC}$. The total capacity of the parallel transformer is $V_{PT\text{-}UPFC}$. The unit capacity cost of the parallel transformer is $d_{PT\text{-}UPFC}$. The cost of MMC is C_{MMC}. The total capacity of MMC is V_{MMC}. The cost of unit capacity of MMC is d_{MMC}. Then the total cost of UPFC can be expressed as

$$C_{UPFC} = C_{ST\text{-}UPFC} + C_{PT\text{-}UPFC} + C_{RS\text{-}UPFC} \quad (12)$$

The cost of each major component can be expressed as

$$\begin{cases} C_{ST,UPFC} = V_{ST,UPFC} \times d_{ST,UPFC} \\ C_{PT,UPFC} = V_{PT,UPFC} \times d_{PT,UPFC} \\ C_{MMC} = V_{MMC} \times d_{MMC} \end{cases} \quad (13)$$

Based on Equations (12) and (13) above, the solution matrix of C_{PST} of total PST cost regarding component capacity $V_{PT\text{-}UPFC}$, $V_{ST\text{-}UPFC}$, and V_{MMC} and unit capacity cost $d_{PT\text{-}UPFC}$, $d_{ST\text{-}UPFC}$, and d_{MMC} can be established, as shown in Equation (12).

$$C_{UPFC} = [V_{ST\text{-}UPFC}, V_{ET\text{-}UPFC}, V_{MMC}]\begin{bmatrix} d_{ST\text{-}UPFC} \\ d_{PT\text{-}UPFC} \\ d_{MMC} \end{bmatrix} \quad (14)$$

According to the UPFC engineering application of Suzhou Power Grid and the UPFC engineering case of Northwest China power grid DOJ-XL line in [26], the engineering cost of setting up UPFC of 330 kV voltage grade line with a primary transmission capacity of 680 MW on the DOJ-XL line is 158.29 million yuan. The cost of a single UPFC used in the Huizhou triangle ring network is estimated at 103.12 million yuan, out of which the serial transformer $C_{ST\text{-}UPFC}$ is 24.66 million yuan. The cost of the parallel transformer $C_{PT\text{-}UPFC}$ is 24.16 million yuan, and the total cost C_{MMC} is 49 million yuan.

According to the above analysis, the cost comparison between 220 kV PST and UPFC is shown in Table 3 below. The total cost of a single PST decreases by 49.86% compared with UPFC. For economic reasons, PST is more suitable for engineering applications.

Table 3. Cost comparison between 220 kV PST and UPFC.

	Cost of Each Unit	Total Cost
UPFC	10.312 million yuan	206 million yuan
PST	5.170 million yuan	129 million yuan
PST compared with UPFC	Reduced by 49.86%	Reduced by 49.86%

6. Experimental Verification

On the basis of theoretical analysis and simulation verification, a 380 V (±4 gear) prototype is designed and manufactured for experimental verification of its power flow regulation performance. The experimental schematic diagram is shown in Figure 18. The three-phase power line used in the experiment platform has a voltage of 380 V and a frequency of 50 Hz. The three-phase power supply is adjusted by a voltage regulator of 220 V/(0~430)V to obtain the system's head voltage of 220 V phase voltage.

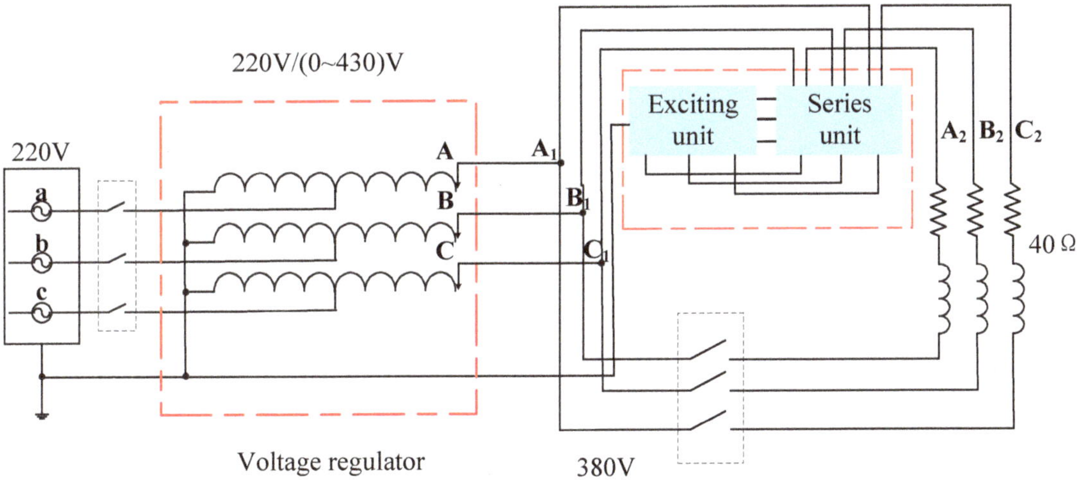

Figure 18. Experimental schematic diagram.

In this experiment, the voltage at both ends of the system is the same. In the absence of PST, there is no phase difference between the first and the end of the system, the line current is 0, and the active power and reactive power flow are 0. The experimental wiring diagram is shown in Figure 19, and the experimental results are shown in Table 4.

Figure 19. Experimental wiring diagram.

Table 4. Experimental test results.

Gear	U_s (V)	U_L (V)	θ (°) Theoretical	θ (°) Measured	θ_{PST} (°) Theoretical	θ_{PST} (°) Measured	P (W)
1	224	223	6.2	7.2	93.1	93.6	17.43
2	224	223	12.4	13.9	96.2	97.2	64.99
3	223	223	18.7	20.6	99.4	100.3	141.39
4	223	222	25.0	25.5	102.5	102.6	240.71
0	223	223	0	0	/	/	0
−1	223	223	−6.2	−7.4	−93.1	−93.6	−13.74
−2	223	223	−12.4	−14.6	−96.2	−96.4	−57.48
−3	223	223	−18.7	−20.6	−99.4	−100.8	−127.72
−4	223	223	−25.0	−26.1	−102.5	−103.5	−224.12

In Table 4, U_s is the voltage before the phase shift, U_L is the voltage after the phase shift, and θ represents the phase difference before and after phase shifting; θ_{PST} represents the phase angle of the output voltage of the PST. P is the active power transmitted by the line.

The variation of line transmission power with the phase shift angle of PST output is shown in Figure 20. We can conclude from the experimental results that the designed PST phase shift angle ranges from −25.5° to 26.17°. The adjustment range of P is −224.12 W~240.17 W by changing the angle of phase shift.

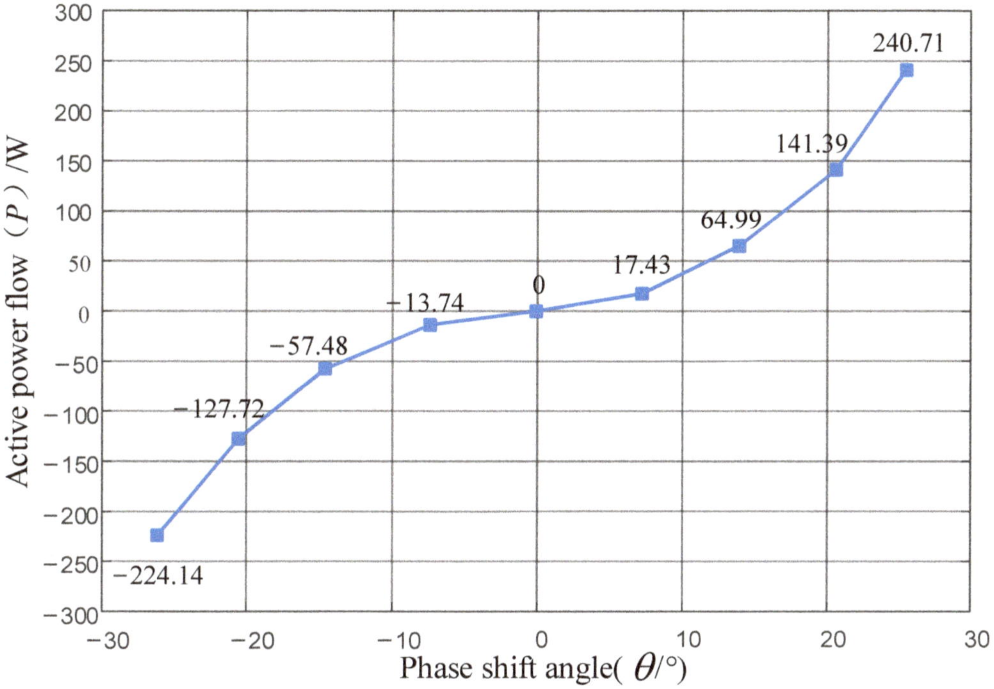

Figure 20. Relationship between line active power flow and phase shift angle.

By comparing the above experimental data with the theoretical values, the following conclusions can be drawn:

(1) When the PST is switched to +4 gear, the maximum leading phase shift angle is output, and the angle is $-25.5°$. When the PST is switched to -4, the maximum lag phase shift angle is output, and the angle is $26.17°$. The error between the measured value and the theoretical value of each gear is small, and the reason for the error is the leakage reactance of transformer winding.

(2) With the change of phase shift angle, the adjustment range of line transmission power is -224.12 W~240.17 W. When the on-load voltage regulating switch of the PST is turned to different gears, the active power of the line can be adjusted without changing the voltage amplitude before and after the phase shift, and thus, we can realize the independent control of the active power of the line.

7. Conclusions

In this paper, the working principle, topological structure, selection research, mathematical model, power flow control function, and economic analysis of high voltage phase shift transformer were studied, and the following conclusions were drawn:

(1) Comparing the characteristics of various types of PST, a high voltage of 220 kV and above was more suitable for the selection of double core symmetrical PST. The influence of winding leakage reactance and saturation effect should be considered in the design of PST parameters; the iron core area and winding turns should be reasonably selected, and a certain phase shifting angle margin should be reserved.

(2) For the actual 220 kV ring network of the Guangdong Power Grid, when PST was installed in the Huizhou–Sandong line and Huizhou–Yongyuan line, the transmission limit of the two lines sections could be increased to 1645 MW under normal working conditions, and the load rate of the four transmission channels was close to 100%. The transmission limit of both lines and planes could be increased to 1219 MW under

N−1 condition. However, the phase shift angle required to adjust the power flow was smaller (θ = 8°), and the cost was lower when it was installed on the Huizhou–Third line. Therefore, the installation site chosen was the Huizhou–Sandong line.

(3) As FACTS elements, UPFC had faster response speed and stronger power flow regulation ability than PST and could provide reactive power compensation for the system, with certain low−voltage ride through capability. However, PST had simple structure and better economy and reliability. In the 220 kV ring network scheme, the total cost of a single PST was reduced by about 49.86%. On the premise of ensuring the effect of power flow regulation, the investment cost was significantly reduced, and the economic benefit was remarkable.

(4) Through the prototype experiment, the regulation ability of dual core symmetrical PST to line power flow was verified, and the controllable regulation of line power flow was realized on the premise of constant voltage amplitude. However, due to the influence of measurement error and background harmonic, there were some errors in the experimental results. This paper did not consider the switching process and transient regulation performance of PST, so it was only suitable for steady−state power flow analysis. If the transient performance needed to be analyzed, it would be necessary to do further research in combination with power electronic devices.

Author Contributions: Conceptualization, M.Y.; methodology, J.Y.; validation, J.Y.; formal analysis, W.Z.; investigation, X.Y.; data curation, S.X.; writing—original draft preparation, M.Y.; writing—review and editing, W.Z. and J.M.; visualization, W.Z.; supervision, J.Y.; project administration, F.L. and M.Y.; funding acquisition, Z.L. All authors have read and agreed to the published version of the manuscript.

Funding: This research and the APC were funded by China Southern Power Grid Corporation's science and technology project "Research and demonstration application of key technologies of economic power grid power flow controller based on controllable phase shifter", grant number 037700KK52190009.

Institutional Review Board Statement: Not applicable.

Informed Consent Statement: Not applicable.

Data Availability Statement: Not applicable.

Conflicts of Interest: The authors declare no conflict of interest.

References

1. Song, T.; Yi, L.; Rui, Z. Study on optimal allocation of waste mine pumping and storage power station based on high proportion of renewable energy. *Power Eng. Eng. Technol.* **2020**, *39*, 87–95.
2. Belivanis, M.; Bell, K.R.W. Coordination of phase shifting transformers to improve transmission network utilization. In Proceedings of the 2010 IEEE PES Innovative Smart Grid Technologies Conference Europe (ISGT Europe), Gothenburg, Sweden, 11–13 October 2010; pp. 1–6.
3. Jinxing, W.; Qing, L. Transmission control technology of large power grid new energy system with UPFC. *Electr. Meas. Instrum.* **2018**, *55*, 51–57.
4. Rong, R.; Zirong, Z.; Zhinong, W. Multi−stage and Multi−Objective Reactive Power Optimization Algorithm of power system considering UPFC. *Power Eng. Technol.* **2020**, *39*, 76–85.
5. Hussain, S.; Yang, X.; Aslam, M.K.; Shaheen, A.; Javed, M.S.; Aslam, N.; Aslam, B.; Liu, G.; Qiao, G. Robust TiN nanoparticles polysulfide anchor for Li–S storage and diffusion pathways using first principle calculations. *Chem. Eng. J.* **2020**, *391*, 123595. [CrossRef]
6. Wang, L.; Wang, Z.; Li, H. Asymmetrical Duty Cycle Control and Decoupled Power Flow Design of a Three−port Bidirectional DC−DC Converter for Fuel Cell Vehicle Application. *IEEE Trans. Power Electron.* **2019**, *27*, 891–904. [CrossRef]
7. Munisamy, V.; Sundarajan, R.S. Hybrid technique for load frequency control of renewable energy sources with unified power flow controller and energy storage integration. *Int. J. Energy Res.* **2021**, *45*, 17834–17857. [CrossRef]
8. Zhou, L.; Swain, A.; Ukil, A. Reinforcement learning controllers for enhancement of low voltage ride through capability in hybrid power systems. *IEEE Trans. Ind. Inform.* **2019**, *16*, 5023–5031. [CrossRef]
9. Walter, S. *Phase Shifting Transformers Discussion of Specific Characteristics*; CIGRE: Paris, France, 1998; pp. 12–306.

10. Wang, Q.; Zhu, X.; Chen, Y.F. A Novel Zinc Oxide Sensor: Application for Transformer Fault Diagnosis. *J. Nanoelectron. Optoelectron.* **2018**, *13*, 1789–1792. [CrossRef]
11. Qiu, H.; Jiang, C.; Li, G.; Hao, D.; Yu, X.; Sun, Y. Design of the Voltage Transformer Based on Nanocrystalline Alloys and Its Application in Intelligent Detection of Secondary Polarity. *J. Nanoelectron. Optoelectron.* **2021**, *16*, 1501–1509. [CrossRef]
12. Zhiwei, C.; Zhiwei, L.; Wenping, L. Brief discussion on design of large power phase shifting transformer. *Transformer* **2014**, *51*, 1–4. (In Chinese)
13. Siddiqui, A.S.; Khan, S.; Ahsan, S.; Khan, M.I.; Annamalai. Application of phase shifting transformer in Indian Network. In Proceedings of the International Conference on Green Technologies, Trivandrum, India, 18–20 December 2012; pp. 186–191.
14. Zenghui, Y.; Yong, C.; Dechang, W. Classification phase shifting transformer control strategy research. *East China Electr. Power* **2013**, *41*, 2233–2236.
15. Hou, C.H.; Dai, C.B.; Sun, Y. Status quo and feature of thyristor controlled phase shifting transformer. *Smart Grid* **2014**, *2*, 18–20. (In Chinese)
16. Enrico, M.C.; Gabriele, M.; Dietrich, B. *Power Flow Control on the Italian Network by Means of Phase Shifting Transformers*; A2–206; CIGRE: Paris, France, 2006.
17. Khan, U. *Modeling and Protection of Phase Shifting Transformers*; University of Western Ontario: Ontario, ON, Canada, 2013.
18. Li, D.; Ding, J.; Wang, Z.F.; Dai, C.C.; Song, Y.T.; Shen, X.H. The research situation and engineering application of phase shifting transformer. *Smart Grid* **2015**, *7*, 9.
19. Liexin, W.U.; Mengze, Y.U.; Zuohong, L.I. Electromagnetic unified power flow controller and its application in loop power flow regulation. *High Volt. Technol.* **2018**, *44*, 3241–3249.
20. Qun, L.; Ningyu, Z.; Shan, G. RTDS modeling and application scenario analysis of power grid phase shifting transformer. *Power Eng. Technol.* **2021**, *40*, 53–58.
21. Heng, C. *Steady State Analysis of Power System*; China Electric Power Press: Beijing, China, 2007; pp. 104–110.
22. Verboomen, J.; Van Hertem, D.; Schavemaker, P.H.; Kling, W.L.; Belmans, R. Phase shifting transformers: Principles and applications. In Proceedings of the International Conference on Future Power Systems, Amsterdam, The Netherlands, 6 March 2006; pp. 1–6.
23. Yu, J.L.; Zhuo, L.; Zhang, K.; Feng, J. Treatment of longitudinal and transversal regulations in power flow calculation. *Autom. Electr. Power Syst.* **1992**, *5*, 31–33. (In Chinese)
24. Liguo, Z. *Research on Key Technologies of 220kV Phase Shift Transformer*; Shandong University: Shandong, China, 2016.
25. Jiying, Z. *Electric Field Calculation and Analysis of High Voltage and Large Capacity Single Core Phase Shifting Transformer*; North China Electric Power University: Beijing, China, 2015.
26. Yong, C.; Yinghui, Y.; Zenghui, Y. Ultra−high voltage grid and controllable phase shifting transformer demonstration engineering project evaluation. *East China Electr. Power* **2013**, *41*, 2237–2240.
27. Jie, M.; Chunding, G.; Yinghui, Y. Application of phase shifting transformer technology to EHV power grid. *East China Electr. Power* **2013**, *41*, 2066–2067.
28. Fei, G.; Xin, L.; Litong, W. Study on calculation method of maximum stage voltage of on load tap changer of symmetrical double core phase shifting transformer. *Chin. J. Electr. Eng.* **2017**, *37*, 2110–2119.
29. Yinghao, Z.; Dazhong, S. *Electrical Mechanism of on−Load Tap Changer*; China Electric Power Press: Beijing, China, 2011.
30. Litong, W. *Research on Key Problems of Hybrid on Load Tap Changer in Phase Shifting Transformer*; North China Electric Power University: Bejing, China, 2017.
31. Xin, L.; Guishu, L.; Litong, W. Determination method of technical parameters of on load tap changer in double core phase shifting transformer. *High Volt. Technol.* **2017**, *43*, 838–844.
32. Parvathy, S.; Thampatty, K. Dynamic modeling and control of UPFC for power flow control. *Procedia Technol.* **2015**, *21*, 581–588. [CrossRef]
33. Rajabi−Ghahnavieh, A.; Fotuhi−Firuzabad, M.; Shahidehpour, M.; Feuillet, R. UPFC for enhancing power system reliability. *IEEE Trans. Power Deliv.* **2010**, *25*, 2881–2890. [CrossRef]

energies MDPI

Article

Application of Block Sparse Bayesian Learning in Power Quality Steady-State Data Compression

Wenjian Hu [1,2,*], Mingxing Zhu [1,2] and Huaying Zhang [3]

1 College of Electrical Engineering and Automation, Anhui University, Hefei 232000, China; zhumingxing@ahu.edu.cn
2 Power Quality Engineering Research Center, Ministry of Education, Hefei 232000, China
3 New Smart City High-Quality Power Supply Joint Laboratory of China Southern Power Grid, Shenzhen Power Supply Co., Ltd., Shenzhen 518000, China; zhyszpower@163.com
* Correspondence: z21514036@stu.ahu.edu.cn; Tel.: +86-137-0559-2785

Abstract: In modern power systems, condition monitoring equipment generates a great deal of steady-state data that are too large for data transmission and, thus, data compression is needed. Therefore, there is a balance to strike between compression quality and data accuracy. Greedy algorithms are effective but suffer from low data reconstruction accuracy. This paper proposes a block sparse Bayesian learning (BSBL)-based data compression method. Based on the prior distribution and posterior probability of the sparse signals, it uses the Bayesian formula to excavate the block structure of these signals. This paper also adds two indicators to the evaluation process to validate the proposed method. The proposed method is effective in terms of signal-to-noise ratio (SNR), relative root mean square error (RRMSE), amplitude error, energy recovery percentage (ERP), and angle error. The first three indicate better performance of the proposed method than the traditional method by giving the same compression ratio. Therefore, the method validates the possibility of a more accurate and economical solution to power quality assurance.

Keywords: block sparse Bayesian learning; compressed sensing; data compression; evaluation indicators; power quality steady-state data

Citation: Hu, W.; Zhu, M.; Zhang, H. Application of Block Sparse Bayesian Learning in Power Quality Steady-State Data Compression. *Energies* **2022**, *15*, 2479. https://doi.org/10.3390/en15072479

Academic Editors: Vítor Monteiro and J. C. Hernandez

Received: 3 February 2022
Accepted: 25 March 2022
Published: 28 March 2022

1. Introduction

As more power electronic equipment and impact loads are connected to the power system, more power quality problems also emerge. When investigating and solving these problems, the data from the time of the accident are needed. When data cannot be stored indefinitely, methods to achieve better reconstruction performance at a low compression ratio become a key issue [1,2].

The current methods for power quality data compression are divided into the traditional methods and those based on compressed sensing (CS). The traditional method is based on wavelet theory. Fundamentally, it is to keep the coefficients larger than the set threshold and discard the coefficients smaller than the set threshold after a wavelet transformation of the data [3,4]. It is found that the magnitudes of wavelet-transformed coefficients associated with disturbance events are much larger than those of disturbance-free coefficients; this fact leads to an effective method to compress power quality disturbance data [3]. The compressed data size is one-sixth to one-third of that of the original data. Ref. [5] uses two-dimensional lifting wavelet transformations based on a one-dimensional wavelet to improve the computational efficiency of the compression algorithm. However, this method is based on the Nyquist theorem, and its high sampling rate and high hardware cost will hinder the efficiency of data transmission and analysis.

Based on the above problems, Donoho proposed the compressed sensing theory [6]. It is known from the CS theory that after the signal is sparsely transformed, the signal can be sampled, compressed, and reconstructed far below the standard of the Nyquist sampling frequency. Presently, many methods for dealing with power quality problems are based on CS theory.

Greedy algorithms represented by the orthogonal matching pursuit (OMP), sparsity adaptive matching pursuit (SAMP), and compressed sampling matching pursuit (CoSAMP) algorithms are widely used, which could compress power quality data. The core idea is to go step-by-step and make the best choice based on an optimization measure according to the current situation [7–11].

Sparse Bayesian learning (SBL) and its improved algorithms are also based on CS theory. They combine existing data to make short-term predictions, compress and reconstruct data, or classify and identify events. Ref. [12] combines the sparse Bayesian learning and extreme learning machine (ELM) algorithms to propose an improved algorithm for power quality event classification, and the test data proves that the algorithm has a faster calculation speed and generalization performance. Ref. [13] proposes a hyperparameter adaptive group Bayesian learning algorithm, which can locate the harmonic source with a small number of measured values and estimate the magnitude and angle of the harmonic source. Ref. [14] proposes a joint clustering and sparse Bayesian learning algorithm when the clustering partition is unknown. The method can adaptively divide clusters, and different clusters can learn different time correlation coefficients, thereby reducing the interference between clusters. Ref. [15] proposes a complex spatiotemporal sparse Bayesian learning (CSTSBL) algorithm to reconstruct angular positions using the spatiotemporal correlation structure of the solution matrix, which can achieve high accuracy and resolution performance at low signal-to-noise ratios.

In addition to statistical features, general sparse vectors also have certain structural features that can provide additional information for algorithm convergence. Based on this idea, the block SBL (BSBL) algorithm is applied to regression and classification in machine learning. Ref. [16] uses the variational Bayesian block sparse method to estimate the radar target scattering center and achieves good results in combination with simulation. Ref. [17] proposes a clustered block sparse Bayesian learning (CBSBL) algorithm for millimeter wave channel estimation by exploiting the correlation between mm wave channels. The algorithm improves the accuracy and robustness of mm wave channel estimation.

BSBL can not only recover signals with block structure but also consider intra-block correlation. Therefore, its recovery of sparse signals can be good. It has been successfully applied to monitor fetal electrocardiograms and electroencephalograms via wireless body area networks [18]. However, there is not much research on the related field of power quality signals. Based on this, this paper proposes a reconstruction method that combines BSBL and sparsity in the transform domain. Different transform domains are used to sparsely transform different power quality steady-state signals. The experimental results using the constructed signal and the measured signal show that the BSBL algorithm outperforms the traditional CS-based algorithms in relative performance indicators.

However, the composition of actual power quality is complex. If it is compressed using a greedy algorithm, it is difficult to pre-estimate the sparsity of the data before finding a suitable dictionary matrix. This causes larger errors in the reconstruction after data compression. Aiming to solve this problem, this paper proposes a BSBL method for compressing steady-state power quality data. It uses the known prior distribution and posterior probability of the signal to mine the correlation within the block and improve the reconstruction accuracy of the signal. For indicators of the reconstructed signal performance comparison, SNR (signal-to-noise ratio), ERP (energy recovery percentage), and RRMSE (relative root mean square error) are more commonly used. However, there is usually a lack of indicators established by combining its signal characteristics to analyze the reconstruction performance. In this paper, two performance indicators are proposed according to the characteristics of the power quality steady-state signal and the relative standard, and their practicability is demonstrated. The Section 2 briefly introduces the BSBL framework and the concept of power quality steady-state data. The Section 3 introduces the flow chart of BSBL and the proposed performance indicators. The Section 4 provides case studies and performance comparisons with other algorithms. The Section 5 gives the recommended transform domain and CR values for the proposed algorithm in different engineering signals.

2. BSBL Algorithm and Power Quality Steady-State Data

2.1. CS Theory

CS theory points out that when a signal $f \in R^{N\times 1}$ of length N is sparse or its sparse representation on a certain orthogonal transform domain Ψ is x, $f = \psi x$. Using a measurement matrix $\Phi \in R^{M\times N}$, the measurement signal of f is

$$y = \Phi f = \Phi\psi x = Ax \tag{1}$$

where M is the signal observation dimension, and M << N. Furthermore, $A = \Phi\psi$ is the sensing matrix, and y is the observation value of the sparse x under A. Ref. [18] used mathematics to prove that the measurement matrix Φ must meet the requirements of the restricted isometry property (RIP):

$$(1-\delta)\left|\left|y\right|\right|_2^2 \leq \left|\left|\Phi y\right|\right|_2^2 \leq (1+\delta)\left|\left|y\right|\right|_2^2 \tag{2}$$

where δ is the restricted equidistant constant and $\delta \in (0,1)$ and $||y||_2$ are the norms of y. The equivalent condition of RIP is that y is not related to Φ. The Gaussian random matrix, Bernoulli random matrix, sparse binary matrix, and other matrices are nearly irrelevant to any sparse matrix at present.

2.2. BSBL Framework

For the sparse x in Equation (1), its non-zero elements often have block structure characteristics [19]. In this structure, suppose it is divided into s continuous blocks, the number of elements in the i-th block is t_i, and there are

$$N = \sum_{i=1}^{s} t_i \tag{3}$$

When the number of elements in each block is equal, there are

$$x = [\underbrace{x_1, x_2, \ldots x_t}_{\text{t datas}}, \underbrace{x_{t+1}, x_{t+2}, \ldots x_{2t}}_{\text{t datas}}, \ldots, \underbrace{x_{(s-1)t+1}, x_{(s-1)t+2}, \ldots x_{st}}_{\text{t datas}}] \tag{4}$$

That is, $t_1 = t_2 = \ldots = t_s$ and $N = st$, assuming that there are k blocks in these blocks that are non-zero ($k << s$), then this vector is called block k-sparse.

Assuming that the blocks are independent of each other and obey Gaussian distribution,

$$p(x_i; \gamma_i, B_i) \sim N(0, \gamma_i B_i), i = 1, 2, \ldots, s \tag{5}$$

where γ_i is a hyperparameter. Most γ_i tend to zero, and $\gamma_i \geq 0$. B_i is a positive definite matrix. The prior distribution with γ_i and B_i as parameters are

$$p(x; \{\gamma_i, B_i\}_{i=i}^{s}) \sim N(0, \Sigma_0) \tag{6}$$

$$\sum\nolimits_0 = diag\{\gamma_1 B_1, \gamma_2 B_2, \ldots, \gamma_s B_s\} = \begin{pmatrix} \gamma_1 B_1 & \cdots & 0 \\ \vdots & \ddots & \vdots \\ 0 & \cdots & \gamma_s B_s \end{pmatrix} \tag{7}$$

Therefore, the compression sampling process for signal x is

$$y = Ax + \varepsilon \tag{8}$$

where ε is the noise signal in the sampling process and satisfies the Gaussian distribution with a mean value of 0 and a variance of λ ($\lambda > 0$),

$$p(\varepsilon_i) \sim N(0, \lambda E) \tag{9}$$

where E is the identity matrix. According to the Bayesian formula, the posterior probability distribution of x is obtained as a Gaussian distribution with mean μ_x and variance Σ_x:

$$p(x|y; \lambda, \{\gamma_i, B_i\}_{i=1}^{s}) \sim N(\mu_x, \Sigma_x) \tag{10}$$

$$\mu_x = \frac{1}{\lambda}\Sigma_x A^{\mathrm{T}} y \tag{11}$$

$$\Sigma_x = (\Sigma_0^{-1} + \frac{1}{\lambda}A^{\mathrm{T}}A)^{-1} = \Sigma_0 - \Sigma_0 A^{\mathrm{T}}(\lambda E + A\Sigma_0 A^{\mathrm{T}})^{-1}A\Sigma_0 \tag{12}$$

Thus far, the maximum posterior probability estimate of x is $\hat{x}$, and $\widehat{x} = \mu_x$.

2.3. Steady-State Power Quality Signal

The steady-state indicators of power quality include voltage deviation, frequency deviation, harmonics, three-phase voltage unbalance, voltage fluctuation and flicker, etc. [20]. In this paper, common and harmful harmonics, voltage fluctuations, and flicker are used to construct the steady-state signal of power quality.

After performing a discrete Fourier transform (DFT) on a periodic non-sinusoidal AC signal, the components greater than integer multiples of the fundamental frequency are called harmonics, and the non-integer multiples are called interharmonics. The analytical expression is

$$x(t) = \sum_{h=0.1}^{M} A_h \cos(2\pi f_0 h t + \varphi_h) \tag{13}$$

where h is the (inter)harmonic order, f_0 is the power frequency, A_h and φ_h are the amplitude and phase of the corresponding order, respectively.

Voltage fluctuation is the phenomenon of a series of changes or continuous changes in the root mean square value of the voltage. Flicker is the cumulative effect of voltage fluctuations over a period of time, which is reflected by the visual perception caused by unstable light illumination. Flicker becomes a kind of harm caused by voltage fluctuations. Because the voltage fluctuation period is generally greater than the power frequency period, the power frequency signal can be regarded as the carrier wave, and the fluctuation signal can be regarded as the modulating wave. The analytical expression of the voltage fluctuation is as follows:

$$x(t) = U_0[1 + m\cos(2\pi f_s t + \varphi_s)]\cos(2\pi f_0 t + \varphi_0) \tag{14}$$

where U_0 is the power frequency voltage amplitude, m is the modulation indicator, and $m < 1$, f_s and φ_s are the modulation wave frequency and initial phase, respectively.

3. Algorithm Implementation Process and Evaluation Indicators

3.1. Parameter Learning Rules under the BSBL Algorithm

In order to determine the value of the hyperparameter $\Theta = (\gamma_i, B_i, \lambda)$ ($i = 1, 2, 3, \ldots, s$), expectation-maximization (EM) can be used to obtain the maximum value of $p(y; \Theta)$ [21,22]. The problem of obtaining the maximum value of $p(y; \Theta)$ can also be transformed into obtaining the minimum value of $L(\Theta)$.

$$L(\Theta) = y^T \Sigma_y^{-1} y + \log\left|\Sigma_y\right| \tag{15}$$

$$\Sigma_y = \lambda E + A\Sigma_0 A^T \tag{16}$$

For signals with no noise or noise close to 0, λ can be set to 0 directly. When the signal-to-noise ratio (SNR) is low, the learning rule is [23]

$$\lambda \leftarrow \frac{\|y - A\mu_x\|_2^2 + \sum_{i=1}^{s} Tr(\Sigma_x^i (A^i)^T A^i)}{M} \tag{17}$$

where the symbol "$\leftarrow$" means assigning the result on the right to the symbol on the left. Furthermore, Tr is the trace of the matrix, A^i is the sub-matrix corresponding to the i-th block of x, and Σ_x is the i-th main diagonal block matrix of the Σ_x. In other cases, the learning rule of λ is

$$\lambda \leftarrow \frac{||y - A\mu_x||_2^2 + Tr(\Sigma_x A^T A)}{M} \tag{18}$$

The learning rules of B_i are

$$B_i \leftarrow \frac{\Sigma_x^i + \mu_x^i (\mu_x^i)^T}{\gamma_i} \tag{19}$$

However, for different block numbers i, if different B_i are used, over-fitting will occur. Therefore, for any i, the learning rule of setting the average value of the parameters is

$$B \leftarrow \frac{1}{m} \sum_{i=1}^{s} \frac{\Sigma_x^i + \mu_x^i (\mu_x^i)^T}{\gamma_i} \tag{20}$$

where μ_x^i is the i-th block of μ_x.

At the same time, in order to better find the intra-block correlation, a first-order auto-regressive model can be used. At this time, the matrix form in the corresponding block is

$$B = \text{Toeplitz}([1,\ r, r^2 \ldots, r^{t-1}]) = \begin{pmatrix} 1 & r & \ldots & r^{t-1} \\ r & 1 & \ldots & r^{t-2} \\ \ldots & \ldots & \ldots & \ldots \\ r^{t-1} & r^{t-2} & \ldots & 1 \end{pmatrix} \tag{21}$$

where Toeplitz$([1,\ r, r^2 \ldots, r^{t-1}])$ is the corresponding Toeplitz matrix, and r is the autoregressive coefficient.

The learning rules of γ_i are

$$\gamma_i \leftarrow \frac{1}{t_i} Tr[B_i^{-1}(\Sigma_x^i + \mu_x^i (\mu_x^i)^T)] \tag{22}$$

3.2. The BSBL Algorithm Process

The BSBL algorithm process calculates μ_x and Σ_x, according to Equations (11) and (12), as well as the initialization parameters or return values. Then, the parameters λ, B, and γ_i are continuously iterated according to the method shown in Section 2.3 until the μ_x and μ_x' calculated twice in a row are less than a certain threshold μ_{th}, the final μ_x is output, and the iteration ends. The flowchart is shown in Figure 1.

It can be seen from the flowchart that BSBL will perform better due to the exploration of the intra-block correlation of the signal and continuous parameter iteration. At the same time, since each iteration will find the inverse of the Σ_x (Equation (12)), the running time will inevitably increase.

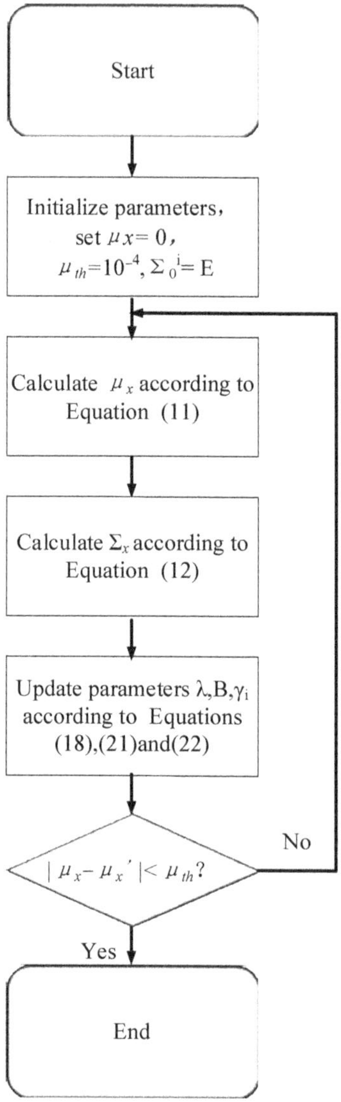

Figure 1. Flowchart of the BSBL algorithm.

3.3. *Evaluation Indicators of Power Quality Data Compression Performance*

The compression performance evaluation indicator is an objective evaluation of the performance of the compression algorithm, which can judge the approximation degree of the reconstructed signal to the original signal. The compression performance evaluation indicators generally include the compression ratio (CR), SNR, ERP, and RRMSE.

(1) The calculation equation of the CR is

$$\mathrm{CR} = \frac{N_R}{N_O} \times 100\% \tag{23}$$

where N_R and N_O are the reconstructed signal and the original signal, respectively.

(2) The equation for calculating the SNR is:

$$\mathrm{SNR} = 10\lg \frac{\sum\limits_{i=1}^{N} [x(i)]^2}{\sum\limits_{i=1}^{N} [x(i) - \overline{x}(i)]^2} \tag{24}$$

where $x(i)$ is the original signal without white noise, and $\overline{x}(i)$ is the reconstructed signal; the unit of the SNR is the decibel (dB). The larger the value, the smaller the noise contained in the reconstructed signal.

(3) The calculation equation of the ERP is

$$\mathrm{ERP} = \frac{\sum\limits_{i=1}^{N} [\overline{x}(i)]^2}{\sum\limits_{i=1}^{N} [x(i)]^2} \times 100\% \tag{25}$$

The closer the ERP is to 1, the closer the energy recovery is to the original signal.

(4) The calculation equation of relative root mean square error is

$$\mathrm{RRMSE} = \sqrt{\frac{\sum\limits_{i=1}^{N} [x(i) - \overline{x}(i)]^2}{\sum\limits_{i=1}^{N} [x(i)]^2}} \times 100\% \tag{26}$$

The smaller the RRMSE, the smaller the relative error of the reconstructed signal.

The above indicators are only proposed for compression and reconstruction errors of general signals. However, the voltage harmonics of the power quality steady-state signal will cause voltage deviation; the amplitude and phase of the harmonics will affect the analysis of the three-phase voltage imbalance. In addition to the error between the reconstructed and original signals, we must also evaluate the characteristics of the power quality signal. For example, Section 5.2 of *IEC41000-4-30 Testing and measurement techniques—Power quality measurement methods* stipulates that the measurement uncertainty of the supply voltage magnitude should not exceed 0.1% of U_{din} (declared input voltage). Meanwhile, Section 5.1 of *IEC41000-4-7 Testing and measurement techniques—General guide on harmonics and interharmonics measurements and instrumentation, for power supply systems and equipment connected thereto* stipulates that the measurement error of the harmonic voltage/current of the Class I harmonic measuring instrument shall not exceed 5%. According to the requirements of these standards and the characteristics of the power quality signal, two new evaluation indicators are proposed: the relative error of the harmonic amplitude (hereafter referred to as the amplitude error) and the relative error of the harmonic phase angle (hereafter referred to as the phase angle error). The calculation equations are

$$\text{Amplitude error} = \sqrt{\frac{\sum\limits_{h=0.1}^{N} (A_h - \hat{A}_h)^2}{\sum\limits_{h=0.1}^{N} A_h^2}} \times 100\% \tag{27}$$

$$\text{Angle error} = \sqrt{\frac{\sum\limits_{h=0.1}^{N} (\varphi_h - \overset{\frown}{\varphi}_h)^2}{\sum\limits_{h=0.1}^{N} \varphi_h^2}} \times 100\% \tag{28}$$

where A_h and φ_h are divided into the amplitude and phase angle of each harmonic and interharmonic of the original data after the DFT. The values of $\hat{A}_h$ and $\hat{\varphi}_h$ are divided into the amplitude and phase angle of each harmonic and interharmonic of the reconstructed signal after the DFT. Thus, amplitude and phase angle errors can reflect the degree of harmonic distortion as the judgment basis after signal compression and reconstruction and meet the relevant national standards, thereby increasing the credibility of the algorithm performance evaluation results and engineering applications.

4. Case Study

4.1. Construction of the Steady-State Power Quality Signal

This article takes the signal length N = 2560 (N is the number of sampling points) and the signal sampling frequency of 12.8 kHz, and Gaussian white noise is added to the signal to make the *SNR* = 30 dB. For the choice of the two parameters, μ_{th} and block length, their being too large or too small will degrade the relative performance of the results. Therefore, a sensitivity analysis was carried out on these parameters by using synthetic data and measured data. The results are shown in Figures 2 and 3. It can be seen from Figure 2 that the value of the SNR gradually decreases with the increase of μ_{th}. When $\mu_{th} = 10^{-4}$, the value of the SNR can be greater than 30 dB. It can be seen from the comparison in Figure 3 that the value of the SNR first increases and then decreases with the increase of block length. When block length = 128, the SNR of both data can be greater than 30 dB. Therefore, we set $\mu_{th} = 10^{-4}$ and block length = 128. The measurement matrix was selected as a sparse binary matrix, and the transform domains used by the signal were the DFT domain and the discrete cosine transform (DCT) domain. Suppose that the harmonic signal contains harmonics of 250 Hz, 350 Hz, 1900 Hz, 2100 Hz, 3950 Hz, and 4050 Hz and inter harmonics of 85 Hz and 115 Hz. In the voltage fluctuation signal, we set m = 0.1, $f_s = 20\text{Hz}$, $\varphi_s = \frac{\pi}{3}$. The signal waveforms and expressions are shown below (normalized basic amplitude to 1 p.u.) and in Figure 4.

$$\begin{aligned} x_1(t) = {} & \sin(2\pi \times 50t) + 0.1\sin(2\pi \times 250t - 2\pi/3) + 0.1\sin(2\pi \times 350t + 2\pi/3) \\ & +0.08\sin(2\pi \times 1900t + 5\pi/6) + 0.08\sin(2\pi \times 2100t + 5\pi/6) \\ & +0.05\sin(2\pi \times 3950t + \pi/3) + 0.05\sin(2\pi \times 4050t + \pi/3) \\ & +0.06\sin(2\pi \times 85t) + 0.06\sin(2\pi \times 115t) \end{aligned} \tag{29}$$

$$x_2(t) = [1 + 0.1\cos(10\pi t) + \pi/3] \times \cos(100\pi t) \tag{30}$$

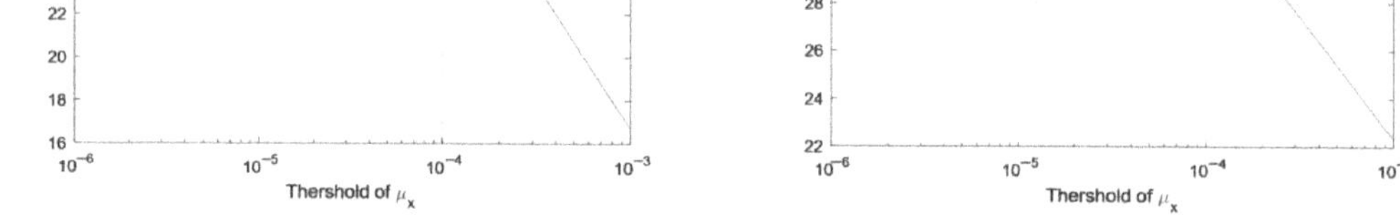

(a) (b)

Figure 2. Sensitivity curve of μ_{th} in different data. (**a**) Sensitivity curve of μ_{th} in measured data. (**b**) Sensitivity curve of μ_{th} in synthetic data.

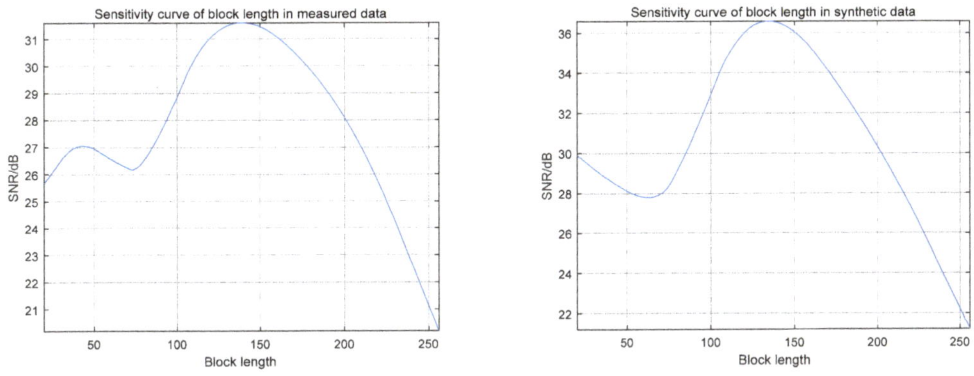

(a) (b)

Figure 3. Sensitivity analysis of block length in different data. (**a**) Sensitivity curve of block length in measured data. (**b**) Sensitivity curve of block length in synthetic data.

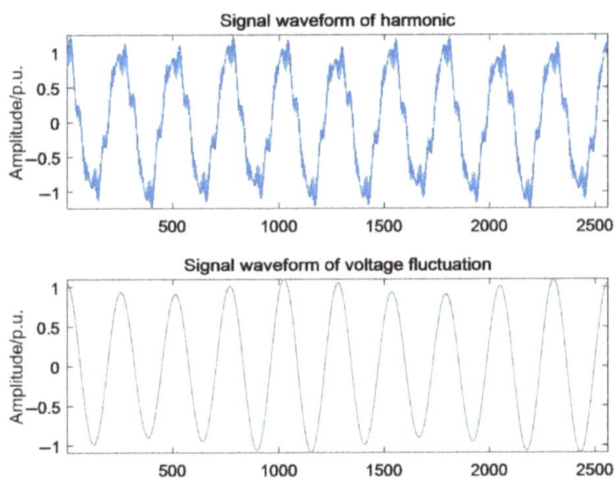

Figure 4. Signal waveform of harmonic and voltage fluctuation.

For harmonic signals and voltage fluctuation signals, the DFT and DCT were used to perform sparse transformations and then compressed and reconstructed. The coefficients in the DFT and DCT domains are shown in Figures 5 and 6, respectively.

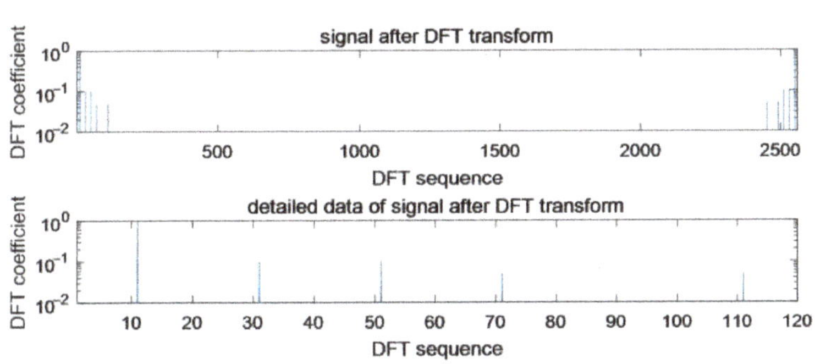

Figure 5. *Cont.*

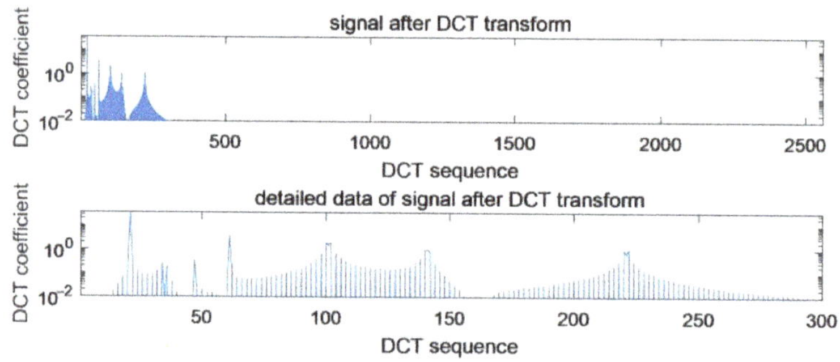

Figure 5. Different transform basis coefficients of the harmonic signal.

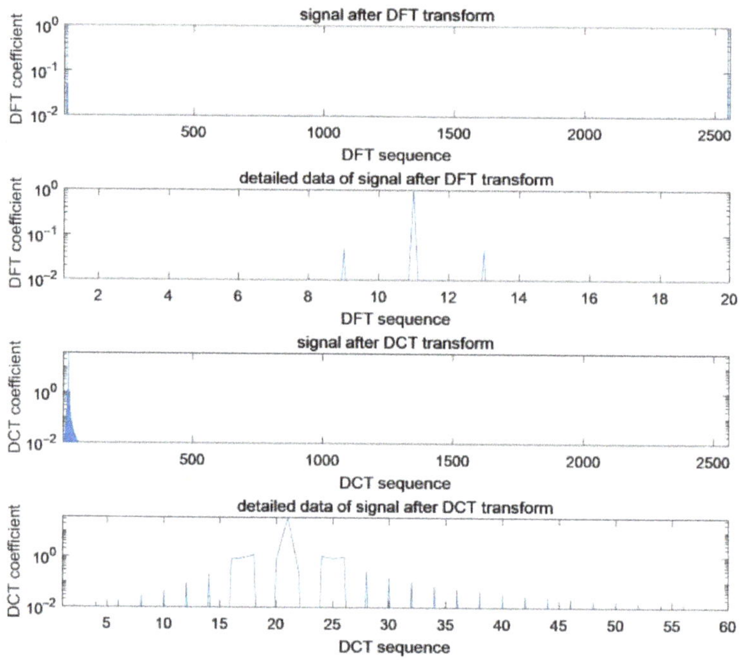

Figure 6. Different transform basis coefficient of voltage fluctuation signal.

It can be seen from Figures 5 and 6 that after the transformation of the above two signals, the non-sparse components in the DCT domain were more concentrated in several blocks. Therefore, after compression and reconstruction by the BSBL algorithm, its performance should be slightly better than that under the DFT.

4.2. *BSBL Algorithm Reconstruction Performance Analysis*

The amplitude error was set up to ensure that the performance of the reconstructed signal meets Section 5.1 of *IEC 41000-4-7*. SNR is an important indicator for signal compression and reconstruction. However, there is no standard for this indicator at present. In the following cases, when the SNR of the reconstructed signal is more than 30 dB and the amplitude error is less than 5%, the minimum CR value is considered to be the best.

4.2.1. Performance Analysis of the Harmonic Signal

The various indicators of the harmonic signal in the DFT and DCT domains when different CR compressions and reconstructions were used are shown in Tables 1 and 2.

Table 1. Reconstruction results of harmonic signals in the DFT domain.

Transform Domain	DFT			
CR (%)	30	40	50	60
SNR (dB)	23.74	28.91	37.65	52.47
RRMSE (%)	6.5	3.58	1.31	0.24
Amplitude error (%)	6.44	3.52	1.25	0.19
ERP (%)	101.49	100.49	100.15	100.03
Angle error (%)	1.46	1.42	1.4	1.22

Table 2. Reconstruction results of harmonic signals in the DCT domain.

Transform Domain	DCT			
CR (%)	30	40	50	60
SNR (dB)	25.06	32.5	40.58	48.29
RRMSE (%)	5.58	2.37	0.93	0.38
Amplitude error (%)	5.5	2.28	0.85	0.3
ERP (%)	101.27	100.45	100.16	100.07
Angle error (%)	1.45	1.43	1.39	1.32

It can be seen from Tables 1 and 2 that in the harmonic signal, when the CR was 30% to 50%, the reconstruction effect of the DCT was better than that of the DFT. In the DFT domain, the value of the ERP differed from 1 by less than 0.5%, the amplitude error was less than 5%, and the phase error was less than 2%. The amplitude error meets the allowable error standard for the measurement of Class I instruments in *IEC41000-4-7*, and the situation is similar in the DCT domain. After the DCT, when CR = 40%, the result of the SNR is more than 30 dB. Therefore, for the constructed harmonic signal CR = 40% is an ideal choice under the DCT, and CR = 50% is a good choice under the DFT.

4.2.2. Performance Analysis of Voltage Fluctuation Signal

The various indicators of the voltage fluctuation signal in the DFT and DCT domains when different CR compressions and reconstructions were used are shown in Tables 3 and 4.

Table 3. Reconstruction results of voltage fluctuation signal in the DFT domain.

Transform Domain	DFT			
CR (%)	30	40	50	60
SNR (dB)	49.57	59.53	61.35	62.02
RRMSE (%)	0.33	0.11	0.09	0.08
Amplitude error (%)	0.29	0.08	0.07	0.06
ERP (%)	100.52	100.08	100.02	100.01
Angle error (%)	1.1	0.94	0.92	0.87

Table 4. Reconstruction results of voltage fluctuation signal in the DCT domain.

Transform Domain	DCT			
CR (%)	30	40	50	60
SNR (dB)	54.62	58.45	59.14	59.72
RRMSE (%)	0.19	0.12	0.11	0.1
Amplitude error (%)	0.14	0.08	0.07	0.07
ERP (%)	100.2	100.03	100.01	100
Angle error (%)	1.29	1.15	1.11	1.07

For the constructed voltage fluctuation signal, the reconstruction performance after the DCT and DFT is good, the SNR can reach 50 dB or more, the amplitude error is less than

0.5%, and the difference between the ERP and 1 approaches 0.5%. Therefore, compression reconstruction with CR = 30% is an ideal choice for this case.

4.2.3. Performance Analysis When the Signal Is Not Sparsely Transformed

The BSBL algorithm also has a good effect on non-sparse signals. The signal does not use any sparse transform domain and directly performs compression and reconstruction. However, the performance cannot meet the requirement of SNR > 30dB when $\mu_{th} = 10^{-4}$, so we set $\mu_{th} = 10^{-5}$. The performance indicators and reconstruction error of each point are shown in Table 5.

Table 5. Reconstruction results of the harmonic signals without a sparse transform.

Transform Domain	No Transform Domain			
CR (%)	30	40	50	60
SNR (dB)	25.40	26.83	28.10	30.05
RRMSE (%)	5.37	4.55	3.93	3.17
Amplitude error (%)	4.26	3.56	3.05	2.43
ERP (%)	100.62	100.15	100.03	100.02
Angle error (%)	1.09	1.02	0.96	0.91

It can be seen from Table 5 that the performance of direct compression and reconstruction of the signal is worse than the result in the DFT or DCT domain. When the CR = 60%, the amplitude error can still be less than 5%, the value of the SNR could be more than 30 dB, the value of the ERP differs from 1 by less than 0.5%, and the phase error is less than 2%. Therefore, in the case of low accuracy and lack of corresponding hardware facilities, the proposed algorithm can directly compress and reconstruct harmonic signals.

4.3. Comparison of Reconstruction Performance with Other Algorithms

In order to judge the performance of the proposed algorithm more scientifically and objectively, this paper compares the performance of this algorithm with other data compression algorithms. For the performance comparison between different algorithms, the priorities of comparison were amplitude error, SNR, RRMSE, ERP, and angle error. This experiment selected the harmonic signal; the CR was selected as 40%; the DCT transform domain was selected; and the CoSaMP, SbOMP [24], and SAMP-SD [25] algorithms were compared. The sparsity (K) was set to 16, and the number of iterations was set to 40. Its comparative performance is shown in Table 6.

Table 6. Performance comparison of reconstruction algorithms (the harmonic signal).

	Traditional Methods			Proposed Method
Algorithms	SAMP-SD	CoSaMP	SbOMP	BSBL
SNR (dB)	39.41	38.01	35.94	40.58
RRMSE (%)	1.11	1.15	1.53	0.93
Amplitude error (%)	1.08	1.12	1.47	0.85
ERP (%)	100.06	100.09	100.13	100.16
Angle error (%)	1.39	1.43	1.40	1.39

From Table 6, it can be seen that the SNR of the proposed algorithm was better than that of the traditional algorithm. Both the RRMSE and amplitude error were less than 1%, while other algorithms were more than 1%. This shows that the proposed algorithm can improve the first three indicators.

4.4. Compression and Reconstruction of Measured Signals

4.4.1. Signals with Harmonics

In order to further verify the practicability of the BSBL algorithm, the real-world data of the capacitive current of a wind farm retrieved from Ningxia Province, China were used for compression and reconstruction. It can be seen from Figure 7 that the signal segment contains rich (inter)harmonic content. The various indicators of reconstruction are shown in Tables 7 and 8.

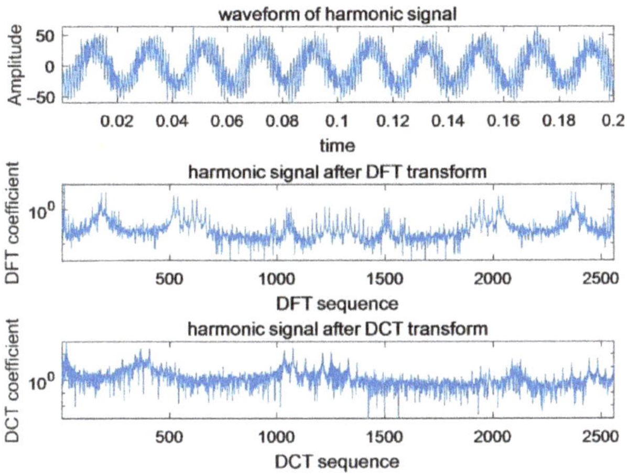

Figure 7. Measured original signal waveform and its DFT and DCT coefficients.

Table 7. Reconstruction results of the harmonic signal in the DFT domain.

Transform Domain	DFT			
CR (%)	30	40	50	60
SNR (dB)	22.24	25.10	29.66	37.53
RRMSE (%)	7.73	5.56	3.29	1.33
Amplitude error (%)	7.62	5.49	3.24	1.31
ERP (%)	103.78	101.63	100.84	100.28
Angle error (%)	1.34	1.36	1.35	1.32

Table 8. Reconstruction results of the harmonic signal in the DCT domain.

Transform Domain	DCT			
CR (%)	30	40	50	60
SNR (dB)	25.91	28.25	30.51	32.28
RRMSE (%)	6.10	4.72	2.95	2.85
Amplitude error (%)	5.83	4.33	3.16	2.51
ERP (%)	102.33	101.03	100.30	100.30
Angle error (%)	1.15	1.13	1.07	1.03

It can be seen from Tables 7 and 8 that for the measured harmonic data, no matter which transformation was adopted, its sparsity was much lower than that of the constructed signal. Therefore, under the same conditions, its performance indicators were lower than those of the constructing signals. When the CR was $\geq$ 50%, the amplitude error met the *IEC40000-6-7* standard. The SNR after the DCT was more than 30 dB, and the SNR after the DFT was 30 dB. In this case, CR = 50% was selected.

The following compares the performance of the proposed algorithm and the traditional algorithm applied to real-world data. The traditional algorithm was set to CR = 40%. Since the sparsity of the measured data cannot be estimated, the sparsity was cycled from 40 to 300. The final result of the traditional algorithm is the result of the largest SNR under the DFT and DCT and the sparsity cycle as the experimental result to compare with the proposed algorithm. The proposed algorithm takes CR = 40%, the result in the DCT domain.

It can be seen from Table 9 that the first three indicators of the proposed algorithm are far better than the traditional ones and because there is no need to predict the sparsity in advance. The process of the proposed algorithm is also simpler. The comparison result shows that the proposed algorithm can be applied to engineering.

Table 9. Performance comparison of reconstruction algorithms (real-world harmonic signal).

	Traditional Methods			Proposed Method
Algorithms	SAMP-SD	CoSaMP	SbOMP	BSBL
SNR (dB)	23.46	21.56	20.98	28.25
RRMSE (%)	6.34	8.35	7.43	4.72
Amplitude error (%)	6.28	8.23	7.36	4.33
ERP (%)	100.48	100.60	100.92	101.03
Angle error (%)	1.19	1.22	1.26	1.13

4.4.2. Fluctuant Signal

Figure 8 shows the fluctuant signal. The application scenario of these data is the sub-synchronous oscillation caused by the super-synchronous interharmonic of the port power electronic converter group. The various indicators of reconstruction are shown in Tables 10 and 11.

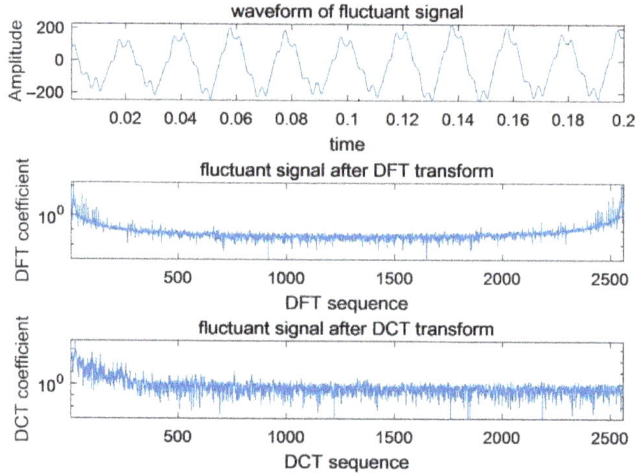

Figure 8. Measured fluctuant signal waveform and its DFT and DCT coefficients.

Table 10. Reconstruction results of the fluctuant signal in the DFT domain.

Transform Domain	DFT			
CR (%)	30	40	50	60
SNR (dB)	23.59	36.56	42.13	43.76
RRMSE (%)	6.61	1.49	0.78	0.65
Amplitude error (%)	6.55	1.48	0.76	0.64
ERP (%)	111.51	102.03	100.34	100.04
Angle error (%)	1.28	1.30	1.35	1.37

For the measured fluctuation signal, the performance of the DFT was better than that of the DCT. Taking CR = 40% under the DFT can be satisfied.

Table 12 shows the performance comparison between the proposed and traditional algorithms with parameter settings the same as those in Section 4.4.1.

Table 11. Reconstruction results of the fluctuant signal in the DCT domain.

Transform Domain	DCT			
CR (%)	30	40	50	60
SNR (dB)	23.52	26.58	35.75	40.76
RRMSE (%)	6.67	4.69	1.63	0.92
Amplitude error (%)	5.98	4.14	1.33	0.71
ERP (%)	103.71	100.81	100.20	100.03
Angle error (%)	1.18	1.16	1.02	0.90

Table 12. Performance comparison of reconstruction algorithms (real-world fluctuant signal).

	Traditional Methods			Proposed Method
Algorithms	SAMP-SD	CoSaMP	SbOMP	BSBL
SNR (dB)	35.23	34.72	35.33	36.56
RRMSE (%)	1.85	1.83	1.72	1.49
Amplitude error (%)	1.80	1.67	1.68	1.48
ERP (%)	100.10	100.04	100.1	102.03
Angle error (%)	1.36	1.22	1.29	1.30

It can be seen from the corresponding coefficients in Figures 7 and 8 that the sparsity of the fluctuant signal was much smaller than that of the harmonic signal, so the performance of each algorithm was good, and the performance of the proposed algorithm was slightly better than the traditional algorithm. Furthermore, from the comparison of these two experiments, the reconstruction performance of the proposed algorithm was better than other algorithms for signals with more harmonic content.

4.5. Analysis of Results

The reconstruction performance of the voltage fluctuation signal and harmonic signal (Tables 1–4) shows that the smaller the sparsity of the original signal after sparse transformation, the higher the reconstruction performance. For the harmonic signal, when the compression ratio increases, the performance of each indicator also increases. For the above constructed and measured data, the trend was consistent with CR. However, the measured signal performed worse than the construction signal because of its rich harmonic content. By comparing the constructed signal with the measured signal, with any transformation and harmonic signal with CR = 50%, the amplitude error met the requirements of *IEC41000-4-7*; the SNR was more than 30 dB; and the performance results of the other indicators, such as RRMSE, ERP, and angle error, were also good. There was little difference between the two transformation domains. For fluctuation signals, when the CR = 40%, the amplitude error met the requirements, but other indicators under the DCT were lower, so using the DFT is recommended.

When compared with the performance of greedy algorithms (SbOMP, SAMP-SD, CoSAMP), the BSBL algorithm does not need to estimate the sparsity of the steady-state power quality signal in advance. Furthermore, the BSBL algorithm has a more prominent effect when the harmonic component in the signal is large. The proposed algorithm can be used normally without sparse transformation, but its performance is slightly worse than with sparse transformation. Therefore, when a large number of steady-state power quality data need to be recorded, sampled, and transmitted in engineering, BSBL is a good solution.

5. Conclusions

This paper studies the application of BSBL in steady-state power quality data compression, converts data into sparse data through different transformation domains (DFT, DCT), and compresses and reconstructs it using BSBL. Based on the relative standard, two indicators of power quality data characteristics, namely relative amplitude error and relative phase angle error, are also presented in this paper. Then, the algorithm's validity is verified by constructing and measuring signals. The results show that CR = 50% can meet the standard for harmonic signals in engineering, no matter which transformation is used. For fluctuation signals, using the CR = 40% selection after the DFT is recommended to obtain a better result. The results also show that when $\mu_{th} = 10^{-5}$ and CR = 60% are selected, the standard requirement can be met without sparse transformation of signals. Finally, compared with other greedy algorithms (SbOMP, SAMP-SD, CoSAMP), we show that the proposed algorithms are superior to other greedy algorithms in terms of relevant indicators. Therefore, in the case of conforming to the standard, the proposed algorithm can meet the requirements of steady-state signal compression of power quality in engineering. Meanwhile, in the performance comparison with other algorithms, the two indicators of the ERP and angle error are not improved. This will be the direction of our next research study.

Author Contributions: Writing—original draft, W.H.; Writing—review and editing, M.Z.; Resources, H.Z. All authors have read and agreed to the published version of the manuscript.

Funding: This research was funded by the China Southern Power Grid Corporation Science and Technology Project [No. 090000KK52190169; SZKJXM2019669].

Acknowledgments: We thank Shenzhen Power Supply Co., Ltd. for the data support they provided.

Conflicts of Interest: The authors declare no conflict of interest.

References

1. Minh Khoa, N.; van Dai, L. Detection and classification of power quality disturbances in power system using modified-combination between the Stockwell transform and decision tree methods. *Energies* **2020**, *13*, 3623. [CrossRef]
2. Wang, S.; Chen, H.; Wu, L.; Wang, J. A novel smart meter data compression method via stacked convolutional sparse auto-encoder. *Int. J. Electr. Power Energy Syst.* **2020**, *118*, 105761. [CrossRef]
3. Santoso, S.; Powers, E.; Grady, W. Power quality disturbance data compression using wavelet transform methods. *IEEE Trans. Power Deliv.* **1997**, *12*, 1250–1257. [CrossRef]
4. Wang, K.; Zhang, M.; Zhang, S.; Xu, Z. A PQ Data Compression Algorithm Based on Wavelet Domain Principal Component Analysis. In Proceedings of the 2020 Asia Energy and Electrical Engineering Symposium (AEEES), Chengdu, China, 29–31 May 2020; pp. 347–350. [CrossRef]
5. Bao, W.; Zhou, R.; Liu, J.F. A Periodical Data Compression Method Based on 2-D Lifting Wavelet Transform in Thermal Power Plant. *Proc. CSEE* **2007**, *29*, 98–103.
6. Donoho, D.L. Compressed sensing. *IEEE Trans. Inf. Theory* **2006**, *52*, 1289–1306. [CrossRef]
7. Mallat, S.; Zhang, Z. Matching pursuits with time-frequency dictionaries. *IEEE Trans. Signal Process.* **1993**, *41*, 3397–3415. [CrossRef]
8. Goklani, H.S.; Sarvaiya, J.N.; Fahad, A.M. Image reconstruction using Orthogonal Matching Pursuit (OMP) algorithm. In Proceedings of the 2014 2nd International Conference on Emerging Technology Trends in Electronics, Communication and Networking, Surat, India, 26–27 December 2014; pp. 1–5. [CrossRef]
9. Ahmed, M.M.; Bedour, H.; Hassan, S.M. Design and Implementation of a Multistage Image Compression and Reconstruction System Based on the Orthogonal Matching Pursuit Using FPGA. In Proceedings of the 2019 14th International Conference on Computer Engineering and Systems (ICCES), Cairo, Egypt, 17–18 December 2019; pp. 174–179. [CrossRef]
10. Song, S.; Zhang, X.; Hao, Q.; Wang, Y.; Feng, N.; Shen, Y. An improved reconstruction method based on auto-adjustable step size sparsity adaptive matching pursuit and adaptive modular dictionary update for acoustic emission signals of rails. *Measurement* **2022**, *189*, 110650. [CrossRef]
11. Sun, M.; Zhao, A.; Li, B.; Zhang, J.; He, Q.; Song, X. Virtual compressed sensing photoacoustic imaging using CoSaMP algorithm based on k-wave. In Proceedings of the Sixteenth National Conference on Laser Technology and Optoelectronics, International Society for Optics and Photonics, Shanghai, China, 3–6 June 2021; Volume 11907, p. 1190739.
12. Ucar, F.; Cordova, J.; Alcin, O.F.; Dandil, B.; Ata, F.; Arghandeh, R. Bundle extreme learning machine for power quality analysis in transmission net-works. *Energies* **2019**, *12*, 1449. [CrossRef]
13. Zhou, W.; Wu, Y.; Huang, X.; Lu, R.; Zhang, H.-T. A group sparse Bayesian learning algorithm for harmonic state estimation in power systems. *Appl. Energy* **2021**, *306*, 118063. [CrossRef]
14. Wang, S.; Li, D.; Liu, M.; Huang, W.; Chen, H.; Cen, Y. Clustered-Sparse Bayesian Learning for Channel Estimation in Underwater Acoustic OFDM Systems. In Proceedings of the 2020 International Conference on Wireless Communications and Signal Processing, Nanjing China, 21–23 October 2020; IEEE: Piscataway, NJ, USA, 2020; pp. 546–551. [CrossRef]
15. Ni, Z.; Huang, B.; Cao, M. Angular Positions Estimation of Spatially Extended Targets for MIMO Radar Using Complex Spatiotemporal Sparse Bayesian Learning. *IEEE Access* **2019**, *7*, 94473–94480. [CrossRef]

16. Chen, D.; Luo, X.; Tao, S.; Chen, R. Multi radar imaging based on variational Bayesian block sparse method. In Proceedings of the 2018 IEEE International Conference on Computational Electromagnetics (ICCEM), Chengdu, China, 26–28 March 2018.
17. Liu, J.; Li, X.; Fang, K.; Fan, T. Millimeter wave channel estimation based on clustering block sparse bayesian learning. In Proceedings of the 2019 11th International Conference on Wireless Communications and Signal Processing (WCSP), Xi'an, China, 23–25 October 2019.
18. Zhang, Z.; Jung, T.-P.; Makeig, S.; Rao, B.D. Compressed Sensing of EEG for Wireless Telemonitoring with Low Energy Consumption and Inexpensive Hardware. *IEEE Trans. Biomed. Eng.* **2012**, *60*, 221–224. [CrossRef] [PubMed]
19. Candes, E.J.; Romberg, J.K.; Tao, T. Stable signal recovery from incomplete and inaccurate measurements. *Commun. Pure Appl. Math.* **2006**, *59*, 1207–1223. [CrossRef]
20. Xiao, X.N. *Power Quality Analysis and Control*; China Electric Power Press: Beijing, China, 2010.
21. Wipf, D.P.; Rao, B.D. An Empirical Bayesian Strategy for Solving the Simultaneous Sparse Approximation Problem. *IEEE Trans. Signal Process.* **2007**, *55*, 3704–3716. [CrossRef]
22. Eldar, Y.C.; Kuppinger, P.; Bolcskei, H. Block-Sparse Signals: Uncertainty Relations and Efficient Recovery. *IEEE Trans. Signal Process.* **2010**, *58*, 3042–3054. [CrossRef]
23. Zhang, Z.; Rao, B.D. Sparse Signal Recovery with Temporally Correlated Source Vectors Using Sparse Bayesian Learning. *IEEE J. Sel. Top. Signal Process* **2011**, *5*, 912–926. [CrossRef]
24. Hang, L.; Bo, W.; Daizhi, L.; Rongqiang, H. Processing Optical Fiber Sensing Signals with Improved Compressed Sensing Algorithm. In Proceedings of the 2020 5th International Conference on Computer and Communication Systems (ICCCS), Shanghai, China, 15–18 May 2020; pp. 660–664. [CrossRef]
25. Zhiyuan, S.; Qianqian, W.; Xinmiao, C. A sparsity adaptive compressed signal reconstruction based on sensing dictionary. *J. Syst. Eng. Electron.* **2021**, *32*, 1345–1353. [CrossRef]

energies

MDPI

Article

A Novel Hybrid Transfer Learning Framework for Dynamic Cutterhead Torque Prediction of the Tunnel Boring Machine

Tao Fu, Tianci Zhang and Xueguan Song *

School of Mechanical Engineering, Dalian University of Technology, No.2 Linggong Road, Ganjingzi District, Dalian 116024, China; futao_cam@126.com (T.F.); tczhang@mail.dlut.edu.cn (T.Z.)
* Correspondence: sxg@dlut.edu.cn

Abstract: A tunnel boring machine (TBM) is an important large-scale engineering machine, which is widely applied in tunnel construction. Precise cutterhead torque prediction plays an essential role in the cost estimation of energy consumption and safety operation in the tunneling process, since it directly influences the adaptable adjustment of excavation parameters. Complicated and variable geological conditions, leading to operational and status parameters of the TBM, usually exhibit some spatio-temporally varying characteristic, which poses a serious challenge to conventional data-based methods for dynamic cutterhead torque prediction. In this study, a novel hybrid transfer learning framework, namely TRLS-SVR, is proposed to transfer knowledge from a historical dataset that may contain multiple working patterns and alleviate fresh data noise interference when addressing dynamic cutterhead torque prediction issues. Compared with conventional data-driven algorithms, TRLS-SVR considers long-ago historical data, and can effectively extract and leverage the public latent knowledge that is implied in historical datasets for current prediction. A collection of in situ TBM operation data from a tunnel project located in China is utilized to evaluate the performance of the proposed framework.

Keywords: tunnel boring machine (TBM); cutterhead torque prediction; operation parameters; transfer learning

Citation: Fu, T.; Zhang, T.; Song, X. A Novel Hybrid Transfer Learning Framework for Dynamic Cutterhead Torque Prediction of the Tunnel Boring Machine. *Energies* **2022**, *15*, 2907. https://doi.org/10.3390/en15082907

Academic Editor: Manoj Khandelwal

Received: 14 March 2022
Accepted: 11 April 2022
Published: 15 April 2022

1. Introduction

Tunnel boring machines (TBM) are widely applied in various tunnel construction projects, such as subways, mining ores, railways, etc., due to advantages of higher reliability, safety, and environmental friendliness [1]. Figure 1 illustrates a typical structure of the TBM, which contains multiple sub-systems, such as the cutterhead driving system, thrust system, cutterhead system, etc. In real-world applications, TBMs generally work in heterogeneous and complicated geological environments, such as spalling, faulting, fracturing, rock bursting, squeezing, swelling, and high water in the flow [2], that pose severe challenges to the operation of TBMs. A schematic illustration of the geological conditions of a tunnel is demonstrated in Figure 2. To ensure construction safety and reduce energy consumption, it is desirable to accurately predict the dynamic load (generally referring to the cutterhead torque) under spatio-temporally varying geological conditions and to dynamically adjust the TBM control parameters during excavation.

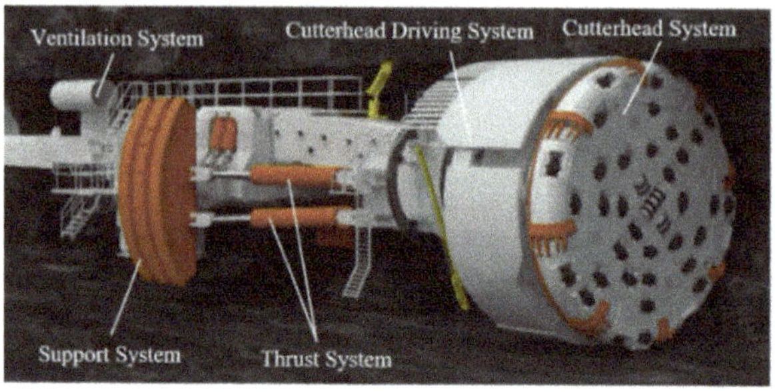

Figure 1. A typical diagram of the TBM. Reproduced with permission from [3], *Springer Nature*, 2016.

In general, the prediction methods for cutterhead torque can be roughly grouped into three types: rock–soil mechanics methods, empirical methods (combined with experiments), and soft computing methods. The rock–soil mechanics method establishes a model according to the force balance among rock, cutters, and internal machinery [4,5]. The empirical models are based on engineering experience involving a large amount of laboratory tests, field measurements, and construction records [6,7]. The soft computing methods are developed as data-based solutions for predicting the TBM's load through mathematical mapping. Rostami [8] elaborated theoretical and empirical methods in a recent review. S. K. Shreyas [9] and Shahrour Isam [10] provided a brief retrospect of recent application of soft computing methods to predict various parameters in tunneling and underground excavations.

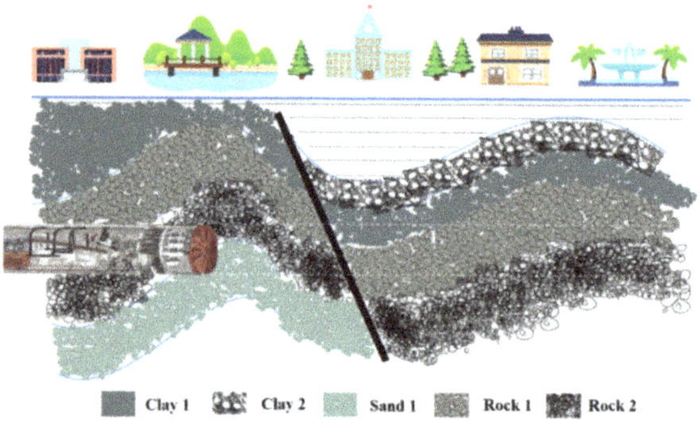

Figure 2. Longitudinal geological profile of a tunnel. Reproduced with permission from [11], IEEE, 2019.

By dividing the tunnel alignment into three general sections in terms of geological and geotechnical conditions, Avunduk et al. [12] proposed an empirical model for predicting excavation performance of TBM. Through the mechanical decoupling method for analyzing the cutterhead–ground interaction, Zhang et al. [13] proposed an approximate calculation method for determining the load acting on the cutterhead. Based on the interaction between the TBM and excavated material, Faramarzi et al. [14] applied the discrete element method (DEM) to evaluate the TBM torque and thrust. Rock–soil mechanics methods and empirical models are both based on the premise that the geological information is known. However, the accurate prediction of a geological profile before excavation is a hard and challenging task. In tunneling and underground excavation, the geological information is obtained through borehole sampling, and the stratum between sampling points are usually estimated by linear fitting. The distance between the sampling points is typically considerable, and the relevant result is often different from the real distribution, which may affect the accuracy of the rock-soil mechanics methods and the empirical models [15].

Assisted by the advancement of sensor and measurement technology, modern TBMs can record series of operation parameters closely related to dynamic load, which provides a basis for the practical application of soft computing methods. Sun et al. [16] utilized the random forest (RF) algorithm to design a predictor for TBM load. Kong et al. [17] took geological conditions and operational data as inputs to build a prediction model based on the RF for predicting driving forces of a TBM in a soil–rock, mixed-face ground. Li et al. [18] used the one-dimensional convolutional neural networks and long short-term memory network (CNN-LSTM) to predict cutterhead speed and penetration rate (PR). Qin et al. [19] applied a deep neural network-based method to predict dynamic cutterhead torque based on operating data and status parameters. Suwansawat et al. [20] applied the multi-layer perceptron (MLP) to determine the correlation among TBM operational data, groundmass characteristics, and surface movements. Lau et al. [21] used a radial basis function (RBF) to estimate tunneling production rates of successive cycles. Gao et al. [22] used three kinds of recurrent neural networks (RNNs) to deal with TBM operating parameters' real-time prediction. Soft calculation methods usually involve the optimization of many parameters, and the selection of parameters based on experience will reduce the accuracy of the analysis results. To deal with this problem, there have been many hybrid methods proposed in the literature. For example, Zhou et al. [23] applied three optimization algorithms to optima of the hyper-parameters of the support vector machine (SVM) technique in forecasting the advance rate (AR) of TBMs. Armaghani et al. [24,25] proposed two hybrid, intelligent systems, namely the particle swarm optimization (PSO)-artificial neural network (ANN) and the imperialism competitive algorithm (ICA)-ANN, to approximate the PR and AR of TBMs, respectively.

Although relatively accurate prediction results can be achieved by soft computing approaches, most of them generally assume that training samples and future test samples have identical distribution characteristics, and their practicability still has room for improvement. During the excavation process, TBMs encounter varying geological and working conditions, such as accelerating, turning, jamming releasing, etc., resulting in considerable changes in the underlying pattern of operation data over space and time. So, historical datasets behave as a non-stationary time series that makes the correlation among parameters in a high degree of complicated, changeable, and challenging conditions to be described by simple or fixed mathematical expressions. Hence, it is a serious challenge to extract common knowledge from historical datasets to assist in building an adaptive model which dynamically changes with geological conditions and operating parameters, for implementing dynamic cutterhead torque prediction at the current moment. To a certain degree, this problem is similar to the paradigm of transfer learning [26,27], which addresses this problem by utilizing experiences gained from source tasks to improve the learning of new related tasks. Hu et al. [28] applied the concept of transfer learning for efficient wind speed prediction. The prediction model was trained on samples from older data-rich farms to extract wind speed patterns, and then finely tuned with samples from newly built farms. Rui et al. [29] constructed a novel transfer learning paradigm for time series prediction, and the principle of transfer learning is employed. However, TBM's historical data contains a variety of geological information and working modes. So, directly adopting the most intuitive transfer learning method without distinguishing all the working modes in the historical data may result in negative transfer problems.

Herein, a novel hybrid data-mining framework based on clustering, multitask learning (MTL), transfer learning, and least-squares support vector regression machines (LS-SVR), abbreviated as TRLS-SVR, is proposed for dynamic cutterhead torque forecasting of TBMs. In this framework, LS-SVR is selected as a baseline model, which has a powerful capability to capture underlying nonlinear relationships for a complex system. The underlying patterns in historical data are effectively divided according to the relationship among attributes [30]. To take advantage of the knowledge contained in different working modes and to eliminate the damage from dataset bias, we adopt the idea of MTL [31], which explicitly exploits commonalities and differences across multiple working modes by learning them simultaneously rather than individually, to improve knowledge extraction ability. Based on the common knowledge extracted from historical data, we utilize the newly collected operation data to continuously update the pattern-specific biases parameters for adapting to the changing geological and working conditions. This study offers the following innovations and contributions. (1) The unsupervised learning algorithm for data clustering is combined with the MTL paradigm to explore and exploit the correlations among multiple working modes by learning simultaneously rather than individually, which enhances the ability of extracting public knowledge from a diversely recorded TBM historical dataset. (2) It employs a transfer learning paradigm to reuse the public knowledge that is contained in the historical dataset to supply new data, and it alleviates random noise interference and fits the varying geological and working conditions well. (3) The TRLS-SVR performs superior performance in geologically complex and changeable locations, compared with that of conventional data-driven algorithms.

The rest of this study is organized as follows. Section 2 presents details of the proposed framework. In Section 3, the experimental verification is presented. In Section 4, some discussions on experimental results are provided. Section 5 concludes the whole study and provides future work.

2. The Proposed Dynamic Cutterhead Torque Prediction Framework

2.1. Overall Framework

The framework of dynamic cutterhead torque prediction proposed in this paper draws inspiration from various machine learning methods, including clustering, MTL, and transfer learning. The overall framework of the TRLS-SVR mainly consists of four components, namely data pre-processing, dividing of typical working modes based on unsupervised clustering algorithm, extracting implicit common knowledge by MTL algorithm, and knowledge reuse based on transfer learning, as described in Figure 3. In the first step, a large number of historical datum that have a long-time span with current sample are extracted from the database. In the second step, a clustering algorithm was used to effectively divide working modes in the historical dataset according to the relationship among attributes. Next, the MTL paradigm was used to exploit representative knowledge from multiple working modes. Based on the transfer learning paradigm, experiences extracted from the historical dataset were retained and utilized to train a fresh model. The detailed descriptions of each component are introduced below.

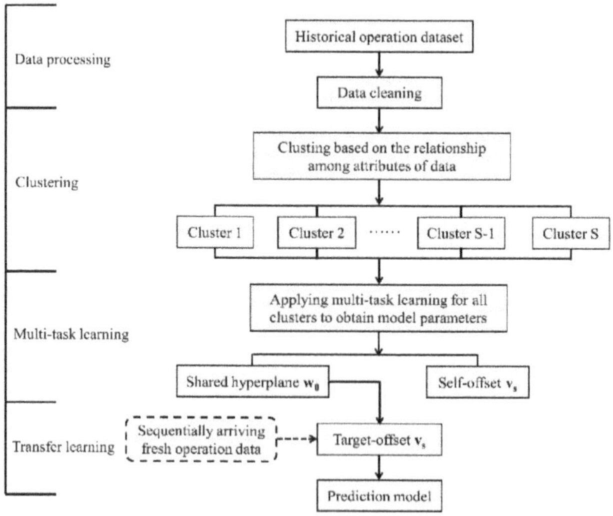

Figure 3. The proposed dynamic cutterhead torque prediction framework.

2.2. Clustering Based on the Relationship among Attributes

Due to continuous changes in geological conditions and work patterns, the historical dataset may contain multiple modes. In order to better extract public knowledge under different patterns, the first step is to divide the historical dataset into different clusters. Clustering as a pre-processing algorithm to uncover the underlying patterns and find natural partitioning within a dataset is widely utilized in engineering data analyses, such as fault detection, pattern recognition, and risk analysis. Currently, widely used clustering algorithms such as K-Nearest Neighbor (K-NN) and the fuzzy c-means algorithm (FCM) are mostly based on the spatial distribution to classify the dataset. However, the spatial distribution of different categories of TBM operation data is often similar, and conventional data clustering methods might not partition it effectively. The relationship among attributes varies considerably under different working and geological conditions, which can be used to improve clustering performance [32]. Thus, in this paper, we employ the modified FCM algorithm, namely, SVR-FCM, presented by Shi et al. [30] for TBM operation data clustering, which is designed under the architecture of FCM, but it partitions the data based on the relationship among attributes rather than their spatial distribution. The distance D_{ik} is defined as follows:

$$D_{ik} = (x_{obj,k} - SVR(x_{1,k}, \ldots, x_{obj-1}, x_{obj+1,k}, \ldots, x_{s,k})_i)^2 \quad (1)$$

The clustering objective function modified as follows:

$$J_{SVR-FCM} = \sum_{i=1}^{c}\sum_{k=1}^{n} u_{ik}^{m}(x_{obj,k} - SVR(x_{1,k}, \ldots, x_{obj-1,k}, x_{obj+1,k}, \ldots, x_{s,k})_i)^2 \quad (2)$$

The necessary conditions for minimizing (2) result in the following partition matrix:

$$u_{ik} = \left[\sum_{t=1}^{c}\left(\frac{(x_{obj,k} - SVR(x_{1,k}, \ldots, x_{obj-1,k}, x_{obj+1,k}, \ldots, x_{s,k})_i)^2}{(x_{obj,k} - SVR(x_{1,k}, \ldots, x_{obj-1,k}, x_{obj+1,k}, \ldots, x_{s,k})_t)^2}\right)^{\frac{1}{m-1}}\right]^{-1} \quad (3)$$

A more detailed description of the algorithm architecture can be seen in [22].

2.3. Extracting Public Knowledge from Historical Dataset

The clustering categories correspond to typical working modes, which are combined of representative working and geological conditions. It should be noted that the data distribution is distinct but similar in different working modes. To extract the public knowledge contained in typical working modes, we adopt the paradigm of MTL, which explicitly exploits commonalities and differences across multiple working modes by learning them simultaneously rather than individually to improve knowledge extraction ability. MTL reinforces each task by using the interconnections between tasks, considering both the relevance and the difference between tasks to enhance the generalization performance. There has been abundant literature on MTL, showing that learning various related tasks simultaneously can be advantageous in predictive performance relative to learning these tasks independently [33,34]. This study adopts the MTL method based on the minimization of the regularization function similar to LS-SVR, which has been successfully utilized for single-task learning [35]. The LS-SVR can be formulated as Equation (4), which solves the regression problem by optimizing the output weight vector, $\mathbf{w}$, and bias term, b, by minimizing a cost function with constraint, as shown in Equation (5).

$$\mathbf{y} = \mathbf{w}^T \varphi(\mathbf{x}) + b \cdot 1 \quad (4)$$

where $\varphi(\bullet)$ denotes a features map.

$$\begin{aligned} &\min J(\mathbf{w},\mathbf{e}) = \tfrac{1}{2}\mathbf{w}^T\mathbf{w} + \rho\mathbf{e}^T\mathbf{e} \\ &s.t.\mathbf{y} = \mathbf{w}^T\varphi(\mathbf{x}) + b\cdot 1 + \mathbf{e} \end{aligned} \tag{5}$$

$\mathbf{e}$ is a vector consisting of slack variables, and the hyper-parameter ρ controls the relative weight of each term. Herein, the output weight vector of different working modes, noted as $\mathbf{w}_s$, can be divided into the common vector $\mathbf{w}_0$, shared by all working modes and working-mode-specific bias vectors, $\mathbf{v}_s$, which can be formulated as follows:

$$\mathbf{w}_s = \mathbf{w}_0 + \mathbf{v}_s, \forall s \in S \tag{6}$$

We estimate all $\mathbf{v}_s$ as well as the (common) $\mathbf{w}_0$ simultaneously. To this end, we solve the following optimization problem, which is analogous to the LS-SVR used for single-task learning:

$$\begin{aligned} &\min_{w_0, v_s, \xi_{s,i}, \rho_{s,i}} \left\{ J(\mathbf{w}_0, \mathbf{v}_s, \xi_{s,i}, \rho_{s,i}) := \gamma \cdot \tfrac{1}{2}\sum_{s=1}^{S}\sum_{i=1}^{ns} \xi_{s,i}{}^2 + \eta \cdot \tfrac{1}{2}\sum_{s=1}^{S}\sum_{i=1}^{ns} \rho_{s,i}{}^2 + \tfrac{1}{2}\cdot\tfrac{\lambda}{S}\sum_{s=1}^{S} ||\mathbf{v}_s||^2 + \tfrac{1}{2}||\mathbf{w}_0||^2 \right\} \\ &s.t.\phi(\mathbf{x}_{s,i})^T \cdot (\mathbf{w}_s) + b_s = y_{s,i} - \xi_{s,i} \\ &\phi(\mathbf{x}_{s,i})^T \cdot \mathbf{w}_0 + b_{s_0} = y_{s,i} - \rho_{s,i} \end{aligned} \tag{7}$$

The number of tasks is S, which is equal to the number of clustering results. Specifically, $\mathbf{x}_{s,i}$ represents the ith sample of the sth task, λ is the constraint coefficient, γ and η are penalty coefficient, and $\xi_{s,i}$ and $\rho_{s,i}$ represent the training error vector of the sth task. According to the Lagrangian multiplier method, to solve Equation (7) is equivalent to solving the corresponding Lagrangian problem:

$$\begin{aligned} &L_D = \tfrac{1}{2}||w_0||^2 + \tfrac{1}{2}\cdot\tfrac{\lambda}{S}\sum_{s=1}^{S}||v_s||^2 + \gamma\cdot\tfrac{1}{2}\sum_{s=1}^{S}\sum_{i=1}^{ns}\xi_{s,i}{}^2 + \eta\cdot\tfrac{1}{2}\sum_{s=1}^{S}\sum_{i=1}^{ns}\rho_{s,i}{}^2 - \\ &\sum_{s=1}^{S}\sum_{i=1}^{ns}\alpha_{s,i}\times\left\{(w_0+v_s)^T\cdot\phi(x_{s,i}) + b_s + \xi_{s,i} - y_{s,i}\right\} - \sum_{s=1}^{S}\sum_{i=1}^{ns}\beta_{s,i}\times\left\{w_0{}^T\cdot\phi(x_{s,i}) + b_{s_0} + \rho_{s,i} - y_{s,i}\right\} \end{aligned} \tag{8}$$

where $\alpha_{s,i}$ and $\beta_{s,i}$ are the ith Lagrangian multiplier for the sth task. Based on the Karush–Kuhn–Tucker (KKT) conditions, setting the first partial derivatives of L_D to zero,

$$\begin{cases} \frac{\partial L_D}{\partial \mathbf{w}_0} = 0 \Rightarrow \mathbf{w}_0 = \sum_{s=1}^{S}\sum_{i=1}^{ns}(\alpha_{s,i}+\beta_{s,i})\cdot\phi(\mathbf{x}_{s,i}) \\ \frac{\partial L_D}{\partial \mathbf{v}_s} = 0 \Rightarrow \mathbf{v}_s = \frac{S}{\lambda}\sum_{i=1}^{ns}\alpha_{s,i}\cdot\phi(\mathbf{x}_{s,i}), \forall_s \in S \\ \frac{\partial L_D}{\partial b_s} = 0 \Rightarrow \sum_{i=1}^{ns}\alpha_{s,i} = 0, \forall_s \in S \\ \frac{\partial L_D}{\partial b_{s_0}} = 0 \Rightarrow \sum_{s=1}^{S}\sum_{i=1}^{ns}\beta_{s,i} = 0 \\ \frac{\partial L_D}{\partial \xi_{s,i}} = 0 \Rightarrow \alpha_{s,i} = \gamma\cdot\xi_{s,i}, \forall_s \in S \\ \frac{\partial L_D}{\partial \rho_{s,i}} = 0 \Rightarrow \beta_{s,i} = \eta\cdot\rho_{s,i}, \forall_s \in S \\ \frac{\partial L_D}{\partial \alpha_{s,i}} = 0 \Rightarrow (\mathbf{w}_0+\mathbf{v}_s)^T\cdot\phi(\mathbf{x}_{s,i}) + b_s + \xi_{s,i} - y_{s,i} = 0, \forall_s \in S \\ \frac{\partial L_D}{\partial \beta_{s,i}} = 0 \Rightarrow \mathbf{w}_0{}^T\cdot\phi(\mathbf{x}_{s,i}) + b_{s_0} + \rho_{s,i} - y_{s,i} = 0, \forall_s \in S \end{cases} \tag{9}$$

Eliminating $\mathbf{w}_0$, $\{\mathbf{v_i}\}_{i=1}^{S}$, $\{\xi_{s,i}\}_{s=1,i=1}^{S,n_i}$, and $\{\rho_{s,i}\}_{s=1,i=1}^{S,n_i}$ results in the solution of (9), being $\boldsymbol{\alpha}^* = (\boldsymbol{\alpha}_1^{*T}, \boldsymbol{\alpha}_2^{*T}, \cdots, \boldsymbol{\alpha}_S^{*T})^T$ and $\boldsymbol{\beta}^* = (\boldsymbol{\beta}_1^{*T}, \boldsymbol{\beta}_2^{*T}, \cdots, \boldsymbol{\beta}_S^{*T})^T$, where

$\boldsymbol{\alpha}_s^* = \left(\alpha_{s,1}^*, \alpha_{s,2}^*, \cdots, \alpha_{s,ns}^*\right)^T$ and $\boldsymbol{\beta}_s^* = \left(\beta_{s,1}^*, \beta_{s,2}^*, \cdots, \beta_{s,ns}^*\right)^T$. The working mode-specific bias vectors can be mathematically formulated as follows:

$$\mathbf{v}_s = \frac{S}{\lambda} \sum_{i=1}^{ns} \alpha^*{}_{s,i} \cdot \phi(\mathbf{x}_{s,i}), \forall_s \in S \tag{10}$$

The extracted public knowledge is denoted as the following vector:

$$\mathbf{w}_0 = \sum_{s=1}^{S} \sum_{i=1}^{ns} (\alpha^*{}_{s,i} + \beta^*{}_{s,i}) \cdot \phi(\mathbf{x}_{s,i}) \tag{11}$$

2.4. Dynamic Cutterhead Torque Prediction Based on Transfer Learning

Transfer learning is an emerging framework that aims to provide a paradigm to utilize previously acquired experience to solve new but similar problems faster and more effectively [33]. There are some commonalities and associations between transfer learning and MTL. Both of them aim to improve the performance of learners via knowledge transfer. Transfer learning has been studied extensively for different applications in recent years, providing many opportunities for applying data-based methods to assist in design and analysis of complex engineering systems.

During the excavation process, geological information and operating parameters generally change continuously, so operation data around the excavation point have more reference significance for subsequent dynamic cutterhead torque prediction. In addition, vibration and shock often occur during excavation, and random noise interference inevitably exists in the measurement of fresh data, which may have a substantial impact on the prediction performance. Hence, training a new model by utilizing the knowledge contained in the historical dataset to reduce the requirement of number of new samples and alleviate the interference of random noise is always considered advisable. To leverage experiences extracted from the historical dataset, the output weight vector of the fresh model, noted as $\mathbf{w}_t$, is feasible to minimize the difference with the public vector, $\mathbf{w}_0$, that can be regarded as the public knowledge transferred from the historical dataset. We intend to train an approximator which has the minimal norm parameter vector and training errors for available fresh samples, that can be written as,

$$\begin{gathered} \min L = \tfrac{1}{2}||\mathbf{w}_t||^2 + \tfrac{1}{2}\mu||\mathbf{w}_t - \mathbf{w}_0||^2 + \tfrac{C}{2}\sum_{j=1}^{m_t} \xi_j^2 \\ s.t. \phi(\mathbf{x}_j)^T \mathbf{w}_t + b_t = y_j - \xi_j \end{gathered} \tag{12}$$

where $\mathbf{w}_t$ is the output weight vector over the fresh data, μ denotes the penalty parameter, C is the regularization parameter, ξ_j is the training error, and m_t is number of fresh training sets around the excavation point. According to Lagrangian multiplier method, to solve Equation (12) is equivalent to solving the corresponding Lagrangian problem:

$$L_D = \frac{1}{2}||\mathbf{w}_t||^2 + \frac{1}{2}\mu||\mathbf{w}_t - \mathbf{w}_0||^2 + \frac{C}{2}\sum_{j=1}^{m_t} \xi_j^2 - \sum_{j=1}^{m_t} \alpha_j(\phi(\mathbf{x}_j)^T \mathbf{w}_t + b_t - y_j + \xi_j) \tag{13}$$

where α_j is the *jth* Lagrangian multiplier, and based on the KKT conditions, the problem can be solved with the Lagrangian multiplier method,

$$\begin{cases} \frac{\partial L_D}{\partial \mathbf{w}_t} = 0 \Rightarrow \mathbf{w}_t = \frac{1}{1+\mu}\left((\mu \mathbf{w}_0 + \sum_{j=1}^{m_t} \alpha_j \phi(\mathbf{x}_j)\right) \\ \frac{\partial L_D}{\partial \xi_j} = 0 \Rightarrow \alpha_j = C\xi_j \\ \frac{\partial L_D}{\partial b_t} = 0 \Rightarrow \sum_{j=1}^{m_t} \alpha_j = 0 \\ \frac{\partial L_D}{\partial \alpha_j} = 0 \Rightarrow \phi(\mathbf{x}_j)^T \cdot \mathbf{w}_t + b_t - y_j + \xi_j = 0 \end{cases} \tag{14}$$

On analysis of Equation (14), it can be concluded that:

$$\begin{cases} \frac{1}{1+\mu}\left(\phi(\mathbf{x}_j)^T \cdot \mu \cdot w_0 + \sum_{k=1}^{m_t} \alpha_k \phi(\mathbf{x}_k)^T \cdot \phi(\mathbf{x}_j)\right) + b_t - y_j + \frac{\alpha_j}{C} = 0 \\ \sum_{j=1}^{mt} \alpha_j = 0 \end{cases} \tag{15}$$

Plugging Equations (12) and (13) into Equation (15) can we obtain:

$$\begin{gathered} \frac{1}{1+\mu}\left(\mu \sum_{s=1}^{S}\sum_{i=1}^{ns} (\alpha_{s,i}+\beta_{s,i}) \cdot \phi(\mathbf{x}_{s,i}) \cdot \phi(\mathbf{x}_j{}^T) + \sum_{k=1}^{m_t} \alpha_k \phi(\mathbf{x}_k) \cdot \phi(\mathbf{x}_j{}^T)\right) + b_t - y_j + \frac{\alpha_j}{C} = 0 \\ \Rightarrow \left(\frac{1}{1+\mu}\begin{bmatrix} \phi(\mathbf{x}_1)\cdot\phi(\mathbf{x}_1{}^T) & \cdots & \phi(\mathbf{x}_1)\cdot\phi(\mathbf{x}_{m_t}{}^T) \\ \vdots & \ddots & \vdots \\ \phi(\mathbf{x}_{m_t})\cdot\phi(\mathbf{x}_1{}^T) & \cdots & \phi(\mathbf{x}_{m_t})\cdot\phi(\mathbf{x}_{m_t}{}^T) \end{bmatrix} + \frac{1}{C}\right)\begin{bmatrix} \alpha_1 \\ \alpha_2 \\ \vdots \\ \alpha_{m_t} \end{bmatrix} + b_t = \begin{bmatrix} \frac{-\mu}{1+\mu}\cdot\sum_{s=1}^{S}\sum_{i=1}^{ns}(\alpha_{s,i}+\beta_{s,i})\cdot\phi(\mathbf{x}_{s,i})\cdot\phi(\mathbf{x}_1{}^T) + y_1 \\ \frac{-\mu}{1+\mu}\cdot\sum_{s=1}^{S}\sum_{i=1}^{ns}(\alpha_{s,i}+\beta_{s,i})\cdot\phi(\mathbf{x}_{s,i})\cdot\phi(\mathbf{x}_2{}^T) + y_2 \\ \vdots \\ \frac{-\mu}{1+\mu}\cdot\sum_{s=1}^{S}\sum_{i=1}^{ns}(\alpha_{s,i}+\beta_{s,i})\cdot\phi(\mathbf{x}_{s,i})\cdot\phi(\mathbf{x}_{mt}{}^T) + y_{m_t} \end{bmatrix} \\ \Rightarrow \left(\frac{1}{1+\mu}\begin{bmatrix} \phi(\mathbf{x}_1)\cdot\phi(\mathbf{x}_1{}^T) & \cdots & \phi(\mathbf{x}_1)\cdot\phi(\mathbf{x}_{m_t}{}^T) & 1+\mu \\ \vdots & \ddots & \vdots & \vdots \\ \phi(\mathbf{x}_{m_t})\cdot\phi(\mathbf{x}_1{}^T) & \cdots & \phi(\mathbf{x}_{m_t})\cdot\phi(\mathbf{x}_{m_t}{}^T) & 1+\mu \\ 1 & \cdots & 1 & \frac{-1-\mu}{C} \end{bmatrix} + \frac{1}{C}\right)\begin{bmatrix} \alpha_1 \\ \alpha_2 \\ \vdots \\ \alpha_{m_t} \\ b_t \end{bmatrix} = \begin{bmatrix} \frac{-\mu}{1+\mu}\cdot\sum_{s=1}^{S}\sum_{i=1}^{ns}(\alpha_{s,i}+\beta_{s,i})\cdot\phi(\mathbf{x}_{s,i})\cdot\phi(\mathbf{x}_1{}^T) + y_1 \\ \frac{-\mu}{1+\mu}\cdot\sum_{s=1}^{S}\sum_{i=1}^{ns}(\alpha_{s,i}+\beta_{s,i})\cdot\phi(\mathbf{x}_{s,i})\cdot\phi(\mathbf{x}_2{}^T) + y_2 \\ \vdots \\ \frac{-\mu}{1+\mu}\cdot\sum_{s=1}^{S}\sum_{i=1}^{ns}(\alpha_{s,i}+\beta_{s,i})\cdot\phi(\mathbf{x}_{s,i})\cdot\phi(\mathbf{x}_{mt}{}^T) + y_{m_t} \\ 0 \end{bmatrix} \end{gathered} \tag{16}$$

Let the solution of (16) be $\boldsymbol{\alpha}_t^* = (\alpha_1^*, \alpha_2^*, \cdots, \alpha_{m_t}^*)^T$ and $\mathbf{b}_t^*$. In addition, the dynamic cutterhead torque prediction of fresh data can be mathematically formulated as follows:

$$\begin{aligned} f_t(\mathbf{x}) &= \phi(\mathbf{x})^T \mathbf{w}_t^* + b_t^* \\ &= \frac{\mu}{1+\mu}\sum_{s=1}^{S}\sum_{i=1}^{ns}(\alpha_{s,i}^* + \beta_{s,i}^*)\phi(\mathbf{x}_{s,i})\phi(\mathbf{x}^T) + \frac{1}{1+\mu}\sum_{k=1}^{m_t}\alpha_k^*\phi(\mathbf{x}_k)\phi(\mathbf{x}^T) + b_t^* \end{aligned} \tag{17}$$

3. Numerical Experiments

In this section, a collection of real-world operational and status parameters of TBM is utilized to demonstrate the superiority and applicability of the framework.

3.1. Experimental Settings

The tunnel project studied in this study is located in Shenzhen, China, which is about 2000 m long and 6.4 m in diameter. As described in Figure 4a, from the ground surface to the tunnel floor, various geological layers, such as clay, sand, and rock, are unevenly distributed. The tunneling equipment used in this tunnel is shown in Figure 4b, and has an earth pressure balance shield TBM with 500 T of total mass and 120 knives on its cutter head. The basic equipment parameters are listed in Table 1. During the tunneling process, the operational and state data of the TBM were recorded by a PLC, which was further read by an industrial computer at regular intervals and stored in the database. Thus, the fresh data in the database were added in batches during the tunneling process. The collected operation dataset represents the operational information and status parameters along the length of the tunnel, which contains about 44 attributes, such as cutterhead torque, chamber pressure, and advance velocity, etc. Please refer to the appendix for a detailed list of these attributes (see Table A1). In the process of dynamic cutterhead torque prediction, data come in batches. We selected five sets of sequence data to construct the test datasets, covering various working and geological conditions. Each collection of data contained approximately 5000 rows and 44 columns; the first 80% of the dataset were used as training samples and the last 20% were used as test samples. Each row of data represents the data of all physical quantities at a certain moment, and each column of data represents the data of a physical quantity at any moment.

(a) (b)

Figure 4. Geological sampling results and the TBM used. (**a**) Geological sampling results. (**b**) The TBM used.

Table 1. Basic parameters of the TBM used.

Parameters	Value	Unit
Cutterhead diameter	6680	mm
Maximum torque	8322	kNm
Rated power of drive motor	160	kW
Number of drive motors	8	1

To improve prediction accuracy, in this paper, we first normalized the samples with a normalization method, which is an essential pre-processing step in the field of machine learning. It is commonly referred to simply as "normalization," or sometimes as "feature scaling," and can be formulated as:

$$\min - \max = \frac{x - X_{\min}}{X_{\max} - X_{\min}} \tag{18}$$

where x is the current value and $X_{\min}$ and $X_{\max}$ are the minimum and maximum values of the entire dataset, respectively. The min–max method rescales values and confines samples to an interval between 0 and 1.

The operational data modeling was conducted with a personal computer (CPU: Intel Core i7-10700; RAM: 32 G). The framework was coded by the author with Matlab and set as follows: the clustering algorithm parameter set refers to the setting of references [30], where the fuzzification parameter, m, was 2, threshold value, ε, was 10^{-6}, number of clusters was 4, and maximum iteration was 1000. The radial basis function (RBF) was selected as a kernel function for LS-SVR. Compared with ordinary LS-SVR models, the framework proposed in this paper has more hyper-parameters, such as, η, λ, and μ, that determine the information extracted from historical data and knowledge transferred for constructing a new model. In this section, we set $\eta = 100$, $\lambda = 1$, and μ was determined according to the forecast accuracy of the previous batch, varying with the value of μ as {1, 5, 10, 15}. Other hyper-parameters were set to the same values with the baseline model LS-SVR, i.e., $\gamma = C = 100$.

3.2. Experiments and Results

To verify the efficacy and superiority, the performance of TRLS-SVR was compared to that of existing data-driven methods, such as RF, SVR, Lasso, and deep neural networks, i.e., long short-term memory (LSTM) networks [22] and online learning methods (i.e., online support vector regression (OSVR) [36]). The fitness of these prediction models was evaluated with four error criterions, i.e., the coefficient of determination (R^2), mean absolute error (*MAE*), root mean square error (*RMSE*), and mean absolute percentage error (*MAPE*). These metrics have the following formula.

$$R^2 = \frac{\sum_{i=1}^{n} (y_i' - \overline{y})^2}{\sum_{i=1}^{n} (y_i - \overline{y})^2} \quad (19)$$

$$MAE = \frac{1}{n}\sum_{i=1}^{n} |y_i - y_i'| \quad (20)$$

$$RMSE = \sqrt{\frac{1}{n}\sum_{i=1}^{n} (y_i - y_i')^2} \quad (21)$$

$$MAPE = \frac{1}{n}\sum_{i=1}^{n} \frac{|y_i - y_i'|}{y_i} \times 100\% \quad (22)$$

where n is total number of samples, $\overline{y}_i$ is the average of all actual values, and y_i' is the predicted value of y_i. The closer R^2 is to 1, the better the performance is. The *MAE* and *RMSE* measure the disparity between actual values and predicted values, which reflects the dispersion of models. The *RMSE* is more sensitive to large errors than *MAE* because the errors are squared, and the large errors are amplified further. *MAPE* is the ratio between errors and actual values. It can be considered as a relative error function, and the smaller the value, the higher the prediction accuracy. These four error criterions can be applied to evaluate the fitness of these prediction models from various viewpoints.

The evaluation results of the proposed TRLS-SVR and other five data-driven models on the five test datasets are shown in Tables 2–5. In general, the results show that three indicators of TRLS-SVR, i.e., *MAE*, *RMSE*, and *MAPE*, are lower than the other five data-driven models, and the coefficient of determination, R^2, is higher than others. The average value of R^2, *MAE*, *RMSE*, and *MAPE* in the five datasets are 0.83, 63.34, 95.11, and 3.61% for the proposed TRLS-SVR; −3.03, 321.60, 385.72, and 16.38% for RF; −0.391, 199.86, 235.22, and 10.58% for LSTM; −0.48, 194.75, 236.27, and 10.52% for SVR; −0.146, 160.55, 211.86, and 8.89% for Lasso; 0.56, 88.97, 148.40, and 5.28% for OSVR, respectively. Hence, the average *MAE* of TRLS-SVR is 80.3% less than RF; 68.31% less than LSTM; 67.48% less than SVR; 60.55% less than Lasso; 28.8% less than OSVR, respectively. In addition, the average *RMSE* of TRLS-SVR is 75.34% less than RF; 59.56% less than LSTM; 59.74% less than SVR; 55.11% less than Lasso; 35.91% less than OSVR, respectively. Moreover, the prediction precision of TRLS-SVR is 77.95% higher than RF; 65.85% higher than LSTM; 65.67% high than SVR; 59.35% higher than Lasso; 31.61% higher than OSVR, respectively.

Table 2. R^2 of different methods in five datasets.

Datasets	RF	LSTM	SVR	Lasso	OSVR	TRLS-SVR
1	0.43	0.68	0.73	0.55	0.58	0.83
2	−9.86	−1.60	0.31	−1.33	0.55	0.85
3	−2.10	−1.21	−3.27	−0.84	0.46	0.85
4	−0.43	0.56	0.46	0.74	0.77	0.85
5	−3.17	−0.39	−0.64	0.15	0.45	0.77

Table 3. *MAE* of different methods in five datasets.

Datasets	RF	LSTM	SVR	Lasso	OSVR	TRLS-SVR
1	172.49	123.69	116.54	135.93	98.61	81.37
2	493.13	231.81	108.57	213.93	71.23	38.00
3	264.91	218.95	298.31	186.23	77.73	49.12
4	215.99	152.91	163.10	93.97	75.85	58.91
5	461.49	271.95	287.26	172.69	121.41	89.31

Table 4. *RMSE* of different methods in five datasets.

Datasets	RF	LSTM	SVR	Lasso	OSVR	TRLS-SVR
1	205.35	153.85	143.10	183.64	176.36	111.59
2	549.17	268.61	138.01	254.49	111.35	65.46
3	309.10	260.87	362.76	238.04	128.90	67.80
4	311.40	173.49	190.99	133.39	124.72	101.36
5	553.55	319.28	346.48	249.76	200.65	129.36

Table 5. *MAPE* of different methods in five datasets.

Datasets	RF	LSTM	SVR	Lasso	OSVR	TRLS-SVR
1	9.18%	6.53%	6.51%	7.85%	6.31%	4.45%
2	22.19%	10.45%	5.04%	9.69%	3.38%	1.82%
3	15.89%	14.05%	18.18%	11.58%	5.14%	3.09%
4	13.56%	10.53%	10.64%	7.26%	5.78%	4.65%
5	21.09%	11.33%	12.22%	8.05%	5.79%	4.04%

For visual comparison, the real cutterhead torque values and predicted cutterhead torque values with these models are also provided in Figures 5–9. It can be observed that the prediction accuracy of existing data-driven models, i.e., RF, LSTM, SVR, and Lasso, is relatively low, and can only predict the average value and changing trend but cannot achieve prediction dynamically and accurately. The main reason may lie in that the cutterhead torque sequence is nonlinear and non-stationary, and it may contain several different working conditions simultaneously. Therefore, it is not advisable to describe the cutterhead torque sequence data by a simple or fixed mathematical formula. The in situ monitoring data are spatio-temporally coupled, and the data close to the excavation point have more reference significance for subsequent load prediction. Using these fresh data to update the model parameters dynamically can capture the load data sequence's changing trend with the geological parameters and the working parameters. Therefore, online learning-based methods' prediction accuracy is higher than traditional statistical data-driven models. In addition, in spite of online learning-based methods, OSVR has high prediction accuracy in some samples; its accuracy is still less than TRLS-SVR on the entire dataset, mainly because there is random noise interference in the measurement of cutterhead torque data. Only using a small amount of fresh data that are close to the excavation point to update model parameters will inevitably overfit random noises and introduce model bias, which leads to performance degradation.

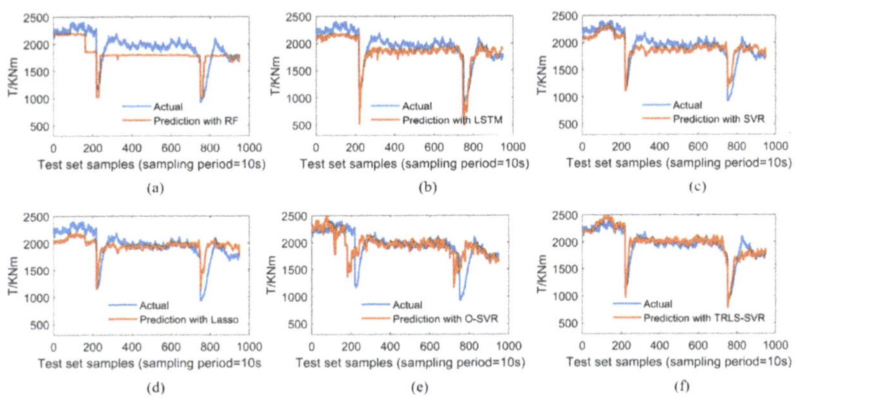

Figure 5. Comparisons between real and predicted cutterhead torque for dataset 1. (**a**) Prediction result of RF. (**b**) Prediction result of LSTM. (**c**) Prediction result of SVR. (**d**) Prediction result of Lasso. (**e**) Prediction result of OSVR. (**f**) Prediction result of TRLS-SVR.

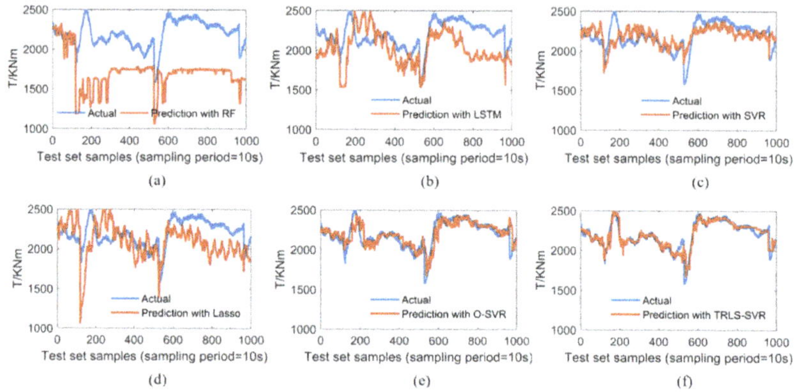

Figure 6. Comparisons between real and predicted cutterhead torque for dataset 2. (**a**) Prediction result of RF. (**b**) Prediction result of LSTM. (**c**) Prediction result of SVR. (**d**) Prediction result of Lasso. (**e**) Prediction result of OSVR. (**f**) Prediction result of TRLS-SVR.

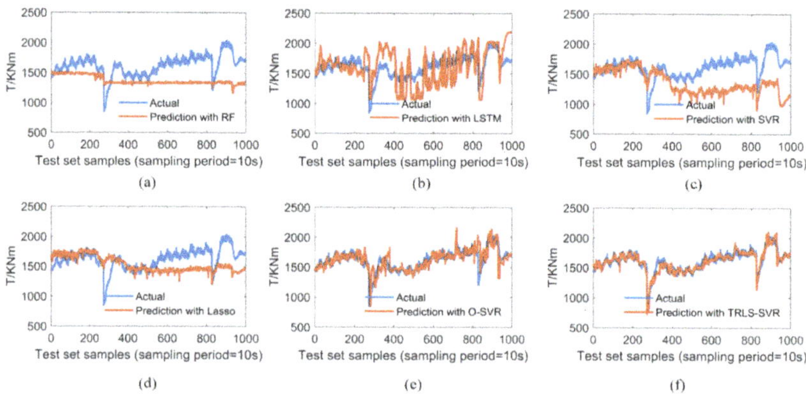

Figure 7. Comparisons between real and predicted cutterhead torque for dataset 3. (**a**) Prediction result of RF. (**b**) Prediction result of LSTM. (**c**) Prediction result of SVR. (**d**) Prediction result of Lasso. (**e**) Prediction result of OSVR. (**f**) Prediction result of TRLS-SVR.

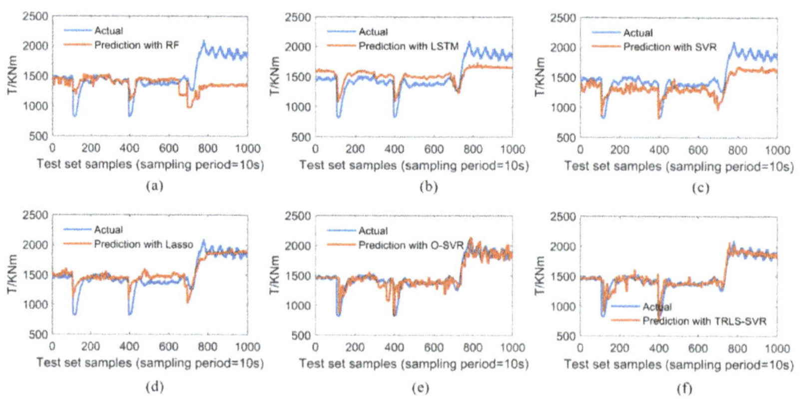

Figure 8. Comparisons between real and predicted cutterhead torque for dataset 4. (**a**) Prediction result of RF. (**b**) Prediction result of LSTM. (**c**) Prediction result of SVR. (**d**) Prediction result of Lasso. (**e**) Prediction result of OSVR. (**f**) Prediction result of TRLS-SVR.

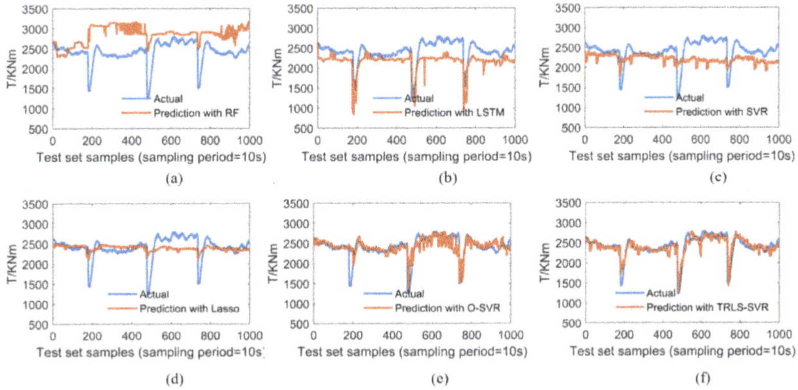

Figure 9. Comparisons between real and predicted cutterhead torque for dataset 5. (**a**) Prediction result of RF. (**b**) Prediction result of LSTM. (**c**) Prediction result of SVR. (**d**) Prediction result of Lasso. (**e**) Prediction result of OSVR. (**f**) Prediction result of TRLS-SVR.

The TRLS-SVR can effectively divide different working and geological conditions of historical data, and learn the cutterhead torque sequence's changing rule under different working modes. When the new coming data are disturbed by random noises or the excavation section's geological conditions, the implicit knowledge contained in historical data is explicitly transferred to reduce over-fitting of random noise, and to avoid introducing model bias. As a result, the proposed TRLS-SVR can achieve better prediction performance than that of existing data-driven methods.

4. Discussion

Compared with those of the baseline data-driven method, LS-SVR, the TRLS-SVR has more hyper-parameters, for example, η, λ, μ, and the number of fresh training sets, m_t. These hyper-parameters determine the amount of information extracted from historical data and the proportion of this information in the model update, which may affect the performance of the algorithm. As mentioned in Section 3.1, regularization parameter μ is determined according to the prediction accuracy of the previous batch. In this section, we focus on how the hyper-parameters η, μ, and number of fresh training sets, m_t, influence the prediction accuracy of the TRLS-SVR framework.

4.1. Analysis of the Number of Fresh Training Sizes

In these experiments, we select 10, 20, 50, 100, 200, 300, and 400 of the datum which are near the excavation point as fresh training sets. The prediction accuracy of the different number of fresh training sets is compared in Figure 10. It can be seen that when the number of training sizes, m_t, is small, the performance of TRLS-SVR improves faster as the number of samples increases, and when the number of training size, m_t, is relatively high, the performance decreases as the number of samples increases. When the number of training sizes, m_t, is 50, the proposed framework tends to provide the best prediction performance. This is because too little training data cannot reduce the interference of noise, which will lead to over-fitting of the noise and affect the prediction accuracy, while too much training data will smooth the changing characteristics of the continuous data to obtain average statistical characteristics and reduce the prediction accuracy.

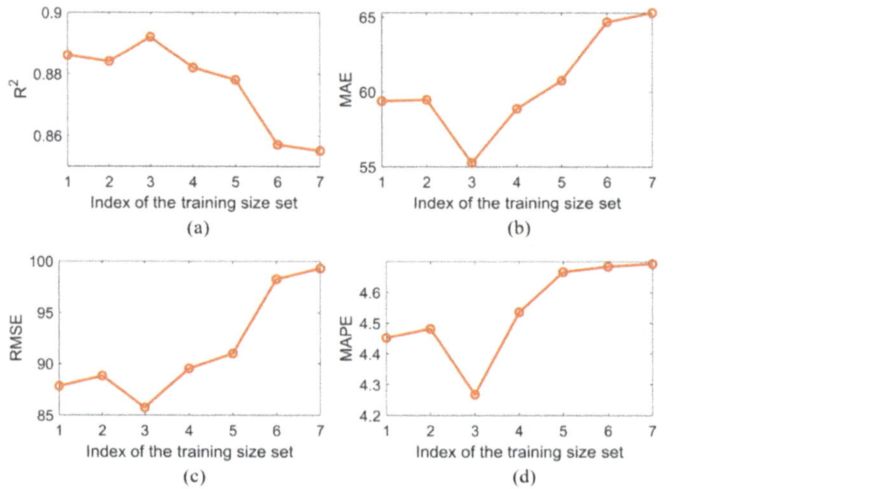

Figure 10. Sensitivity analysis on training size. (**a**) R2. (**b**) MAE. (**c**) RMSE. (**d**) MAPE.

4.2. Analysis of Regularization Parameters

We conduct experiments on the TBM dataset to discuss the sensitivity of the two regularization parameters η and λ. We fix the number of fresh training sets, m_t, as 50, hyper-parameters as $C = \gamma = 100$, and regularization parameter μ is determined according to the prediction accuracy of the previous batch. For the sensitivity analysis of the regularization parameter, η, we fix $\lambda = 1$ and vary the value of η as $\{10^{-3}, 10^{-2}, 10^{-1}, 1, 10, 100, 200, 500, 1000, 2000\}$. For the sensitivity analysis of the regularization parameter, λ, we fix $\eta = 100$ and vary the value of λ as $\{10^{-5}, 10^{-4}, 10^{-3}, 10^{-2}, 10^{-1}, 1, 10, 100, 1000\}$. The prediction accuracy of different values of regularization parameters are compared in Figures 11 and 12. In Figure 11, it can be seen that the optimal prediction accuracy by TRLS-SVR is achieved by setting $\eta = 100$ when $\lambda = 1$ is fixed. From Figure 12, it can be seen that the optimal prediction accuracy by TRLS-SVR is achieved by setting the value of λ as a small value. In addition, the prediction accuracy of TRLS-SVR changes slightly when the value of λ is in the range of $[10^{-5}, 1]$.

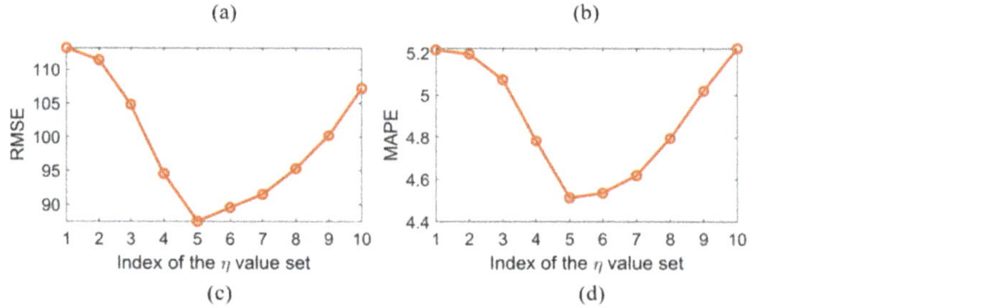

Figure 11. Sensitivity analysis on the regularization parameter, η. (**a**) R2. (**b**) MAE. (**c**) RMSE. (**d**) MAPE.

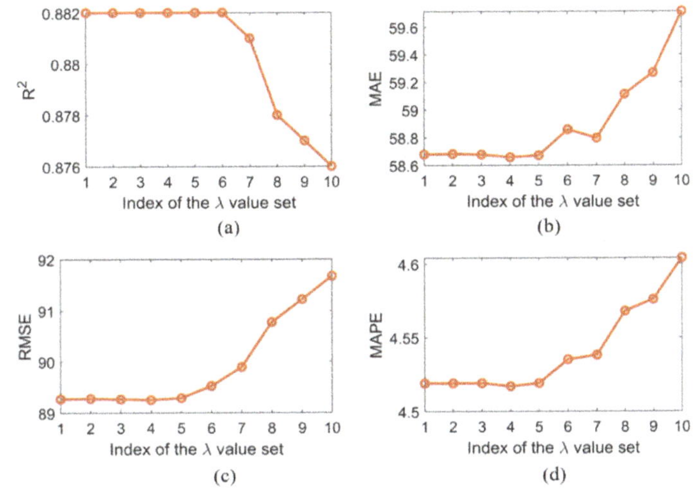

Figure 12. Sensitivity analysis on the regularization parameter, λ. (**a**) R2. (**b**) MAE. (**c**) RMSE. (**d**) MAPE.

4.3. Limitations and Recommendations

The heterogeneous in situ data of the TBM include not only numerical data but also categorical data, such as the geological data. The heterogeneous in situ data have one special characteristic that is different for the sizes of the geological data and the operation data, which limits the application of data-driven techniques on them. Thus, in this paper, we only consider the operational data and ignore the geological data. In the future, to further improve the prediction accuracy of the framework, it is necessary to integrate geological data through multi-source heterogeneous data fusion.

5. Conclusions

In this study, a novel hybrid transfer learning framework named TRLS-SVR, that aims to enhance the accuracy of TBM dynamic cutterhead torque prediction, is proposed. In the proposed framework, the underlying patterns in historical datasets were effectively divided according to the relationship among attributes. The idea of MTL was adopted to exploit commonalities and differences across various working modes by learning them simultaneously rather than individually, to capture the public knowledge from historical datasets. In order to cope with the changing geological and working conditions, the idea of transfer learning was adopted and the newly collected operation data were utilized to continuously update the parameters of the forecasting model as a supplement. Real-world, in situ operational and status parameters from a tunnel located in Shenzhen, China, were utilized to evaluate the efficacy and superiority of the proposed framework. Experimental results demonstrated that the TRLS-SVR alleviated the shortcoming of traditional statistical data-driven methods, which can only predict the average value and changing trend of the cutterhead torque but cannot achieve dynamically and accurately the prediction of the load. Additionally, compared with the method of an online learning paradigm, which puts more attention to data closer to the excavation point, the framework has stronger robustness. This is because the model can use the knowledge contained in historical data to reduce the impact of random noise and alleviate over-fitting issues. In summary, the major novelty of this study is to provide a first test of merging MTL and transfer learning for TBM dynamic cutterhead torque prediction. Though the framework is presented in the context of dynamic cutterhead torque prediction of TBM, it can be easily extended to the status monitoring of other engineering systems, such as wind power equipment, automobiles, etc. In the near future, we plan to further investigate the adaptable adjustment of TBM's operating status based on the proposed framework, which is of great significance to the operation safety and energy consumption.

Author Contributions: Conceptualization, T.F.; Funding acquisition, X.S.; Investigation, T.Z.; Methodology, T.F. and T.Z.; Project administration, X.S.; Resources, X.S.; Software, T.F. and T.Z.; Writing—original draft, T.F. All authors have read and agreed to the published version of the manuscript.

Funding: This research was funded by the National Key R&D Program of China (Grant No 2018YFB1702502) and the National Natural Science Foundation of China (Grant No. 52075068).

Institutional Review Board Statement: Not applicable.

Informed Consent Statement: Not applicable.

Data Availability Statement: Not applicable.

Conflicts of Interest: The authors declare no conflict of interest. The funders had no role in the design of the study; in the collection, analyses, or interpretation of data; in the writing of the manuscript, or in the decision to publish the results.

Appendix A

Table A1. List of operating parameters.

Parameter (Unit)	Parameter (Unit)
Temperature of oil tank (°C)	Temperature of gear oil (°C)
Rotation speed of cutterhead(r/min)	Cutter power (kw)
Propelling pressure (bar)	Propelling pressure of A group (bar)
Propelling pressure of B group (bar)	Propelling pressure of C group (bar)
Propelling pressure of D group (bar)	Pressure of equipment bridge (bar)
Pressure of articulation system (bar)	Pressure of shield tail seal at top right front (bar)
Pressure of shield tail seal at right front (bar)	Pressure of shield tail seal at left front (bar)
Pressure of shield tail seal at top right back (bar)	Pressure of shield tail seal at right back (bar)
Pressure of shield tail seal at bottom back (bar)	Pressure of shield tail seal at left front (bar)
Pressure of shield tail seal at top left front (bar)	Pressure of shield tail seal at left back (bar)
Pressure of shield tail seal at top left back (bar)	Pressure of shield tail seal at right back (bar)
Rolling angle (°)	Pressure of screw pump at back (bar)
Pressure of chamber at top left (bar)	Pressure of chamber at bottom left (bar)
Pressure of chamber at bottom right (bar)	Bentonite pressure (bar)
Temperature of screw conveyor (°C)	Pitch angle (°)
Thrust of cutterhead (kN)	Advance velocity (mm/min)
Torque of cutterhead (kNm)	Displacement of A group of thrust cylinders (mm)
Displacement of B group of thrust cylinders (mm)	Displacement of C group of thrust cylinders (mm)
Displacement of D group of thrust cylinders (mm)	Displacement of articulated system at top right (mm)
Displacement of articulated system at left (mm)	Displacement of articulated system at top left (mm)
Displacement of articulated system at right (mm)	Bentonite pressure of shield shell (bar)
Pressure of screw conveyor at front (bar)	Pressure of screw pump (bar)

References

1. Zheng, Y.L.; Zhang, Q.B.; Zhao, J. Challenges and opportunities of using tunnel boring machines in mining. *Tunn. Undergr. Space Technol.* **2016**, *57*, 287–299. [CrossRef]
2. Delisio, A.; Zhao, J.; Einstein, H.H. Analysis and prediction of TBM performance in blocky rock conditions at the Lötschberg Base Tunnel. *Tunn. Undergr. Space Technol.* **2013**, *33*, 131–142. [CrossRef]
3. Sun, W.; Wang, X.; Wang, L.; Zhang, J.; Song, X. Multidisciplinary design optimization of tunnel boring machine considering both structure and control parameters under complex geological conditions. *Struct. Multidiscip. Optim.* **2016**, *54*, 1073–1092. [CrossRef]
4. Wang, L.; Gong, G.; Shi, H.; Yang, H. Modeling and analysis of thrust force for EPB shield tunneling machine. *Autom. Constr.* **2012**, *27*, 138–146. [CrossRef]
5. Hassanpour, J.; Rostami, J.; Khamehchiyan, M.; Bruland, A. Developing new equations for TBM performance prediction in carbonate-argillaceous rocks: A case history of Nowsood water conveyance tunnel. *Geomech. Geoeng.* **2009**, *4*, 287–297. [CrossRef]
6. Delisio, A.; Zhao, J. A new model for TBM performance prediction in blocky rock conditions. *Tunn. Undergr. Space Technol.* **2014**, *43*, 440–452. [CrossRef]
7. Yagiz, S. New equations for predicting the field penetration index of tunnel boring machines in fractured rock mass. *Arab. J. Geosci.* **2017**, *10*, 33. [CrossRef]

8. Rostami, J. Performance prediction of hard rock Tunnel Boring Machines (TBMs) in difficult ground. *Tunn. Undergr. Space Technol.* **2016**, *57*, 173–182. [CrossRef]
9. Shreyas, S.K.; Dey, A. Application of soft computing techniques in tunnelling and underground excavations: State of the art and future prospects. *Innov. Infrastruct. Solut.* **2019**, *4*, 46. [CrossRef]
10. Shahrour, I.; Zhang, W. Use of soft computing techniques for tunneling optimization of tunnel boring machines. *Undergr. Space* **2021**, *6*, 233–239. [CrossRef]
11. Zhao, J.; Shi, M.; Hu, G.; Song, X.; Zhang, C.; Tao, D.; Wu, W. A Data-Driven Framework for Tunnel Geological-Type Prediction Based on TBM Operating Data. *IEEE Access* **2019**, *7*, 66703–66713. [CrossRef]
12. Avunduk, E.; Copur, H. Empirical modeling for predicting excavation performance of EPB TBM based on soil properties. *Tunn. Undergr. Space Technol.* **2018**, *71*, 340–353. [CrossRef]
13. Zhang, Q.; Hou, Z.; Huang, G.; Cai, Z.; Kang, Y. Mechanical characterization of the load distribution on the cutterhead-ground interface of shield tunneling machines. *Tunn. Undergr. Space Technol.* **2015**, *47*, 106–113. [CrossRef]
14. Faramarzi, L.; Kheradmandian, A.; Azhari, A. Evaluation and Optimization of the Effective Parameters on the Shield TBM Performance: Torque and Thrust—Using Discrete Element Method (DEM). *Geotech. Geol. Eng.* **2020**, *38*, 2745–2759. [CrossRef]
15. Leng, S.; Lin, J.R.; Hu, Z.Z.; Shen, X. A Hybrid Data Mining Method for Tunnel Engineering Based on Real-Time Monitoring Data from Tunnel Boring Machines. *IEEE Access* **2020**, *8*, 90430–90449. [CrossRef]
16. Sun, W.; Shi, M.; Zhang, C.; Zhao, J.; Song, X. Dynamic load prediction of tunnel boring machine (TBM) based on heterogeneous in-situ data. *Autom. Constr.* **2018**, *92*, 23–34. [CrossRef]
17. Kong, X.; Ling, X.; Tang, L.; Tang, W.; Zhang, Y. Random forest-based predictors for driving forces of earth pressure balance (EPB) shield tunnel boring machine (TBM). *Tunn. Undergr. Space Technol.* **2022**, *122*, 104373. [CrossRef]
18. Li, L.; Liu, Z.; Zhou, H.; Zhang, J.; Shen, W.; Shao, J. Prediction of TBM cutterhead speed and penetration rate for high-efficiency excavation of hard rock tunnel using CNN-LSTM model with construction big data. *Arab. J. Geosci.* **2022**, *15*, 280. [CrossRef]
19. Qin, C.; Shi, G.; Tao, J.; Yu, H.; Jin, Y.; Lei, J.; Liu, C. Precise cutterhead torque prediction for shield tunneling machines using a novel hybrid deep neural network. *Mech. Syst. Signal Process.* **2021**, *151*, 107386. [CrossRef]
20. Suwansawat, S.; Einstein, H.H. Artificial neural networks for predicting the maximum surface settlement caused by EPB shield tunneling. *Tunn. Undergr. Space Technol.* **2006**, *21*, 133–150. [CrossRef]
21. Lau, S.C.; Lu, M.; Ariaratnam, S.T. Applying radial basis function neural networks to estimate next-cycle production rates in tunnelling construction. *Tunn. Undergr. Space Technol.* **2010**, *25*, 357–365. [CrossRef]
22. Gao, X.; Shi, M.; Song, X.; Zhang, C.; Zhang, H. Recurrent neural networks for real-time prediction of TBM operating parameters. *Autom. Constr.* **2019**, *98*, 225–235. [CrossRef]
23. Zhou, J.; Qiu, Y.; Zhu, S.; Armaghani, D.J.; Li, C.; Nguyen, H.; Yagiz, S. Optimization of support vector machine through the use of metaheuristic algorithms in forecasting TBM advance rate. *Eng. Appl. Artif. Intell.* **2021**, *97*, 104015. [CrossRef]
24. Armaghani, D.J.; Mohamad, E.T.; Narayanasamy, M.S.; Narita, N.; Yagiz, S. Development of hybrid intelligent models for predicting TBM penetration rate in hard rock condition. *Tunn. Undergr. Space Technol.* **2017**, *63*, 29–43. [CrossRef]
25. Armaghani, D.J.; Koopialipoor, M.; Marto, A.; Yagiz, S. Application of several optimization techniques for estimating TBM advance rate in granitic rocks. *J. Rock Mech. Geotech. Eng.* **2019**, *11*, 779–789. [CrossRef]
26. Zhuang, F.; Qi, Z.; Duan, K.; Xi, D.; Zhu, Y.; Zhu, H.; Xiong, H.; He, Q. A Comprehensive Survey on Transfer Learning. *Proc. IEEE* **2021**, *109*, 43–76. [CrossRef]
27. Lu, J.; Behbood, V.; Hao, P.; Zuo, H.; Xue, S.; Zhang, G. Transfer learning using computational intelligence: A survey. *Knowl.-Based Syst.* **2015**, *80*, 14–23. [CrossRef]
28. Hu, Q.; Zhang, R.; Zhou, Y. Transfer learning for short-term wind speed prediction with deep neural networks. *Renew. Energy* **2016**, *85*, 83–95. [CrossRef]
29. Ye, R.; Dai, Q. A novel transfer learning framework for time series forecasting. *Knowl.-Based Syst.* **2018**, *156*, 74–99. [CrossRef]
30. Shi, M.; Zhang, L.; Sun, W.; Song, X. A fuzzy c-means algorithm guided by attribute correlations and its application in the big data analysis of tunnel boring machine. *Knowl.-Based Syst.* **2019**, *182*, 104859. [CrossRef]
31. Zhang, Y.; Yang, Q. A Survey on Multi-Task Learning. *IEEE Trans. Knowl. Data Eng.* **2021**, *4347*, 1–20. [CrossRef]
32. Song, X.; Shi, M.; Wu, J.; Sun, W. A new fuzzy c-means clustering-based time series segmentation approach and its application on tunnel boring machine analysis. *Mech. Syst. Signal Process.* **2019**, *133*, 106279. [CrossRef]
33. Li, Y.; Tian, X.; Liu, T.; Tao, D. On better exploring and exploiting task relationships in multitask learning: Joint model and feature learning. *IEEE Trans. Neural Netw. Learn. Syst.* **2018**, *29*, 1975–1985. [CrossRef] [PubMed]
34. Xu, S.; An, X.; Qiao, X.; Zhu, L. Multi-task least-squares support vector machines. *Multimed. Tools Appl.* **2014**, *71*, 699–715. [CrossRef]
35. Evgeniou, T.; Pontil, M. Regularized Multi–Task Learning. In Proceedings of the Tenth ACM SIGKDD International Conference on Knowledge Discovery and Data Mining, Washington, DC, USA, 22–25 August 2004.
36. Ma, J.; Theiler, J.; Perkins, S. Accurate On-line Support Vector Regression. *Neural Comput.* **2003**, *15*, 2683–2703. [CrossRef]

energies

Article

Experimental Interharmonic Sensitivity Evaluation of LED Lamps Based on the Luminous Flux Flicker Model

Hongtao Li [1], Yifan Song [1], Mingxing Zhu [2] and Yadong Jiao [2,*]

[1] State Grid Beijing Electric Power Research Institute, State Grid Beijing Electric Power Company, Beijing 100075, China; lihongtao@bj.sgcc.com.cn (H.L.); songyifan@bj.sgcc.com.cn (Y.S.)
[2] School of Electrical Engineering and Automation, Anhui University, Hefei 230601, China; zhumingxing@ahu.edu.cn
* Correspondence: jiao12345ya@163.com

Abstract: LED lamps have gradually replaced other lighting sources and have become mainstream in the lighting industry. The research on interharmonic sensitivity affecting their lighting quality cannot be ignored. By deconstructing the lamp-eye-brain module in the International Electrotechnical Commission (IEC) flicker model, a luminous flux flicker model without the constraints of a specific light source was proposed. The test system and corresponding analysis method of the interharmonic-luminous flux transfer coefficient in the model were described in detail, and the accuracy of the test results of the system was verified via incandescent lamp heat balance model simulations. Based on the test results, the conversion method of the interharmonic ratio of LED lamps under the flicker limit based on the interharmonic-flicker limit curve of incandescent lamps was deduced. By testing and comparing the differences in interharmonic-flicker limit curves of different driving types of LED lamps, the experimental evaluation of their sensitivity was completed, and the reference for LED lamp selection, driver design, and compatibility standard formulation in different application scenarios was provided.

Keywords: LED lamps; luminous flux; flicker; interharmonics; sensibility

Citation: Li, H.; Song, Y.; Zhu, M.; Jiao, Y. Experimental Interharmonic Sensitivity Evaluation of LED Lamps Based on the Luminous Flux Flicker Model. *Energies* **2022**, *15*, 2990. https://doi.org/10.3390/en15092990

Academic Editors: Nicu Bizon and S. M. Muyeen

Received: 22 February 2022
Accepted: 18 April 2022
Published: 19 April 2022

1. Introduction

Lighting is an indispensable part of the normal operation of modern society. With people becoming more concerned with lighting quality, energy conservation, and environmental protection, LED lamps relying on semiconductor technology have gradually replaced incandescent lamps, fluorescent lamps, and compact energy-saving lamps and have become the new mainstream lighting source [1,2]. Presently, the lighting change as a result of voltage fluctuations caused by interharmonics in the power grid is one of the most important factors affecting lighting quality [3–5]. The evaluation of LED lamp sensitivity to interharmonics is an essential part of product application selection, drive design optimization, and compatibility standard formulation.

Flicker quantifies the impact of light changes on human vision, and the measurement method for flicker is given by the International Electrotechnical Commission (IEC) 61000-4-15 standard based on the flicker effect of amplitude modulation voltage on tungsten incandescent lamps (referred to as incandescent lamps) [6–9]. Based on this, researchers have made various improvements, studied the flicker effect of interharmonics, and obtained the interharmonic level corresponding to the flicker limit [10,11]. However, these studies are based on the thermal radiation balance of incandescent lamps. This model is not applicable to the study of light sources with different luminous principles [12]. In this regard, the interharmonic sensitivity of fluorescent lamps has been studied by establishing the interharmonic flux flicker test platform and using the iterative approximation method, and the interharmonic-flicker limit curve of fluorescent lamps has been drawn one by one [13,14]. However, this method has some problems such as low efficiency and complexity; yet, there is little research on LED flicker. In [15], an actual case where the flicker meter could not detect exceedance of the standard was introduced, but the human eye could observe the light and shade changes of LED lights. The problem was solved by replacing different brands of LED lights. In [16], an incandescent lamp as the control group of the existing light source was used, and under the condition of the same voltage fluctuation level, the relative change in illuminance of different light sources was compared and analyzed to measure the degree of influence. In [17], the flicker effect of LED lamps with different color temperatures was studied; however, in the comparison between cool white and warm white lamps, the flicker value was measured via FULKE 435, the algorithm of which belongs to the IEC flicker model based on incandescent lamps. Therefore, its measurement results are not referential. In [18–21], the response distribution of different types of LED lamps under the same interharmonic conditions was analyzed with the percent flicker and flicker index from the perspective of luminous flux as the bases for evaluating the quality of lighting products. These studies lack the quantitative analysis of the corresponding relationship between the light and dark changes of LED lights caused by interharmonics. However, [22] proposed a flicker meter based on the general model of the lamp. The flicker test for different types of light sources was based on the changes in three parameters derived from the lamp type and its circuit, but sufficient experimental data to demonstrate the rationality and accuracy of the derived parameters are still needed. Although LED lamps are widely used, it is still necessary to study a new flicker model that is separate from incandescent lamps and take the flicker limit as a constraint to realize the evaluation of the interharmonic sensitivity of LED lamps.

By deconstructing the lamp-eye-brain module in the IEC flicker model, this paper proposed a luminous flux flicker model that is free from the constraints of specific light sources and describes in detail the test system and corresponding analysis method of lamp interharmonic luminous flux transfer coefficient in the model. The accuracy of the test results of the system was verified via incandescent lamp heat balance model simulations. Based on the results, the conversion method of the interharmonic ratio of LED lamps based on the interharmonic-flicker limit curve of incandescent lamps under the flicker limit was deduced. The experimental evaluation of its sensitivity was completed by testing and comparing the differences between interharmonic-flicker limit curves of different driving types of LED lamps.

The main contributions of this paper could be summarized as follows:

1. The lamp-eye-brain module in the IEC flicker model was deconstructed, and the luminous flux flicker model, free from the constraints of incandescent lamps, was obtained, providing a general model for the flicker effect measurement of any light source.
2. Through simulations and measurements of a specific incandescent lamp, the test system and corresponding analysis method of the interharmonic-luminous flux transfer coefficient composed of a programmable power supply, an illuminance sensor, a shading box, a multi-channel data acquisition device, and a data analysis system were verified. The luminous flux flicker model accuracy provided a test platform for its application.
3. Using the incandescent lamp given by IEC as a reference, and by applying the transfer coefficient of interharmonic voltage and luminous flux fluctuation of different light sources to calculate the corresponding interharmonic-flicker limit curve of the light source, the proposed method simplified the test operation steps while overcoming the programmable power limitations.
4. By comparing the magnitude of the interharmonic-flicker limit curves of different light sources at different frequencies, the sensitivity to interharmonics was measured.

The remainder of this paper is organized as follows. The luminous flux flicker model, derived based on the IEC flicker model, is described in Section 2. In Section 3, the core interharmonic-luminous flux testing system in the luminous flux flicker model is established and verified. The interharmonic sensitivity evaluation is presented in Section 4 via LED lamp interharmonics-flicker limit curve calculation. Section 5 discusses and analyzes the interharmonic sensitivity test results for LED lamps classified according to drive. The conclusions are presented in Section 6.

2. Luminous Flux Flicker Model

Flicker is the subjective visual perception of human eyes on the change in light brightness. Considering differences between individuals, to obtain a relatively objective flicker evaluation value, it is necessary to statistically analyze the observation data of human eyes under specific lamp conditions. Presently, the flicker measurement model provided by IEC 61000-4-15 is widely used. It is based on the test results of 60 W incandescent lamps at 120 V/60 Hz and 230 V/50 Hz. It is mainly composed of two parts: simulating the response of lamp-eye-brain to fluctuating voltage and statistical data processing, as shown in Figure 1a [23].

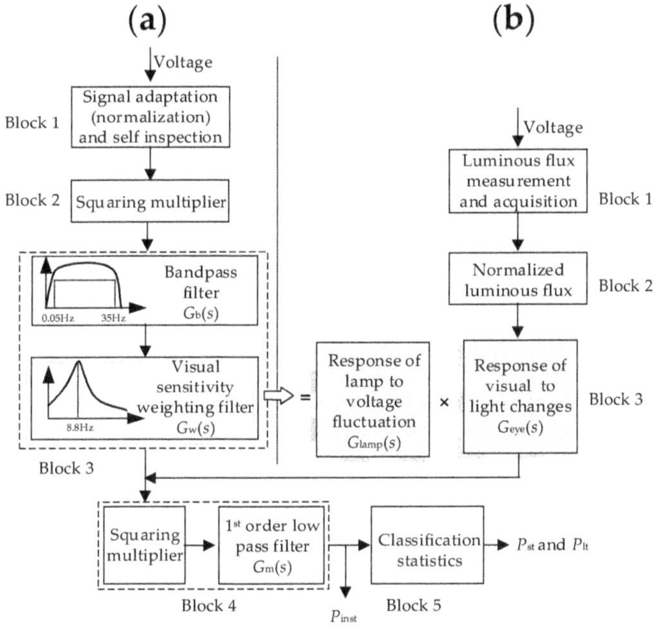

Figure 1. Relationship between IEC and the luminous flux flicker model: (**a**) IEC and (**b**) luminous flux flicker meter models.

Block 1 adapts (normalizes) the input voltage based on the half-wave RMS value. Blocks 2, 3, and 4 form the analogue part of the lamp-eye-brain link. Block 2 adopts the square detection method to measure the voltage fluctuation of the modulated power frequency and simulate the behavior of incandescent lamps. Block 3 consists of a 0.05–35 Hz bandpass filter $G_b(s)$ that extracts the frequency components that cause flicker, and a visual sensitivity weighting filter $G_w(s)$ that simulates the response of the human eye to light changes. Block 4 simulates the reflection and memory effects of the optic nerve in the human brain on the eyes. The squaring multiplier simulates the nonlinear reflection process of the human brain, and the first order low-pass filter $G_m(s)$ simulates the human brain's visual perception and memory effect and outputs the instantaneous flicker value P_{inst}. Finally, Block 5 obtains the short-term flicker value P_{st} (10 min) and the long-term flicker value P_{lt} (2 h) through the classification statistics of the evaluation indicators [6].

The visual sensitivity weighting filter is the core of the IEC flicker model. The transfer function is shown in (1), which is expressed with parameters k, λ, ω_1, ω_2, ω_3, and ω_4, and is closely related to the specific parameters of incandescent lamps, and the details are shown in Table 1 [6], which in turn makes it impossible for other types of light sources to use this model for correct flicker measurement.

$$G_W(s) = \frac{k\omega_1 s}{s^2 + 2\lambda s + \omega_1^2} \times \frac{(1 + \frac{s}{\omega_2})}{(1 + \frac{s}{\omega_3})(1 + \frac{s}{\omega_4})} \quad (1)$$

Table 1. Parameters of the visual sensitivity weighted filter.

Variable	60 W Incandescent Lamp (230 V/50 Hz)	60 W Incandescent Lamp (120 V/60 Hz)
k	1.74802	1.6357
λ	$2\pi \times 4.05981$	$2\pi \times 4.167375$
ω_1	$2\pi \times 9.15494$	$2\pi \times 9.077169$
ω_2	$2\pi \times 2.27979$	$2\pi \times 2.939902$
ω_3	$2\pi \times 1.22535$	$2\pi \times 1.394468$
ω_4	$2\pi \times 21.9$	$2\pi \times 17.31512$

The transfer function of the visual sensitivity weighting filter can be disassembled into two parts: the lamp response to voltage fluctuation $G_{\text{lamp}}(s)$ and the visual response to light change $G_{\text{eye}}(s)$ [8,24,25], as shown in Figure 1b. $G_{\text{lamp}}(s)$ completely depends on the light source parameters. When the transfer function of the response of the incandescent lamp to voltage fluctuation is obtained in an effective way, the visual response part, independent of the light source, can be separated according to (2). Through light change measurements, the corresponding flicker level can be obtained, making it possible to test the flicker effect of other light sources.

$$G_{\text{eye}}(s) = \frac{G_W(s)}{G_{\text{lamp}}(s)} \quad (2)$$

The flicker model oriented to light change consists of two parts: the light parameters inputted by the model and the processing of light change signals. Luminous flux is a parameter describing the ability of light source radiation to generate a visual response in unit time, and it is the best choice for model input. In addition, to reduce the differential impact of different spectral luminous efficiencies and the complexity of the human eye perception system [7], the relative change in the luminous flux obtained by processing is the key to the model. Therefore, referring to Figure 1a, the luminous flux flicker model can be divided into five parts, as shown in Figure 1b: Block 1 tests the changes of voltage and luminous flux through an appropriate system; Block 2 normalizes the signal to obtain the relative change of luminous flux and tests the behavior of the lamp, which is only related to the light source parameters; Block 3 strips the visual response of the light source to light changes to simulate the behavior of the human eye; and Blocks 4 and 5 function in the same way as the IEC standard flicker model.

The luminous flux flicker model is suitable for measuring the flicker effect of light sources with different luminous principles, which solves the light source limitation of the IEC flicker model. In addition, because the band-pass filter is removed from the model, it is possible to test the flicker effect of interharmonics on different light sources.

3. Construction and Verification of the Test System

3.1. Interharmonics-Luminous Flux Test System

To realize the luminous flux flicker model, the key is to obtain an accurate visual response $G_{\text{eye}}(s)$ to light changes. An interharmonic luminous flux test system based on the frequency sweep principle could be constructed to realize Block 1 of the luminous flux flicker model, obtain the $G_{\text{lamp}}(s)$ of the incandescent lamp in the IEC flicker model, and calculate $G_{\text{eye}}(s)$ according to (2).

As the model measures flicker using the relative change of luminous flux, the linear relationship between illuminance and luminous flux makes the test system use illuminance instead of luminous flux. As shown in Figure 2, the interharmonic-luminous flux test system could be divided into five main parts:

1. Programmable power supply: Used to output "fundamental + interharmonic set frequency" voltage, which could be realized via Chroma 61860. At the same time, considering factors such as data processing volume and observation error, the minimum interharmonic frequency interval was set to not exceed 1 Hz (1 Hz is recommended).
2. Illuminance sensor: Used to test the lamp light brightness change caused by interharmonic voltage and simulate the response to light changes as seen by human eyes. Konica Minolta T-10A with an analogue output function (0–3 V) was selected because its measurement accuracy and reliability meet DIN 5032 Class B requirements and its spectral luminous efficiency deviation does not exceed 6%.
3. Shading box: Used as the confined space where the lamp was tested and the illuminance sensor was placed. It has the ability to isolate external light changes and eliminate internal irregular reflection interference. In this study, the shading box was a 0.7 × 0.7 × 1.0 m (L× W × H) box the inner wall of which was evenly covered with black flannelette.
4. Multi-channel data acquisition device: Used for multi-channel data synchronous acquisition and recording, it recorded the illuminance data of the programmable power supply output interharmonic voltage and illuminance sensor test. The device sampling frequency was not lower than 12.8 kHz.
5. Data analysis system: Based on MATLAB coding, it could realize DFT data analysis in different time windows.

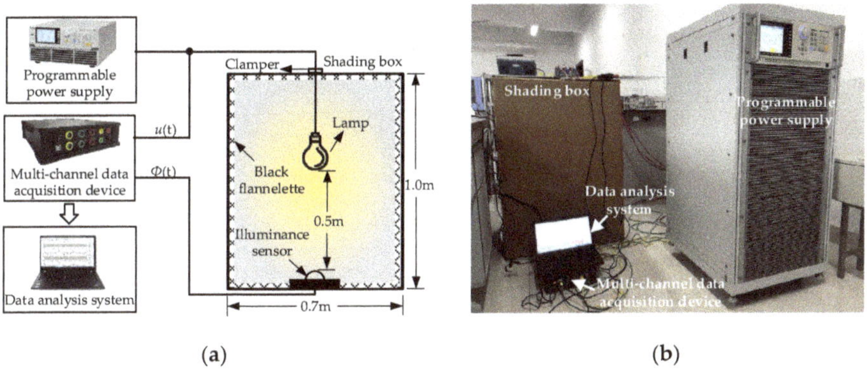

Figure 2. Interharmonics-luminous flux test system: (**a**) schematic and (**b**) actual layout.

During the test, a clamp was used to fix the lamp to be tested in the middle of the shading box and the illuminance sensor was placed horizontally directly below the lamp, with the distance between them maintained at 0.5 m, and we then set the illuminance sensor to be in the appropriate range. Then, the shading box was closed. Finally, the programmable power supply connected to the voltage test line was started and controlled to generate a 230 V/50 Hz fundamental wave voltage. A frequency sweep interval of 1 Hz was used, the interharmonics with a duration of no less than 1 s and a constant ratio (5% is recommended) were stacked, and a multi-channel data acquisition device was used to synchronously measure the test voltage and the corresponding luminous flux fluctuation data.

The measured data were analyzed by fast Fourier transform (FFT) at a time interval of 50 cycles to obtain the corresponding luminous flux fluctuation component of the lamp to be tested under different interharmonic frequencies. The interharmonic voltage and luminous flux fluctuation transfer coefficient of the lamp were calculated using (3).

$$\left|G_{\text{lamp}}(f_{\text{IH}})\right| = \frac{\frac{\Phi(f_{\text{M}})}{\Phi_{\text{DC}}}}{\frac{U(f_{\text{IH}})}{U_1}} \tag{3}$$

where U_1 is the fundamental wave amplitude, $U(f_{\text{IH}})$ is the interharmonic wave amplitude with a frequency of f_{IH}, Φ_{DC} is the DC component of the luminous flux, and $\Phi(f_{\text{M}})$ is the luminous flux amplitude with a fluctuation frequency of f_{M}. The relationship between this frequency and the interharmonic frequency is shown in (4).

$$f_{\text{M}} = |f_{\text{IH}} - h \times f_1| \tag{4}$$

where f_1 is the fundamental voltage frequency, h is the odd harmonic number closest to f_{IH}, $h = 1, 3, 5, \ldots$, and $f_{\text{IH}} \neq h \times f_1$; thus, $f_{\text{M}} < 50$ Hz.

Taking a 60 W incandescent lamp of 230 V/50 Hz as an example, the programmable power output was superposed with a 45 Hz interharmonic voltage and the analysis results of the measured data in the time and frequency domains are shown in Figure 3. At this time, the illuminance fluctuation frequency (i.e., light flux) was 5 Hz, which corresponds to the relationship expressed in (4).

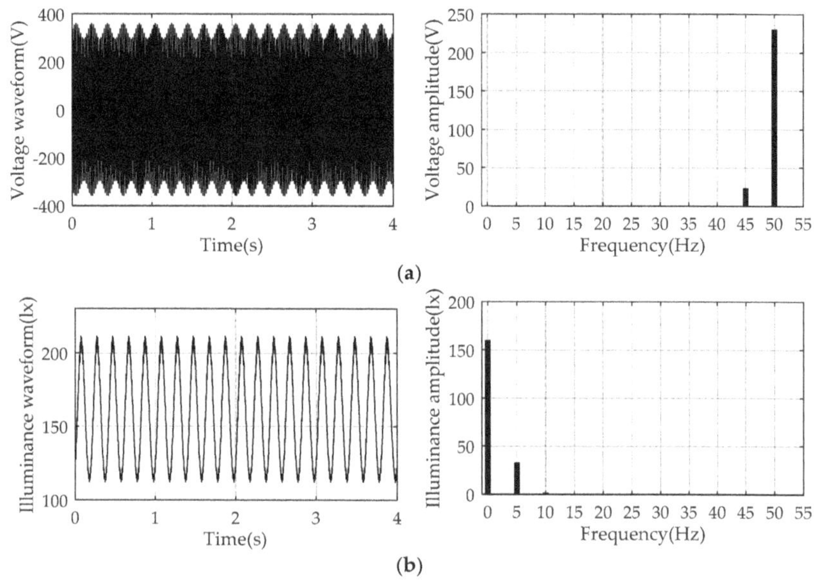

Figure 3. Test results of an incandescent lamp response to 45 Hz interharmonics: (**a**) voltage and (**b**) illuminance.

3.2. Verification of the Test System Based on an Incandescent Lamp

An incandescent lamp is a typical thermal radiation light source. The relationship between voltage and filament temperature can be deduced by using the heat balance equation as follows:

$$\frac{u^2(t)}{R} = C_\theta \frac{dT(t)}{dt} + \frac{T(t)}{R_\theta} \tag{5}$$

where $T(t)$ is the temperature of the filament of the incandescent lamp that changes with time t, C_θ is the heat capacity of the filament, R_θ is the thermal resistance of the filament,

$u(t)$ is the voltage across the incandescent lamp that changes with time t, and R is the filament resistance.

Using (5), the filament temperature $T(t)$ can be expressed as (6), where T_c is room temperature.

$$T(t) = e^{-\frac{t}{C_\theta R_\theta}} \left[\frac{1}{C_\theta}\int_0^t \frac{u^2(t)}{R} \cdot e^{\frac{t}{C_\theta R_\theta}} dt + T_c\right] \tag{6}$$

In [7], it was pointed out that the luminous flux produced by incandescent lamps is a nonlinear function determined by the temperature of the filament. The physical properties of the material and the geometric shape, with a small temperature variation range, can be approximately described as an exponential function:

$$\Phi(t) = \beta \times T(t)^{\alpha} \tag{7}$$

In (7), α and β were constants related to the temperature of the filament (usually $\alpha = 1.7$).

By substituting (6) into (7), the quantitative relationship between voltage and luminous flux under incandescent lamp conditions could be determined, and the simulation model could be established in MATLAB/Simulink. For the 230 V/50 Hz 60 W incandescent lamp in the IEC flicker model, under an average operating temperature of 2700 K, its thermal time constant was $R_\theta C_\theta$ = 19 ms and the filament resistance was R = 883 Ω. These parameters were brought into the simulation model, and the relative fluctuation of luminous flux at different frequencies was obtained by changing the input interharmonic voltage frequency of the model. As the IEC flicker model is aimed at voltage modulation, interharmonics could be used for simulation, the frequency relationship meets (4), and the $|G_{lamp}|$ value was calculated with the corresponding frequency according to (3). In addition, the test system in Figure 2 was used to test the 230 V/50 Hz 60 W incandescent lamp, and the results of the two were verified and compared.

The experimental and simulated curves of a 230 V/50 Hz 60 W incandescent lamp were highly coincident, which verified that the constructed interharmonic-luminous flux test system could accurately measure the lamp's response to voltage fluctuations. Observing the differential relationship between the power supply voltage and luminous flux of the incandescent lamp described by (6) and (7), it was determined that it is equivalent to a first-order system in the frequency domain. Thus, data fitting was performed based on the experimental test analysis results, as shown in Figure 4, and the response transfer function of the test incandescent lamp to voltage fluctuation was obtained as in (8), which is consistent with the results of [8].

$$G_{lamp}(s) = \frac{191.79}{s + 52.66} \tag{8}$$

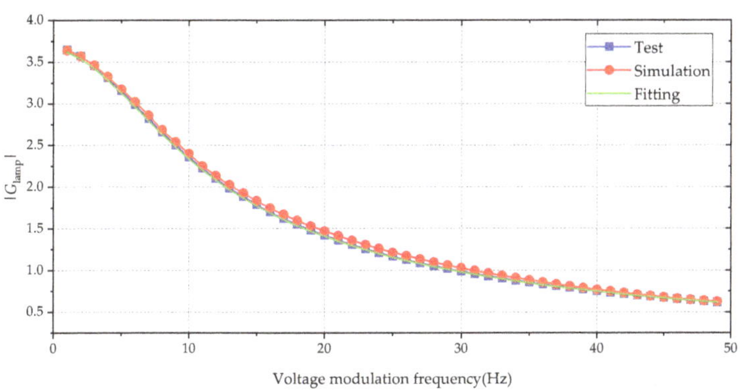

Figure 4. Response of the 60 W incandescent lamp at 230 V/50 Hz under different voltage modulation frequencies.

The result obtained using (2) was used to calculate the visual response G_{eye} (s) to light changes, as shown in (9).

$$G_{eye}(s) = \frac{38.77s^3 + 2597s^2 + 2.925 \times 10^4 s}{s^4 + 196.3s^3 + 1.178 \times 10^4 s^2 + 5.348 \times 10^5 s + 3.505 \times 10^6} \quad (9)$$

Here, the relative fluctuation data of luminous flux of any light source under the action of interharmonic voltage could be obtained through the test system, which could be input into the luminous flux flicker model to quantify the flicker level of the light source caused by the luminous flux fluctuation.

4. Interharmonic Sensitivity Evaluation of LED Lamps

Using the luminous flux flicker model, the interharmonic level of different light sources under the flicker limit can be obtained by repeatedly adjusting the amplitude of interharmonics at different frequencies to approach the flicker limit in the interharmonic-luminous flux test system. However, there are significant differences in the flicker effect of different light sources, and the method lacks a reference initial value, which greatly increases the number of test repetitions, the tediousness of operation steps, and the complexity of the data analysis. In addition, the accuracy of the measurement results is also limited by the minimum adjustment step size of the programmable power supply amplitude. Therefore, taking the incandescent lamp as the reference, using the transmission coefficient of interharmonic voltage and luminous flux fluctuation of different light sources, the calculation method of the interharmonic-flicker limit curve was proposed to simplify the test ratio and overcome the programmable power supply limitation simultaneously.

According to the luminous flux flicker model in Figure 1b, the short-term flicker of any light source caused by interharmonics of a certain frequency could be expressed by (10).

$$P_{st} = 0.714 \times \frac{\Phi(f_M)}{\Phi_{DC}} \times G_{eye}(f_M) \times \sqrt{G_m(f_M)} \quad (10)$$

The fluctuation of the luminous flux in (10) could be obtained by using the interharmonic-luminous flux testing system. Assuming that there are two lamps, A and B, the difference in the short-term flicker measurement could be expressed by (11).

$$\frac{P_{st,A}}{P_{st,B}} = \frac{\frac{\Phi_A(f_M)}{\Phi_{DC,A}}}{\frac{\Phi_B(f_M)}{\Phi_{DC,B}}} \quad (11)$$

By substituting (3) into (11), (12) could be obtained.

$$\frac{P_{st,A}}{P_{st,B}} = \frac{\left|G_{lamp,A}(f_{IH})\right| \times \frac{U_A(f_{IH})}{U_{1,A}}}{\left|G_{lamp,B}(f_{IH})\right| \times \frac{U_B(f_{IH})}{U_{1,B}}} \quad (12)$$

Based on (12), if the transfer coefficient $|G_{lamp,B}|$ of the interharmonic voltage of lamp B and the luminous flux fluctuation and the ratio of the interharmonic to the fundamental of different frequencies caused by $P_{st,B} = 1$ are known, then based on the transfer coefficient $|G_{lamp,A}|$ of the interharmonic voltage of lamp A and the luminous flux fluctuation, the ratio of the interharmonic to the fundamental corresponding to different frequencies of lamp A when $P_{st,A} = 1$ could be calculated.

Based on (8), the transfer coefficient $|G_{\text{lamp,ref}}|$ of the interharmonic voltage and luminous flux fluctuation of a 230 V/50 Hz 60 W incandescent lamp is given by:

$$\left|G_{\text{lamp,ref}}(f_{\text{IH}})\right| = \frac{191.79}{\sqrt{4\pi^2(f_{\text{IH}} - hf_1)^2 + 52.66^2}} \tag{13}$$

In addition, $s = \text{j}2\pi \times |f_{\text{IH}} - h \times f_1|$, and the transfer function of the visual sensitivity weighting filter of the 230 V/50 Hz 60 W incandescent lamp brought into (1) could calculate the modulus corresponding to the transfer function at different frequencies to obtain the visual sensitivity weighting coefficient $|G_{\text{W}}(f_{\text{IH}})|$. Then, according to the corresponding relationship between single interharmonic and flicker given in [26], the ratio of the interharmonics to the fundamental (interharmonic ratio, expressed as a percentage) of different frequencies of the incandescent lamp with $P_{\text{st}} = 1$ was calculated according to (14), as shown in Figure 5.

$$\left.\frac{U(f_{\text{IH}})}{U_1}\right|_{\text{ref},P_{\text{st}}=1} = \frac{0.25\%}{0.714 \times 2 \times |G_{\text{W}}(f_{\text{IH}})|} \tag{14}$$

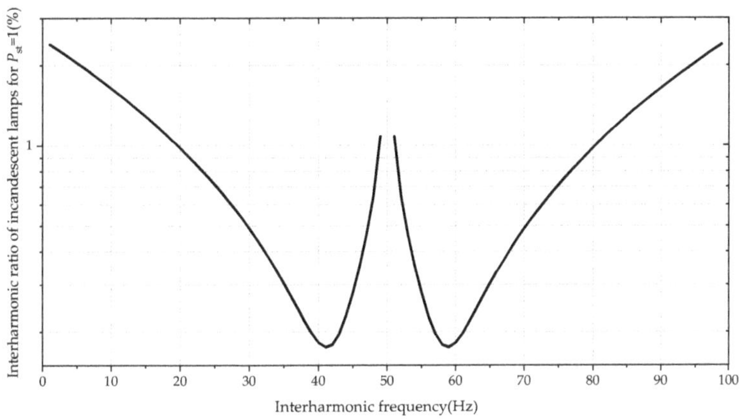

Figure 5. Interharmonic-flicker limit curve of the incandescent lamp.

The interharmonic ratio takes the 230 V/50 Hz 60 W incandescent lamp as the reference standard, uses the interharmonic-luminous flux test system, superimposes the non-specific constant interharmonic amplitude value within the allowable voltage range of the lamp, and sweeps the frequency to obtain the transmission coefficient $|G_{\text{lamp,A}}|$ between the interharmonic voltage and luminous flux fluctuation of any light source. Then, according to (15), the interharmonic ratio of different frequencies corresponding to this lamp for $P_{\text{st}} = 1$ could be calculated, and the interharmonic-flicker limit curve could be drawn.

$$\left.\frac{U(f_{\text{IH}})}{U_1}\right|_{P_{\text{st}}=1} = \frac{\left|G_{\text{lamp,ref}}(f_{\text{IH}})\right|}{\left|G_{\text{lamp,a}}(f_{\text{IH}})\right|} \times \left.\frac{U(f_{\text{IH}})}{U_1}\right|_{\text{ref},P_{\text{st}}=1} \tag{15}$$

The interharmonic-flicker limit curve of an LED lamp could be calculated by using the interharmonic-luminous flux test system, taking the incandescent lamp as the benchmark. If the interharmonic ratio of the power supply voltage of the LED lamp is in the area above the curve, the interharmonic ratio causes the flicker value to exceed the limit. This means that the LED lamp with a lower curve level is more sensitive to interharmonics. Therefore, the evaluation of the sensitivity of the LED lamp to interharmonics could be completed by comparing the interharmonic ratio at the same frequency.

5. Results and Discussion

As the light-emitting body of an LED lamp is a semiconductor device, it can work normally only with a DC power supply. Facing the AC low-voltage power grid, it needs a driving device to convert the AC-DC power supply. This is the core component of LED lamps and has the most significant impact on their working performance. Therefore, in the face of a wide variety of LED lamp products, LED lamps are orderly classified according to the type of power drive to compare and study the influence of interharmonics on LED lamps with different driving modes.

Presently, LED drives are mainly divided into two types: constant voltage and constant current [27–31]. An RC buck drive is a representative constant voltage drive. Its structure is shown in Figure 6a. This circuit forms a voltage-dividing circuit through two capacitors to reduce the power supply voltage amplitude and limits the current flowing through the LED lamp through a resistance. It has low cost, simple circuit structure, and high universality. Therefore, it is widely used in some low-end LED products. To improve energy efficiency, a constant current drive (Figure 6b) is favored in the commercial market. It uses the principle of switching power supply for DC-DC conversion to achieve a stable LED current.

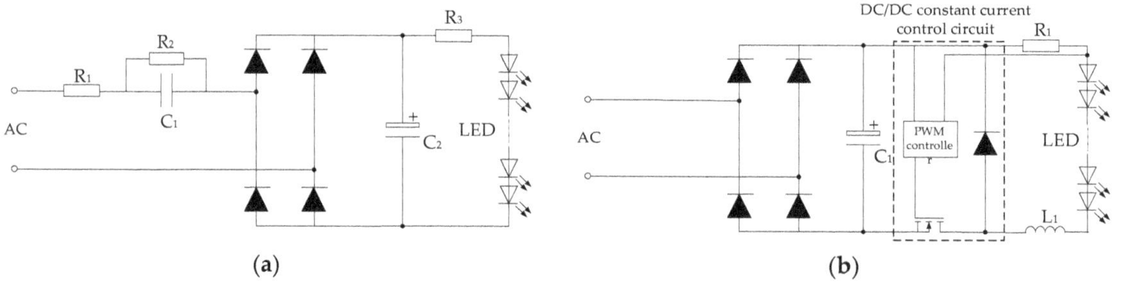

Figure 6. Schematic of common LED lamp driving types: (**a**) type I: RC buck drive; (**b**) type II: constant current drive.

According to the type of driving circuit, LED lamps with representative brands are selected, and the interharmonic sensitivity test evaluation is carried out. The test object information is listed in Table 2.

Table 2. Performance index parameters of the test LED lamp.

Drive Type	Brand Code	Power (W)	Luminous Flux (lm)	Color Temperature (K)
Type I	A	3	170	3000
	A	12	1000	5000
	B	16	1800	6500
Type II	B	6.5	600	6500
	C	3.5	249	6500
	C	7	550	5000
	D	4.5	380	6500
	D	8.5	860	6500

Each LED lamp in Table 2 was tested in the interharmonic-luminous flux testing system, and the transfer coefficient analysis of interharmonic voltage and luminous flux fluctuation was carried out according to (3). The results are shown in Figure 7.

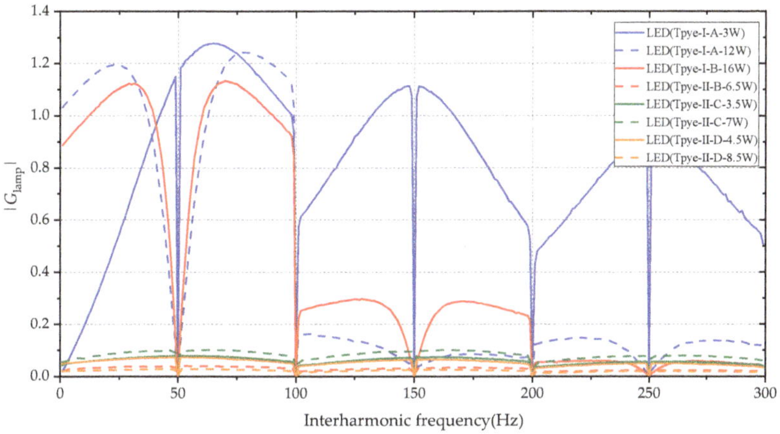

Figure 7. Interharmonic voltage and luminous flux fluctuation transfer coefficient of test LED lamps.

Figure 7 shows that the responses of different LED lamps to the interharmonic voltage are different, especially between LED lamps with different drivers. Combining Figures 5 and 7, when $P_{st} = 1$, the interharmonic ratio of eight LED lamps tested at different frequencies could be calculated according to (15), and the interharmonic-flicker limit curve of each LED lamp could be drawn, as shown in Figure 8.

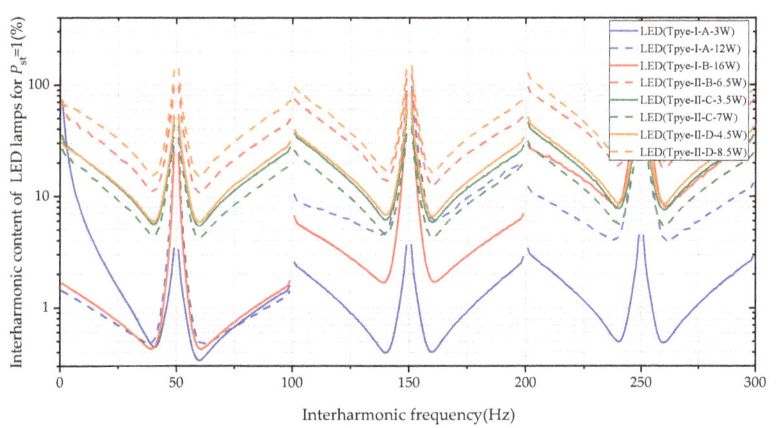

Figure 8. Interharmonic-flicker limit curve of test LED lamps.

According to the LED lamp test results in Figure 8, the following were inferred:

1. In contrast with incandescent lamps, high-frequency interharmonics (more than 100 Hz) can cause flicker of LED lamps. Taking the odd harmonic frequency as the center, it follows that the higher the frequency is, the higher the interharmonic-flicker limit curve is, indicating that the LED is less sensitive to interharmonics with higher frequency.
2. The interharmonic-flicker limit curve of the RC-buck-driven LED lamp is significantly lower than that of the constant-current-driven LED lamp, which shows that the former lacks a control circuit and is more vulnerable to external voltage fluctuation; thus, the RC-buck-driven LED lamp is more sensitive to interharmonics. For LED lamps with different drives, the minimum interharmonic ratio causing flicker out-of-limit is shown in Figure 9. The minimum interharmonic ratio of RC-buck-driven LEDs is below 1%, while the minimum interharmonic ratio of constant-current-driven LEDs

is more than 4%, and can even reach approximately 15%. In the current power grid environment, constant-current-driven LEDs have low probability of light intensity changes that can be detected by the naked eye. Therefore, in scenes with high lighting quality, it is recommended to use constant-current-driven LED lights.

3. The higher the power of LED lamps of the same brand and driving type, the higher the overall level of the interharmonic-flicker limit curve, which indicates that higher-power LEDs are less sensitive to interharmonics.

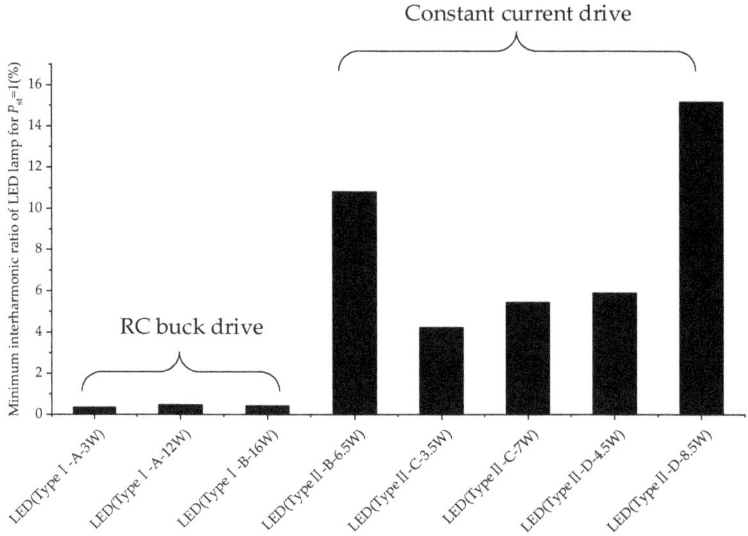

Figure 9. Minimum interharmonic ratio of out-of-limit flicker.

6. Conclusions

The influence of interharmonics on the lighting quality of LED lamps widely used in the market cannot be ignored. The "weakness" of LED lamps in interharmonics can be quantified in terms of flicker. Through the deconstruction of the IEC flicker model, this paper proposed a luminous flux flicker model that is free from the constraints of specific light sources. In the interharmonic luminous flux test system in the model, the calculation method of the LED interharmonic-flicker limit curve based on incandescent lamps was deduced, and the experimental evaluation of LED interharmonic sensitivity was completed. Based on the results, the following conclusions were drawn:

1. The interharmonic luminous flux test system composed of a programmable power supply, shading box, illuminance sensor, data acquisition device, and data analysis system could accurately eliminate the part that depends on incandescent lamp parameters in the visual sensitivity weighting filter of the IEC flicker model and realize a luminous flux flicker model suitable for any light source.
2. Taking a 230 V/50 Hz 60 W incandescent lamp as the reference, using the interharmonic-luminous flux test system, superimposing the non-specific constant interharmonic amplitude, and sweeping the frequency to obtain the transfer coefficient of the interharmonic voltage and luminous flux fluctuation of any light source, the interharmonic-flicker limit curves could be quickly drawn, improving the efficiency of existing methods and eliminating the limitations of programmable power hardware.
3. The interharmonic-flicker limit curve could describe the interharmonic sensitivity of LED lamps. The lower the curve level, the more sensitive the LED lamps were to interharmonics, and vice versa.
4. The test results showed that the driving power supply is an important factor affecting the sensitivity of interharmonics. The RC-buck-driven LED lamp was more sensitive

to interharmonics than the constant-current-driven LED lamp. In a scene with high lighting quality, the constant-current-driven LED lamp is recommended. In addition, high-frequency interharmonics (more than 100 Hz) also caused flicker of LED lamps. The higher the frequency centered on odd harmonics, the lower the interharmonic sensitivity, while the higher the power, the lower the interharmonic sensitivity of LED lamps.

Owing to the relatively small number of brands and types of LED lamps tested in this study and the focus on the influence of driving types on interharmonic sensitivity, the evaluation results are not widely representative. In future work, it is necessary to apply the luminous flux flicker model to various other tests. Further interharmonic sensitivity evaluations for LED lamps can be performed to investigate aspects such as illuminance, appearance, light irradiation angle, and dimming. This should provide a test basis for determining the electromagnetic compatibility limits and aid the iterative improvement of LED lamp product technology.

Author Contributions: Conceptualization, H.L. and M.Z.; methodology, H.L., Y.S., M.Z., and Y.J.; validation, Y.S., M.Z., and Y.J.; investigation, H.L., Y.S., M.Z., and Y.J.; writing—original draft preparation, Y.S. and Y.J.; writing—review and editing, H.L., Y.S., M.Z., and Y.J.; project administration, H.L. and M.Z. All authors have read and agreed to the published version of the manuscript.

Funding: This research was funded by the Science and Technology Project Foundation of State Grid Beijing Electric Power Company, grant number 52022320006D.

Institutional Review Board Statement: Not applicable.

Informed Consent Statement: Not applicable.

Data Availability Statement: The data presented in the study are available in the article.

Conflicts of Interest: The authors declare no conflict of interest.

References

1. Kar, A. New Generation Illumination Engineering-An Overview of Recent Trends in Science & Technology. In Proceedings of the 2014 First International Conference on Automation, Control, Energy and Systems (ACES), Adisaptagram, India, 1–2 February 2014; pp. 1–6.
2. Lokesh, J.; Kini, S.G.; Mahesha, M.G.; Padmasali, A.N. A study on the performance characteristics of light emitting diodes from reliability prospective. In Proceedings of the 2021 IEEE 2nd International Conference on Smart Technologies for Power, Energy and Control (STPEC), Bilaspur, India, 19–22 December 2021; pp. 1–6.
3. Tayjasanant, T.; Wang, W.; Li, C.; Xu, W. Interharmonic-flicker curves. *IEEE Trans. Power Deliv.* **2005**, *20*, 1017–1024. [CrossRef]
4. Jing, Y.; Sun, C.; Xu, W. Voltage flicker caused by two interharmonics. In Proceedings of the 2006 Canadian Conference on Electrical and Computer Engineering, Ottawa, ON, Canada, 7–10 May 2006; pp. 239–242.
5. Yong, J.; Sun, C.X.; Jian-Bo, L.I.; Zhou, H.B.; Chen, L. Light flicker characteristics caused by interharmonics and flicker limit curve. *Proc. CSEE* **2008**, *28*, 88–93.
6. *Standard 61000-4-15*; Electromagnetic Compatibility (EMC)-Part 4: Testing and Measurement Techniques Section 15: Flicker Meter-Functional and Design Specifications. International Electrotechnical Commission (IEC): Geneva, Switzerland, 2010.
7. Peretto, L.; Emanuel, A.E. A theoretical study of the incandescent filament lamp performance under voltage flicker. *IEEE Trans. Power Deliv.* **1997**, *12*, 279–288. [CrossRef]
8. Cho, S.H.; Jung, J.A.; Jang, G.; Kwon, S.H.; Kang, M.H. Experimental verification of 220V incandescent lamp characteristics for the IEC flickermeter. In Proceedings of the 2008 13th International Conference on Harmonics & Quality of Power, Wollongong, NSW, Australia, 28 September–1 October 2008; pp. 1–4.
9. Hooshyar, A.; El-Saadany, E.F. Development of a flickermeter to measure non-incandescent lamps flicker. *IEEE Trans. Power Deliv.* **2013**, *28*, 2103–2115. [CrossRef]
10. Gallo, D.; Landi, C.; Langella, R.; Testa, A. IEC flickermeter response to interharmonic pollution. In Proceedings of the 2004 11th International Conference on Harmonics and Quality of Power, Lake Placid, NY, USA, 12–15 September 2004; pp. 489–494.
11. Jiao, P.; Ma, C. Study on IEC flickermeter applicability based on flux parameters. *Power Syst. Prot. Control.* **2015**, *43*, 85–90.
12. Hui, S.Y.; Qin, Y.X. A General photo-electro-thermal theory for light emitting diode (LED) systems. *IEEE Trans. Power Electron.* **2009**, *24*, 1967–1976. [CrossRef]
13. Lin, Y.; Yang, Y.; Lan, C. Formulation of flicker curve for inter-harmonic of fluorescent lamps. *Northeast. Electr. Power Technol.* **2021**, *42*, 56–59.

14. Chen, S.; Heah, M.Y.; Then, A.B.; Foo, M.K. Automatic evaluation of flickering sensitivity of fluorescent lamps caused by interharmonic voltages. In Proceedings of the 2008 13th International Conference on Harmonics & Quality of Power, Wollongong, NSW, Australia, 28 September–1 October 2008; pp. 1–6.
15. Drapela, J.; Langella, R.; Testa, A.; Grappe, J.E. A real life light flicker case-study with LED lamps. In Proceedings of the 2018 18th International Conference on Harmonics and Quality of Power, Ljubljana, Slovenia, 13–16 May 2018; pp. 1–6.
16. Chmielowiec, K. Flicker effect of different types of light sources. In Proceedings of the 2011 11th International Conference on Electrical Power Quality and Utilisation, Lisbon, Portugal, 17–19 October 2011; pp. 1–6.
17. Kumar, K.J.; Sudhir Kumar, R. Voltage flicker from warm and cool white LED bulbs. In Proceedings of the 2020 IEEE PES Innovative Smart Grid Technologies Europe (ISGT-Europe), The Hague, The Netherlands, 26–28 October 2020; pp. 934–938.
18. Petrinska, I.; Ivanov, D. Estimation of flicker of contemporary LED lamps and luminaires. In Proceedings of the 2019 11th Electrical Engineering Faculty Conference (BulEF), Varna, Bulgaria, 11–14 September 2019; pp. 1–4.
19. Lehman, B.; Wilkins, A.; Berman, S.; Poplawski, M.; Miller, N.J. Proposing measures of flicker in the low frequencies for lighting applications. In Proceedings of the 2011 IEEE Energy Conversion Congress and Exposition, Phoenix, AZ, USA, 17–22 September 2011; pp. 2865–2872.
20. Chaki, R.; Anand, V.; Ghosh, S.; Ghosh, M. A Comparative Analysis of Flicker Index and Modulation Depth on Few Existing Dimmable LED Drivers. In Proceedings of the 2018 2nd International Conference on Power, Energy and Environment: Towards Smart Technology (ICEPE), Shillong, India, 1–2 June 2018; pp. 1–5.
21. In, C.; Kim, J.; Oh, Y.; Cho, K. Flicker Characteristics of a Multi-Channel Current-Controlled PWM Dimming Method of LED Lightings. In Proceedings of the 2019 IEEE 60th International Scientific Conference on Power and Electrical Engineering of Riga Technical University (RTUCON), Riga, Latvia, 7–9 October 2019; pp. 1–7.
22. Drapela, J.; Langella, R.; Slezingr, J.; Testa, A. A tunable flickermeter to account for different lamp technologies. *IEEE Trans. Power Deliv.* **2016**, *32*, 872–880. [CrossRef]
23. Wang, J.; Liu, H. A calculation method for interharmonics-caused flicker and flicker source identification. *Autom. Electr. Power Syst.* **2011**, *35*, 57–63, 80.
24. Cai, R.; Cobben, J.; Myrzik, J.; Blom, J.H.; Kling, W.L. Flicker responses of different lamp types. *IET Gener. Transm. Distrib.* **2009**, *3*, 816–824. [CrossRef]
25. Wang, C.S.; Devaney, M.J.; Yang, S.W. Decomposition of the IEC flickermeter weighting curves [light flicker due to fluctuating current]. In Proceedings of the 21st IEEE Instrumentation and Measurement Technology Conference, Como, Italy, 18–20 May 2004; pp. 1378–1382.
26. Zeng, Q.; Yang, H.; Wang, Z.; Hongping, L.I. Application of all-phase spectrum method to flicker detection. *Proc. CSU-EPSA* **2017**, *29*, 66–71.
27. Abdalaal, R.M.; Man Ho, C.N. Characterization of commercial LED lamps for power quality studies. *IEEE Can. J. Electr. Comput. Eng.* **2021**, *44*, 94–104. [CrossRef]
28. Padmavathi, P.; Natarajan, S. A survey on efficient converter driver techniques for LED lighting applications. In Proceedings of the 2019 Innovations in Power and Advanced Computing Technologies (i-PACT), Vellore, India, 22–23 March 2019; pp. 1–7.
29. Gupta, N.; Verma, D. A review and analysis of topologies of light emitting diode (LED) drivers. In Proceedings of the 2021 IEEE 2nd International Conference on Electrical Power and Energy Systems (ICEPES), Bhopal, India, 10–11 December 2021; pp. 1–6.
30. Luo, Q.; Huang, J.; He, Q.; Ma, K.; Zhou, L. Analysis and design of a single-stage isolated AC–DC LED driver with a voltage doubler rectifier. *IEEE Trans. Ind. Electron.* **2017**, *64*, 5807–5817. [CrossRef]
31. Petrinska, I.; Ivanov, D.; Pavlov, D.; Kassev, K. Photometric flicker of LED luminaires according to driver type. In Proceedings of the 2020 21st International Symposium on Electrical Apparatus & Technologies (SIELA), Bourgas, Bulgaria, 3–6 June 2020; pp. 1–4.

energies

MDPI

Article

A Single-Phase Transformerless Nine-Level Inverter and Its Control Strategy

Yuan Li [1], Wenwen Huang [2], Zhen Li [3,4], Yue Sun [2] and Bi Liu [2,*]

1 School of Electronic and Electrical Engineering, Shangqiu Normal University, Shangqiu 476000, China; liyuan8330@126.com
2 School of Electrical Engineering and Automation, Anhui University, Hefei 230601, China; w_huang_w@163.com (W.H.); sunyue5678@163.com (Y.S.)
3 School of Electrical Engineering, Shandong University, Jinan 250061, China; zhenli0901@sdu.edu.cn
4 National Engineering Laboratory of Efficiency-Saving Motor & Control Technology, Heifei 230601, China
* Correspondence: b_liu_2014@163.com

Abstract: To reduce the device number per unit level of the existing nine-level inverters, a topology of single-phase transformerless nine-level inverter was proposed. The proposed topology consists of only 10 switching devices and 4 capacitors. Firstly, the working principle of the proposed topology was analyzed in detail, and the comparison with conventional nine-level topologies in terms of device switching loss and conduction loss was presented. Then, combined with one-dimensional space vector modulation, a floating capacitor voltage estimation method was presented by the analysis of the operation mode before the arrival of redundant switching state. On this basis, a floating capacitor voltage sensorless control scheme was proposed, to achieve the balance of the floating capacitor voltages. Finally, simulations and experiments verified the effectiveness and correctness of the proposed topology and floating capacitor voltage sensorless control method.

Keywords: nine-level inverter; one-dimensional space vector modulation; floating capacitor voltage estimation; voltage balancing control

Citation: Li, Y.; Huang, W.; Li, Z.; Sun, Y.; Liu, B. A Single-Phase Transformerless Nine-Level Inverter and Its Control Strategy. *Energies* **2022**, *15*, 3418. https://doi.org/10.3390/en15093418

Academic Editor: Tek Tjing Lie

Received: 28 March 2022
Accepted: 5 May 2022
Published: 7 May 2022

1. Introduction

In recent years, high-order multi-level inverters have attracted the attention of many scholars in low-voltage applications such as new energy power generation, microgrids, and motor drives, due to their advantages of high output power quality, low current harmonics, and low voltage stress of switching devices [1–5]. At present, multi-level inverter circuits mainly include neutral-point-clamped (NPC), flying capacitors (FC), cascaded H-bridge (CHB), stacked multicell converter (SMC) and other topologies. A nine-level active NPC (ANPC) inverter topology is proposed to expand the inverter capacity while reducing the filter size by interleaving and paralleling two five-level ANPC inverters [6]. However, this topology requires 24 switching devices, and the circuit hardware cost is high. A new approach is developed in [7], which connects four flying capacitor H-bridges in series with one two-level half bridge, effectively reducing the number of switching devices in the nine-level topology to 18, but the flying capacitor voltage control of this topology is relatively complex. The nine-level inverter topology presented in [8] uses a switched capacitor circuit combined with an H-bridge. This circuit only needs one voltage source, and fewer switching devices. Although the voltage stress of switching devices is relatively small, the energy of this circuit can only be transmitted in one direction. The stacked multi-unit multi-level circuit is proposed in [9,10], which not only reduces the energy storage of the inverter, but also improves the voltage withstand capability of the circuit. The above-mentioned multi-level topology is not suitable for low-voltage application, due to the large number of devices and complex control. In order to solve the above problems, four nine-level topologies suitable for low-voltage application are proposed in the literature [11–13].

However, these topologies require more DC power sources in practical applications, which greatly increases the system cost.

The unbalance of floating capacitor voltages has always been one of the key problems to be solved in high-order multilevel inverters. Thus, several floating capacitor voltage balance control methods have been proposed. A capacitance voltage estimation method is proposed in [14]. By connecting a voltage sensor between two capacitor anodes and analyzing the current flow path, only half of the voltage sensors can estimate the overall capacitance voltage. The SVPWM modulation strategy is adopted in [15,16], and the balance between the dc-link capacitor voltages and the floating capacitor voltages is achieved by reasonably selecting the redundant vector. However, due to the large number of voltage vectors, the calculation is very complicated. Compared with the SVPWM method, the SPWM is simple and has been widely used in high-order multi-level topologies. A simple logic-form equation based PWM is designed in [17], which can make the floating capacitors voltage around its reference value. An inverter capacitor voltage balancing method based on optimal zero sequence voltage injection is proposed in [18]. This method is realized by hybrid carrier pulse width modulation, which can balance and adjust the voltage of DC link capacitor and floating capacitors in the topology. A new carrier-based modulation scheme was adopted in [19], which solves the problem of floating capacitors voltage balance by clamping the modulation reference safely to the pole voltage level conducive to floating capacitors voltage balance. A SPWM-based floating capacitor voltage control method is proposed in [20], which achieves the balance of the floating capacitor voltage by reasonably selecting redundant switch states during phase switching. By adjusting the duty cycle, the dwell time of the redundant switch state is changed, and then the floating capacitor voltages are controlled to be balanced [21]. A one-dimensional space vector modulation method is proposed in [22], which can achieve the voltage balance of the floating capacitors while reducing the switching loss and the control complexity. Compared with SPWM, the one-dimensional space vector can increase the control degree of freedom of the inverter, which is more conducive to the high-frequency output of the inverter.

However, the realization of the above-mentioned floating capacitor voltage balance methods requires the use of voltage sensor to obtain a voltage signal. Since the actual operating environment of the inverter is complex, humidity, vibration, noise, electromagnetic interference, etc. will affect the measurement accuracy of the sensor, and even lead to sensor failure. Therefore, how to realize sensorless control or less sensor control is the key to ensure the reliable operation of the inverter system. However, most of the existing sensorless control strategies are proposed to cancel the grid voltage or current sensors [23–25], while the floating capacitor voltage sensorless control is relatively rarely reported.

To tackle the above problems, this paper proposes a low-voltage single-phase non-isolated nine-level inverter topology with simple structure and few switching devices, and it presents a floating capacitor voltage sensorless control strategy based on one-dimensional space vector modulation. First, the working principle of the proposed nine-level inverter is analyzed in detail, and the comparison with the existing nine-level topologies is presented. Then, combined with the one-dimensional space vector modulation algorithm, a control strategy without floating capacitor voltage sensor is proposed, and the voltage balance is realized by estimating and predicting the floating capacitor voltages. Finally, simulation and experimental results have verified the validity and feasibility of the proposed topology and control algorithm.

2. The Proposed Nine-Level Inverter Topology and Its Working Principle

2.1. Topology

The proposed single-phase non-isolated nine-level inverter topology is shown in Figure 1. The circuit is mainly composed of 10 switching devices and 4 capacitors. S_2, S_3, and S_7 are composed of two switching devices, which are bidirectional blocking switches, and S_1, S_4, S_5 and S_6 are unidirectional blocking switches. Assuming that the total dc-link voltage V_{DC} is $8E$, the voltage of the dc-link capacitors C_1 and C_2 is $4E$, and the voltage of the floating capacitors C_3 and C_4 is E. I_{DC} and I_O are the dc-link current and load current, respectively, while I_{C1}, I_{C2}, I_{C3} and I_{C4} are the currents flowing through the capacitors C_1, C_2, C_3 and C_4, respectively.

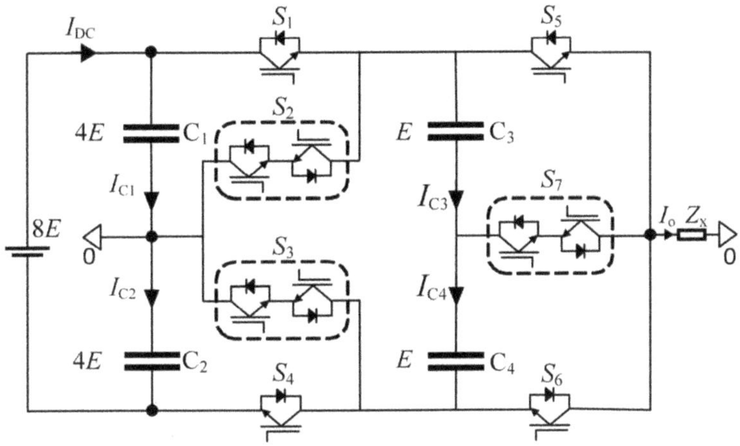

Figure 1. The topology of the proposed 9-level converter.

2.2. *Working Modal Analysis*

Figure 2 shows the 12 working states of the proposed topology. The solid arrows in the figure represent the reference positive direction of the current on each path, and the dashed arrows represent the actual flow direction of the current. The detailed switch status analysis is as follows:

(1) State P4 (Figure 2a): The switches S_1 and S_5 are turned on, the current flows from the positive terminal of the dc-link through the switches S_1, S_5 and the load, and then returns to the neutral point, the inverter output voltage $V_O = V_{DC}/2$. The current flowing through capacitors C_3 and C_4 is $I_{C3} = I_{C4} = 0$.

(2) State P3 (Figure 2b): The switches S_1 and S_7 are turned on, the current starts from the positive end of the dc-link and returns to the neutral point after passing through S_1, C_3, S_7 and the load, and the output voltage $V_O = 3V_{DC}/8$. The current through C_3 is I_O and the current through C_4 is 0.

(3) State P2P (Figure 2c): the switches S_1 and S_6 are turned on, the current starts from the positive terminal of the dc-link and returns to the neutral point through S_1, C_3, C_4, S_6 and the load. The output voltage $V_O = V_{DC}/4$, and the currents $I_{C3} = I_{C4} = I_O$.

(4) State P2N (Figure 2d): the switch S_3 and S_5 are turned on, the current starts from the neutral point and returns to the neutral point after passing through S_3, C_4, C_3, S_5 and the load. The output voltage $V_O = V_{DC}/4$, and the currents $I_{C3} = I_{C4} = -I_O$.

(5) State P1 (Figure 2e): the switches S_3 and S_7 are turned on, the current starts from the neutral point and returns to the neutral point after passing through S_3, C_4, S_7 and the load. The output voltage $V_O = V_{DC}/8$, while the currents $I_{C3} = 0$ and $I_{C4} = -I_O$.

(6) State OP (Figure 2f): The switches S_2 and S_5 are turned on, and the current starts from the neutral point and returns to the neutral point after passing through S_2, S_5 and the load. At this time, $V_O = 0$, $I_{C3} = I_{C4} = 0$.

(7) The state is ON (Figure 2g): the switches S_3 and S_6 are turned on, the current starts from the neutral point and then returns to the neutral point after passing through S_3, S_6 and the load, and the voltage $V_O = 0$, and the currents $I_{C3} = I_{C4} = 0$.

(8) State N1 (Figure 2h): The switches S_2 and S_7 are turned on, the current starts from the neutral point and returns to the neutral point after passing through S_2, C_3, S_7 and the load. The output voltage $V_O = -V_{DC}/8$, and the currents $I_{C3} = I_O$, $I_{C4} = 0$.

(9) State N2P (Figure 2i): The switch tubes S_2 and S_6 are turned on, the current starts from the neutral point and returns to the neutral point after passing through S_2, C_3, C_4, S_6 and the load. The voltage $V_O = -V_{DC}/4$, and the currents $I_{C3} = I_{C4} = I_O$.

(10) State N2N (Figure 2j): The switches S_4 and S_5 are turned on, and the current starts from the negative end of the dc-link and returns to the neutral point through S_4, C_4, C_3, S_5 and the load. The output voltage $V_O = -V_{DC}/4$, and the currents $I_{C3} = I_{C4} = -I_O$.

(11) State N3 (Figure 2k): the switches S_4 and S_7 are turned on, the current starts from the negative end of the dc-link and then returns to the negative end of the dc-link through S_4, C_4, S_7 and the load. The voltage $V_O = -3V_{DC}/8$, and the current $I_{C4} = -I_O$.

(12) State N4 (Figure 2l): The switches S_4 and S_6 are turned on, the current starts from the negative end of the dc-link and returns to the neutral point after passing through S_4, S_6 and the load. The voltage $V_O = -V_{DC}/2$, and the currents $I_{C3} = I_{C4} = 0$.

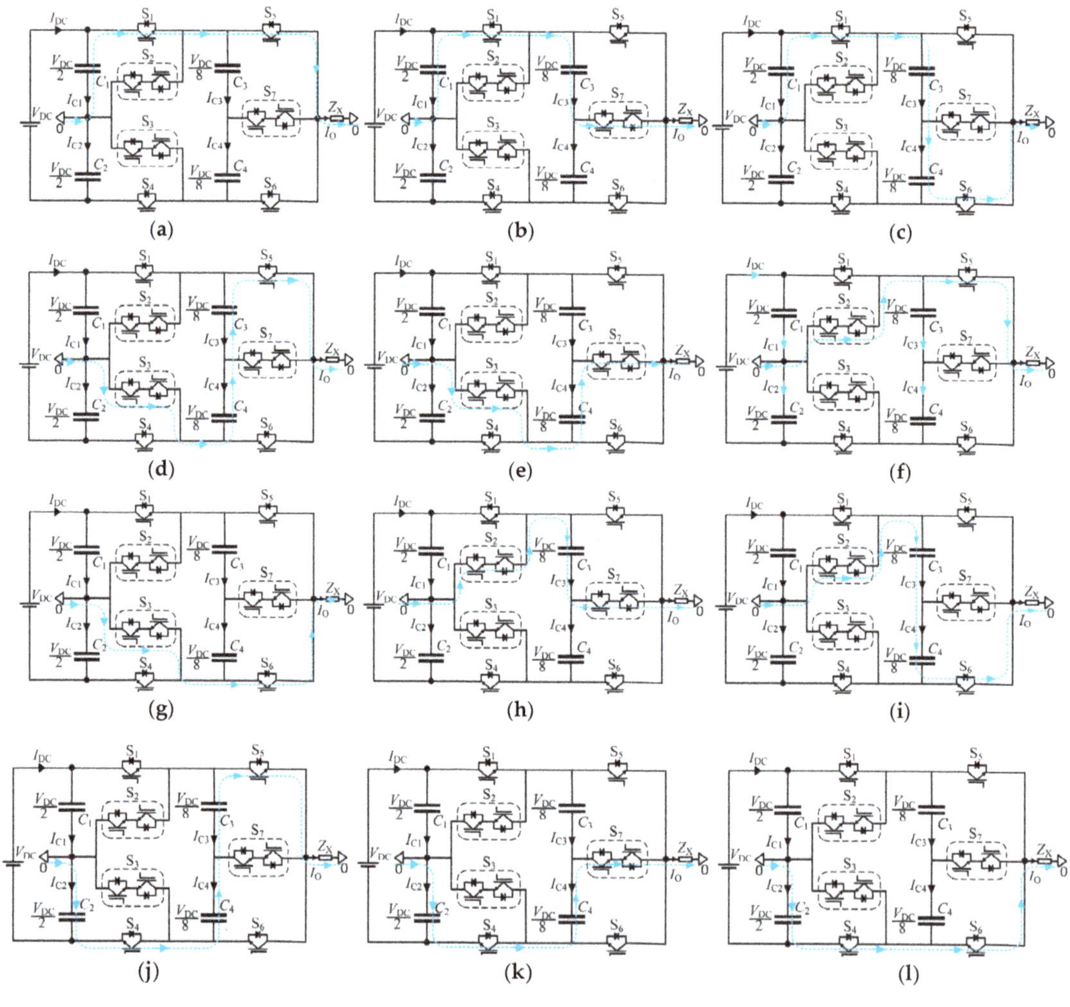

Figure 2. Operating states of the proposed 9-level topology. (**a**) $V_O = 4E$; (**b**) $V_O = 3E$; (**c**) $V_O = 2E$; (**d**) $V_O = 2E$; (**e**) $V_O = E$; (**f**) $V_O = 0$; (**g**) $V_O = 0$; (**h**) $V_O = -E$; (**i**) $V_O = -2E$; (**j**) $V_O = -2E$; (**k**) $V_O = -3E$; (**l**) $V_O = -4E$.

It can be seen from Figure 2 that the proposed nine-level topology has a total of 12 switching states, as shown in Table 1.

Table 1. Switching states of the proposed 9-level converter.

S_1–S_7	States	V_{C3}		V_{C4}		V_O
		$I_O > 0$	$I_O < 0$	$I_O > 0$	$I_O < 0$	
1000100	P4	-	-	-	-	4E
1000001	P3	↑	↓	-	-	3E
1000010	P2P	↑	↓	↑	↓	2E
0010100	P2N	↓	↑	↓	↑	2E
0010001	P1	-	-	↓	↑	E
0100100	OP	-	-	-	-	0
0010010	ON	-	-	-	-	0
0100001	N1	↑	↓	-	-	E
0100010	N2P	↑	↓	↑	↓	−2E
0001100	N2N	↓	↑	↓	↑	−2E
0001001	N3	-	-		↑	−3E
0001010	N4	-	-	-	-	−4E

Note: '↑' means capacitor charging; '↓' means capacitor discharge.

2.3. Comparison with Existing Nine-Level Inverter Topologies

In order to explain the advantages of the proposed topology more intuitively, this paper selects the traditional ANPC and SMC nine-level topology as the comparison objects. Figure 3 shows the above circuit topologies.

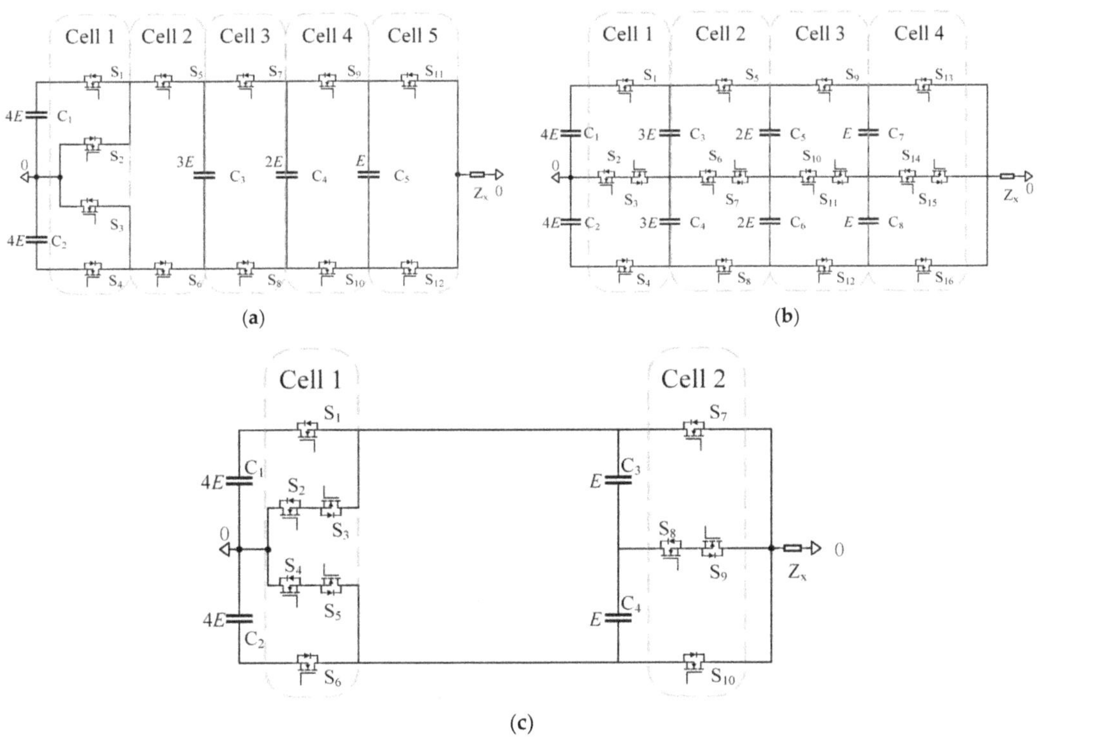

Figure 3. The topology of the three nine-level converters. (**a**) ANPC-9L; (**b**) SMC-9L; (**c**) The proposed topology.

Table 2 presents the main circuit parameter comparison between the proposed topology and two typical nine-level topologies. In terms of the number of switching devices, the number of switching devices in the proposed topology is 5/8 than that of the SMC-9L topology and 5/6 than that of the ANPC-9L topology. The proposed topology can effectively reduce the difficulty of circuit design, reduce the system cost, and improve the reliability of the inverter. Compared with ANPC-9L, the number of floating capacitors in the proposed topology is reduced by 1, and 4 less than that of SMC-9L, and the withstand voltage is lower, which can reduce the volume of floating capacitors. In addition, the number of sensors of the proposed topology is also less than other topologies.

Table 2. Parameters comparison of the main circuit.

Parameters	ANPC-9L	SMC-9L	Proposed Topology
number of switching devices	12	16	10
switching devices withstand voltage	$4E \times 4 + E \times 8$	$E \times 16$	$4E \times 6 + E \times 4$
capacitive voltage sensing number of devices	5	8	4
number of floating capacitors	3	6	2
suspension capacitance withstand voltage	$3E \times 1 + 2E \times 1 + E \times 1$	$3E \times 2 + 2E \times 2 + E \times 2$	$E \times 2$

In order to compare the switching losses of the above three topologies, the ANPC-9L is divided into 5 units, the SMC-9L is divided into 4 units, and the proposed topology is divided into 2 units, as shown in Figure 3. It can be seen that the ANPC-9L unit 1 has the same structure as the unit 1 of the proposed topology, and the voltages of the devices are both $4E$. Therefore, under the same switching frequency, the switching losses of the devices in the two topological units 1 are the same. The devices of ANPC-9L units 2, 3, 4, and 5 and the devices in the proposed topology unit 2 are both subjected to voltage E, but the proposed topology unit 2 has only 4 switching devices, while the ANPC-9L units 2, 2, There are a total of 8 switching devices in 3, 4 and 5. Obviously, at the same switching frequency, the proposed topology has smaller switching losses.

Compared with the SMC-9L, the proposed topology unit 2 and SMC-9L unit 4 have the same structure, and the voltages of the devices are both E. Under the same switching frequency, the switching loss of the proposed topology unit 2 and SMC-9L unit 4 is the same. There are 6 switching devices in the proposed topology unit 1, the withstand voltage is $4E$, the number of switching devices in SMC-9L units 1, 2, and 3 is 12, and the device withstand voltage is E; thus, under the same switching frequency, the device switching loss of the unit 1 in proposed topology is slightly larger than that of the sum of the units 1, 2, 3 in SMC-9L.

Table 3 shows the comparison results of the total number of devices in the current path when the three nine-level topologies output different voltage levels. It can be seen that when the proposed topology outputs each level, the number of devices through which the current flows is the smallest, so the conduction loss is also the smallest. Generally, the switching frequency of a multilevel inverter is low, and the conduction loss accounts for the main part of the device loss. Therefore, the topology proposed in this paper has the lowest total device loss.

Table 3. Comparison of the total number of devices in the current path.

V_O	Total Number of Devices in the Current Path		
	ANPC-9L	SMC-9L	Proposed Topology
$4E$	5	4	2
$3E$	5	5	3
$2E$	5	6	2
E	5	7	4
0	5	8	3
$-E$	5	7	4
$-2E$	5	6	2
$-3E$	5	5	3
$-4E$	5	4	2

3. Control Strategy of Voltage Sensor without Floating Capacitance

The control objectives of the proposed nine-level inverter are: output voltage V_O, dc-link capacitor voltages V_{C1}, V_{C2}, and flying capacitor voltages V_{C3}, V_{C4}. Since the one-dimensional space vector modulation algorithm has the advantages of high degree of freedom and can control multiple targets at the same time [26,27], the one-dimensional space vector modulation algorithm is adopted in this paper. On this basis, by analyzing the circuit state before the redundant switch state arrives, the estimation method of the floating capacitor voltage is given, and then the appropriate value is selected according to the difference between the estimated value of the floating capacitor voltage and the reference value and the positive and negative current of the current. Redundant switching state, a voltage sensor control algorithm without floating capacitance is proposed.

3.1. One-Dimensional Space Vector Modulation Algorithm

Let the output voltage vector of the single-phase nine-level inverter be $\boldsymbol{V}_O$, and the reference voltage vector be the modulating wave vector $\boldsymbol{u}_r$. Figure 4 shows the one-dimensional space vector diagram of the single-phase nine-level inverter.

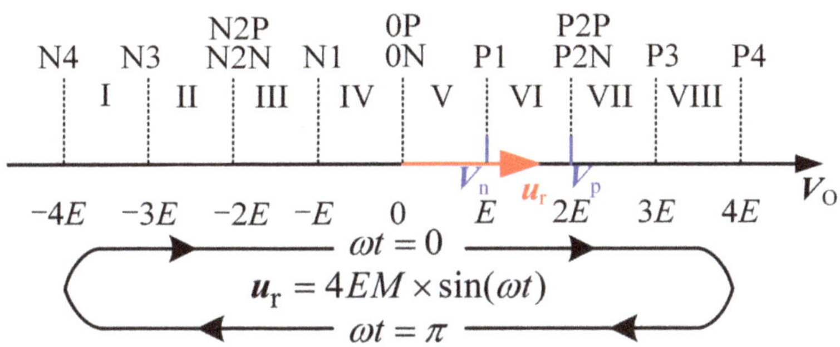

Figure 4. One-dimensional SVM of the proposed 9-level converter.

The expression of the modulation signal u_r is defined as

$$u_r = \frac{1}{2} M V_{DC} \sin(\omega t), \tag{1}$$

where M is the modulation degree, and $0 \leq M \leq 1$. In the one-dimensional vector space, select the most recent two voltage vectors to synthesize the reference voltage $\boldsymbol{u}_r$. Let the vector with the larger modulus value be V_p, and the vector with the smaller modulus value be V_n. It can be noticed that the interval where $\boldsymbol{u}_r$ is located can be determined according to the sign and modulus value of $\boldsymbol{u}_r$. If $\boldsymbol{u}_r$ is in the interval VI, the vector $2E$ and E are used for synthesis.

According to volt-second balance principle, the dwell time of the vectors V_p and V_n satisfies Equation (2), where T_s is the switching period.

$$\begin{cases} u_r T_s = \boldsymbol{V}_p T_p + \boldsymbol{V}_n T_n \\ T_s = T_p + T_n \end{cases}, \tag{2}$$

3.2. Control Strategy of Voltage Sensor without Floating Capacitance

From Table 1, the redundant switch states P2P, P2N, N2P and N2N can affect the floating capacitors C_3 and C_4 at the same time. Therefore, the floating capacitor voltage can be adjusted by reasonably selecting the redundant switch states corresponding to the $\pm 2E$ level. In order to cancel the floating capacitor voltage sensor, it is necessary to accurately estimate the floating capacitor voltage before outputting the $\pm 2E$ level.

Figure 5 shows the switch sequence diagram in which the output levels $2E$ and $-2E$ participate. Before the output voltage reaches the $2E$ level, there are two level states, $3E$

and E. Therefore, it is necessary to perform voltage estimation in these two states. Taking Figure 5a as an example, the floating capacitor voltage value is estimated when the $3E$ level is output, and the redundant switch state selection is performed when the $2E$ level is output. Before the output voltage reaches the $-2E$ level, there are two level states, $-3E$ and $-E$, and the voltage needs to be estimated in these two states. In Figure 5c, when the output voltage is at the $-E$ level, the floating capacitor voltage value is estimated, and the redundant switch state selection is performed when the output $-2E$ level.

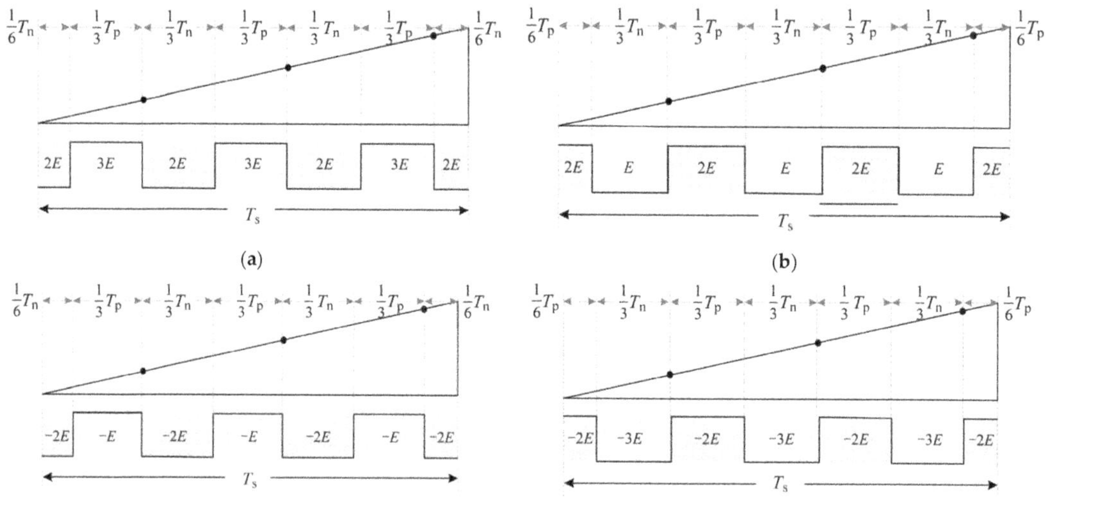

Figure 5. The switching sequences that contain the $2E$, $-2E$ voltage vector. (**a**) $N = 3$; (**b**) $N = 1$; (**c**) $N = -1$; (**d**) $N = -3$.

In order to distinguish the four level states of $3E$, E, $-E$, and $-3E$, define the inverter output voltage to round up the level N:

$$N = \text{ceil}(\frac{V_O}{E}), \quad (3)$$

where ceil(x) is the round-up function.

As can be seen from Figure 5, there are 3 sampling data reading points in one switching sequence cycle, and the values of V_{C1}, V_{C2}, V_O, and I_O can be read at the sampling points. In Figure 2b, when the level state is $3E$, the current flows through C_1, C_3 and the load and returns to the neutral point. According to Kirchhoff's voltage law, the floating capacitor voltage V_{C3} can be expressed as $V_{C1} - V_O$. In Figure 2h, when the level state is $-E$, the current flows through C_3 and the load and returns to the neutral point, and the floating capacitor voltage V_{C3} can be expressed as $-V_O$. Therefore, when the switch states are P3 and N1, the voltage of C_3 can be estimated. In Figure 2e, when the level state is E, the current flows through C_4 and the load and returns to the neutral point, and the floating capacitor voltage V_{C4} can be expressed as V_O. When the flat state is $-3E$, the current flows through C_2, C_4 and the load and returns to the neutral point, and the floating capacitor voltage V_{C4} can be expressed as $V_O + V_{C2}$. Therefore, when the switching states are P1 and N3, the voltage of C_4 can be estimated. Therefore, the voltage estimation formula of capacitors C_3 and C_4 can be expressed as:

$$V_{FC\_est} = \begin{cases} V_{C1} - V_O, & V_O = 3E \\ V_O, & V_O = E \\ -V_O, & V_O = -E \\ V_O + V_{C2}, & V_O = -3E \end{cases}, \quad (4)$$

However, when voltage estimation is performed, there will be cases where the estimated value changes abruptly as the switch state changes. Therefore, it is necessary to deal with this sudden change in the control algorithm: use the estimated value at the previous moment to select the redundant switch state, so as to avoid the sudden change in the estimated value causing the estimated voltages of C_3 and C_4 to be unbalanced.

Since the proposed topology and modulation signal u_r are both symmetrical, in theory, the charge and discharge capacities of capacitors C_3 and C_4 are not much different, and the redundant switch states at $2E$ and $-2E$ levels have the same effect on the charge and discharge of C_3 and C_4. Therefore, it is not necessary to set the priority of voltage regulation of C_3 and C_4 in the proposed control algorithm, and unified tracking regulation can be performed according to the estimated floating capacitor voltage value corresponding to the moment before the arrival of the output level $\pm 2E$. Let ΔV_{CF} be the voltage offset state between the estimated value of C_3 or C_4 and the reference value, the polarity of the voltage offset of C_3 and C_4 can be expressed as:

$$D_C = \begin{cases} 1, & \Delta V_{CF} > 0 \\ 0, & \Delta V_{CF} \leq 0 \end{cases}, \quad (5)$$

Let the direction of the current flowing out of the inverter be the positive direction, and the current direction is represented by D_I:

$$D_I = \begin{cases} 1, & i_O > 0 \\ 0, & i_O \leq 0 \end{cases}, \quad (6)$$

It can be seen from the literature [28] that the selection of the reference value of the floating capacitor voltage determines the storage and transmission of capacitive energy in the topology, and the transfer relationship of the capacitive energy is:

$$\Delta Q_{C1} + \Delta Q_{FC\_P} = \Delta Q_{C2} + \Delta Q_{FC\_N}, \quad (7)$$

In the formula, ΔQ_{C1} and ΔQ_{C2} are the energy transferred from C_1 and C_2 to the load in one modulating wave cycle, respectively; ΔQ_{FC_P} and ΔQ_{FC_N} are the energy transferred from C_3 and C_4 to the load in the positive and negative half cycles. Therefore, the voltage reference value of the positive and negative half-cycle floating capacitors can be set as:

$$V^*_{FC\_P} = E + K(2E - \frac{V_{C2}}{2}), \quad (8)$$

$$V^*_{FC\_N} = E + K(2E - \frac{V_{C1}}{2}), \quad (9)$$

In the formula, K is the voltage correction coefficient, and the value of K affects the balance degree of V_{C1} and V_{C2}.

When the estimated voltage of the floating capacitor C_3 or C_4 is greater than or less than the reference voltages $V_{FC\_P^*}$ and $V_{FC\_N^*}$, it needs to be discharged or charged, and then the switch state at the next moment is selected according to the direction of the output current I_O at this time. When the inverter outputs $2E$ and $-2E$ levels, the switching state selection principle for balancing the voltage of the floating capacitor is as follows:

$$S_{P2} = [D_C \odot D_I] * \text{P2P} + [D_C \oplus D_I] * \text{P2N}, \quad (10)$$

$$S_{N2} = [D_C \odot D_I] * N2P + [D_C \oplus D_I] * N2N. \quad (11)$$

3.3. Implementation of the Proposed Algorithm

The implementation process of the floating capacitor voltage estimation is shown in Figure 6. By judging the size of the output voltage rounding up the level N, the estimation formula of the floating capacitor voltage is selected.

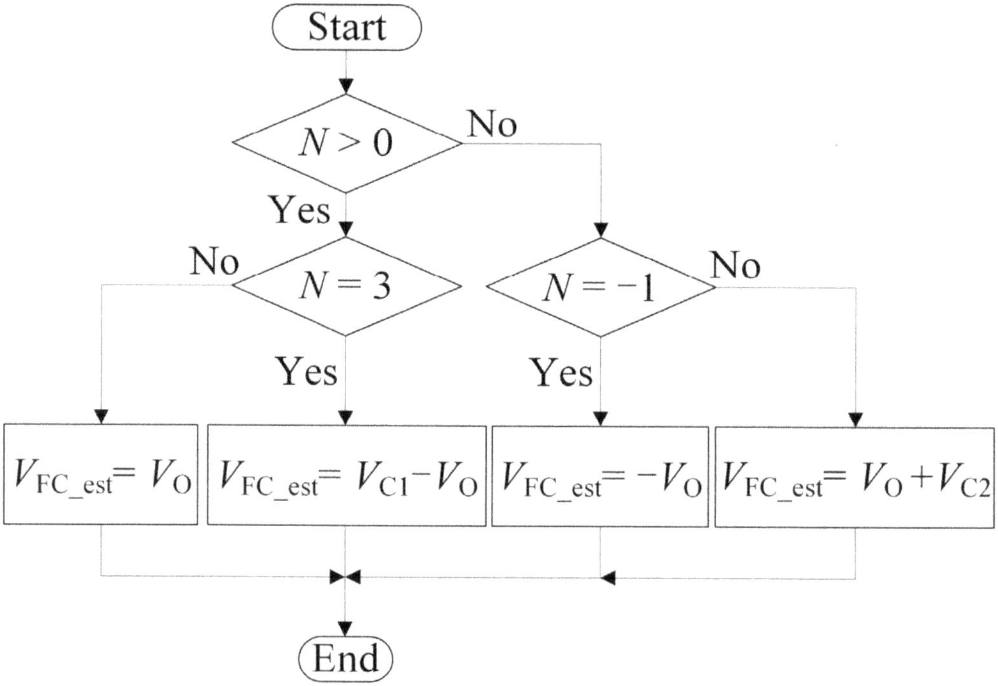

Figure 6. The flow chart of floating capacitor voltage estimation.

The flowchart of redundant switch state selection is shown in Figure 7. Combined with Table 1, if the sampled output current $I_O > 0$ and the output voltage V_O is positive, when the estimated value of the floating capacitor voltage is greater than the reference value, select the P2N switch state to discharge; when the estimated value of the floating capacitor voltage is less than the reference value, select the P2P switching state for charging, so that the voltage of the floating capacitor returns to a balanced state. Similarly, other redundant switch states can be selected according to Figure 7.

The block diagram of the proposed control strategy is shown in Figure 8. The nine-level inverter adopts one-dimensional space vector modulation, obtains the real-time operating parameters V_{C1}, V_{C2}, V_O, I_O of the inverter through sampling, selects the calculation formula of the floating capacitor voltages according to the redundant switch state, judges the positive and negative output current. Then, calculate the difference between the estimated value of the floating capacitor voltage and the reference value, and select the redundant switch state according to the charging and discharging requirements of the floating capacitor, and finally output the driving pulse of the switch tube S_k ($k = 1, 2, \ldots 7$).

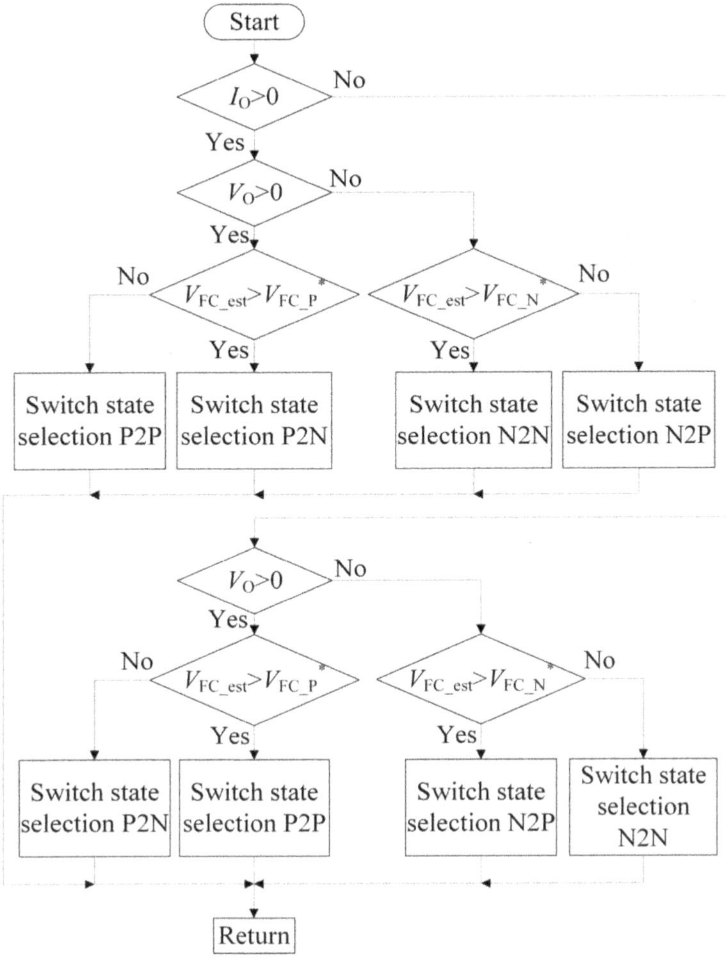

Figure 7. The flow chart of redundant switching states selection.

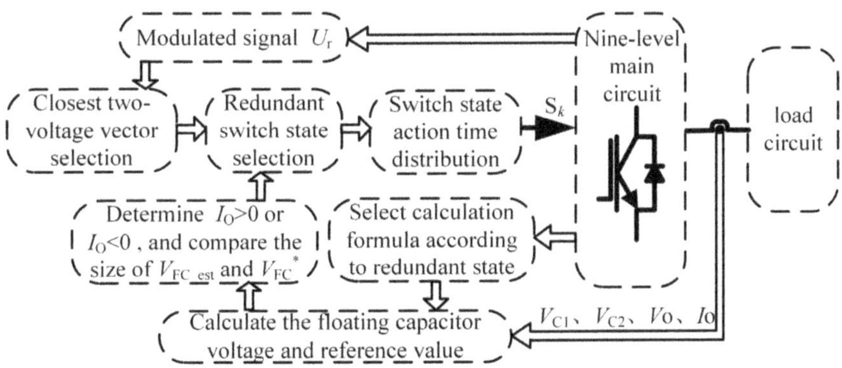

Figure 8. The block diagram of the proposed floating capacitor voltage sensorless control scheme.

4. Simulation and Experimental Analysis

4.1. Simulation Analysis

In order to verify the validity of the proposed topology and the feasibility of the control strategy, a single-phase nine-level inverter simulation model was built in Matlab/Simulink. The simulation parameters are listed as follows: dc-link voltage reference value is 400 V, bus capacitor $C_1 = C_2 = 6720$ uF, flying capacitor $C_3 = C_4 = 6$ mF, output voltage fundamental frequency is 50 Hz, load $L_f = 30$ mH, $R_f = 10\ \Omega$. The voltage correction coefficient K in Equations (8) and (9) is tuned as 0.1.

Figure 9 shows the steady-state simulation waveforms of the proposed topology. It can be seen from Figure 9a,b that the output voltage of the proposed topology is a nine-level staircase wave, the output current has a high sine degree, and the current THD is 0.28%. In Figure 9c, the bus capacitor voltage is stable around 200 V, with a fluctuation range of ±3.8 V, and the floating capacitor voltage is stable around 50 V, with an up and down fluctuation range of ±2.2 V.

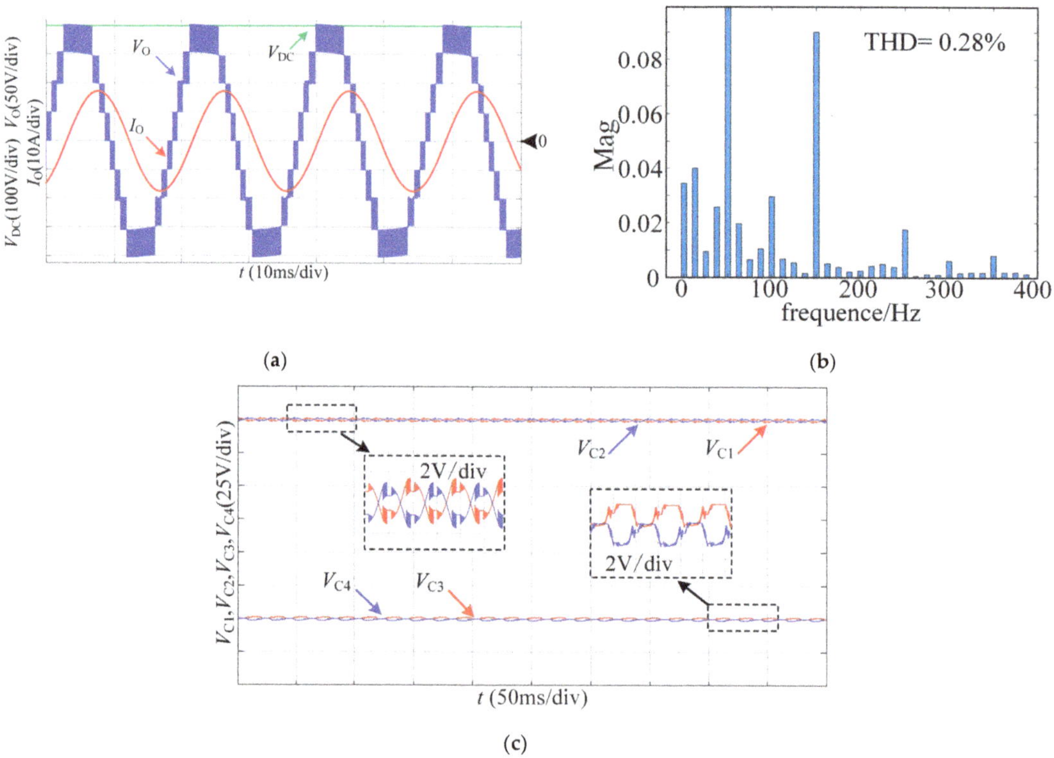

Figure 9. Simulation waveforms of the proposed topology in the steady-state. (**a**) Bus voltage output voltage and output current simulation waveform; (**b**) Output current FFT analysis results; (**c**) Simulation Waveforms of dc-link Capacitor Voltage and Suspended Capacitor Voltage.

Figure 10 shows the simulation waveforms of the dc-link voltage, bus capacitor voltage, floating capacitor voltage, output voltage and output current when the modulation factor $M = 0.7$, $M = 0.85$ and $M = 1$. THD_V and THD_I are the THD values of the output voltage and output current, respectively. It can be seen from Figure 10 that when the modulation degree is small, the output voltage presents a seven-level staircase wave. When the modulation degree changes, the output voltage can achieve a smooth transition. Under different modulations, both the bus capacitor voltage and the floating capacitor voltage are stable near their reference values, where the bus capacitor voltage ripple is 7 V, and the floating capacitor voltage ripple is 5 V.

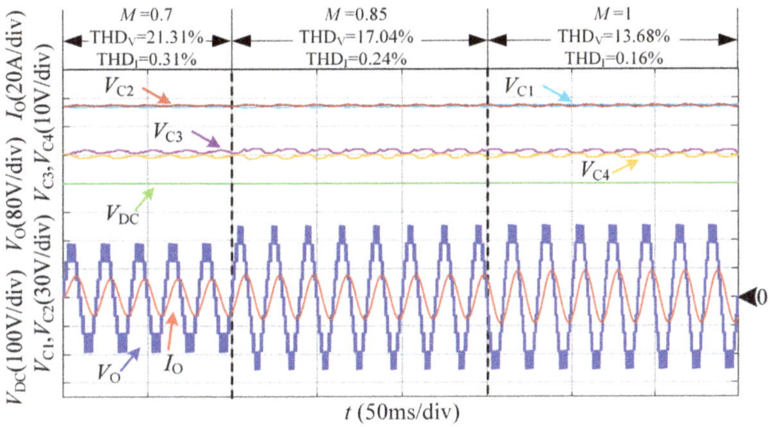

Figure 10. Simulation results under different modulation degree M.

Figure 11 shows the simulation waveforms of the proposed voltage sensorless control strategy. It can be seen from Figure 11a that the proposed control method can track the voltage of the floating capacitor quickly and accurately under different modulation degrees, and the error rate is less than 2%. Comparing Figures 10 and 11b, the proposed method has the same output performance as the traditional sensor measurement method, but its capacitor voltage ripple is larger than that of the sensor measurement, the bus capacitor voltage ripple is 13 V, and the floating capacitor voltage ripple is 7 V. Because the floating capacitor voltage estimation method is affected by the sampling time and the system calculation time, a slight delay leads to a larger estimation error.

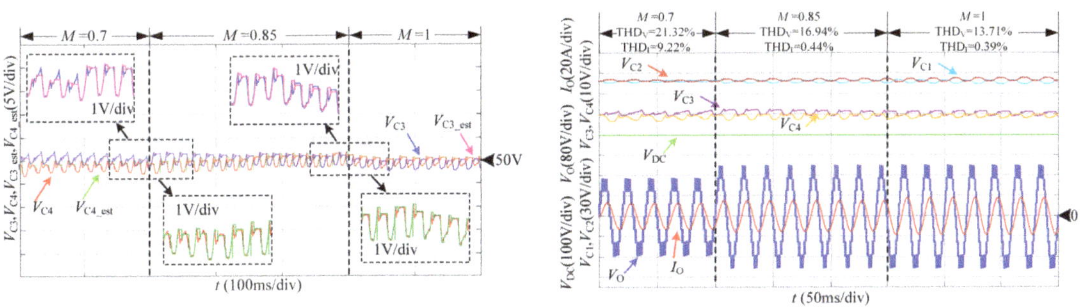

(**a**) (**b**)

Figure 11. Simulation waveforms of the proposed floating capacitor voltage sensorless control scheme. (**a**) Simulation waveform of actual and estimated value of floating capacitor voltage; (**b**) Capacitor voltage, output voltage and output current simulation waveforms.

4.2. Experimental Analysis

To further verify the effect of the circuit topology and control method proposed in this paper, an experimental platform is built, as shown in Figure 12. The experimental parameters are as follows: dc-link voltage reference value is 400 V, dc-link capacitor $C_1 = C_2 = 6720$ uF, floating capacitor $C_3 = C_4 = 6$ mF, output voltage fundamental frequency $f_g = 50$ Hz, load $L_f = 30$ mH, $R_f = 6\ \Omega$, and system switch Frequency $f_s = 2$ kHz. The switch device model used in this platform is IPW60R070C6, and its driver is QC962-8A.

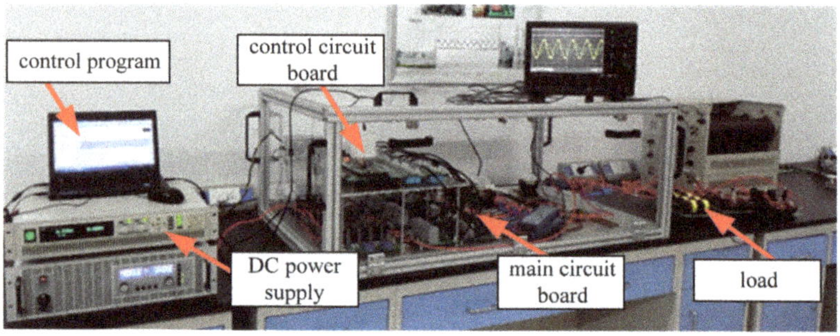

Figure 12. Scaled-down experimental platform of the proposed single-phase nine-level inverter.

Figure 13 shows the experimental waveforms of the proposed topology at different modulation degrees. In Figure 13a, the experimental waveform is basically consistent with the simulation result, the output voltage is a nine-level staircase wave, and the output current has a high sine degree. Due to the influence of the load inductance L_f, the current lags the voltage. It can be seen that the nine-level topology has good output performance. It can be seen from Figure 13b that V_{C1} and V_{C2} are both stable at about 200 V, and the voltage fluctuation range is ±12 V. It can be seen from Figure 13c that V_{C3} and V_{C4} are both stable at about 50 V, and the voltage fluctuation range is ±4 V.

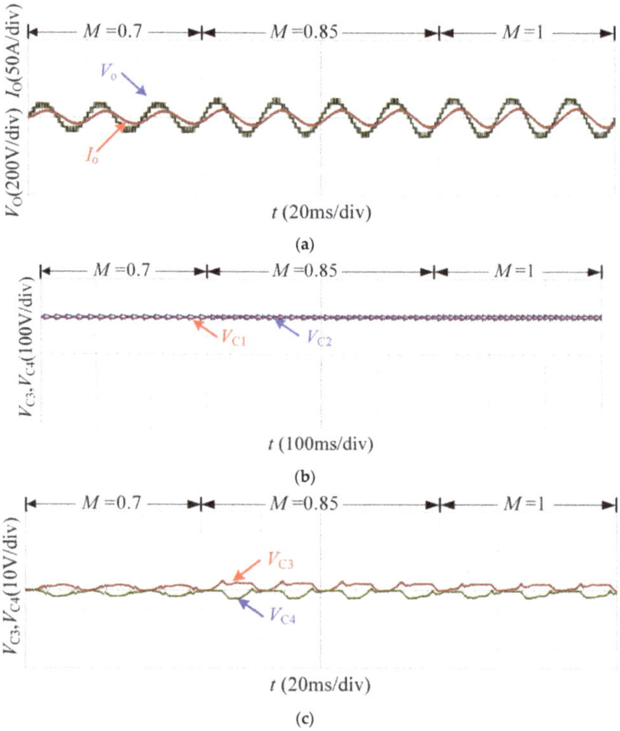

Figure 13. Experimental results of the proposed inverter with different M. (**a**) Output voltage V_O and output current I_O waveform; (**b**) dc-link capacitor voltage V_{C1}, V_{C2} waveform; (**c**) Suspension capacitor voltage V_{C3}, V_{C4} waveform.

Figure 14 shows the experimental waveforms of the proposed voltage sensorless control strategy. It can be seen from Figure 14a that under different modulation degrees, the proposed method has the same effect as using the floating capacitor voltage sensor, and both can achieve stable output of the inverter. In Figure 14b, V_{C1} and V_{C2} have the smallest ripple when $M = 0.85$, and the voltage ripple range is ±14 V. From Figure 14c, it can be noticed that the voltage fluctuation range of V_{C3} and V_{C4} is ±8 V, which is slightly larger than that of the method with floating capacitor voltage sensor. It is because the voltage ripple is affected by factors such as the control frequency of the system, the correction coefficient K, and the number of estimated sampling points. Although the voltage ripple is large, the proposed sensorless control strategy can reduce the hardware cost of the system, since there is no need to design the sampling circuit of the floating capacitor voltages.

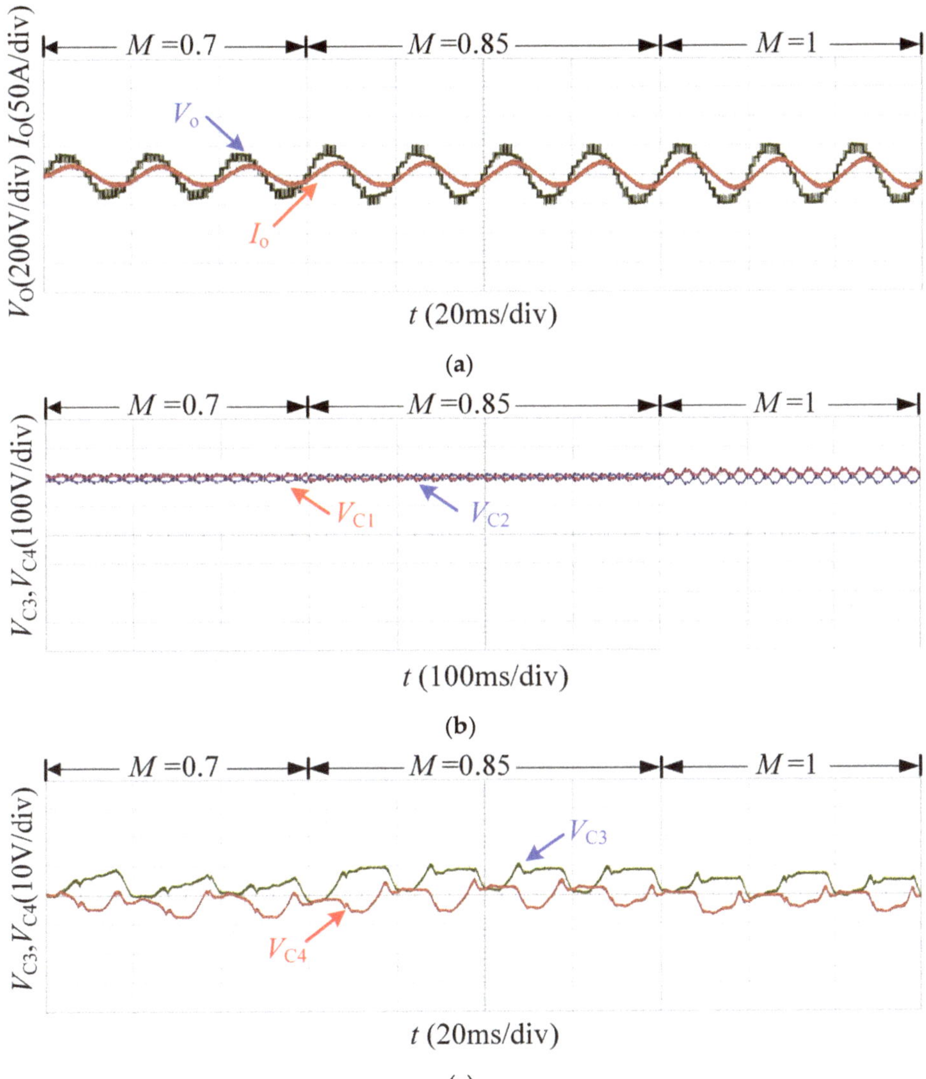

Figure 14. Experimental results of the floating capacitor voltage sensorless control scheme. (**a**) Waveforms of the output voltage V_O and output current I_O; (**b**) Waveforms of the dc-link capacitor voltages V_{C1}, V_{C2}; (**c**) Waveforms of the capacitor voltage V_{C3}, V_{C4}.

5. Conclusions

This paper proposes a single-phase non-isolated nine-level inverter topology. Compared with the traditional topology, this topology has the advantages of fewer switching devices and fewer floating capacitors. Additionally, a floating capacitor voltage sensorless control algorithm is presented. The algorithm realizes the estimation of the floating capacitor voltages by sampling the output voltage and the bus capacitor voltage, and it achieves the balance of the floating capacitor voltages combined with the one-dimensional space vector modulation method. Finally, the proposed algorithm is verified by computer simulation and experiments. Theoretical analysis and simulation experimental results show that the proposed topology and control method have the following advantages:

(1) Compared with the traditional nine level inverter topology, the proposed nine level inverter has obvious advantages in the number of switching devices and the number of flying capacitors. The most prominent is that the number of switching devices through which the current flows is the least, which reduces the device switching loss in the inverter.

(2) The proposed method reduces 50% capacitor voltage sampling circuit, which can greatly reduce the cost of the system, and it can realize the accurate estimation of the floating capacitor voltage and its balance control at a specific moment.

Author Contributions: Conceptualization, Y.L. and Y.S.; methodology, Y.L., W.H., B.L. and Y.S.; software, W.H. and B.L.; validation, Y.L., Z.L. and W.H.; formal analysis, W.H. and B.L.; writing—original draft preparation, Y.L. and Y.S.; writing—review and editing, W.H. and B.L. All authors have read and agreed to the published version of the manuscript.

Funding: This research was funded by the National Science Foundation of China (No.61273251) and open Project of National and Local Joint Engineering Laboratory of High Energy-saving Motor and Control Technology of Anhui University (KFKT202104).

Institutional Review Board Statement: Not applicable.

Informed Consent Statement: Not applicable.

Data Availability Statement: Not applicable.

Acknowledgments: The authors gratefully appreciate the reviewers for their constructive comments and suggestions.

Conflicts of Interest: The authors declare no conflict of interest.

References

1. Wang, Y.Q.; Ku, R.H.; Zhou, C.L.; Wang, Z.; Wang, M.D. Hybrid T-type multilevel inverter and its modulation strategy. *High Volt. Eng.* **2020**, *46*, 3220–3228.
2. Yu, H.Y.; Chen, B.; Yao, W.X.; Liu, Z.Y. Hybrid seven-level converter based on t-type converter and h-bridge cascaded under SPWM and SVM. *IEEE Trans. Power Electron.* **2018**, *33*, 689–702. [CrossRef]
3. Zhang, Y.L.; Wang, Q.J.; Hu, C.G.; Zhou, Y.F. A novel low voltage seventeen-level converter and control strategy. *Proc. CSEE* **2020**, *40*, 1095–1105.
4. Yousofi-Darmian, S.; Barakati, S.M. A new asymmetric multilevel inverter with reduced number of co-mponents. *IEEE J. Emerg. Sel. Top. Power Electron.* **2020**, *8*, 4333–4342. [CrossRef]
5. Hu, C.G.; Rui, T.; Ma, D.J.; Wang, Q.J.; Luo, F.L. SVM control strategy for neutral point voltage balance and switching loss reduction of three-level ANPC inverter. *Proc. CSEE* **2016**, *36*, 3598–3608.
6. Cao, Y.; Yang, J.T.; Li, R.; Cai, X.; Yu, W.; Xu, J.; Du, C.R.; Huang, L.; Zhang, T.; Ma, Z.Y.; et al. Circulating current suppression strategy of ANPC nine-level inverter. *Proc. CSEE* **2020**, *40*, 232–242.
7. Viju, N.R.; Arun, R.S.; Kaarthik, R.S. Generation of higher number of voltage levels by stacking inverters of lower multilevel structures with low voltage devices for drives. *IEEE Trans. Power Electron.* **2017**, *32*, 52–59.
8. Liu, J.F.; Wu, J.L.; Zeng, J.; Guo, H.F. A novel nine-level inverter employing one voltage source and reduced components as high-frequency AC power source. *IEEE Trans. Power Electron.* **2017**, *32*, 2939–2947. [CrossRef]
9. Ghias, A.M.Y.M.; Pou, J.; Agelidis, V.G. An active voltage-balancing method based on phase-shifted PWM for stacked multicell converters. *IEEE Trans. Power Electron.* **2016**, *31*, 1921–1929. [CrossRef]

10. Ghias, A.M.Y.M.; Pou, J.; Acuna, P.; Ceballos, S.; Heidari, A.; Agelidis, V.G.; Merabet, A. Elimination of low-frequency ripples and regulation of neutral-point voltage in stacked multicell converters. *IEEE Trans. Power Electron.* **2017**, *32*, 164–175. [CrossRef]
11. Lee, S.S.; Sidorov, M.; Lim, C.S. Hybrid cascaded multilevel inverter (HCMLI) with improved symmetrical 4-Level submodule. *IEEE Trans. Power Electron.* **2018**, *33*, 932–935. [CrossRef]
12. Sandeep, N.; Yaragatti, U.R. Design and implementation of a sensorless multilevel inverter with reduced part count. *IEEE Trans. Power Electron.* **2017**, *33*, 6677–6683. [CrossRef]
13. Sandeep, N.; Yaragatti, U.R. Operation and control of a nine-level modified ANPC inverter topology with reduced part count for grid-connected applications. *IEEE Trans. Ind. Electron.* **2018**, *65*, 4810–4818. [CrossRef]
14. Liu, C.K.; Deng, F.J.; Wang, Q.J.; Wang, Y.B.; Blaabjerg, F.; Wang, Z. Double half-bridge submodule-based modular multilevel converters with reduced voltage sensors. *IEEE Trans. Power Electron.* **2021**, *36*, 3643–3648. [CrossRef]
15. Li, J.J.; Jiang, J.G. A novel space vector pulse width modulation for five-level nested neutral point piloted converter. *Proc. CSEE* **2018**, *38*, 3306–3315.
16. Hu, C.G.; Dong, H.; Zhang, Y.L.; Li, J. Hybrid active neutral clamped seven level inverter and its control strategy. *Electr. Mach. Control* **2020**, *24*, 40–49.
17. Sandeep, N.; Yaragatti, U.R. A switched-capacitor-based multilevel inverter topology with reduced components. *IEEE Trans. Power Electron.* **2018**, *33*, 5538–5542. [CrossRef]
18. Cheng, Q.; Wang, C.; Chen, Z.; Li, Z. A capacitor-voltage-balancing method based on optimal zero-sequence voltage injection in stacked multicell converter. *IEEE J. Emerg. Sel. Top. Power Electron.* **2021**, *9*, 4700–4714. [CrossRef]
19. Davis, T.T.; Joseph, T.; Dey, A. A capacitor voltage balancing scheme for single-source fed switch optimized three-phase nine-level inverter. *IEEE Trans. Ind. Electron.* **2021**, *68*, 3652–3661. [CrossRef]
20. Tian, K.; Wu, B.; Narimani, M.; Xu, D.; Cheng, Z.Y.; Zargari, N.R. A capacitor voltage-balancing method for nested neutral point clamped (NNPC) inverter. *IEEE Trans. Power Electron.* **2016**, *31*, 2575–2583. [CrossRef]
21. Wang, K.; Xu, L.; Zheng, Z.D.; Li, Y.D. Voltage balancing control of a four-level hybrid-clamped converter based on zero-sequence voltage injection using phase-shifted PWM. *IEEE Trans. Power Electron.* **2016**, *31*, 5389–5399. [CrossRef]
22. Li, W.C.; Ma, W.M.; Wang, G.G.; Lin, C.M.; Lie, S.X.; Guo, D.H. Novel space vector pulse width modulation method for neutral point clamped H-bridge cascaded single-phase inverter. *Proc. CSEE* **2014**, *34*, 5313–5319.
23. Liu, B.; Feng, X.Y.; Deng, R.; Xia, W.J.; Song, W.S. Grid voltage sensorless control strategy of single-phase PWM rectifiers with model reference adaptive system. *Proc. CSEE* **2019**, *39*, 6065–6074.
24. Zhang, L.H.; Born, R.; Gu, B.; Chen, B.F.; Zheng, C.; Zhao, X.N.; Lai, J.S. A sensorless implementation of the parabolic current control for single-phase stand-alone inverters. *IEEE Trans. Power Electron.* **2016**, *31*, 3913–3921. [CrossRef]
25. Viswadev, R.; Mudlapur, A.; Ramana, V.V.; Venkatesaperumal, B.; Mishra, S. A novel AC current sensorless hysteresis control for grid tie inverters. *IEEE Trans. Circuits Syst. II* **2020**, *67*, 2577–2581. [CrossRef]
26. Leon, J.I.; Portillo, R.; Vazquez, S.; Padilla, J.J.; Franquelo, L.G.; Carrasco, J.M. Simple unified approach to develop a time-domain modulation strategy for single-phase multilevel converters. *IEEE Trans. Ind. Electron.* **2008**, *55*, 3239–3248. [CrossRef]
27. Liu, F.; Wu, X.Z.; Zhao, Y.M.; Liu, J.D.; Zhang, P.; Zhao, Y.X.; Wang, J. Research on common mode voltage suppression technology of multilevel inverter based on one-dimensional space vector modulation. *Power Syst. Technol.* **2020**, *44*, 3972–3982.
28. Teymour, H.R.; Sutanto, D.; Muttaqi, K.M.; Cinfo, P. A novel modulation technique and a new balancing control strategy for a single-phase five-level ANPC converter. *IEEE Trans. Ind. Appl.* **2015**, *51*, 1215–1227. [CrossRef]

energies

MDPI

Article

Impact of the COVID-19 Pandemic Crisis on the Efficiency of European Intraday Electricity Markets

Jan Niklas Buescher [1], Daria Gottwald [2,*], Florian Momm [3] and Alexander Zureck [2]

1 Automatisierungstechnik, RWTH Aachen University, Templergraben 55, 52062 Aachen, Germany; jan.buescher@rwth-aachen.de
2 ISF Institute for Strategic Finance, FOM University of Applied Sciences for Economics and Management, 45127 Essen, Germany; alexander.zureck@fom.de
3 Energie- und Wasseroekonomik, Hochschule Ruhr West, Duisburger Str. 100, 45479 Muelheim an der Ruhr, Germany; florian.momm@stud.hs-ruhrwest.de
* Correspondence: daria.gottwald@fom-net.de

Abstract: Our goal is to examine the efficiency of different intraday electricity markets and if any of their price prediction models are more accurate than others. This paper includes a comprehensive review of Germany, France, and Norway's (NOR1) day-ahead and intraday electricity market prices. These markets represent different energy mixes which would allow us to analyze the impact of the energy mix on the efficiencies of these markets. To draw conclusions about extreme market conditions, (i) we reviewed the market data linked to COVID-19. We expected higher volatility in the lockdowns than before and therefore decrease in the efficiency of the prediction models. With our analysis, (ii) we want to draw conclusions as to whether a mix based mainly on renewable energies such as that in Norway implies lower volatilities even in times of crisis. This would answer (iii) whether a market with an energy mix like Norway is more efficient in highly volatile phases. For the analysis, we use data visualization and statistical models as well as sample and out-of-sample data. Our finding was that while the different price and volatility levels occurred, the direction of the market was similar. We could find evidence that our expectations (i–iii) were met.

Keywords: energy efficiency; energy mix; energy markets; COVID-19; out-of-sample data

Citation: Buescher, J.N.; Gottwald, D.; Momm, F.; Zureck, A. Impact of the COVID-19 Pandemic Crisis on the Efficiency of European Intraday Electricity Markets. *Energies* **2022**, *15*, 3494. https://doi.org/10.3390/en15103494

Academic Editor: Jesús Manuel Riquelme-Santos

Received: 12 April 2022
Accepted: 5 May 2022
Published: 10 May 2022

1. Introduction

While digitalization and climate change continue, renewable electricity generation, such as wind and solar power, will be further expanded [1–3]. At the same time, SARS-CoV-2 (COVID-19) has a major impact on the entire global economy. Changes are particularly noticeable in the energy sector, as both supply and demand are affected. These effects were striking regarding WTI and BRENT oil prices, which in some cases fell into negative values in April 2020 [4]. Since the focus of this paper will be Europe, here are a few examples of European energy prices: The BRENT oil price dropped by 59% compared to its last peak on the 17th of February. Also, gas prices dropped to their lowest since 1995 at the same time [5]. Since the COVID-19 crisis had a major impact on other commodities, such as oil and gas, the question now arises as to the situation in the whole-sale electricity markets. Unlike other commodities, electricity cannot be stored to the same extent, which is why there are special features regarding price forecasts [6]. In contrast to the markets described, the electricity markets have a lower market extent of liquidity, leading to a wider bid-ask spread [7]. The motivation of this paper is to analyze and compare different European short-term electricity markets by applying statistics to corresponding market data. We want to draw conclusions about the impact of the energy mix on prices and volatilities during periods of extreme market conditions caused by the effects of the COVID-19 pandemic crisis on the short-term electricity markets. The motivation is also to analyze if the extreme market conditions had an immediate impact on energy efficiency. Three markets were selected because each market represents a different main energy source. In other words, this research aims to point out the impact on a market relying mainly on wind energy and lignite (Germany), a market relying on nuclear power (France), and a market relying on hydropower (Norway) [8].

Like what has already been done with other commodities, one aim of this paper is to divide the period of 01.01.2020 to 01.02.2021 into different phases and highlight the impacts of the lockdowns. The paper by J. Ali and W. Kahn (2020) serves as a model for this procedure. They looked at the agricultural commodity prices for various products in India and examined the influences on prices in the individual phases of the local lockdowns. They found that the weighted average of prices fell during the lockdown [9]. Some papers focus on the links between the financial and commodity markets. One example is the paper of A. Elsayed et al. (2020). They investigate co-movements between the energy market and the financial markets. Furthermore, they analyze time patterns of volatility spillovers [10]. The research of Bompard et al. (2020) focuses on the immediate impact of COVID-19 on European Electricity systems. This paper considers the quantified impact of strategic decisions on regulation and system operations during the lockdown periods [11].

The paper by O. B. Adekoya and J. A. Oliyide (2021) highlights the link between financial and commodity markets. In addition, it refers to the main price drivers. These linkages are mainly captured by analyzing price movements and volatilities [12]. Our paper aims to analyze price and volatility movements as well. Among all financial assets, spot electricity prices belong to the most volatile asset classes. One of the reasons for the high volatility is the non-storable nature of volatility. In their study L. Han et al. point out the risks for the market participants caused by volatilities and extreme price outcomes. For their analysis, they also looked at the different market regions in Australia individually and then compared them with each other. We take a similar strategic approach in this paper, as we examine and compare various European markets based on their volatilities and prices [13]. In our paper, we aim to determine whether there is a link between the type of energy generation and the price or volatility movement in an extreme economic situation. In doing so, we will follow the approach of S. Halbrügge et al. (2021) and their comprehensive analysis of the German electricity market, as well as C. Fezzi and V. Fanghella (2020), who analyzed the Italian Electricity market [4,14]. In addition, Kuppelwieser and Wozabal (2021) followed a similar approach as it is done in this kind of research. They also made use of intraday power data considering out-of-sample data sets. In contrast to our research, they consider weather forecasting and algorithmic trading. Furthermore, they also took order data and market liquidity into account [7].

In the course of their research, C. Kath and F. Ziel (2018, 2021) examined the forecast accuracy by applying modern regression techniques [15] and conformal predictions [16] in short-term electricity markets [15,16]. The focus and novelty of our paper are not to create a forecast. Rather, we have applied statistics and visualization techniques to test the predictive accuracy in the volatile market. B. Finnah et al. used different approaches to visualizing short-term electricity data in combination with a statistical model. Their focus was on the German day-ahead and the intraday auction electricity markets [17]. K. Maciejowska et al. (2019) performed a region-based comparison between German and Polish short-term electricity data. They also made use of visualizations comparing the different datasets [18]. Kramer and Kiesel (2021) focused on a data visualization regarding the buy and sell order data for German short-term electricity markets (Figure 1). In their paper, they also pointed out how the day-ahead and intraday markets are structured. This graph also supports the understanding of our research [19]. Other factors that can influence electricity prices are extreme weather and bidding behavior. Ghosh et al. (2021) investigated the freeze in Texas in February 2021 and analyzed the effects of extreme weather conditions on electricity prices [20]. Xiao et al. (2021) conducted an analysis of the bidding behavior using virtual bidding. They focused on wind power producers in the course of their research [21].

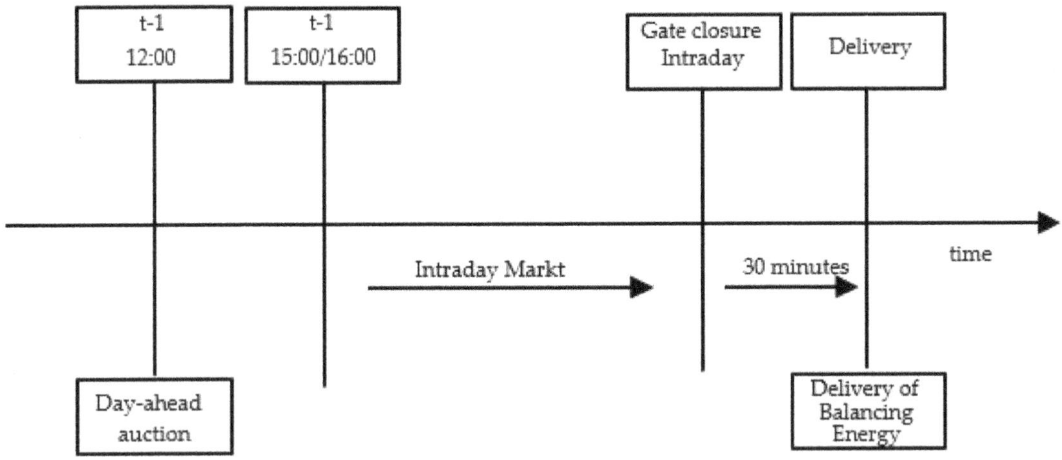

Figure 1. This figure represents the German electricity spot market. This figure functions to improve understanding on the market design. This figure has been presented according to Kramer and Kiesel (2021) [19].

To use (i) out-of-sample data to improve forecasting accuracy, which is defined as the price difference between intraday and day-ahead markets in a similar way to traditional asset management, we have examined two types of prices in extreme situations for this paper [16,22]. We used intraday prices as well as day-ahead prices. One of our goals is to determine whether the forecasting methods have been improved by using this data. The periods of March to May 2020 and December to February 2020/21 are crucial. In addition, countries in Europe reacted with different measures to the pandemic. Considering the COVID-19 measures and the (ii) different energy mixes, it leads to the question if it had an impact on the prices [13,23–25]. In other words, did the energy mix influence the prices during high volatility periods caused by lockdowns? To investigate the question, we looked at the energy markets in Germany, France, and Norway (NO1). Our overall research problem is whether (iii) the forecasting accuracy in the German intraday market has improved in the second lockdown.

1.1. Contributions

Following the guidelines presented in our research design (see Figure 2), we apply a rigorous, transparent, and reproducible methodology for our models and visualizations. All four types of figures follow the same coding methodology using different datasets. We will provide our readers with further details on our scripts when contacting the corresponding author. We are transparent about our data sources and codes being used regarding Bloomberg. We present how different energy mixes react to extreme market conditions in terms of market efficiency by applying our model to three different short-term electricity markets.

1.2. Paper Structure

The paper is structured as follows: we first present common methods regarding volatility and explain our calculation method with regard to confidence intervals. After that, we looked at the data we used. Our results are structured as follows: A general introduction to our approach, followed by a structured analysis looking first at intraday and then at day-ahead prices. In the case of Germany and France, we have also analyzed the volumes. This is followed by a discussion, conclusion, and outlook. To present our results, we refer to (i) extreme market conditions, (ii) the energy mix, and (iii) energy efficiency.

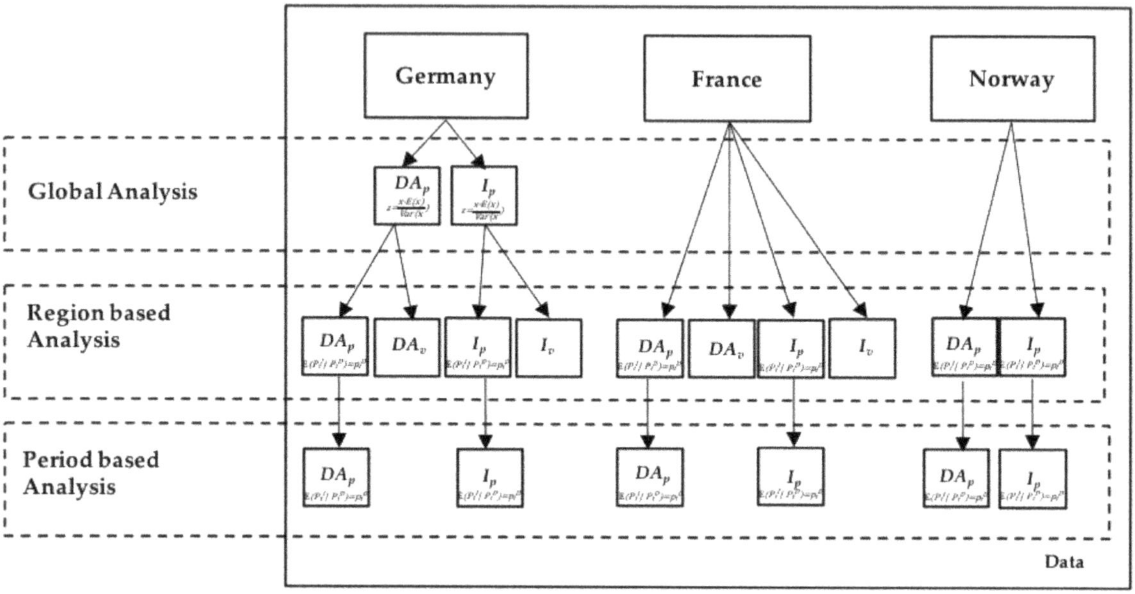

Figure 2. This figure represents the research design including the different stages of research the data went through in the course of the analysis. The extent at which each kind of data got analyzed has been linked to the data availability in Bloomberg. DA = day-ahead, I = intraday, p = Price, v = volume.

2. Materials and Methods

2.1. Price Analysis

We present the data and the corresponding statistical results using various visualization methods to draw a conclusion about the forecast accuracy in a volatile market, following an order that starts with a general overview via graphs. These graphs include all price data for German intraday and day-ahead electricity products. Afterwards, we divided the data into the observed regions (Germany, France, and Oslo—Norway) and added the 30-day volatility and confidence intervals to the analysis. The confidence intervals present the uncertainty estimates regarding the analyzed data. In addition, electricity generation was also taken into consideration. Thereby, we pointed out different energy sources (renewables and fossil fuels). The following statistical values are determined: Mean, median, standard deviation, and a 95% confidence interval. To calculate volatility, the following formula is used in literature for spot prices observed from historical data. Price returns are used in this example to obtain the volatility estimates [26].

$$\sigma^2 dt = E\left[\left(\widetilde{dS/S}\right)^2\right] \tag{1}$$

where:

S = Spot Price
σ = Spot Price Volatility
dt = variance

Regarding the volatility figures, we made use of the 30- and 90-day volatility. Thereby, we retrieved the corresponding data from Bloomberg using the following codes: VOLATILITY_30D and VOLATILITY_90D.

To calculate the 95% confidence interval, which will be used in the visualizations, α is set to 0.05 using the standard normal approximation [27].

$$\beta_k = [b_k - 1.96se(b_k), b_k + 1.96se(b_k)] \quad (2)$$

This type of analysis is carried out to draw connections between the lockdown periods, the price development, and the energy production. The focus is on the German market. The data for France and Norway (NO1) serves as support, as both countries/regions have different energy mixes.

$$z = \frac{x - E(x)}{Var(x)} \quad (3)$$

To perform the standardization according to the z-transformation, Formula (3) was applied. The standardization was carried out for the energy mixes and partly for German electricity prices [27].

$$\mathbb{E}\left(P_t^I \middle| P_t^D\right) = p_t^D \; for \; t = 1,2 \quad (4)$$

Formula (4) represents the underlying message of the research. It states that the expected price of the intraday price is the day-ahead price. Our research incorporates this statement through the visual comparisons and continues to refer to it in the Results and Conclusion. In doing so, we want to highlight the extent to which the forecasts made were accurate in the tense market environment [28].

2.2. *Volume Analysis*

A volume analysis is included for Germany to prove that the price developments during the lockdown periods had the COVID-19 pandemic crisis as the main driver. For France, the volume data are used to support hypotheses. For NO1, this analysis is not performed as no data are available in Bloomberg. Similar to prices, the following statistical quantities are determined for German and French intraday and day-ahead volumes: Mean, Median, and Standard Deviation. For the German intraday and day-ahead volumes, an additional graph was created that also includes the 95% confidence interval. However, this type of analysis only serves to support the statements, whereby the focus is on the German spot market for electricity.

2.3. *Programming Language*

To run our calculations and create the visualizations, we used the programming language python. We used the packages pandas and scripy stats for the statistics and matplotlib and seaborn for our statistical graphics. For the standardization applied to some of the figures, we employed the standardscaler. This package removes the mean and scales of each variable to unit variance. The underlying mathematical approach is made transparent in Formula (3). The style used to set the colors of the graphs is the "darkgrid" style. We made use of visual studio code to create the graphs and calculate the statistics.

2.4. *Data*

2.4.1. Electricity Prices

For the price analysis, Bloomberg data for the German, French, and Norway (NO1) day-ahead and intraday prices have been used. As stated before, the 30-day-volatility and 90-day-volatility were retrieved from Bloomberg. The confidence intervals are based on their own calculations in Python. In the case of the German and French electricity prices, EPEX Spot prices were used. In the case of the Norwegian price data, prices from Nord Pool were used. All calculations are based on hourly data and their mean values. The prices for Germany, France, and NO1 prices were provided in euros. The analysis is based on the daily closing prices of the traded hours. In addition to the prices, the 30- and 90-day volatility was also subtracted for the same data set.

Table 1 gives an overview of all relevant statistical data regarding the prices used More than 5000 data points were used in each case.

The lockdown periods (Table 2) may differ. On the one hand, the time series of data extends to 1 February 2021. On the other hand, various measures or news items have ushered in a period of higher volatility in the markets. Table 2 represents the data periods that have been used for the statistical calculations.

Table 1. General Statistics across all Price Categories in MWh/EUR.

Statistics	German Intraday	German Day-Ahead	France Intraday	France Day-Ahead	NO1 Intraday	NO1 Day-Ahead
count	6356	6532	5367	6532	6532	6248
mean	38.23	34.67	37.49	36.16	14.17	12.72
std	35.93	17.02	18.54	16.97	17.65	14.82
median	35.00	33.58	36.50	35.40	8.84	8.08
min	−150.00	−83.94	−25.20	−8.65	−1.73	0.02
max	1000.00	189.25	328.20	189.25	205.68	152.25

Table 2. Data for the statistical evaluation on electricity prices.

Observation Period	German Intraday	German Day -Ahead	France Intraday	France Day-Ahead	NO1 Intraday	NO1 Day-Ahead
January 2020	1 January 2020–31 January 2020	1 January 2020–31 January 2020	1 January 2020–31 January 2020	1 January 2020–31 January 2020	1 January 2020–31 January 2020	1 January 2020–31 January 2020
January 2021	1 January 2021–31 January 2021	1 January 2021–31 January 2021	1 January 2021–31 January 2021	1 January 2021–31 January 2021	1 January 2021–31 January 2021	1 January 2021–31 January 2021
First Lockdown	3 March 2020-4 May 2020	3 March 2020-4 May 2020	3 March 2020-4 May 2020	3 March 2020-4 May 2020	2 March 2020-1 April 2020	2 March 2020-1 April 2020
Second Lockdown	2 November 2020-1 February 2021	2 November 2020-1 February 2021	15 October 2020-15 December 2020	15 October 2020-15 December 2020	2 November 2020-1 February 2021	2 November 2020-1 February 2021
Summer Months	4 May 2020-30 September 2020	4 May 2020-30 September 2020	1 July 2020-30 September 2020	1 July 2020-30 September 2020	4 May 2020-30 September 2020	4 May 2020-30 September 2020

2.4.2. Electricity Volumes

For the volume analysis, the daily closing volume of the individual hours was used for further analysis. The same data set was also used regarding German and French volumes (Table 3).

Table 3. General Statistics across German and French Volumes in MWh/EUR.

Statistics	German Intraday	German Day-Ahead	France Intraday	France Day-Ahead
count	5689	6816	6392	6486
mean	5689.82	24,379.26	169.43	14,082.17
std	2844.92	4164.75	288.43	2814.81
median	5805.00	23,881.50	40.00	13,963.10
min	0.00	14,441.00	0.00	6892.00
max	61,234.00	43,600.00	2917.00	25,013.00

The entire record refers to five business days in a week and does not include weekends and national bank holidays. If there was no data for a specific product in the record due to a bank holiday or if the product was not traded on that day due to low liquidity, the record will have a blank field.

2.4.3. Electricity Generation

The electricity generation data was gathered from the ENTSO-E Transparency Platform. The observation period starts on 1 January 2020 and ends on 31 December 2020. To get a quick overview of the main electricity sources, we have plotted the main electricity generation sources per country. Figures 3–5 show the standardized values on the left side and the non-standardized values in MWh on the right side [8].

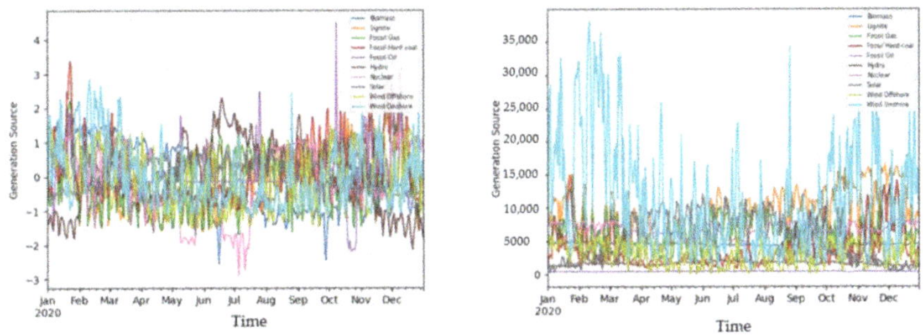

(a) (b)

Figure 3. This figure represents the German electricity generation. (**a**) Shows an approach using data standardization which has been defined in Formula (3); (**b**) Shows the electricity generation sources in MWh. Figures 3–5 got generated using the same code in Python.

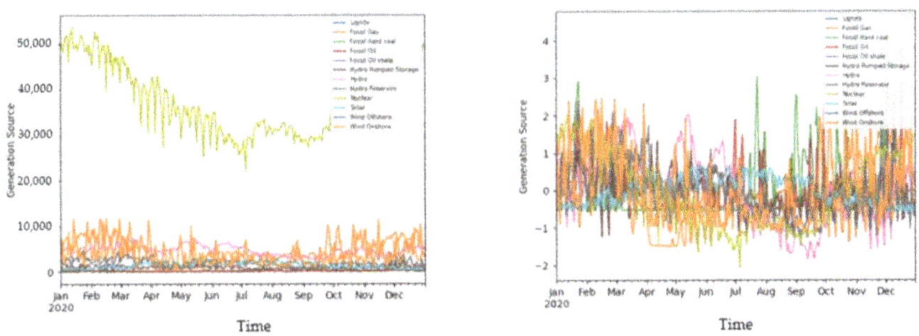

(a) (b)

Figure 4. This figure represents the French electricity generation. (**a**) Shows an approach using data standardization which has been defined in Formula (3); (**b**) Shows the electricity generation sources in MWh. Figures 3–5 got generated using the same code in Python.

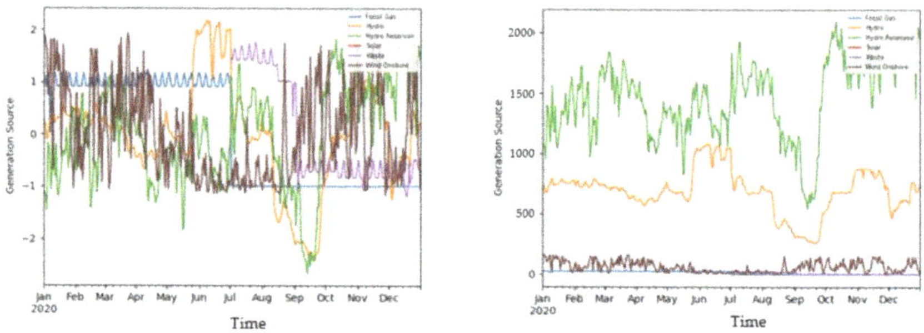

(a) (b)

Figure 5. This figure represents the Norway (NO1) electricity generation. (**a**) Shows an approach using data standardization which has been defined in Formula (3); (**b**) Shows the electricity generation sources in MWh. Figures 3–5 got generated using the same code in Python.

3. Results

3.1. Results on the Overall Investigation

We translated our statistical results into graphs which enabled us to draw conclusions about the energy price and volatility situation during and between the two lockdowns in the three countries. Our summary statistics are presented in Table 1. We looked at intraday prices and day-ahead prices. The data is divided into a country/region-based analysis and two different lockdown periods. In addition, the 30-day volatility and the confidence intervals are considered as well. The following data is not standardized. The focus of the analysis will also be on the influence and crisis-proofing of renewable energies. Therefore, the electricity price development in France and the NO1 region in Oslo, Norway, will also be considered. For all three countries/regions, the following periods were considered: Price and volatility level in January 2020 as the pre-COVID-19 phase compared to the price level in January 2021. The respective first and second lockdown periods. The periods differ due to national regulations. In addition, the summer months are considered because the scope of the measures was small in all three countries.

The observation period from January 2019 to February 2021 was considered. The graph also takes into consideration the 90-day-volatility. One weakness of these graphs is that the overview is not always given (see Figure 6).

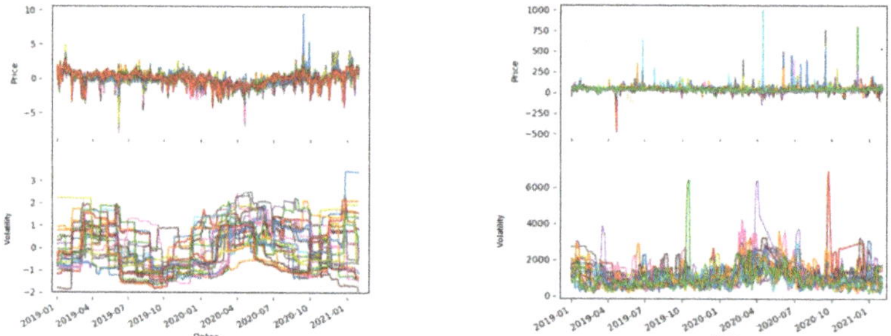

(a) (b)

Figure 6. These figures show an example of the German day-ahead prices (**a**) and intraday prices (**b**). The visualization applies standardized data (see Formula (3)). Figure 6 got generated using the same code in Python.

3.2. German Electricity Prices

In this part of the analysis, the German market is considered first. The graphs for French and Norwegian short-term power follow the same structure. On the left, there is always a presentation of the price level. The right sight displays the volatility development. The last two graphs include the mean as well as a 95% confidence level. A wide confidence level indicates more dispersion and, thus an uncertainty around the actual mean value. Germany had two lockdown periods during the observation period. The first lockdown period started in March 2020 and ended in May 2020. The second lockdown period started as a so-called soft lockdown in November and was then tightened in December. At the time of the data withdrawal in February 2021, the second lockdown in Germany had not yet ended [29]. In terms of the energy mix, Germany drew most of its energy from onshore wind power when it comes to the mean value analysis. Following the same mean value, approach lignite came in second place. It is striking with regard to the energy mix that none of the electricity generation sources make up the majority, i.e., have a share of over 50% [8].

3.2.1. German Intraday Prices

Figure 7 shows a clear increase in volatility in April 2020. The confidence level also shows greater volatility between April and May. In the second lockdown, volatility does not increase as much again. The reasons for the volatility increase in April 2020 could be many and varied. On the one hand, it could be related to the drop in oil prices, as there are correlations, or to weather data. This is particularly evident regarding the 30-day volatility, where a rise is visible after approximately fourteen days. However, this paper considers only the analysis of price, volatility, and production data. Correlations with other commodities are omitted [4,13]. Before April 2020, both prices and volatilities were at a stable level. After showing high outliers in April 2020, the price level recovered again. Therefore, in May, it was back at the level before April. The sample regarding the second lockdown starts on the first of November 2020 and ends on the first of February 2020 and indicates any similarities to the first lockdown. In contrast to the first lockdown, the price level remained at a higher level. Volatility, on the other hand, was lower. It recovered in May 2020, after the first lockdown, and remained at a constantly low level. Except for the numerous outliers highlighted in the price range based on confidence levels, the price movement shows an upward trend. For example, the price in January 2020 at 37.81 EUR is below the price in January 2021 at 54.51 EUR. The lowest price in the data set here is −150.00 EUR, and the highest price is 1000.00 EUR.

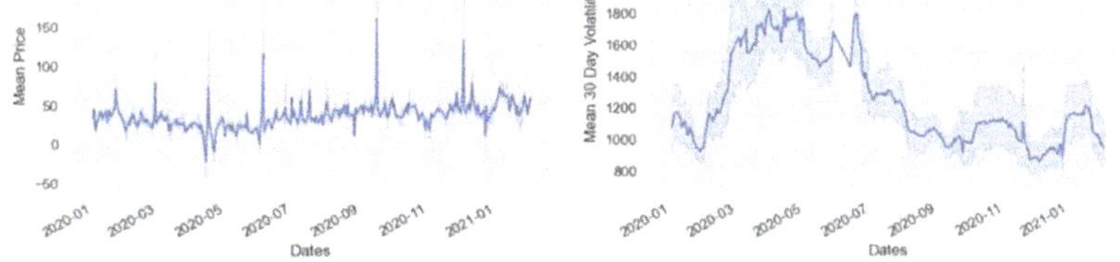

Figure 7. This figure represents the intraday price levels for German power including its' confidence levels on the left side and the 30 day volatility including its' confidence levels on the right side. Figures 7 and 8 got generated using the same code in Python.

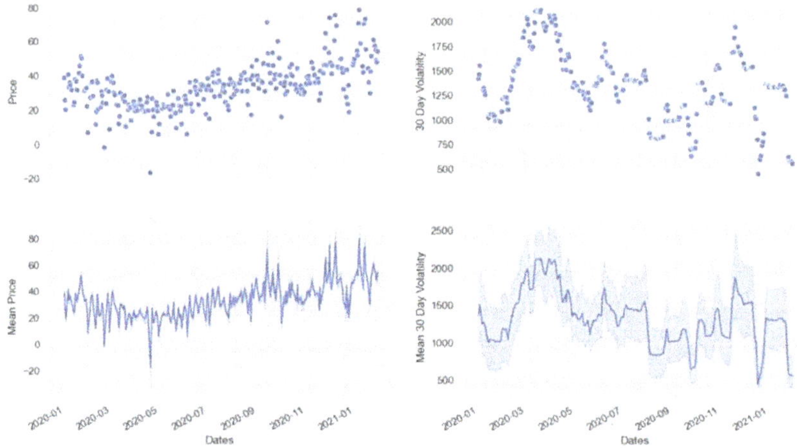

Figure 8. This figure represents the day-ahead price levels for German power including its' confidence levels on the left side and the 30-day volatility including its confidence levels on the right side. Figures 7 and 8 were generated using the same code in Python.

3.2.2. German Day-Ahead Prices

Figure 8 shows that the impact on volatility was greater on the day-ahead market than on the intraday market. Volatility shows the highest level during the first lockdown and decreases towards the end of the lockdown (Table 4). The prices show a clear drop in April 2020, which is the exact opposite development of the intraday market. After the first lockdown, the price level recovers. In April 2020, the average price under consideration dropped significantly into negative territory. As Valitov (2020) demonstrated, negative prices have been possible in the day-ahead market in Germany since 2008. The reason for this is the high feed-in of RES, for example, in times of low demand [30]. The lowest price in the data set is −83.94 EUR, whereas the highest price is 189.25 EUR which is also an indicator of high outliers. Outliers are also characterized by higher confidence intervals in the price graph. At the beginning of December, the price level increased and showed high swings. During the bank holidays around Christmas, the price level decreased again before moving up in January. Also, the 30-day volatility moved down during the bank holidays before it increased around the 15th of November. In January, the volatility increased again, and by the end of January, there was a drop in volatility. The price level in December 2021 is also higher than the price level in December 2020. Compared to the intraday area, the confidence level in the day-ahead area is constantly wider.

Table 4. General Statistics about German Prices and Volatilities in EUR.

Observation Period	Germany Intraday (Mean Price; Mean Vola)	Germany Intraday (Std Price; Std Vola)	Germany Day-Ahead (Mean Price; Mean Vola)	Germany Day-Ahead (Std Price; Std Vola)
January 2020	37.81; 1045.02	21.40; 308.11	33.91; 1138.85	13.54; 1066.01
January 2021	54.51; 1110.88	26.48; 404.37	54.72; 1110.18	17.34; 944.49
First Lockdown	24.01; 1692.26	39.88; 531.82	21.60; 1842.29	13.03; 984.57
Second Lockdown	48.71; 1023.19	33.32; 399.93	48.31; 1226.78	17.71; 1047.86
Summer Months	38.10; 1146.86	39.42; 336.66	32.99; 1214.32	14.15; 916.56

The average price was 38.23 EUR for intraday and 34.67 EUR for day-ahead prices. The standard deviation was 35.93 for intraday and 17.02 for day-ahead.

- Compared with France and NO1, Germany has the most diverse energy mix, but is also heavily dependent on the fluctuating energy production from on-shore wind, which accounts for the largest share even before lignite;
- The wide confidence levels could be explained by the strongly fluctuating electricity production via offshore wind;
- Day-ahead and intraday prices diverge sharply in the first half of the year but have converged significantly in the second half of the year;
- The first lockdown shows a fundamentally lower price level than the second. At the same time, the first lockdown is the period with the highest volatility;
- The price level in January 2021 is significantly higher than the price level in January 2020.

3.2.3. German Traded Volumes

Figures 9 and 10 compare the volumes traded on the German intraday (Figure 9) and day-ahead (Figure 10) markets. When looking at the day-ahead volumes this figure does not indicate any abnormalities during the COVID-19 pandemic crisis. Rather, this figure indicates seasonal fluctuations. In the intraday area, however, the situation is different. The confidence levels show large outliers, especially in the summer months. Higher traded volumes on the intraday market are linked to a higher level of flexibility.

- In January 2020, higher volumes were traded than in January 2021;
- In the first lockdown, the standard deviation in the day-ahead area was higher than in later periods, and more average daily volume was traded in the day-ahead market. It does not decrease again until late summer;
- On the day-ahead market, it is noticeable that the confidence level is always quite constant and that the volumes increase in the summer months;
- In the second lockdown, more was traded in the intraday area/less was traded in the day-ahead area (compared to the first lockdown period).

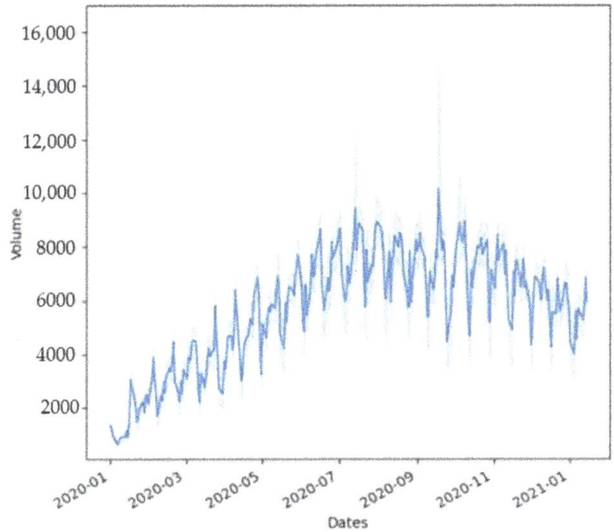

Figure 9. Shows German average traded volumes per day in Intraday markets. The light blue shadow represents the 95% confidence level. Figures 9 and 10 got generated using the same code in Python.

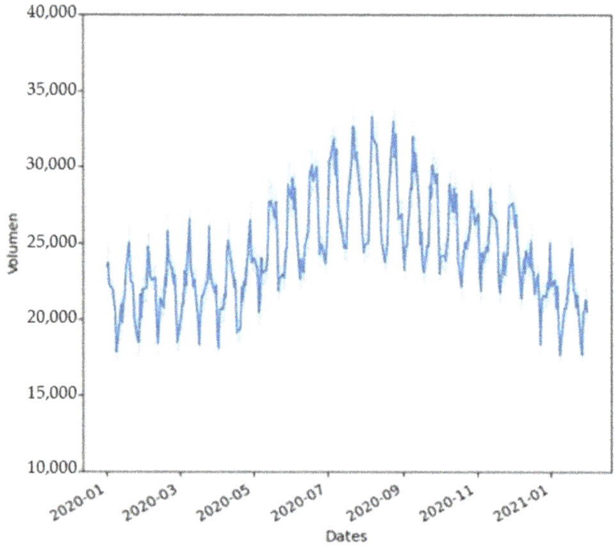

Figure 10. Shows German average traded volumes per day in Day-Ahead markets. The light blue shadow represents the 95% confidence level. Figures 9 and 10 got generated using the same code in Python.

3.3. *French Electricity Prices*

The graphical analysis follows the same pattern for France. The same parameters are superimposed, and day-ahead and intraday prices are also compared. As each country in Europe has its own lockdown rules, the lockdown periods in France differ from those in Germany. In France, the first lockdown period was from March to June/July. This extends the observation period compared to Germany. The second lockdown began in mid-October and was relaxed again in mid-December. Until the end of our observation period on 1 February, a curfew remained in place from 6 p.m. to 6 a.m. CET [31]. France obtains most of its electricity from nuclear energy sources. The calculations based on the average values showed that almost 70% of the electricity produced over the year came from nuclear sources. Hydropower and onshore wind power take second and third place in France. The mean values of both accounted for less than 10% of the energy mix on an annual average [8].

3.3.1. French Intraday Prices

In direct comparison with the German intraday electricity prices, it is noticeable that the price level of the French intraday prices fluctuates even more (Figure 11). However, the general price level falls between March and May 2020 and rises again in June. Volatility initially fell sharply in February 2020 but jumped again in March, coinciding with the first COVID-19 measures. Volatility in June is lower than in the previous month. The confidence interval of volatility from March to June also indicates a higher dispersion and, thus, greater uncertainty regarding forecasts. As volatility decreases and prices rise in June, the confidence level also decreases. At the time of the announcement of the second national lockdown, prices and volatility rose simultaneously. Also, the confidence level of volatility shows greater excesses. At the end of the second lockdown, there was a clear drop in prices and a jump in volatility. However, the situation recovers quickly thereafter. Compared to prices in January 2020, prices in January 2021 are at a higher level. In contrast, intraday volatility is at a lower level. The highest price in the data set is 328.20 EUR, and the lowest is –25.20 EUR. This indicates outliers, which are also evident in the price range via the confidence level.

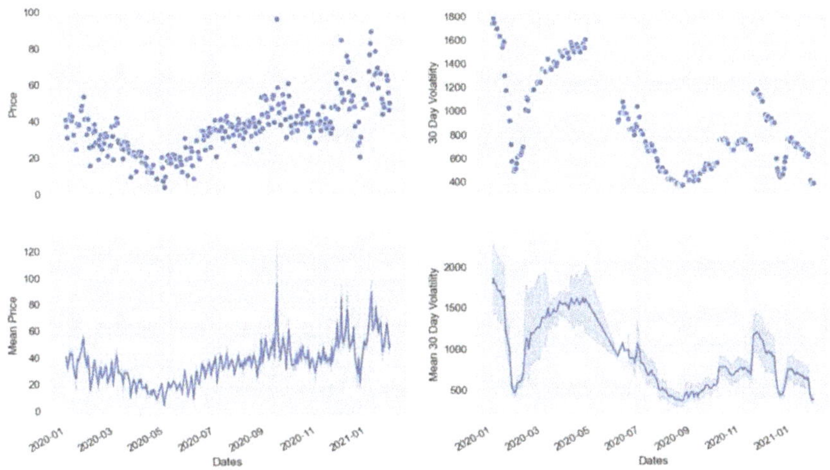

Figure 11. This figure represents the intraday price levels for French power including its confidence levels on the left side and the 30-day volatility including its confidence levels on the right side. Figures 7, 8 and 11–14, got generated using the same code in Python.

3.3.2. French Day-Ahead Prices

The price development in the day-ahead area (Figure 12) hardly differs from the intraday area. Only the confidence intervals are somewhat larger in the intraday area. This becomes visible via the outliers. In the day-ahead area, the lowest price is −8.65 EUR and the highest at 189.25 EUR. Like the intraday area, the prices are lower in summer than winter. A particularly low-price level during the first lockdown should be noted. Compared to the price level in January 2020, the price level in January 2021 is also higher. Differences in intraday prices are visible when it comes to volatility. Volatility is characterized by two peaks during the two lockdowns. First, there was a clear increase in April 2020 and another in June 2020. In the period in between, volatility fell once again. At the same time, the confidence level also decreases during this period. The development of volatility and confidence intervals during the second lockdown shows similarities to the development in the intraday area.

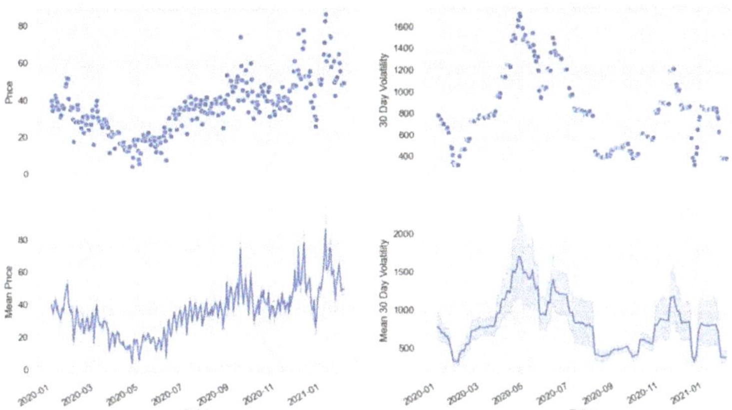

Figure 12. This figure represents the day-ahead price levels for French power including its confidence levels on the left side and the 30-day volatility including its confidence levels on the right side. Figures 7, 8 and 11–14, got generated using the same code in Python.

The average price per day was 37.49 EUR; 36.16 EUR and the standard deviation for the entire period was 18.54; 16.97.

- In France, less nuclear power is produced in the summer months, and a little more use is made of renewable energy sources such as solar. No link between generation and price level is visible;
- Table 5 shows that there are no major fluctuations between day-ahead and intraday. Thus, it can be said for the forecast accuracy that it is higher in France;
- In the first lockdown, prices fell sharply and showed a high volatility. In the second lockdown, prices rose sharply, and volatility was low;
- Day-ahead and intraday prices were always close to each other in the mean and median. However, they also show strong outliers.

Table 5. General Statistics about French Prices and Volatilities in EUR.

Observation Period	France Intraday (Mean Price; Mean Vola)	France Intraday (Std Price; Std Vola)	France Day-Ahead (Mean Price; Mean Vola)	France Day-Ahead (Std Price; Std Vola)
January 2020	38.31; 1221.29	12.21; 717.49	37.55; 546.84	10.73; 321.90
January 2021	62.37; 601.00	18.39; 309.41	61.24; 644.05	17.17; 647.56
First Lockdown	22.00; 1397.17	9.78; 476.57	21.30; 1161.13	9.32; 795.95
Second Lockdown	47.09; 843.06	16.76; 405.42	46.26; 824.63	15.41; 651.17
Summer Months	41.45; 527.49	16.33; 242.01	40.76; 567.71	12.34; 490.50

3.3.3. French Traded Volumes

The following is a brief supporting analysis of the volumes for French electricity prices.

Table 6 represents the general statistics regarding the traded volumes. Thereby, the analysis was conducted on the mean and standard deviation for intraday and day-ahead prices. The average volume traded per day was 169.43 MWh; 14,082.17 MWh and the standard deviation for the entire period was 288.43; 2814.81.

- During the first lockdown, less than normal trading took place on the intraday market;
- During the second lockdown, the most trading took place on the intraday market. The standard deviation was also above average;
- During the summer months, less than average was traded via the day-ahead market;
- In January 2021, more than the average was traded on the day-ahead market and more compared to the previous year;
- The standard deviation of the day-ahead volumes is quite constant.

Table 6. General Statistics about French Volumes in MWh.

Observation Period	France Intraday (Mean)	France Intraday (Std)	France Day-Ahead (Mean)	France Day-Ahead (Std)
January 2020	178.21	292.91	14,724.14	2855.71
January 2021	178.99	257.54	16,449.86	2642.51
First Lockdown	132.13	244.21	14,088.72	2381.12
Second Lockdown	195.73	309.71	14,373.57	2600.35
Summer Months	180.25	288.66	12,450.08	2650.61

3.4. Norwegian Electricity Prices

The graphical analysis follows the same pattern for Norway. It only considers the Oslo region (NO1). The Norwegian government declared a national lockdown in mid-March 2020. At the end of April, the first relaxations were decided, and relaxations were introduced until mid-June [32]. Norway introduced new national restrictions by the end of October 2020. These restrictions were still in place once the observation period ended [33]. In terms of energy production, Norway produces electricity mainly from hydropower. This energy source is divided into the hydro water reservoir and hydro run-of-river and poundage. Together, these two energy sources account for almost 100% on an annual average. Wind onshore or fossil gas accounts for less than 5% of Norway's energy production [8].

3.4.1. NO1 Intraday Prices

The price development in the intraday area (Figure 13) shows a bearish trend, including several outliers until April. At the beginning of June, the price level was relatively stable at EUR 9.00. During the second lockdown period, the prices show volatile movements. The trend is a rising price. Compared to the price level in January 2020, the price level in January 2021 was higher. However, the curve shows a high price level in winter and a low price level in summer. The confidence level of the prices does not show any conspicuous features and is evenly distributed over the year. The lowest price in the data set here is −1.73 EUR, and the highest price is 205.68 EUR. The 30-day volatility also increased during the first lockdown period. Thus, it started being around 350 in the first lockdown and ended up at a level of 770 during the second lockdown. The confidence level of the 30-day volatility increased between March and July and decreased again in September. The second lockdown also shows a widened confidence level.

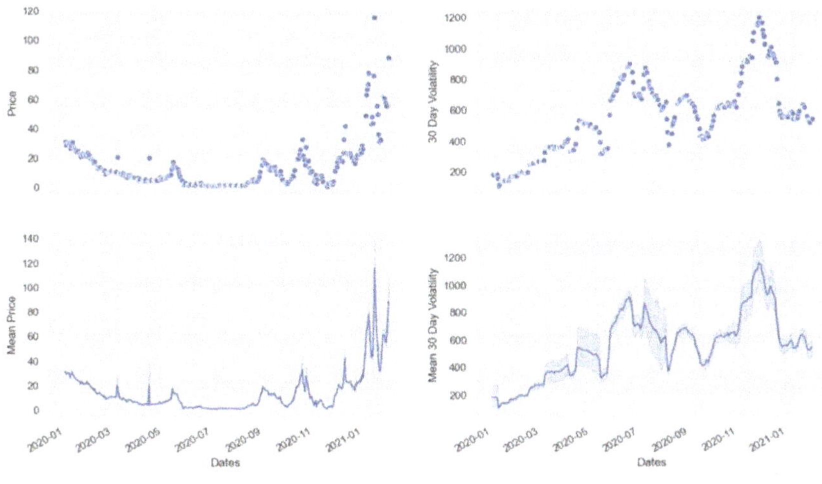

Figure 13. This figure represents the intraday price levels for NO1 power including its confidence levels on the left side and the 30-day volatility including its confidence levels on the right side. Figures 7, 8 and 11–14, got generated using the same code in Python.

3.4.2. NO1 Day-Ahead Prices

The price development in the day-ahead area (Figure 14) also shows a bearish trend until April. Like in the intraday area, the prices reached their lowest point in June. From June, the day remained at nearly the same level. During the summer months, the price level remained low and showed some peaks in September and November. In December, the price level increased again. Compared to January 2020, the price level was higher in January 2021 (Table 7). Regarding the confidence level of the prices, there are no conspicuous features in the day-ahead area. This indicates that the uncertainty regarding prices is less great. In the day-ahead area, the curve also shows a low price level in summer and a high price level in winter. In terms of this shape, there is a similarity to the intraday prices. The lowest price in the data set is 0.02 EUR, and the highest price is 152.25 EUR. At the end of March, volatility started increasing and reached the first peak in the mid of April where it remained relatively stable. The volatility shows more peaks between May and October. In November, it started to rise and ended up at a higher level compared to January 2020. Between November and March, the confidence level widened until it narrowed again in January.

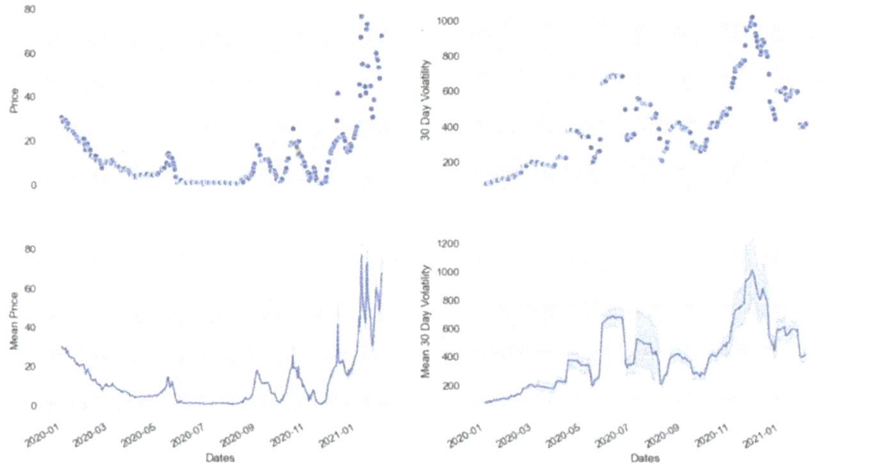

Figure 14. This figure represents the day-ahead price levels for NO1 power including its confidence levels on the left side and the 30-day volatility including its confidence levels on the right side. Figures 7, 8 and 11–14, got generated using the same code in Python.

Table 7. General Statistics about NO1 Prices and Volatilities in EUR.

Observation Period	NO1 Intraday (Mean Price; Mean Vola)	NO1 Intraday (Std Price; Std Vola)	NO1 Day-Ahead (Mean Price; Mean Vola)	NO1 Day-Ahead (Std Price; Std Vola)
January 2020	25.34; 160.33	4.19; 100.58	23.68; 97.94	4.06; 29.08
January 2021	59.16; 575.52	28.79; 124.94	52.41; 539.61	19.25; 162.48
First Lockdown	8.96; 354.71	4.86; 178.76	7.79; 193.10	2.05; 76.19
Second Lockdown	28.86; 773.76	28.10; 332.86	26.10; 688.65	22.98; 365.61
Summer Months	5.47; 620.98	5.20; 299.48	4.71; 419.00	4.65; 272.55

The average price of the daily traded volume was 14.17 EUR for intraday and 12.72 EUR for day-ahead prices. The standard deviation was 17.65 for intraday and 14.82 for day-ahead.

- During the first lockdown, the price in the day-ahead and intraday area fell sharply. This trend was reinforced in the summer months. In the summer months, less electricity was produced by waste and wind power;
- In the second lockdown, the price level and the level of volatility increased. At the beginning of the second lockdown, the production of hydropower by water reservoirs collapsed;

- The price level in January 2021 is significantly higher than the price level in January 2020. The same applies to volatility;
- Intraday and day-ahead prices both have low confidence level shears and have a similar shape (low prices in summer and high prices in winter);
- There are no strong outliers, as is the case with German and French prices. Intraday prices show more outliers than Day-Ahead prices;
- Hydropower as the largest generation source can be stored and is more independent of the weather;
- Both curves show very similar price and volatility developments.

The results confirm the statement that the 2nd lockdown had a stronger impact on the energy markets in the NO1 region.

4. Discussion

4.1. Critical Appraisal

The following section discusses elements of the paper based on the current state of research. Since a further step in analyzing the data would be prediction methods, we will now briefly discuss back testing, which is commonly used in practice. We can refer to the study by L. Han et al. (2020). They pointed out that spot electricity prices are among the financial products with the highest price deviations and the greatest volatility. They attribute this to the non-storable nature of electricity [13]. Due to the many uncertainties caused by the high volatilities and the price outliers, increasingly complex technologies are being used in the energy markets to predict prices [13,34]. This is necessary because errors arise, for example, through back testing. This is the case because back testing cannot detect so-called black swans [35,36]. An example of a so-called black swan event was the negative oil prices in April 2020 [5,36]. Machine learning (ML) can get around this problem. For example, multivariate analyses using ML can be used to find important indicators that point to a crisis [35]. Different energy markets would support the research here. Therefore, we recommend ML techniques for further analysis of the data.

Another challenge regarding the forecasting accuracy of German electricity is the rapidly changing energy mix. The higher volatilities caused by this also require further complex models and automated trading techniques. In addition, there is a growing need for more comprehensive analyses of alternative data types, such as weather [4,6,37]. This alternative data type is an additional element of the analysis compared to the financial markets. O. B. Adekoya and J.A. Oliyide (2021) and A. Elsayed et al. (2020) have highlighted the links between financial markets and other commodities during the COVD-19 crisis. This comparison is not made in this paper, as the focus is on pure price and volatility movements, supported in some cases by volume data [10,12]. J. Ali, W. Kahn (2020) analyzed the markets according to the lockdown periods [9]. The same procedure was used in this paper. However, weaknesses of the method have become apparent since, in a comprehensive comparison of countries, the lockdown periods differ due to national regulations [25,29,31,33]. This leads to more difficult comparability of the data. However, we have deliberately decided to extend the analysis to include different countries and, in contrast to S. Halbrügge et al. (2021) and C. Fezzi and V. Fanghella (2020), to carry out a country comparison [4,14]. Furthermore, this analysis should draw attention to the growing importance of the intraday electricity markets during the expansion of renewable energies [19,38]. In this way, links to the energy mixes of the countries and the response to COVID-19 can be established.

N. Löhndorf and D. Wozabal have shown in the course of their research why the intraday and day-ahead markets are interdependent and why trading only on one of the markets would not make sense. For example, it would not be optimal to fully utilize the capacity of the day-ahead market without leaving capacity for the intraday markets. They point out that the attractiveness of the intraday markets lies in their volatility and information content. They assume that the attractiveness of the day-ahead markets lies in the market department. According to their research, forecast errors in the day-ahead market can be caused by e.g., wrong weather forecasts. These forecast errors then trigger, among other things, price changes in the intraday markets [28,39].

4.2. Research Limitations

This paper refers to forecasting accuracy when the price difference between the intraday and the day-ahead electricity markets are described. However, the prevailing research contradicts the term forecast accuracy is a precise term to describe the price differences between day-ahead and intraday markets. As some bidders perform price arbitrage between day-ahead and intraday markets to maximize profits, the day-ahead price is not often a forecast of the intraday price [40,41]. In order to obtain even more valid results, factors such as weather, seasonal fluctuations and correlations with other commodities, and, if applicable, public holidays would also have to be considered. Furthermore, the data quality of the Norwegian data is limited as there is no volume data available on Bloomberg. For further analyses, it would therefore be helpful to draw on other Nordic markets and more regions in Norway with a high share of renewable energies as well. Sweden or Denmark, for example, could be considered [22]. This paper does not consider different price systems. For instance, in Europe and Australia, a zonal price system is common, whereas the U.S. uses a nodal price system. In the case of a wider selection of countries, these differences need to be considered [42]. In addition to it, we only considered the market closing prices of each day under consideration. In order to draw a conclusion on intraday volatility, hourly price data should be included as it might have had a positive impact on the data quality. We did include weather forecasts in our analysis as done by Kuppelwieser and Wozabal (2021) [7]. An additional factor that could be considered is the COVID-19 case numbers or, in the meantime, the vaccination rate for the countries studied.

5. Conclusions and Outlook

This paper finds that COVID-19 had an impact on the energy markets in Germany, France, and Norway (NO1). The German electricity market is facing changes. Due to the increase in energy production from renewable energies, more and more energy is being traded on the intraday market [39]. This trend is confirmed by our analysis of the volume. Our overall objective was to determine whether (iii) forecasting methods in intraday electricity markets in Germany have improved. To find out, we compared day-ahead prices and intraday prices. We did this because the day-ahead price is the market-clearing price [38,42,43]. In the comparison, we found that prices in France and Norway were consistently close. (ii) The power generation in France, which consists mainly of nuclear electricity, gives a hint here. We conclude that France shows close intraday and day-ahead prices. This may be due to the constant use of nuclear power. We had expected that (ii, iii) Norway would also be highly efficient. Based on our analysis, we can conclude that the intraday and day-ahead prices are close together. Norway seems to be an exception here. Based on our data NO1 data, we can conclude they have low prices and hardly any outliers, which is particularly evident in the first lockdown period when compared to the developments of the German and France day-ahead and intraday markets. Regarding the day-ahead data, the 30-day volatilities were consistently lower than in Germany and France. These findings provide an indication of the link to the increased use of hydropower and hydro reservoirs in Norway (NO1) [8]. In Germany, prices were initially far apart during the first lockdown period. During the second lockdown period, there was a convergence of prices in the mean and median. This convergence is (i) an indication, but not proof, that the use of the out-of-sample data of the first lockdown may have improved the forecast accuracy. In addition, (iii) the constantly changing energy mix in Germany may have played a role. The results for the German market could be used as pioneering data for other countries that want to undertake such an energy transition [39]. Other results are as follows:

- All products considered show a higher price in January 2021 compared to the previous year;

- In the case of German electricity, however, a lower daily average volume was traded in January 2021 compared to the previous year. The volume in France remains constant in the intraday area and is increased in the day-ahead area;
- All prices considered fell in the first lockdown, whereas volatilities rose sharply in some cases;
- In contrast to France and Germany, the electricity prices for NO1 show fewer outliers, which could be explained by the generation from water reservoirs and hydropower;
- All prices increased in the second lockdown. Unlike the other products, the volatility of NO1 increased particularly strongly in the second lockdown. This suggests that NO1 was hit harder by the second lockdown than by the first.

In the general price and volatility analysis, it was determined that the weather factor was omitted in the case of nuclear electricity and electricity from hydropower, which could lead to higher forecast accuracy and to greater crisis security. Higher forecast accuracy is becoming more important because of extreme electricity prices that have occurred with larger magnitude and higher frequency in recent years [44]. In addition, new market regulations and cross-country interconnections need to be considered to understand electricity prices [45]. Macroeconomic events can lead to an increased level of uncertainty in the natural gas and electricity market [46,47]. Furthermore, the aim of our research is to provide statistical evidence for the relationship between energy production and price and volatility movements. We want to do this because this paper only makes assumptions based on the results of the graphical analysis.

To produce more valid results, we plan to expand the dataset and consider further crisis periods. Thus, potential future research would include generalizable indicators from the prevailing time series that could be employed to filter out or predict future crisis situations. Reference data for validation could be price data during the Russian aggression against Ukraine or the subprime crisis. Such a model should be possible even though there are different energy mixes and or other special features regarding the data structure, e.g., different lockdown periods. The overall research goal is to develop methods that can predict the crisis or key indicators such as volatility.

Author Contributions: Conceptualization, J.N.B. and F.M.; methodology, J.N.B. and F.M.; software, J.N.B.; validation, J.N.B., F.M., D.G. and A.Z.; formal analysis, J.N.B., F.M. and D.G.; investigation, D.G.; resources, F.M. and D.G.; data curation, J.N.B.; writing—original draft preparation, D.G.; writing—review and editing, J.N.B., F.M., D.G. and A.Z.; visualization, J.N.B.; project administration, D.G.; Supervision, A.Z. All authors have read and agreed to the published version of the manuscript.

Funding: This research received no external funding.

Institutional Review Board Statement: Not applicable.

Informed Consent Statement: Not applicable.

Data Availability Statement: The price and volume data has been retrieved from Bloomberg. The generation data has been retrieved from: URL https://transparency.entsoe.eu/generation/r2/actualGenerationPerProductionType/show (assessed on 5 June 2021).

Acknowledgments: We would like to thank the 2021 International Joint Conference on Energy, Electrical and Power Engineering (CoEEPE 2021) for giving us the opportunity to present the paper at their first conference in Huangshan, China. The underlying research has been published in the corresponding conference proceedings using the title: How did the COVID-19 crisis affect the efficiency of European intraday electricity markets? The price and volume data has been retrieved via the Bloomberg terminal of the ISF Institute for Strategic finance (FOM University of Applied Sciences for Economics and Management).

Conflicts of Interest: The authors declare no conflict of interest.

References

1. Marshman, D.; Brear, M.; Jeppesen, M.; Ring, B. Performance of wholesale electricity markets with high wind penetration. *Energy Econ.* **2020**, *89*, 104803. [CrossRef]
2. Detemple, J.; Kitapbayev, Y. The value of green energy under regulation uncertainty. *Energy Econ.* **2020**, *89*, 104807. [CrossRef]
3. Härtel, P.; Korpås, M. Demystifying market clearing and price setting effects in low-carbon energy systems. *Energy Econ.* **2020**, *93*, 105051. [CrossRef]
4. Halbrügge, S.; Schott, P.; Weibelzahl, M.; Buhl, H.U.; Fridgen, G.; Schöpf, M. How did the German and other European electricity systems react to the COVID-19 pandemic? *Appl. Energy* **2021**, *285*, 116370. [CrossRef]
5. Ghiani, E.; Galici, M.; Mureddu, M.; Pilo, F. Impact on Electricity Consumption and Market Pricing of Energy and Ancillary Services during Pandemic of COVID-19 in Italy. *Energies* **2020**, *13*, 3357. [CrossRef]
6. Duso, T.; Szücs, F.; Böckers, V. Abuse of dominance and antitrust enforcement in the German electricity market. *Energy Econ.* **2020**, *92*, 104936. [CrossRef]
7. Kuppelwieser, T.; Wozabal, D. Intraday Power Trading: Towards an Arms Race in Weather Forecasting? *Eur. J. Oper. Res.* **2021**, in press.
8. ENTSO-E. Actual Generation per Production Type. Available online: https://transparency.entsoe.eu/generation/r2/actualGenerationPerProductionType/show (accessed on 5 June 2021).
9. Ali, J.; Khan, W. Impact of COVID-19 pandemic on agricultural wholesale prices in India: A comparative analysis across the phases of the lockdown. *Public Aff.* **2020**, *20*, e2402. [CrossRef]
10. Elsayed, A.H.; Nasreen, S.; Tiwari, A.K. Time-varying co-movements between energy market and global financial markets: Implication for portfolio diversification and hedging strategies. *Energy Econ.* **2020**, *90*, 104847. [CrossRef]
11. Bompard, E.; Mosca, C.; Colella, P.; Antonopoulos, G.; Fulli, G.; Masera, M.; Poncela-Blanco, M.; Vitiello, S. The Immediate Impacts of COVID-19 on European Electricity Systems: A First Assessment and Lessons Learned. *Energies* **2020**, *14*, 96. [CrossRef]
12. Adekoya, O.B.; Oliyide, J.A. How COVID-19 drives connectedness among commodity and financial markets: Evidence from TVP-VAR and causality-in-quantiles techniques. *Resour. Policy* **2020**, *70*, 101898. [CrossRef] [PubMed]
13. Han, L.; Kordzakhia, N.; Trück, S. Volatility spillovers in Australian electricity markets. *Energy Econ.* **2020**, *90*, 104782. [CrossRef]
14. Fezzi, C.; Fanghella, V. Real-Time Estimation of the Short-Run Impact of COVID-19 on Economic Activity Using Electricity Market Data. *Environ. Resour. Econ.* **2020**, *76*, 885–900. [CrossRef] [PubMed]
15. Kath, C.; Ziel, F. The value of forecasts: Quantifying the economic gains of accurate quarter-hourly electricity price forecasts. *Energy Econ.* **2018**, *76*, 411–423. [CrossRef]
16. Kath, C.; Ziel, F. Conformal prediction interval estimation and applications to day-ahead and intraday power markets. *Int. J. Forecast.* **2021**, *37*, 777–799. [CrossRef]
17. Finnah, B.; Gönsch, J.; Ziel, F. Integrated day-ahead and intraday self-schedule bidding for energy storage systems using approximate dynamic programming. *Eur. J. Oper. Res.* **2021**, *301*, 726–746. [CrossRef]
18. Maciejowska, K.; Nitka, W.; Weron, T. Day-Ahead vs. Intraday—Forecasting the Price Spread to Maximize Economic Benefits. *Energies* **2019**, *12*, 631. [CrossRef]
19. Kramer, A.; Kiesel, R. Exogenous factors for order arrivals on the intraday electricity market. *Energy Econ.* **2021**, *97*, 105186. [CrossRef]
20. Ghosh, S.; Bohra, A.; Dutta, S. The Texas Freeze of February 2021: Event and Winterization Analysis Using Cost and Pricing Data. In Proceedings of the IEEE Electrical Power and Energy Conference (EPEC), Toronto, ON, Canada, 22–31 October 2021; IEEE: Piscataway, NJ, USA; pp. 7–13. [CrossRef]
21. Xiao, D.; AlAshery, M.K.; Qiao, W. Optimal price-maker trading strategy of wind power producer using virtual bidding. *J. Mod. Power Syst. Clean Energy* **2021**, 1–13. [CrossRef]
22. Snow, D. Machine Learning in Asset Management. *SSRN Electron. J.* **2019**. [CrossRef]
23. Rintamäki, T.; Siddiqui, A.S.; Salo, A. Does renewable energy generation decrease the volatility of electricity prices? An analysis of Denmark and Germany. *Energy Econ.* **2017**, *62*, 270–282. [CrossRef]
24. Cramton, P. Electricity market design. *Oxf. Rev. Econ. Policy* **2017**, *33*, 589–612. [CrossRef]
25. Malec, M.; Kinelski, G.; Czarnecka, M. The Impact of COVID-19 on Electricity Demand Profiles: A Case Study of Selected Business Clients in Poland. *Energies* **2021**, *14*, 5332. [CrossRef]
26. Pilipović, D. *Energy Risk*; McGraw-Hill: New York, NY, USA, 1998; pp. 100–101.
27. Verbeek, M. *Modern Econometrics*, 5th ed.; Wiley Custom: Croydon, UK, 2017; pp. 167–173.
28. Löhndorf, N.; Wozabal, D. The Value of Coordination in Multimarket Bidding of Grid Energy Storage. *Work. Pap.* **2021**. [CrossRef]
29. Robert Koch Institut. Available online: https://www.rki.de/DE/Content/InfAZ/N/Neuartiges_Corona-virus/nCoV_node.html (accessed on 5 June 2021).
30. Valitov, N.; Maier, A. Asymmetric information in the German intraday electricity market. *Energy Econ.* **2020**, *89*, 104785. [CrossRef]
31. Governement. Available online: https://www.gouvernement.fr/en/coronavirus-covid-19 (accessed on 5 June 2021).
32. Ursin, G.; Skjesol, I.; Tritter, J. The COVID-19 pandemic in Norway: The dominance of social implications in framing the policy response. *Health Policy Technol.* **2020**, *9*, 663–672. [CrossRef]
33. Government.no. URL. Available online: https://www.regjeringen.no/en/topics/koronavirus-covid-19/id2692388/ (accessed on 5 June 2021).

34. Wang, B.; Wang, J. Energy futures and spots prices forecasting by hybrid SW-GRU with EMD and error evaluation. *Energy Econ.* **2020**, *90*, 104827. [CrossRef]
35. Se Prado, M.M.L. *Machine Learning for Asset Managers*; Cambridge University Press: Cambridge, UK, 2020; pp. 19–20.
36. Selmi, R.; Bouoiyour, J.; Hammoudeh, S. Negative Oil: Coronavirus, a "Black Swan" Event for the Industry? *2020.* Available online: https://hal.archives-ouvertes.fr/hal-02570614/document (accessed on 5 June 2021).
37. Agnello, L.; Castro, V.; Hammoudeh, S.; Sousa, R.M. Global factors, uncertainty, weather conditions and energy prices: On the drivers of the duration of commodity price cycle phases. *Energy Econ.* **2020**, *90*, 104862. [CrossRef]
38. Glas, S.; Kiesel, R.; Kolkmann, S.; Kremer, M.; Von Luckner, N.G.; Ostmeier, L.; Urban, K.; Weber, C. Intraday renewable electricity trading: Advanced modeling and numerical optimal control. *J. Math. Ind.* **2020**, *10*, 3. [CrossRef]
39. Kiesel, R.; Paraschiv, F. Econometric analysis of 15-minute intraday electricity prices. *Energy Econ.* **2017**, *64*, 77–90. [CrossRef]
40. Hadsell, L. The impact of virtual bidding on price volatility in New York's wholesale electricity market. *Econ. Lett.* **2007**, *95*, 66–72. [CrossRef]
41. Iria, J.; Soares, F.; Matos, M. Optimal supply and demand bidding strategy for an aggregator of small prosumers. *Appl. Energy* **2018**, *213*, 658–669. [CrossRef]
42. Grimm, V.; Rückel, B.; Sölch, C.; Zöttl, G. The impact of market design on transmission and generation investment in electricity markets. *Energy Econ.* **2020**, *93*, 104934. [CrossRef]
43. De Lagarde, C.M.; Lantz, F. How renewable production depresses electricity prices: Evidence from the German market. *Energy Policy* **2018**, *117*, 263–277. [CrossRef]
44. Liu, L.; Bai, F.; Su, C.; Ma, C.; Yan, R.; Li, H.; Sun, Q.; Wennersten, R. Forecasting the occurrence of extreme electricity prices using a multivariate logistic regression model. *Energy* **2022**, *247*, 123417. [CrossRef]
45. Tschora, L.; Pierre, E.; Plantevit, M.; Robardet, C. Electricity price forecasting on the day-ahead market using machine learning. *Appl. Energy* **2022**, *313*, 118752. [CrossRef]
46. Shi, X.; Shen, Y. Macroeconomic uncertainty and natural gas prices: Revisiting the Asian Premium. *Energy Econ.* **2020**, *94*, 105081. [CrossRef]
47. Moutinho, V.; Oliveira, H.; Mota, J. Examining the long term relationships between energy commodities prices and carbon prices on electricity prices using Markov Switching Regression. *Energy Rep.* **2022**, *8*, 589–594. [CrossRef]

energies

MDPI

Article

RLS-Based Algorithm for Detecting Partial Demagnetization under Both Stationary and Nonstationary Conditions

Ze Jiang [1,2], Xiaoyan Huang [1,*] and Wenping Cao [3]

1 Zhejiang Provincial Key Laboratory of Electrical Machine Systems, College of Electrical Engineering, Zhejiang University, Hangzhou 310027, China; jiangze1982@zju.edu.cn
2 Wolong Electric Group Co., Ltd., Shaoxing 312300, China
3 School of Electrical Engineering and Automation, Anhui Unsiversity, Hefei 230039, China; 19122@ahu.edu.cn
* Correspondence: xiaoyanhuang@zju.edu.cn

Abstract: An algorithm was developed detect the partial demagnetization of permanent-magnet synchronous motors (PMSMs) under both stationary and nonstationary conditions. On the basis of the recursive least-squares (RLS) method, the vital component of fault-related harmonics in the current could be extracted on the line, and its proportion to fundamental component could be regarded as the indicator of partial demagnetization faults. The proposed algorithm is fairly easy to realize and could substitute conventional and complicated signal processing methods such as Fourier transform and wavelet transform when detecting partial demagnetization. Experiments with inverter-fed healthy and partially demagnetized PMSMs are carried out to substantiate the effectiveness of proposed algorithm under both stationary and nonstationary conditions. At the end, a way to eliminate the impact of eccentricity fault on the partial demagnetization diagnosis is given.

Keywords: permanent-magnet synchronous motors (PMSMs); partial demagnetization; fault diagnosis; recursive least squares (RLS)

Citation: Jiang, Z.; Huang, X.; Cao, W. RLS-Based Algorithm for Detecting Partial Demagnetization under Both Stationary and Nonstationary Conditions. *Energies* **2022**, *15*, 3509. https://doi.org/10.3390/en15103509

Academic Editors: Lorand Szabo and Federico Barrero

Received: 29 March 2022
Accepted: 9 May 2022
Published: 11 May 2022

1. Introduction

1.1. Motivations

Permanent-magnet synchronous motors (PMSMs) are deployed on a large scale for their inherent merits of high torque and power density, and high efficiency [1]. However, the demagnetization of permanent magnets installed in PMSMs has raised major concerns in the academic and industrial fields. Demagnetization occurs in some harsh operation environments such as high temperatures and large currents [2]. In addition, mechanical failures may cause demagnetization faults [3]. In some applications that require high reliability [4–7], demagnetization faults must be reported early [8].

1.2. Demagnetization Phenomena

In partial demagnetization [9], one or some parts of permanent magnets are demagnetized, which results in the spatial asymmetry of the magnetic field in the air gap. Due to spatial asymmetry, with specific combinations of poles and slots [10,11], back electromotive force (EMF) contains fault-related harmonics that are the source of current harmonics. Many studies proved that this kind of harmonics can be expressed as follows [12–17]:

$$f_{fault} = f_e(1 \pm n/p) \tag{1}$$

where f_{fault} refers to the fault-related harmonic frequency, f_e is the electrical fundamental frequency, p denotes the number of pole pairs, and n is an integer that can be 1, 2, 3, … …

1.3. Related Works

In the literature, many methods emerged to detect partial demagnetization. To directly identify the distribution of magnetic flux, signals measured by the Gaussmeter [18] and Hall sensors [19,20] are analyzed. Since faulted-related harmonics in (1) are introduced in back-EMF, back-EMF could be measured to detect partial demagnetization [21]. However, in most cases, back-EMF cannot be directly measured for PMSMs under operation. Methods based on motor current signature analysis (MCSA) [22–26] were developed and widely commercialized since fault-related harmonics in currents occur simultaneously due to back-EMF harmonics [26]. In addition, information on output torque [27], vibration, and acoustics [28] can also be used to analyze partial demagnetization. In this paper, MCSA was adopted to detect partial online demagnetization due to its feasibility and easy implementation.

According to orders of a fault-related harmonic frequency, partial demagnetization can be recognized by using signal processing methods such as fast Fourier transform (FFT) [22,23], wavelet transform (WT) [24], and Hilbert–Huang transform (HHT) [25]. When PMSMs are operated under stationary conditions, FFT is the most common approach to gain components of fault-related harmonics [26]. However, rigorous conditions must be satisfied when using FFT: (1) integer periods need to be sampled or spectral leakage may happen; (2) sample frequency must be twice higher than the maximal frequency that needs to be analyzed; (3) the number of sample points should be 2N to achieve fast operation of the algorithm [29]. Although the above conditions can be satisfied, there are still some other problems when using FFT, such as complex calculation, and FFT not being suitable when PMSMs are operated under nonstationary conditions since FFT can only reflect the frequency spectrum, but is not able to correlate to time [26]. However, in most applications, such as electrical vehicles and pumps, PMSMs are often operated under nonstationary conditions.

In order to detect partial demagnetization under nonstationary conditions, many time–frequency methods were investigated, including short-time Fourier transform (STFT) [30], continuous and discrete wavelet transform (CWT/DWT) [13,24], and S-transform [21]. The main problem of these time–frequency methods is the resolution of time and frequency [31]. If time resolution is too low, it is difficult to accurately catch the change in frequency during transient processes; if the frequency resolution is too low, the previous frequency cannot be obtained at a specific time. Tme and frequency resolution cannot simultaneously be at a high level according to the Heisenberg uncertainty principle [32]. In order to overcome resolution problems, harmonic order tracking analysis (HOTA) was proposed by means of Gabor transform [33–35]. However, this still requires a window of sample data that must be stored in the memory, and needs complex calculation.

In this paper, an RLS-based algorithm to detect the partial demagnetization of PMSMs under both stationary and nonstationary conditions is presented. Some regression models, such as neural networks [36], ADALINE filters [37], convolutional neural networks (CNNs), the support vector regression method (SVRM) [38], and RLS [39] were applied in most applications, but none of these models was utilized in the field of partial demagnetization detection, where they may present sremarkable merits under stationary and nonstationary conditions. Among these regression models, RLS is comparatively easy to realize and was hence used in this paper. Another advantage of methods such as RLS is the fact that the result is physically interpretable, which is not always true with advanced methods. The proposed RLS-based algorithm can be executed in real time, and magnitudes of fault-related harmonics are established to indicate the health of PMSMs. Moreover, there is no need to store previous data in the proposed algorithm, which enables the use of the algorithm in embedding devices with small-capacity memory. The proposed algorithm could substitute conventional and complicated signal-processing methods, such as Fourier transform and wavelet transform, when detecting partial demagnetization. Experiments with inverter-fed healthy and partially demagnetized PMSMs were carried out to substantiate the effectiveness of proposed algorithm under both stationary and nonstationary conditions. Lastly, a specific part of eccentricity faults and its impact is discussed.

2. Partial Demagnetization Model of PMSMs

To better illustrate the fault-related harmonics introduced in the current due to partial demagnetization, a 9-slot and 8-pole PMSM with star-connected winding, whose structure is presented in Figure 1, is investigated in this paper. The typical frequency spectrum of phase back-EMF when one or more permanent magnets are partially demagnetized is shown in Figure 2, in which magnitude refers to the ratio of the fundamental components. Side frequencies around the fundamental component (2/4th-, 3/4th- and 5/4th-order) were the predominant fault-related harmonics whose magnitudes are major indicators of partial demagnetization faults.

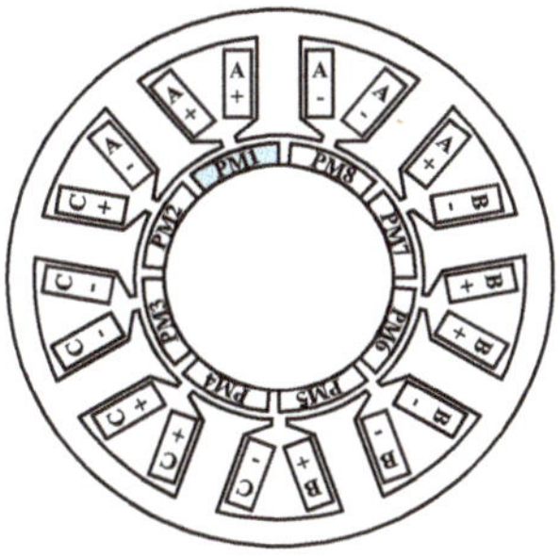

Figure 1. Structure of 9−slot and 8−pole PMSM.

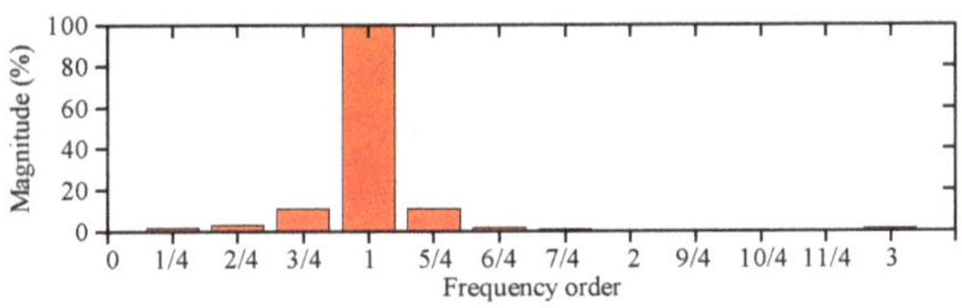

Figure 2. Typical frequency spectrum of phase back-EMF when partial demagnetization happens.

For inverter-fed PMSMs, back-EMF cannot be directly measured, so currents must be sampled and used to detect partial demagnetization faults. To reduce computational resources and use less memory storage, only a few orders of fault-related harmonics are relevant, and phase current i_a when partial demagnetization happens can be written as follows [40]:

$$i_a = \sum_{k=1}^{\infty} A_k \sin(2\pi f_e t \times k/p) + B_k \cos(2\pi f_e t \times k/p) \tag{2}$$

When detecting a partial demagnetization fault, specific values of k are adopted instead of all on the basis of the frequency spectrum. For instance, Figure 2 displays that the other frequencies were near to zero except the 2/4th-, 3/4th-, 5/4th-order and fundamental components; values of k in i_a current expression (2) could be set as 2 (for the 2/4th order), 3 (for the 3/4th order), 4 (for the fundamental components), and 5 (for the 5/4th order). However, the real current content was more complicated than the components above. Some odd harmonics such as the 5th and 7th orders were included in the currents due to magnet saturation, the slot effect, and inverter nonlinearity. Inphase fault-related harmonics disappear in a symmetric three-phase winding; for instance, the 3/4th order harmonic existing in phase back-EMF shown in Figure 2 was absent in the line-to-line back-EMF and may not appear in the currents either. Due to the switching frequency of the inverter, components around the switching and double switching frequencies are also involved in the frequency spectrum of currents, but these high-frequency components could be simply eliminated by a low-pass filter (LPF) with a proper cut-off frequency. Hence, phase current i_a in (2) can be expressed as the combination of selected frequencies:

$$\begin{aligned} i_a = & A_2 \sin(2\pi f_e t \times 2/p) + B_2 \cos(2\pi f_e t \times 2/p) + A_4 \sin(2\pi f_e t \times 4/p) + B_4 \cos(2\pi f_e t \times 4/p) \\ & + A_5 \sin(2\pi f_e t \times 5/p) + B_5 \cos(2\pi f_e t \times 5/p) + A_{20} \sin(2\pi f_e t \times 20/p) + B_{20} \cos(2\pi f_e t \times 20/p) \\ & + A_{28} \sin(2\pi f_e t \times 28/p) + B_{28} \cos(2\pi f_e t \times 28/p) \end{aligned} \quad (3)$$

Magnitude M_k of each component could be established as follows:

$$M_k = \sqrt{A_k{}^2 + B_k{}^2} \quad (4)$$

and phase φ_k of each component is calculated with:

$$\varphi_k = \arctan \frac{B_k}{A_k} \quad (5)$$

The magnitude of fault-related frequency could be deemed to be the indicator of partial demagnetization. The key point is to identify values of A_k and B_k for each selected frequency component, which is discussed in the next sections.

3. Proposed Algorithm of Detecting Partial Demagnetization

The expression of i_a in (3) is the linear combination of selected components; recursive least-squares (RLS) is the simplest way to identify A_k and B_k. Under stationary conditions, motor speed and electrical fundamental frequency f_e are constant, while under nonstationary conditions, motor speed and f_e vary. In order to apply the RLS algorithm, motor speed must be obtained from the position sensors. In this section, RLS is first briefly introduced, and an algorithm of detecting partial demagnetization is proposed that is adaptive to the change in speed and can be utilized under both stationary and nonstationary conditions.

3.1. *Recursive Least Squares (RLS)*

Consider a system with output that can be expressed as the linear combination of inputs, which is presented in (6):

$$y(i) = w_1(i)x_1(i) + w_2(i)x_2(i) + \ldots + w_m(i)x_m(i) = \mathbf{W}(i)^T\mathbf{X}(i) \quad (6)$$

where i is the i-th iteration of RLS algorithm; y is the output of the system; x_m is the m-th input of the system; and w_m is the coefficient of the m-th input that needs to be identified. Auxiliary variable s is inserted to represent the estimated output:

$$s(i) = \hat{w}_1(i)x_1(i) + \hat{w}_2(i)x_2(i) + \ldots + \hat{w}_m(i)x_m(i) = \hat{\mathbf{W}}(i)^T\mathbf{X}(i) \quad (7)$$

where $\hat{w}_m$ is the estimated coefficient of the m-th input. The basic principle of RLS is to minimize the error between estimated output s and real output y; hence, a cost function is given by:

$$C = \frac{1}{2}[y(i) - s(i)]^2 \quad (8)$$

There are many other selections for cost functions, such as symmetric mean absolute percentage error (SMAPE), given by:

$$C = \frac{|y(i) - s(i)|}{(|y(i)| + |s(i)|)/2} \quad (9)$$

To simplify the algorithms, the cost function in (8) was selected in this paper.

By using the gradient descent algorithm to minimize the value of C, the operation process of updating the estimated coefficients could be written as:

$$\hat{\mathbf{W}}(i) = \hat{\mathbf{W}}(i-1) + \mathbf{K}(i)[y(i) - \hat{\mathbf{W}}(i-1)^T\mathbf{X}(i)] \tag{10}$$

$$\mathbf{K}(i) = \frac{\mathbf{P}(i-1)\mathbf{X}(i)}{\lambda + \mathbf{X}(i)^T\mathbf{P}(i-1)\mathbf{X}(i)} \tag{11}$$

$$\mathbf{P}(i) = [\mathbf{I} - \mathbf{K}(i)\mathbf{X}(i)^T] \times \mathbf{P}(i-1)/\lambda \tag{12}$$

where λ is the forgetting factor to deal with data saturation; a larger value of λ always needs longer convergence time, while a smaller λ may cause the vibration of the results; hence, the selection of λ is a trade-off between convergence time and vibration.

In the case of the RLS algorithm, the normal equations are as follows [41–43]:

$$\sum_{i=1}^{n} \lambda^{n-i}\mathbf{X}(i)\mathbf{X}^T(i)\hat{\mathbf{W}}(i) = \sum_{i=1}^{n} \lambda^{n-i}\mathbf{X}(i)y(i) + \sum_{i=1}^{n} \lambda^{n-i}\mathbf{X}(i)v(i) \tag{13}$$

where $v(i)$ is system noise. When the value of λ is close to 1, it is assumed that

$$\frac{1}{n}\sum_{i=1}^{n} \lambda^{n-i}\mathbf{X}(i)v(i) \approx E(\mathbf{X}(n)v(n)) = 0 \tag{14}$$

where E is the mathematical expectation. Then,

$$\sum_{i=1}^{n} \lambda^{n-i}\mathbf{X}(i)\mathbf{X}^T(i)\hat{\mathbf{W}}(i) = \sum_{i=1}^{n} \lambda^{n-i}\mathbf{X}(i)y(i) = \sum_{i=1}^{n} \lambda^{n-i}\mathbf{X}(i)\mathbf{X}^T(i)\mathbf{W}(i) \tag{15}$$

Thus,

$$\hat{\mathbf{W}}(i) = \mathbf{W}(i) \tag{16}$$

which means that the estimated values can be well-converged to the real values when the value of λ is close to 1. When the value of λ is much less than 1,

$$\frac{1}{n}\sum_{i=1}^{n} \lambda^{n-i}\mathbf{X}(i)v(i) \neq 0 \tag{17}$$

Lastly, estimated values are impacted by system noise, leading to the instability of the RLS algorithm.

3.2. Methodology of Partial Demagnetization Detection

By virtue of the RLS algorithm introduced above, the coefficients of each component A_k and B_k can be estimated. Then, their magnitudes can be regarded as indicators of partial demagnetization:

$$y(i) = i_a(i) \tag{18}$$

$$\begin{aligned}\mathbf{X}(i) = \; & [\sin(2\pi f_e t \times 2/p), \cos(2\pi f_e t \times 2/p),\\ & \sin(2\pi f_e t \times 4/p), \cos(2\pi f_e t \times 4/p),\\ & \sin(2\pi f_e t \times 5/p), \cos(2\pi f_e t \times 5/p), \ldots]^T\end{aligned} \tag{19}$$

$$\mathbf{W}(i) = [A_2, B_2, A_4, B_4, A_5, B_5, \ldots]^T \tag{20}$$

The proposed methodology estimates values of A_k and B_k. Figure 3 shows the flowchart of the proposed methodology.

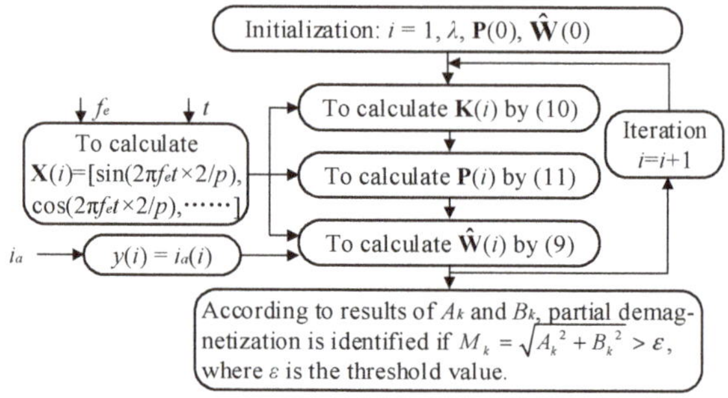

Figure 3. Flowchart of proposed methodology.

Figure 3 shows that f_e and t are used to calculate the selected components in (3), and i_a, 0 measured with current sensors, is system output y. After the system initialization of i, λ, $\mathbf{P}(0)$ and $\hat{\mathbf{W}}(0)$, the values of $\mathbf{K}(i)$, $\mathbf{P}(i)$, and $\hat{\mathbf{W}}(i)$, can be established. Through several iterations, the results of A_k and B_k are able to converge to their real values. Then, the magnitude of each fault-related harmonic component and the fundamental component in the currents are identified through Equation (4). The magnitude values of fault-related harmonic components can help in distinguishing healthy PMSMs from partially demagnetized ones by selecting proper threshold value ε.

Under stationary conditions, motor speed and f_e are fixed, and the proposed algorithm functions are the same as in the FFT method to gain the amplitude of each component. Under nonstationary conditions, motor speed and f_e vary and must be obtained from position sensors; hence, the proposed algorithm is adaptive to speed changes, which allows for dealing with nonstationary situations like in methods of time–frequency analysis. In addition, the proposed algorithm focuses on several dominant components that can avoid the resolution problems of existing time–frequency analysis and reduce computational resources.

4. Experimental and Simulation Validation

In order to substantiate the effectiveness of proposed algorithm, experimental validation was carried out and is described in Figure 4. The main parameters of the tested PMSM are listed in Table 1. The details of RLS algorithms are given in Table 2.

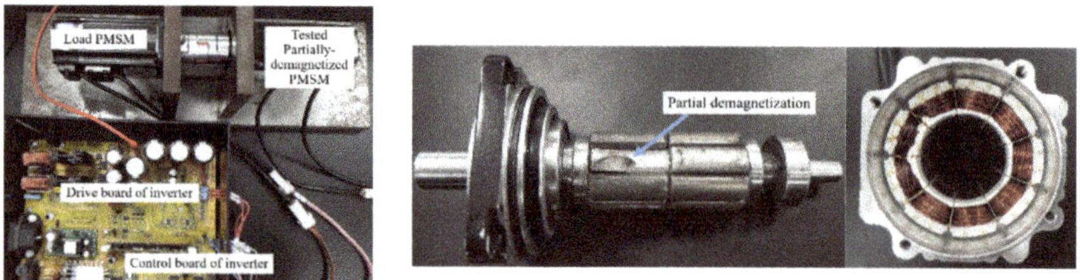

(a) (b)

Figure 4. Partially demagnetized PMSM and experimental setup. (**a**) Experimental setup; (**b**) rotor and stator of partially demagnetized PMSM.

Table 1. Tested PMSM parameters.

Parameter	Value
Number of poles and slots	8 poles and 9 slots
Winding connection	Star-connected winding
Rated speed	1000 rpm
Rated current	2.6A (peak)
Rated voltage	100 V
Stator resistance	2.54 ohm
Stator inductance	7.1 mH
Flux linkage	0.062 Wb

Table 2. Details of RLS algorithms.

Sampling Number	Forgetting Factor λ	Sampling Step
Sampling in each PWM period (0.1 ms)	0.999	0.1 ms

The tested PMSM was partially demagnetized while PMSM load was healthy. Moreover, another healthy PMSM was tested in order to compare the current difference between healthy and partially demagnetized PMSMs. The tested healthy and partially demagnetized PMSMs were both driven by commercial inverter box with the TMS320F28335 DSP and controlled with field oriented control (FOC) strategy, the overall control block is shown in Figure 5:

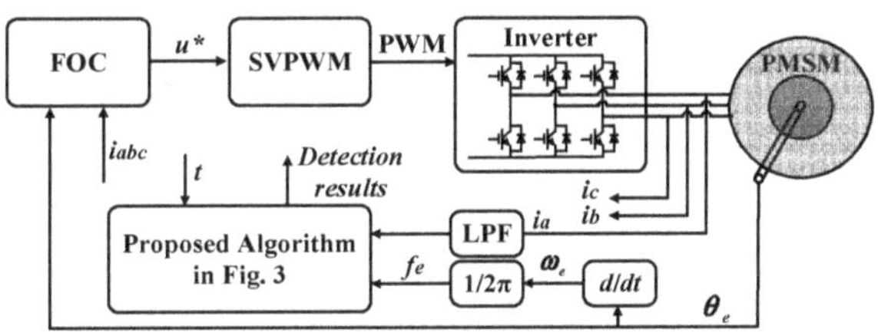

Figure 5. Overall control block.

Figure 5 shows that i_a was measured from the terminal of the motor. Additionally, motor speed was monitored by the encoder installed in the rotor, and fundamental frequency f_e could be established. Values of i_a and f_e were used to execute the proposed algorithm, as shown in Figure 3, to detect partial demagnetization in real time. Both the stationary and nonstationary operations of the tested PMSM were analyzed in the experiments.

4.1. Stationary Conditions

First, the case when the tested PMSMs were operated under stationary conditions was examined. PMSMs are maintained at rated speed of 1000 rpm, and rated phase current i_a had a peak value of 2.6 A. Figure 6 shows the waveforms of i_a for both healthy and demagnetized PMSMs.

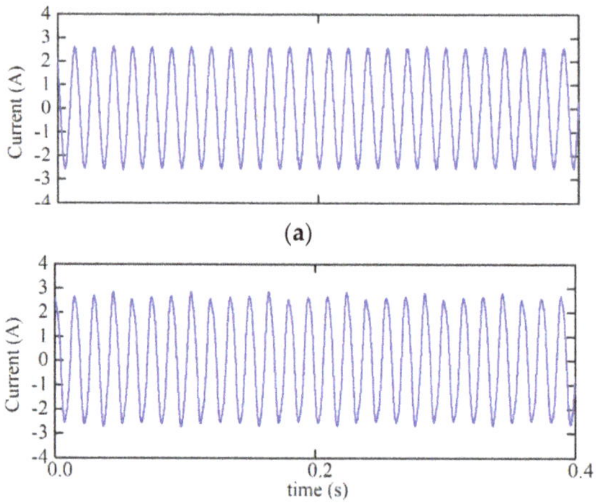

Figure 6. Experimental waveform of current i_a under stationary conditions (1000 rpm, peak current: 2.6A). (**a**) Current waveform for healthy PMSM; (**b**) current waveform for partially−demagnetized PMSM.

By using the proposed algorithm in Figure 3, coefficient values of each selected component in (3) A_k and B_k could be established. Through Equation (4), the magnitude of each fault-related harmonic in current i_a can then be calculated. In order to vividly depict the effectiveness of the proposed algorithm and consider the magnitude change of fundamental component, explicit indicator value η of partial demagnetization can be expressed as:

$$\eta = \frac{M_k}{M_4} \times 100\% = \frac{\sqrt{A_k{}^2 + B_k{}^2}}{\sqrt{A_4{}^2 + B_4{}^2}} \times 100\% \tag{21}$$

which is the magnitude ratio of the $k/4$th-order harmonic component to the fundamental component. Figure 7a,b show the ratio comparisons of 2/4th- and 5/4th-order harmonics, respectively, between partially demagnetized and healthy PMSMs under stationary conditions. Figure 7 shows that the fault-related harmonic ratio of the partially demagnetized PMSM was larger than that of the healthy PMSM. Figure 7a shows that the ratio of the 2/4th-order harmonic to the fundamental component for the partially demagnetized PMSM was 1.4%, while the ratio for the healthy PMSM was less than 0.2%. In terms of the 5/4th-order harmonic, as shown in Figure 7b, the ratio was 2.5% for the partially demagnetized PMSM, and 1.0% for the healthy PMSM. The first 0.2 s in Figure 7 is the convergence time of the proposed algorithm that is related to the value of λ. If the forgetting factor is too small, previous data are forgotten faster, leading to possible numerical instability. On the other hand, data saturation may occur with a larger forgetting factor, resulting in a slow convergence rate. Thus, there is a trade-off when selecting the value of the forgetting factor; in this paper, the final value of the forgetting factor was 0.999. Experimental results with different values of forgetting factor (0.999 and 0.9) are presented in Figure 7.

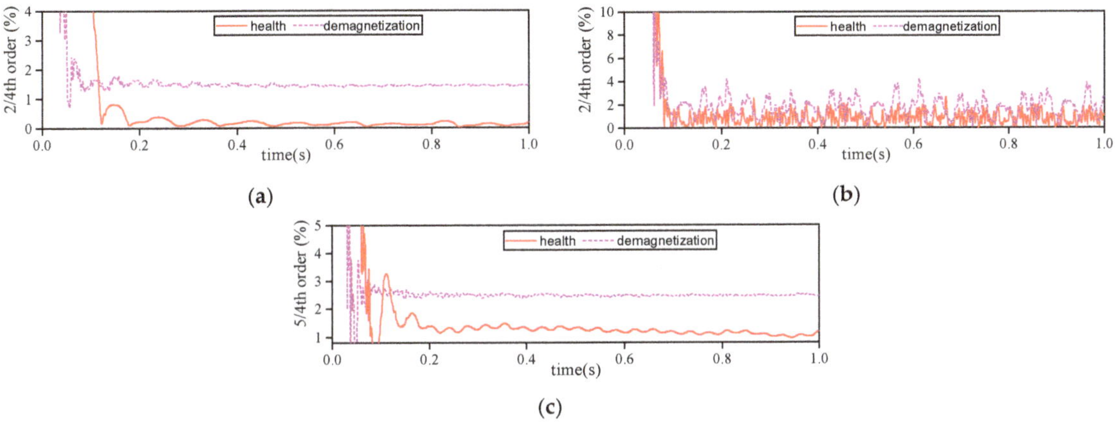

Figure 7. Comparison of harmonics between healthy and partially demagnetized PMSMs under stationary conditions (1000 rpm, 2.6A, in experiment). (**a**) 2/4th-order harmonic, λ = 0.999; (**b**) 2/4th-order harmonic, λ = 0.9; (**c**) 5/4th-order harmonic, λ = 0.9.

Fault-related harmonics in a healthy PMSM are ideally zero, but due to the inherent spatial asymmetry caused by the mechanical installation of rotor shafts and permanent magnets, some small fault-related harmonics exist in the current for healthy PMSMs, which is why current magnitudes of 2/4th- and 5/4th-order harmonics for healthy PMSMs were not absolutely zero in the experiment.

To evaluate the proposed algorithm against a conventional FFT, the magnitudes of corresponding harmonic components in the current with different motor health conditions were also computed through FFT and are listed in Table 3. Important points from Table 3 are as follows:

1. Comparing Columns 2 and 3, and Columns 4 and 5 shows that the magnitudes of each fault-related harmonic component were similar through different algorithms (FFT and RLS) for the partially demagnetized PMSM. The same was true in comparing Columns 4 and 5 for the healthy PMSM, which means the proposed algorithm with RLS could substitute FFT when analyzing the frequency spectrum in real-world use. Minor errors between FFT and RLS may be due to the spectral leakage of FFT, the data saturation of RLS, and other factors.
2. For the partially demagnetized PMSM, the magnitudes of fault-related harmonics components are larger than the ones of healthy PMSM, which can be seen from the column 2 and column 4 as well as the column 3 and column 5 in Table 3. This characteristic can be regarded as the indicator of the partial demagnetization fault.
3. Since the proposed algorithm is in the recursive form, the computational resources can be reduced significantly. The execution time of one iteration in the proposed algorithm is 0.2 ms, while under stationary conditions, the execution time of 512-point FFT in the TMS320F28335 DSP is more than 10 ms. In addition, the FFT requires more memory capacity to store the data points.

Table 3. Magnitude of each fault-related harmonic component in current with different algorithms and different health status (unit: A).

Harmonic Order	Demagnetized PMSM (FFT)	Demagnetized PMSM (RLS)	Healthy PMSM (FFT)	Healthy PMSM (RLS)
2/4th	0.0335	0.0375	0.0007	0.0027
4/4th	2.6299	2.6301	2.5171	2.5197
5/4th	0.0699	0.0662	0.0324	0.0325
5th	0.0139	0.0143	0.0167	0.0199
7th	0.0084	0.0086	0.0090	0.0087

To prove the robustness of the proposed algorithm when the load changes, the PMSMs are operated at the same speed of 1000 rpm with the different peak current of 0.6A. After implementing the proposed algorithm in Figure 3, the extracted 2/4th and 5/4th harmonics are displayed in Figure 8. Similarly, 2/4th and 5/4th harmonic magnitudes when the PMSM was partially demagnetized were larger than those of the healthy PMSM.

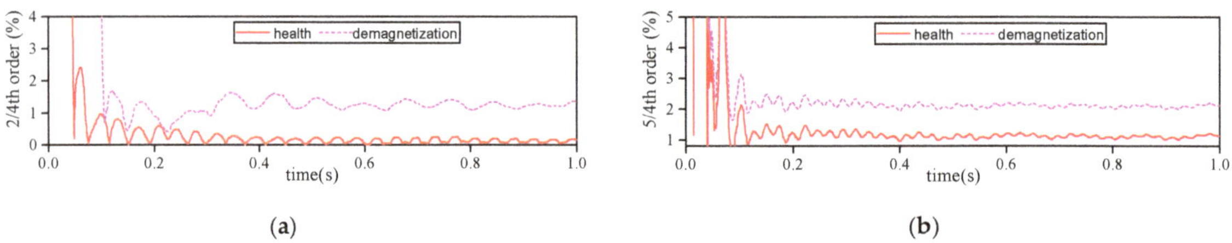

Figure 8. Comparison of harmonics between healthy and partially demagnetized PMSMs under stationary conditions: (**a**) 2/4th-order harmonic (1000 rpm, 0.6 A); (**b**) 5/4th-order harmonic (1000 rpm, 0.6 A).

4.2. Nonstationary Conditions

Under nonstationary conditions, PMSM speed is not constant. Normal FFT cannot analyze the content of harmonic components with the change in time, which hinders recognizing partial demagnetization faults. Many time–frequency analyses were investigated to overcome the problem of FFT under nonstationary conditions, but these methods tend to require complicated calculations, and have poor resolution of time and frequency. Using the proposed algorithm in this paper, it is possible to deal with the above problems. To prove that, the tested partially demagnetized PMSM was operated under nonstationary conditions, and current i_a was measured and analyzed with the proposed algorithm in Figure 3. Under nonstationary conditions, motor speed or fundamental frequency f_e is a crucial factor that greatly impacts the accuracy of the proposed algorithm; hence, motor speed should be precisely sampled through speed sensors.

Waveforms of phase current i_a and speed for the partially demagnetized PMSM under nonstationary conditions are given in Figure 9. During the first 3 s, motor speed was maintained at 900 rpm, and peak current was 0.5A. At 3 s, the motor began to decelerate to 450 rpm, and the current was decreased to 0.4 A. After the transient process, the motor speed was held at 450 rpm.

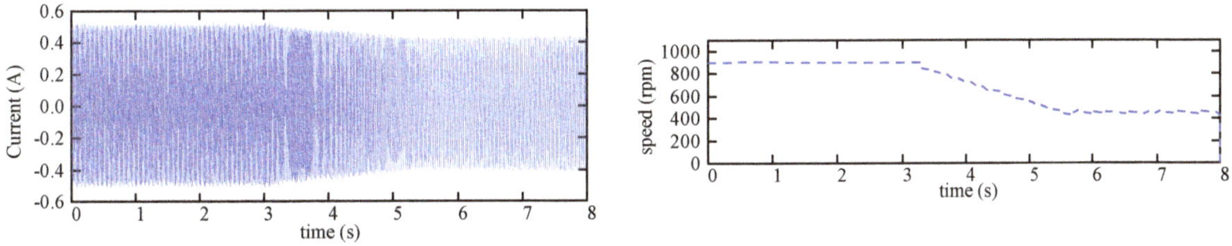

(a) (b)

Figure 9. Experimental waveforms of current i_a and speed for partially−demagnetized PMSM under nonstationary conditions. (**a**) Current waveform of i_a; (**b**) speed waveform.

Using the proposed algorithm in Figure 3 under nonstationary conditions, the magnitudes of fundamental and harmonic components in the current could be extracted, and results are presented in Figure 10, showing that the magnitude of the fundamental component in current i_a began to decrease at 3 s. Results in Figure 10 prove that, with the change in time, the proposed algorithm enables gaining the magnitude of each predominant component in the frequency spectrum of the current under nonstationary conditions, just like other methods of time–frequency analyses.

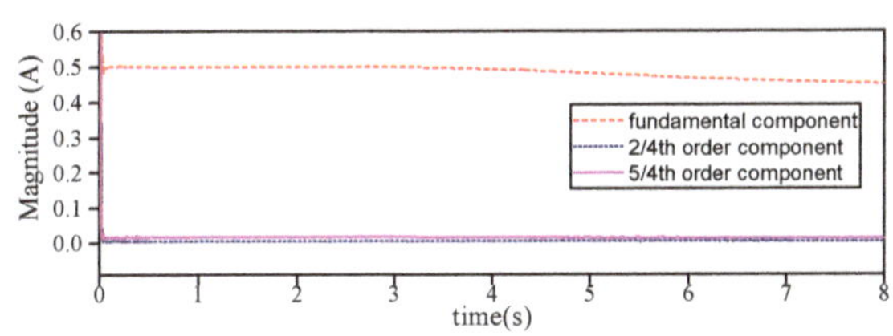

Figure 10. Magnitudes of fundamental and harmonic components in current for partially demagnetized PMSM under nonstationary conditions (experiment).

Figure 11 shows the magnitude ratios of the 2/4th-order component to the fundamental component, which uses indicator value η in (15) to evaluate partial demagnetization. In terms of the 2/4th-order harmonic, the ratio to the fundamental component was 1% for the partially demagnetized PMSM, and less than 0.1% for the healthy PMSM. In addition, Figure 12 displays the magnitude ratios of the 5/4th-order component to the fundamental component; the ratio for the partially demagnetized PMSM was much larger than that of the healthy PMSM. Under nonstationary conditions, magnitudes of fundamental and harmonic components vary, as shown in Figure 10, which hinders setting a threshold value for judging the occurrence of demagnetization. To overcome this problem, the ratio of harmonic component to fundamental component can be regarded to be a unified and per-unit value. When the motor is operated under nonstationary conditions, the ratio is on the same or a similar level, and it is convenient to set threshold value ε to recognize partial demagnetization.

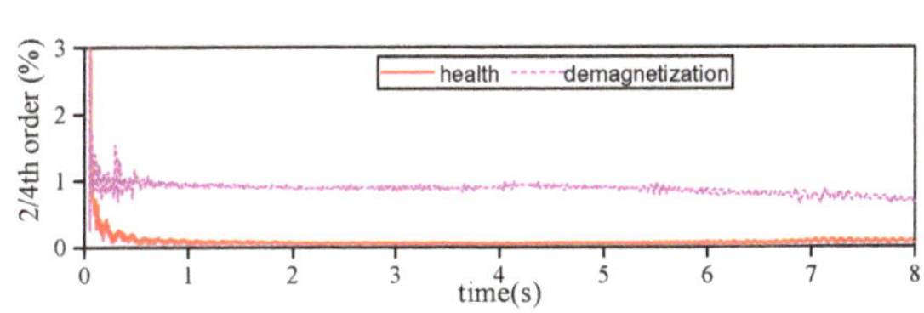

Figure 11. Comparison of 2/4th-order harmonic between partially demagnetized and healthy PMSMs under nonstationary conditions (experiment).

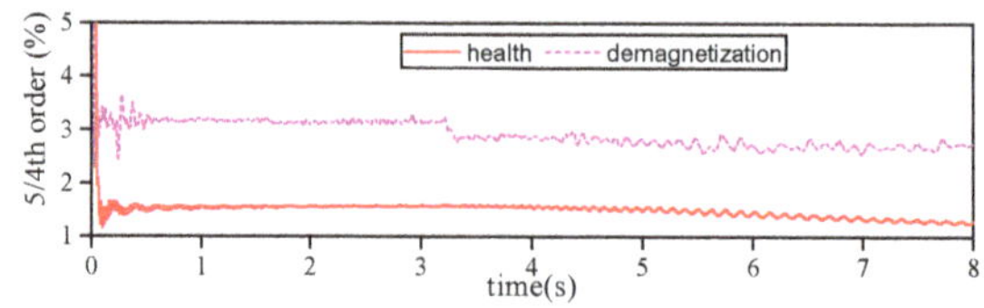

Figure 12. Comparison of 5/4th-order harmonic between partially demagnetized and healthy PMSMs under nonstationary conditions (experiment).

4.3. Comparison with Other Time–Frequency Analyses under Nonstationary Conditions

Furthermore, other tools of time–frequency analysis were compared when the PMSM is operated under nonstationary conditions. As mentioned in Section 1, problems of time and frequency resolution are major concerns when using these tools. For instance, Gabor transform possesses the highest resolution and concentration in the time–frequency domain [35], which can be expressed as:

$$G(t, f) = \sqrt[4]{\sigma} \int_{-\infty}^{\infty} e^{-\sigma\pi(\tau-t)^2} e^{-j2\pi f\tau} x(\tau) d\tau \tag{22}$$

To describe resolution problems in detail, Gabor transform was adopted to track fault-related harmonics in the nonstationary current in Figure 9. Results when $\sigma = 0.5$ and $\sigma = 50$ are shown in Figures 13 and 14., respectively.

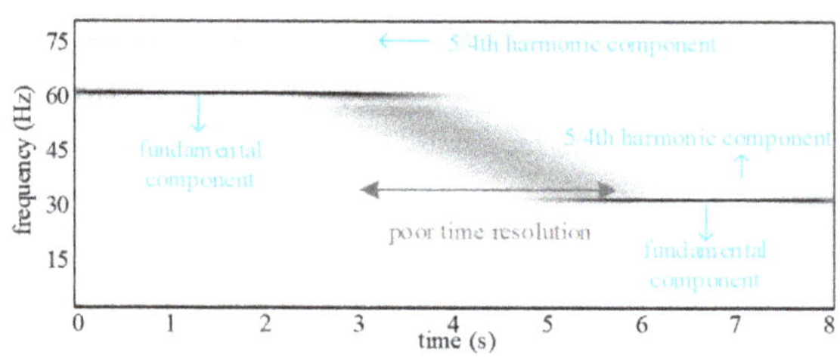

Figure 13. Experimental result using Gabor transform when $\sigma = 0.5$.

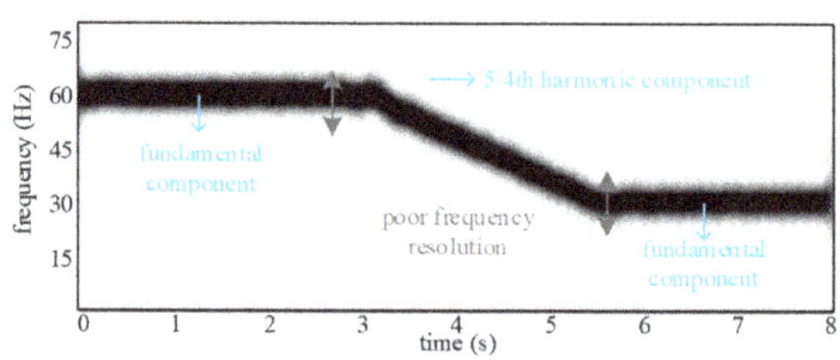

Figure 14. Experimental result using Gabor transform when $\sigma = 50$.

Figure 13 shows that, when σ was smaller ($\sigma = 0.5$), the resolution in the frequency domain was higher, while the resolution in the time domain was not satisfactory, which hindered catching the change in frequency during the transient process in time. Figure 14 shows the opposite: when the value of σ was larger ($\sigma = 50$), resolution in the time domain was higher, so the frequency change was clearer under nonstationary conditions. However, the improvement of resolution in the time domain sacrifices resolution in the frequency domain, which obstructs distinguishing fault-related harmonics in the frequency spectrum, for instance, the 5/4th harmonic component can hardly be seen in Figure 14. Therefore, it is crucial to find a proper value of σ when utilizing Gabor transform. For other tools of time–frequency analysis such as wavelet transform, a similar problem also occurs. By using the proposed algorithm in this paper, the resolution problem in the time and frequency domains could be avoided.

There are also other potential merits of our proposed algorithm compared with other existing methods of time–frequency analysis. The execution time of one iteration in the proposed algorithm is 0.2 ms under both stationary and nonstationary conditions. The execution time of other time–frequency analyses is much longer under nonstationary conditions, for instance, it takes more than 2 ms to complete one-point calculation (for one specific time and one specific frequency) on the time–frequency plane. To finish the whole map of the time–frequency plane, it usually takes several hours when the dataset is too large, which hinders using these time–frequency analyses in embedded devices. In addition, other methods of time–frequency analysis require large memory capacity to store data points within the specific window. The proposed algorithm only needs one data point at the current moment, which pronouncedly reduces the memory burden of embedded devices. The merits and demerits of FFT, time–frequency (TF) analyses, and RLS are listed in Table 4.

Table 4. Merits and demerits of different analytical tools.

Tools	Merits	Demerits
FFT	(1) Most known and mature way to analyze spectrum. (2) No resolution problems, since it can only analyze the frequency domain along with a long stationary time of 0.0335.	(1) Cannot detect the fault under nonstationary conditions. (2) Integer periods need be sampled or spectrum leakage may happen. Sample frequency must be twice higher than the maximal frequency that needs to be analyzed. (3) The number of sample points should be 2N to achieve fast algorithm processing.
TF(Gabor)	(1) Able to detect faults under both stationary and nonstationary conditions.	(1) Resolution problem in time and frequency domains. (2) Complicated calculation (2 ms to complete one-point on time–frequency plane, more time to complete analysis).
RLS	(1) Able to detect faults under both stationary and nonstationary conditions. (2) No resolution problem in time and frequency domains. (3) Execute in recursive form to reduce computational burden (only 0.2 ms for one iteration).	(1) Careful selection of forgetting factor λ. (2) Needs precise speed information.

4.4. Elimination of Eccentricity Faults

Eccentricity faults also introduce the fault harmonics presented in (1) [11,13,17,20], which lead to false diagnoses of partial demagnetization. Some measures should be taken to eliminate the impact of eccentricity faults. Hence, two different kinds of eccentricity, namely, static and dynamic eccentricity (SE and DE), are discussed here. Using finite element analysis (FEA), 3-phase back-EMFs are shown in Figure 15 when the 9-slot and 8-pole PMSM was healthy or suffering from partially demagnetization, and with SE and DE faults.

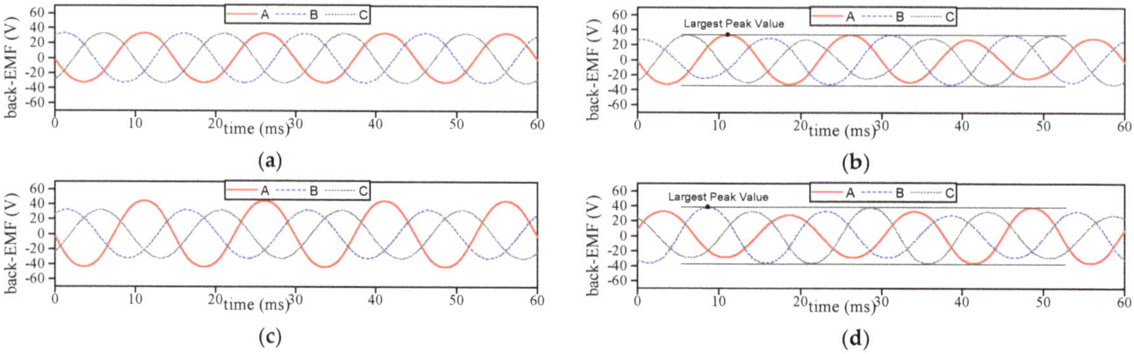

Figure 15. Simulated three−phase back−EMFs for 9−slot and 8−pole PMSM at 1000 rpm. (**a**) Healthy condition; (**b**) partial demagnetization fault; (**c**) static eccentricity fault; (**d**) dynamic eccentricity fault.

On the one hand, SE causes different distances between 3-phase windings on the stator and permanent magnets on the rotor, which obviously resulted in the unbalance of 3-phase back-EMFs in Figure 15c. For one-phase or line-to-line back-EMF, the fault-related harmonics in (1) would not be present, which the frequency spectrum of E_{AB} in Figure 16a shows.

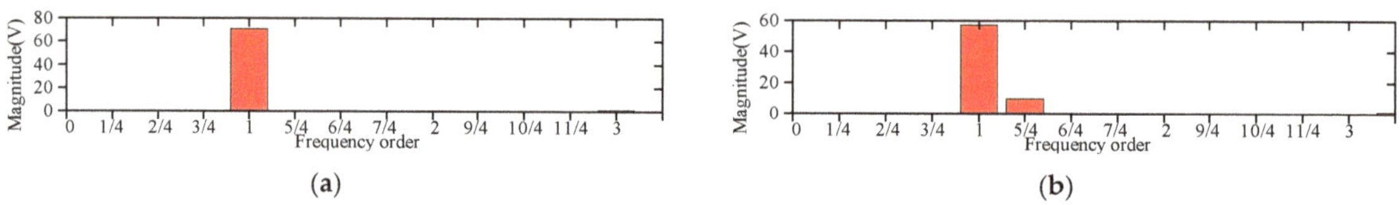

Figure 16. Frequency spectrum of EAB for 9−slot and 8−pole PMSM. (**a**) Static eccentricity fault at 1000 rpm; (**b**) dynamic eccentricity fault at 1000 rpm.

On the other hand, a DE fault introduces the fault-related harmonics shown in Figure 16b. However, the rotor is always dynamically decentered to one direction along with time when the DE fault occurs, which causes the back-EMF amplitude to be much larger than that of healthy motors. Hence, a DE fault can be simply excluded by capturing the largest peak value of E_{AB} in one mechanical period, depicted in Figure 15b,d. As shown in Table 5, the largest peak value of E_{AB} in one mechanical period under different conditions is listed. When DE occurs, the largest peak value is larger than that of healthy motors; when partial demagnetization happens, the largest peak value is always near that of a healthy motor. By using this characteristic, the impact of eccentricity could be eliminated from the partial demagnetization fault. Because manufacturing eccentric PMSMs always takes a long time, and the main focus of this paper is the application of RLS algorithm in the field of partial demagnetization, the experimental validation of eccentricity faults is absent.

Table 5. Largest peak value of E_{AB} in one mechanical period under different conditions (for 9-slot and 8-poles motor at 1000 rpm).

Healthy	DE 25%	DE 100%	DEMAG 1.78%	DEMAG 3.57%	DEMAG 5.35%	DEMAG 7.14%	DEMAG 8.92%	DEMAG 10.71%
58.2 V	65.5 V	88.5 V	58.4 V	58.3 V	58.1 V	58.1 V	58.1 V	58.1 V

5. Conclusions

This paper proposed a RLS-based algorithm to detect partial demagnetization under both stationary and nonstationary conditions. With the use of RLS, the proposed algorithm enables extracting the fault-related harmonic components introduced in the current; magnitudes of these fault-related harmonics can be regarded as an indicator of partial demagnetization. To better set the threshold value to evaluate the occurrence of partial demagnetization, a magnitude ratio of fault-related harmonics to fundamental component is given that can also be used when the magnitude of current changes with time. Experimental results show the effectiveness of the proposed algorithm under stationary and nonstationary conditions. Furthermore, comparisons with other existing methods of time–frequency analysis show that the proposed algorithm is an alternative method to detect partial demagnetization while overcoming time- and frequency-domain resolution problems, reducing computational burden and memory storage. Lastly, the impact of eccentricity faults was discussed. Experimental validation of the impact of eccentricity could and the distinction of different faults could future research directions. Although a method was proposed in the paper to eliminate the impact of eccentricity, the method is still not applicable. In addition, it was impossible to apply all faults in one paper; hence, the impacts of other possible faults on the detection of demagnetization still need further investigation.

Author Contributions: Conceptualization, methodology, validation, writing—original draft preparation, Z.J. and X.H.; writing—review and editing, X.H. and W.C.; supervision, X.H.; funding acquisition, X.H. All authors have read and agreed to the published version of the manuscript.

Funding: This work was supported in part by the National Key R&D Program of China under Grant No. 2019YFE0123500 and in part by the National Natural Science Foundation of China under Grant No. 51922095.

Institutional Review Board Statement: Not applicable.

Informed Consent Statement: Not applicable.

Data Availability Statement: Not applicable.

Conflicts of Interest: The authors declare no conflict of interest.

References

1. Im, S.-H.; Gu, B.-G. A Snubberless Solid-State Tap Changer for Permanent Magnet Synchronous Motors. *IEEE Trans. Power Electron.* **2020**, *35*, 12143–12152. [CrossRef]
2. Choi, G.; Zhang, Y.; Jahns, T.M. Experimental Verification of Rotor Demagnetization in a Fractional-Slot Concentrated-Winding PM Synchronous Machine Under Drive Fault Conditions. *IEEE Trans. Ind. Appl.* **2017**, *53*, 3467–3475. [CrossRef]
3. Moosavi, S.; Djerdir, A.; Amirat, Y.A.; Khaburi, D. Demagnetization fault diagnosis in permanent magnet synchronous motors: A review of the state-of-the-art. *J. Magn. Magn. Mater.* **2015**, *391*, 203–212. [CrossRef]
4. Da, Y.; Shi, X.; Krishnamurthy, M. A New Approach to Fault Diagnostics for Permanent Magnet Synchronous Machines Using Electromagnetic Signature Analysis. *IEEE Trans. Power Electron.* **2013**, *28*, 4104–4112. [CrossRef]
5. Dini, P.; Saponara, S. Design of adaptive controller exploiting learning concepts applied to a BLDC-based drive system. *Energies* **2020**, *13*, 2512. [CrossRef]
6. Bernardeschi, C.; Dini, P.; Domenici, A.; Palmieri, M.; Saponara, S. Formal verification and co-simulation in the design of a synchronous motor control algorithm. *Energies* **2020**, *13*, 4057. [CrossRef]
7. Bernardeschi, C.; Dini, P.; Domenici, A.; Saponara, S. Co-Simulation and Verification of a Non-Linear Control System for Cogging Torque Reduction in Brushless Motors. In Proceedings of the International Conference on Software Engineering and Formal Methods, Oslo, Norway, 16–20 September 2019; Springer: Cham, Switzerland, 2019.
8. Moon, S.; Jeong, H.; Lee, H.; Kim, S.W. Detection and Classification of Demagnetization and Interturn Short Faults of IPMSMs. *IEEE Trans. Ind. Electron.* **2017**, *64*, 9433–9441. [CrossRef]
9. Reigosa, D.; Fernández, D.; Martínez, M.; Park, Y.; Lee, S.B.; Briz, F. Permanent magnet synchronous machine non-uniform demagnetization detection using zero-sequence magnetic field density. *IEEE Trans. Ind. Appl.* **2019**, *55*, 3823–3833. [CrossRef]
10. Zafarani, M.; Goktas, T.; Akin, B. A comprehensive magnet defect fault analysis of permanent-magnet synchronous motors. *IEEE Trans. Ind. Appl.* **2016**, *52*, 1331–1339. [CrossRef]
11. Zafarani, M.; Goktas, T.; Akin, B.; Fedigan, S.E. An investigation of motor topology impacts on magnet defect fault signatures. *IEEE Trans. Ind. Electron.* **2017**, *64*, 32–42. [CrossRef]
12. Wang, C.; Prieto, M.D.; Romeral, L.; Chen, Z.; Blaabjerg, F.; Liu, X. Detection of partial demagnetization fault in PMSMs operating under nonstationary conditions. *IEEE Trans. Magn.* **2016**, *52*, 1–4. [CrossRef]
13. Ruiz, J.R.; Rosero, J.A.; Espinosa, A.G.; Romeral, L. Detection of demagnetization faults in permanent-magnet synchronous motors under nonstationary conditions. *IEEE Trans. Magn.* **2009**, *45*, 2961–2969. [CrossRef]

14. Hong, J.; Park, S.; Hyun, D.; Kang, T.J.; Lee, S.B.; Kral, C.; Haumer, A. Detection and classification of rotor demagnetization and eccentricity faults for PM synchronous motors. *IEEE Trans. Ind. Appl.* **2012**, *48*, 923–932. [CrossRef]
15. Urresty, J.; Riba, J.; Romeral, L. A back-emf based method to detect magnet failures in PMSMs. *IEEE Trans. Magn.* **2013**, *49*, 591–598. [CrossRef]
16. Goktas, T.; Zafarani, M.; Akin, B. Discernment of broken magnet and static eccentricity faults in permanent magnet synchronous motors. *IEEE Trans. Energy Convers.* **2016**, *31*, 578–587. [CrossRef]
17. Mazaheri-Tehrani, E.; Faiz, J.; Zafarani, M.; Akin, B. A fast phase variable abc model of brushless PM motors under demagnetization faults. *IEEE Trans. Ind. Electron.* **2019**, *66*, 5070–5080. [CrossRef]
18. Hong, J.; Hyun, D.; Lee, S.B.; Yoo, J.; Lee, K. Automated Monitoring of Magnet Quality for Permanent-Magnet Synchronous Motors at Standstill. *IEEE Trans. Ind. Appl.* **2010**, *46*, 1397–1405. [CrossRef]
19. Park, Y.; Yang, C.; Lee, S.B.; Lee, D.M.; Fernandez, D.; Reigosa, D.; Briz, F. Online Detection and classification of rotor and load defects in PMSMs based on hall sensor measurements. *IEEE Trans. Ind. Appl.* **2019**, *55*, 3803–3812. [CrossRef]
20. Park, Y.; Fernandez, D.; Lee, S.B.; Hyun, D.; Jeong, M.; Kommuri, S.K.; Cho, C.; Reigosa, D.D.; Briz, F. Online detection of rotor eccentricity and demagnetization faults in PMSMs based on hall-effect field sensor measurements. *IEEE Trans. Ind. Appl.* **2019**, *55*, 2499–2509. [CrossRef]
21. Song, X.; Zhao, J.; Song, J.; Dong, F.; Xu, L.; Zhao, J. Local Demagnetization Fault Recognition of Permanent Magnet Synchronous Linear Motor Based on S-Transform and PSO–LSSVM. *IEEE Trans. Power Electron.* **2020**, *35*, 7816–7825. [CrossRef]
22. le Roux, W.; Harley, R.G.; Habetler, T.G. Detecting rotor faults in low power permanent magnet synchronous machines. *IEEE Trans. Power Electron.* **2007**, *22*, 322–328. [CrossRef]
23. Rajagopalan, S.; Roux, W.l.; Habetler, T.G.; Harley, R.G. Dynamic eccentricity and demagnetized rotor magnet detection in trapezoidal flux (brushless DC) motors operating under different load conditions. *IEEE Trans. Power Electron.* **2007**, *22*, 2061–2069. [CrossRef]
24. Ishikawa, T.; Seki, Y.; Kurita, N. Analysis for Fault Detection of Vector-Controlled Permanent Magnet Synchronous Motor with Permanent Magnet Defect. *IEEE Trans. Magn.* **2013**, *49*, 2331–2334. [CrossRef]
25. Espinosa, A.G.; Rosero, J.A.; Cusidó, J.; Romeral, L.; Ortega, J.A. Fault Detection by Means of Hilbert–Huang Transform of the Stator Current in a PMSM with Demagnetization. *IEEE Trans. Energy Convers.* **2010**, *25*, 312–318. [CrossRef]
26. Faiz, J.; Mazaheri-Tehrani, E. Demagnetization modeling and fault diagnosing techniques in permanent magnet machines under stationary and nonstationary conditions: An overview. *IEEE Trans. Ind. Appl.* **2017**, *53*, 2772–2785. [CrossRef]
27. Ebrahimi, B.M.; Faiz, J. Demagnetization Fault Diagnosis in Surface Mounted Permanent Magnet Synchronous Motors. *IEEE Trans. Magn.* **2013**, *49*, 1185–1192. [CrossRef]
28. Yang, Z.; Shi, X.; Krishnamurthy, M. Vibration monitoring of PM synchronous machine with partial demagnetization and inter-turn short circuit faults. In Proceedings of the 2014 IEEE Transportation Electrification Conference and Expo (ITEC), Dearborn, MI, USA, 15–18 June 2014; pp. 1–6.
29. Oraintara, S.; Chen, Y.J.; Nguyen, T.Q. Integer fast Fourier transform. *IEEE Trans. Signal Process.* **2002**, *50*, 607–618. [CrossRef]
30. Rajagopalan, S.; Aller, J.M.; Restrepo, J.A.; Habetler, T.G.; Harley, R.G. Detection of Rotor Faults in Brushless DC Motors Operating Under Nonstationary Conditions. *IEEE Trans. Ind. Appl.* **2006**, *42*, 1464–1477. [CrossRef]
31. Puche-Panadero, R.; Pineda-Sanchez, M.; Riera-Guasp, M.; Roger-Folch, J.; Hurtado-Perez, E.; Perez-Cruz, J. Improved Resolution of the MCSA Method Via Hilbert Transform, Enabling the Diagnosis of Rotor Asymmetries at Very Low Slip. *IEEE Trans. Energy Convers.* **2009**, *24*, 52–59. [CrossRef]
32. Stanković, L. The Support Uncertainty Principle and the Graph Rihaczek Distribution: Revisited and Improved. *IEEE Signal Process. Lett.* **2020**, *27*, 1030–1034. [CrossRef]
33. Sapena-Bano, A.; Pineda-Sanchez, M.; Puche-Panadero, R.; Perez-Cruz, J.; Roger-Folch, J.; Riera-Guasp, M.; Martinez-Roman, J. Harmonic Order Tracking Analysis: A Novel Method for Fault Diagnosis in Induction Machines. *IEEE Trans. Energy Convers.* **2015**, *30*, 833–841. [CrossRef]
34. Sapena-Bano, A.; Riera-Guasp, M.; Puche-Panadero, R.; Martinez-Roman, J.; Perez-Cruz, J.; Pineda-Sanchez, M. Harmonic Order Tracking Analysis: A Speed-Sensorless Method for Condition Monitoring of Wound Rotor Induction Generators. *IEEE Trans. Ind. Appl.* **2016**, *52*, 4719–4729. [CrossRef]
35. Sapena-Bano, A.; Burriel-Valencia, J.; Pineda-Sanchez, M.; Puche-Panadero, R.; Riera-Guasp, M. The Harmonic Order Tracking Analysis Method for the Fault Diagnosis in Induction Motors Under Time-Varying Conditions. *IEEE Trans. Energy Convers.* **2017**, *32*, 244–256. [CrossRef]
36. Qiu, T.; Wen, X.; Zhao, F. Adaptive-Linear-Neuron-Based Dead-Time Effects Compensation Scheme for PMSM Drives. *IEEE Trans. Power Electron.* **2016**, *31*, 2530–2538. [CrossRef]
37. Wang, L.; Zhu, Z.Q.; Bin, H.; Gong, L. A Commutation Error Compensation Strategy for High-Speed Brushless DC Drive Based on Adaline Filter. *IEEE Trans. Ind. Electron.* **2021**, *68*, 3728–3738. [CrossRef]
38. Maraaba, L.S.; Milhem, A.S.; Nemer, I.A.; Al-Duwaish, H.; Abido, M.A. Convolutional Neural Network-Based Inter-Turn Fault Diagnosis in LSPMSMs. *IEEE Access* **2020**, *8*, 81960–81970. [CrossRef]
39. Sarrafan, K.; Muttaqi, K.M.; Sutanto, D. Real-Time Estimation of Model Parameters and State-of-Charge of Li-Ion Batteries in Electric Vehicles Using a New Mixed Estimation Model. *IEEE Trans. Ind. Appl.* **2020**, *56*, 5417–5428. [CrossRef]

40. Liu, G.; Chen, B.; Wang, K.; Song, X. Selective Current Harmonic Suppression for High-Speed PMSM Based on High-Precision Harmonic Detection Method. *IEEE Trans. Ind. Inform.* **2019**, *15*, 3457–3468. [CrossRef]
41. Leung, S.H.; So, C.F. Gradient-based variable forgetting factor RLS algorithm in time-varying environments. *IEEE Trans. Signal Process.* **2005**, *53*, 3141–3150. [CrossRef]
42. Paleologu, C.; Benesty, J.; Ciochina, S. A Robust Variable Forgetting Factor Recursive Least-Squares Algorithm for System Identification. *IEEE Signal Process. Lett.* **2008**, *15*, 597–600. [CrossRef]
43. Ciochina, S.; Paleologu, C.; Benesty, J.; Enescu, A.A. On the influence of the forgetting factor of the RLS adaptive filter in system identification. In Proceedings of the 2009 International Symposium on Signals, Circuits and Systems, Iasi, Romania, 9–10 July 2009; pp. 1–4. [CrossRef]

energies

MDPI

Article

Industrial Data Denoising via Low-Rank and Sparse Representations and Its Application in Tunnel Boring Machine

Yitang Wang, Yong Pang, Wei Sun and Xueguan Song *

School of Mechanical Engineering, Dalian University of Technology, No. 2 Linggong Road, Ganjingzi District, Dalian 116024, China; yitangwang@mail.dlut.edu.cn (Y.W.); pangy@mail.dlut.edu.cn (Y.P.); sunwei@dlut.edu.cn (W.S.)

* Correspondence: sxg@dlut.edu.cn

Abstract: The operation data of a tunnel boring machine (TBM) reflects its geological conditions and working status, which can provide critical references and essential information for TBM designers and operators. However, in practice, operation data may get corrupted due to equipment failures or data management errors. Moreover, the working state of a TBM system usually changes, which results in patterns of operation data that vary comparatively. This paper proposes a denoising approach to process the corrupted data. This approach is combined with low-rank matrix recovery (LRMR) and sparse representation (SR) theory. The classical LRMR model requires that the noise must be sparse, but the sparsity of noise cannot be fully guaranteed. In the proposed model, a weighted nuclear norm is utilized to enhance the sparsity of sparse components, and a constraint of condition number is applied to ensure the stability of the model solution. The approach is coupled with a fuzzy c-means algorithm (FCM) to find the natural partitioning using the TBM operation data as input. The performances of the proposed approach are illustrated through an application to the Shenzhen metro. Experimental results show that the proposed approach performs well in corrupted TBM data denoising. The different excavation status of the TBM recognition accuracy is improved remarkably after denoising.

Keywords: tunnel boring machine; industrial data denoising; low rank; sparse representation; fuzzy c-means clustering

Citation: Wang, Y.; Pang, Y.; Sun, W.; Song, X. Industrial Data Denoising via Low-Rank and Sparse Representations and Its Application in Tunnel Boring Machine. *Energies* **2022**, *15*, 3525. https://doi.org/10.3390/en15103525

Academic Editors: Adolfo Dannier and Manoj Khandelwal

Received: 25 March 2022
Accepted: 9 May 2022
Published: 11 May 2022

1. Introduction

Tunnel boring machine (TBM), as an important complex engineering system, have been widely used in various forms of tunnel construction, such as in metros, railways, highways, etc. With the application of cyber-physical systems and sensing technologies, massive amounts of operation data have been measured and recorded, which can reflect TBM conditions and working status and promotes the development of data-driven design, analysis, and control of TBM [1–3]. However, the TBM used in mixed ground conditions faces many challenges, such as geological hazards, instrument damage, instability of the excavation face, and change in working state, which may result in noise and data pattern changes. These disturbances cannot be fully avoided in the construction process and seriously reduce the quality and reliability of data, so it is necessary to develop approaches and techniques to remove the noises from recorded TBM parameters and identify the different excavation states of the TBM.

In recent years, data denoising technology has received tremendous attention, with researchers proposing a series of effective methods [4–6]. The common methods of data denoising for data preprocessing include traditional filtering denoising [7,8], wavelet transform (WT) denoising [9,10], and empirical mode decomposition (EMD) denoising [11,12]. The filtering denoising methods, such as median filtering denoising [13], Kalman filtering [14] and Winer filtering [15], are rarely used in TBM data denoising due to the lack of prior knowledge of noise statistical characteristics. With respect to tunneling data denoising, Zhang and Liu [16] calculated the Mahalanobis distance at first, and removed the data outside of the p90 percentile, before finally employing the DB4 wavelet with two-layer decomposition and soft thresholding to denoise the operation data. As a solution, Zhou and Xu [17] also utilized the wavelet transform to remove noise before building a dynamic predictive model of TBM data. However, the pre-divided temporal frequency features and the linear non-adaptive transformation of the WT limit its application ability in nonlinear TBM systems. EMD decompose the signal into a finite number of intrinsic mode functions (IMFs), which has been successfully used in data denoising in various fields such as seismic data [18], electrochemical data [19], and global navigation satellite system (GNSS) position data [20]. Unfortunately, the true physical processes could be obscured by the disruption of a given signal, when the signal contains intermittent processes [21].

In general, there are various kinds of noise in TBM operation data during the tunneling process, and the aforementioned methods are potentially restricted for particularly noisy operation data. As a matter of a fact, some attributes of TBM operational data are highly correlated. In addition, the parameters correlation can effectively boost the denoising results. Rank minimization is a common method for exploring the correlation, and some well-performed denoising methods based on low-rank matrix recovery (LRMR) have been proposed. For instance, Li and Wang [22] proposed an effective sparse representation model for image fusion and denoising, which decomposes noisy images into two different components and exploits a weight nuclear norm sparse regularization term to avoid corruption by noises. With respect to hyperspectral image denoising, Zhao and Yang [23] utilized the local and global redundancy and correlation (RAC) in spatial and spectral domains jointly because it is has particularly global and local RAC in spatial and spectral dimensions. As a solution, Gomes and Costa [24] employed low-rank tensor approximation to denoise data in R-D sensor array processing problems. However, most methods mentioned above are mostly used for computer vision applications. The literature survey showed that the LRMR method has not yet been used in tunneling data denoising.

In this article, a new denoising approach based on LRMR is proposed for TBM data preprocessing. The robust principal component analysis (RPCA) [25] is a typical LRMR framework, which allows two facts that the high correlation of matrix can be captured by low-rank prior, meanwhile the noise possesses sparse characteristics. In fact, the robustness of RPCA is demonstrated by assuming that noise is sparse, regardless of noise intensity. However, the sparsity of noise cannot be fully guaranteed, due to the complexity of the noise sources in the tunneling data. In order to break the above hurdle, the weighted nuclear norm is utilized to improve the ability of the model in sparse decomposition. The weighted nuclear norm can also reduce the influence of singular values in the process of nuclear norms approximate to low-rank matrix. Besides this, we employ a Forbenius-2 norm to improve the stability of model solutions. The proposed LRMR model is a jointly nonconvex problem, and we introduce the inexact augmented Lagrange multiplier (IALM) [26] technique and an alternating minimization strategy to solve it.

Since there are different operation states of the TBM system, the patterns of operation vary greatly as well. It necessary to partition the data into different parts based on characteristics to better help the design and analysis for TBM designers and operators. In recent years, data clustering methods have been developed and widely used in the field of engineering for the partition of operation data. Generally, these methods can be grouped into two categories, hard clustering methods and fuzzy clustering methods. Different from hard clustering, fuzzy clustering methods allows data to belong to multiple clusters with varying degrees of membership. Thus, we utilize the fuzzy c-means algorithm (FCM) [27] to classify the TBM operation status. By doing so, we can make a reasonable partitioning of the TBM operation data.

Our main contributions are as follows:

(1) A new non-convex low-rank approximation that utilizes the weighted nuclear norm and Forbenius norm is introduced for TBM denoising. This method can avoid the

deficiencies of RPCA. To the best of the authors' knowledge, the RPCA model has not yet been used in tunneling data denoising.

(2) An efficient optimization framework based on IALM and alternating minimization is designed to minimize the non-convex proposed model. Experimental results with real-world datasets from a tunnel in China illustrate that the proposed method performs well.

The remainder of the paper is organized as follows: Section 2 presents detailed information on the proposed method and the optimization for this method. Experimental results and analysis are reported in Section 3. Finally, the conclusions are given in Section 4.

2. Proposed Approach

2.1. Preliminary

Assume that the observed data matrix $\boldsymbol{D} \in \mathbb{R}^{m \times n}$ can be decomposed into a low-rank matrix $\boldsymbol{L} \in \mathbb{R}^{m \times n}$ and a sparse matrix accounting for both small Gaussian noise and large scale outliers $\boldsymbol{S} \in \mathbb{R}^{m \times n}$, so that $\boldsymbol{D} = \boldsymbol{L} + \boldsymbol{S}$.

To determine $\boldsymbol{L}$ and $\boldsymbol{S}$, the RPCA algorithm aims to solve the minimization problem, which can be described as follows:

$$\min_{\boldsymbol{L},\boldsymbol{S}} \operatorname{rank}(\boldsymbol{L}) + \gamma \|\boldsymbol{S}\|_0 \text{ s.t. } \boldsymbol{D} = \boldsymbol{L} + \boldsymbol{S} \tag{1}$$

where $\gamma > 0$ is a trade-off parameter, which is a balance of the relative contribution between the rank of matrix $\boldsymbol{L}$ and the ℓ_0 norm of matrix $\boldsymbol{S}$. It is obvious that the optimization problem in (1) is highly non-convex for all matrix variables simultaneously, it cannot be solved directly using the existing methods. By replacing the ℓ_0 norm with the ℓ_1 norm, and the rank with the nuclear norm, then the problem can be relaxed into the following surrogate:

$$\min_{\boldsymbol{L},\boldsymbol{S}} \|\boldsymbol{L}\|_* + \gamma \|\boldsymbol{S}\|_1 \text{ s.t. } \boldsymbol{D} = \boldsymbol{L} + \boldsymbol{S} \tag{2}$$

where $\|\cdot\|_*$ is the nuclear norm defined as $\|\cdot\|_* = \sum_i |\sigma_i(\cdot)|$, $\sigma_i(\cdot)$ is the i-th singular value of $\cdot$, $\|\cdot\|_1$ is the ℓ_1 norm, which can be calculated as the sum of the absolute values in matrix $\cdot$. In the last decades, many effective methods have been proposed for solving the optimization problem of (2). Specifically, the optimization problem (2) can be effectively solved by the alternating direction method (ADM) algorithm [28] or the inexact augmented Lagrange multipliers (IALM) algorithm [26].

2.2. Denoising Method

In this paper, a data denoising method guided by RPCA is proposed. As analyzed above, TBM data denoising aims to reconstruct clean data $\boldsymbol{L} \in \mathbb{R}^{m \times n}$ from its noisy measurement $\boldsymbol{D} \in \mathbb{R}^{m \times n}$, where $\boldsymbol{L} \in \mathbb{R}^{m \times n}$ is low rank and the noise $\boldsymbol{S} \in \mathbb{R}^{m \times n}$ is sparse. In practice, the sparsity of noise cannot be fully guaranteed, due to the complexity of the noise sources in the TBM operation data. Moreover, the nuclear norm minimization (NNM)-being a low-rank matrix treats different rank components equally, which may result in severe shrinkage of the underlying information [29]. Hence, we introduce a weighted nuclear norm, which assigns different weights to different singular values, to regularize the $\boldsymbol{L}$. Thus, we can reformulate the objective function (2) as:

$$\min_{\boldsymbol{L},\boldsymbol{S}} \|\boldsymbol{L}\|_{\omega,*} + \gamma \|\boldsymbol{S}\|_1 \text{ s.t. } \boldsymbol{D} = \boldsymbol{L} + \boldsymbol{S} \tag{3}$$

where $\|\cdot\|_{\omega,*}$ is the weighted nuclear norm defined as $\|\cdot\|_{\omega,*} = \sum_i |\omega_i \sigma_i(\cdot)|$ [30], $\omega_i > 0$ is the i-th weight assigned to $\sigma_i(\cdot)$. In Equation (3), different singular values are differently in the weighted nuclear norm, which is very beneficial to reduce the influence of singular values in the process of nuclear norm approximation of matrix rank.

However, during noise removal, the stability and sparsity of the model solution cannot be guaranteed [30]. As we know, the condition number is an effective measure of the stability of a linear system. From an optimization point of view, the Forbenius norm is conducive to dealing with the problem of matrix inversion when it is ill-conditioned. Thus, we introduce the Forbenius norm constraint on the $\boldsymbol{L}$, through which we force the stability of the matrix to be restored. Then, the above optimization problem (3) can be re-described as:

$$\min_{\boldsymbol{L},\boldsymbol{S}} \|\boldsymbol{L}\|_{\omega,*} + \gamma\|\boldsymbol{S}\|_1 + \lambda\|\boldsymbol{L}\|_F^2 \text{ s.t. } \boldsymbol{D} = \boldsymbol{L} + \boldsymbol{S} \quad (4)$$

where $\lambda > 0$ is a regularization parameter and it is used to adjust the role of $\|L\|_F^2$ in the denoising model.

2.3. Optimization and Algorithm

In this section, we provide a detailed description of solving optimization problem (4). In the present paper, the inexact augmented Lagrange multiplier (IALM) [26] technique is used to solve the optimization problem (4). For further optimization, we first need to determine the value of weight ω_i. According to the literature [31], the weighted term should be inversely proportional to the weighted term. The algorithm steps of weight updating is shown in Algorithm 1.

Algorithm 1 Algorithm for update weight

Input: Initial $\boldsymbol{L}$ and $\boldsymbol{S}$ and $\omega_i(i = 1, 2, \ldots, \kappa)$ and maximum number of $\boldsymbol{\kappa}$.
Initialize: $\omega^{(0)} = 1 * 1^{\mathrm{T}} \in \mathbb{R}^{m \times n}$, $\varepsilon = 0.01$
while $\boldsymbol{\kappa}$ not reach **do**
(a) Update $\boldsymbol{L}$and $\boldsymbol{S}$
(b) Update weight via $\omega_j^{i+1} = \frac{1}{\sigma_j^i + \varepsilon}$
end while if the maximum number of $\boldsymbol{\kappa}$ has been reached.
Output: ω

Then, we write its IALM form as follows:

$$L(\boldsymbol{L}, \boldsymbol{S}, \boldsymbol{X}, \eta) = \|\boldsymbol{L}\|_{\omega,*} + \gamma\|\boldsymbol{S}\|_1 + \lambda\|\boldsymbol{L}\|_{\mathrm{F}}^2 + \langle \boldsymbol{X}, \boldsymbol{D} - \boldsymbol{L} - \boldsymbol{S}\rangle + \frac{\eta}{2}\|\boldsymbol{D} - \boldsymbol{L} - \boldsymbol{S}\|_{\mathrm{F}}^2 \quad (5)$$

where $\boldsymbol{X} \in \mathbb{R}^{m \times n}$ is the Lagrange multiplier matrix, and $\eta > 0$ is the penalty parameter. We introduce an auxiliary variable $\boldsymbol{A}$ and relax (5) into the following problem:

$$\begin{aligned} L(\boldsymbol{L}, \boldsymbol{S}, \boldsymbol{X}, \eta) = {} & \|\boldsymbol{L}\|_{\omega,*} + \gamma\|\boldsymbol{S}\|_1 + \lambda\|\boldsymbol{A}\|_{\mathrm{F}}^2 + \langle \boldsymbol{X}_1, \boldsymbol{D} - \boldsymbol{L} - \boldsymbol{S}\rangle + \tfrac{\eta_1}{2}\|\boldsymbol{D} - \boldsymbol{L} - \boldsymbol{S}\|_F^2 \\ & + \langle \boldsymbol{X}_2, \boldsymbol{A} - \boldsymbol{S}\rangle + \tfrac{\eta_2}{2}\|\boldsymbol{A} - \boldsymbol{S}\|_F^2 \end{aligned} \quad (6)$$

where $\boldsymbol{X} = (\boldsymbol{X}_1, \boldsymbol{X}_2)$ and $\eta = (\eta_1, \eta_2)$. To facilitate the optimization, the optimization problem (6) can be rewritten as:

$$L(\boldsymbol{L}, \boldsymbol{S}, \boldsymbol{X}, \eta) = \|\boldsymbol{L}\|_{\omega,*} + \gamma\|\boldsymbol{S}\|_1 + \lambda\|\boldsymbol{A}\|_{\mathrm{F}}^2 + \frac{\eta_1}{2}\left\|\boldsymbol{D} - \boldsymbol{L} - \boldsymbol{S} + \eta_1^{-1}\boldsymbol{X}_1\right\|_F^2 + \frac{\eta_2}{2}\left\|\boldsymbol{A} - \boldsymbol{S} + \eta_2^{-1}\boldsymbol{X}_2\right\|_F^2 \quad (7)$$

For each of the five matrices $\boldsymbol{S}$, $\boldsymbol{L}$, $\boldsymbol{A}$, $\boldsymbol{X}_1$ and $\boldsymbol{X}_2$ to be solved in problem (7), the cost function is convex with respect to one and fixed others. We use an alternating minimization strategy to solve the problem (7).

(1) Updating $\boldsymbol{S}$

When we update matrix $\boldsymbol{S}$, we fix $\boldsymbol{L}$, $\boldsymbol{A}$, $\boldsymbol{X}_1$ and $\boldsymbol{X}_2$. Thus, the problem is reduced to:

$$\boldsymbol{S} = \arg\min_{\boldsymbol{S}}\left\{\gamma\|\boldsymbol{S}\|_1 + \frac{\eta_1}{2}\left\|\boldsymbol{D} - \boldsymbol{L} - \boldsymbol{S} + \eta_1^{-1}\boldsymbol{X}_1\right\|_F^2\right\} \quad (8)$$

Obviously, this is a typical ℓ_1 minimization problem which can effectively be solved by the iterative shrinkage algorithm (ISA) [32] or the fast iterative shrinkage thresholding (FIST) [33]. By introducing the shrinkage operator S [34], the solution is:

$$\mathbf{S}^{k+1} = S_{\gamma/\eta_1}\left(\mathbf{D} - \mathbf{L}^{k+1} + \mathbf{X}_1^k/\eta_1^{-k}\right) \tag{9}$$

(2) Updating $\boldsymbol{A}$

Similarly, when we update $\boldsymbol{A}$, we assume that $\mathbf{S}$, $\boldsymbol{L}$, $\mathbf{X}_1$ and $\mathbf{X}_2$ are all fixed. So (7) can be reformulated as:

$$\boldsymbol{A} = \arg\min_{\boldsymbol{A}}\left\{\lambda\|\boldsymbol{A}\|_F^2 + \frac{\eta_2}{2}\left\|\boldsymbol{A} - \boldsymbol{L} + \eta_2^{-1}\mathbf{X}_2\right\|_F^2\right\} \tag{10}$$

which is an ordinary least squares problem, we can obtain the following closed-form solution:

$$\boldsymbol{A} = \frac{\eta_2}{2\lambda + \eta_2}\left(\boldsymbol{L} - \eta_2^{-1}\mathbf{X}_2\right) \tag{11}$$

(3) Updating $\boldsymbol{L}$

To update $\boldsymbol{L}$, we keep the other variables fixed. By removing irrelevant items from the objective function (7), $\boldsymbol{L}$ can be obtained by solving:

$$\boldsymbol{L} = \arg\min_{\boldsymbol{L}}\left\{\|\boldsymbol{L}\|_{\omega,*} + \frac{\eta_1}{2}\left\|\mathbf{D} - \boldsymbol{L} - \mathbf{S} + \eta_1^{-1}\mathbf{X}_1\right\|_F^2\right\} \tag{12}$$

Problem (12) is a weighted nuclear norm minimization that can be found using the method developed in [35]. By introducing the singular value thresholding D [36], and the solution is:

$$\boldsymbol{L}^{k+1} = D_{\omega(\eta_1^k+\eta_2^k)^{-1}}\left[\eta_1^k\left(\mathbf{D} - \mathbf{S}^{k+1} + \eta_1^{-k}\mathbf{X}_1^k\right) + \eta_2^k\left(\boldsymbol{A}^{k+1} + \eta_2^{-k}\mathbf{X}_2^k\right)\right] \tag{13}$$

(4) Updating $\mathbf{X}_1$ and $\mathbf{X}_2$

Similarly, fixing other variables, the Lagrange multiplier matrices $\mathbf{X}_1$ and $\mathbf{X}_2$ can be updated by solving:

$$\mathbf{X}_1 = \arg\min_{\mathbf{X}_1}\left\{\frac{\eta_1}{2}\left\|\mathbf{D} - \boldsymbol{L} - \mathbf{S} + \eta_1^{-1}\mathbf{X}_1\right\|_F^2\right\} \tag{14}$$

$$\mathbf{X}_2 = \arg\min_{\mathbf{X}_2}\left\{\frac{\eta_2}{2}\left\|\boldsymbol{A} - \boldsymbol{L} + \eta_2^{-1}\mathbf{X}_2\right\|_F^2\right\} \tag{15}$$

We have closed-form solutions:

$$\mathbf{X}_1^{k+1} = \mathbf{X}_1^k + \eta_1^k(\mathbf{D} - \boldsymbol{L}_{k+1} - \mathbf{S}_{k+1}) \tag{16}$$

$$\mathbf{X}_2^{k+1} = \mathbf{X}_2^k + \eta_2^k(\boldsymbol{A}_{k+1} - \boldsymbol{L}_{k+1}) \tag{17}$$

A detailed optimization produce of problem (7) is provided in Algorithm 2.

Algorithm 2 Data Denoising Algorithm

Input: Initial $\boldsymbol{D}$ and parameter λ.
Initialize: $\boldsymbol{X}_0 = 0, \boldsymbol{S}_0 = 0, \boldsymbol{x}_0 = 0\ k = 1$
while not converged **do**
(a) Update $\boldsymbol{S}$ via Equation (9)
(b) Update $\boldsymbol{A}$ via Equation (11)
(c) Update $\boldsymbol{L}$ via Equation (13)
(d) Update $\boldsymbol{X}_1$ and $\boldsymbol{X}_2$ respectively via Equations (16) and (17)
end while
Output: L and S

A detailed flowchart of the proposed denoising method is described in Figure 1.

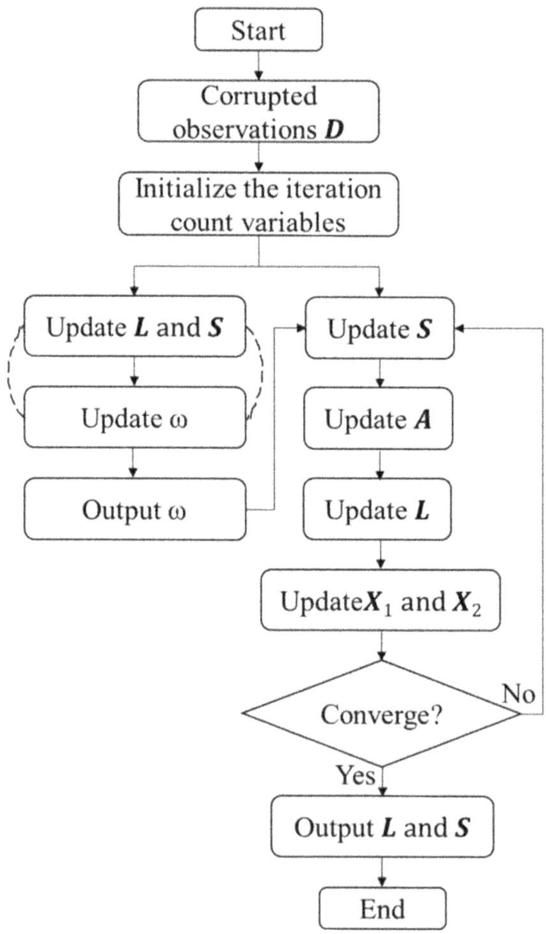

Figure 1. Flowchart of the proposed denoising method.

3. Engineering Application on the Operation Data Analysis of TBM

3.1. Project Review

In this subsection, the proposed method is applied to denoising a real operation dataset from a TBM. The TBM operational data used for analysis in this study belong to a tunnel in Shenzhen, China (as shown in Figure 2). The length of the main tunnel is 2000 m and its diameter is 6.4 m. The ground surface elevation ranges from 0.2–5.8 m, and the depth of the tunnel floor from the ground surface ranges from 11.8–25.4 m. From the ground surface to the tunnel floor, various geological layers, such as clay, sand, and rock, are unevenly distributed, as shown in Figure 3. During the TBM excavation process, an earth pressure balance (EPB) shield TBM is used, as show in Figure 4. The basic equipment parameters are listed in Table 1. This TBM system consists of a cutterhead, chamber, screw conveyor, tail skin and other auxiliary subsystems, which have a diameter of 6.2 m and a total mass of over 500,000 kg. During the tunneling process, the operational and state data of the TBM was recorded by a PLC, which was further read by an industrial computer at regular intervals and stored in the database. Thus, the fresh data in the database are added to in batches during the tunneling process. The collected operation dataset represents the operational information and status parameters along the length of tunnel, which contains about 53 attributes that were continuously measured with a frequency of 1 Hz. Some key attributes are listed in Abbreviations. In the process of dynamic cutterhead torque prediction, the data come in batches. As can be seen from Figure 5, the tunneling data contain extremely large values and extremely small values far from the general level of numerical values. The data acquisition process of TBM is inevitably corrupted by background noise and interference induced by the measuring system. The raw data must be denoised. In addition, the geological properties in tunnels are greatly different from each other, which results in considerable variation in the TBM operating state during tunneling. Two operational datasets of this tunnel are used to validate the denoising algorithm, in which the geological conditions are unlabeled and unknown.

Figure 2. Schematic diagram of research area.

Figure 3. Geological sampling results of the tunnel.

Figure 4. The earth pressure balance shield TBM used.

Table 1. Basic parameters of the TBM used.

Parameters of TBM	Value	Unit
Excavation diameter	6280	mm
Length	75.5	m
Weight	≈500	T
Maximum rotation speed of cutterhead	3.7	r/min
Rated torque of cutterhead	6000	KN·m
Maximum propelling speed	80	mm/min
Number of drive motors	8	1

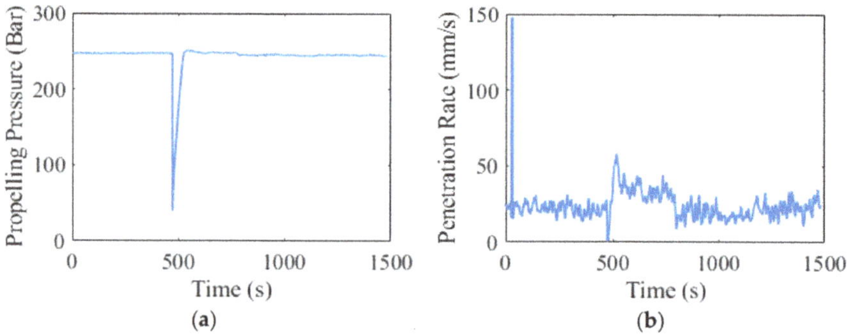

Figure 5. Some attributes with outliers: (**a**) propelling pressure and (**b**) penetration rate.

3.2. Datasets and Experimental Setup

Two datasets of sequence data are used to evaluate the effectiveness of the proposed denoising approach. Some attributes whose values are constant or barely changing should be eliminated first. Then, we eliminated the attributes which are detected independently and have no correlation like temperature with other attributes with the assistance of TBM operators and managers. The two selected datasets have different durations and contain different numbers of attributes, and the details of datasets used in the experiments are described in Table 2.

Table 2. Details of datasets used in the experiments.

Dataset	Attributes Count	Samples Count
Dataset 1	39	1480
Dataset 2	43	860

The proposed approach is compared with other denoising methods to prove its effectiveness. These competing methods include hard threshold (HT) denoising [9], soft threshold (ST) denoising [10], and variational mode decomposition (VMD) denoising [37]. The rate mean-square error (RMSE) and mean absolute error (MAE) are used to quantitatively evaluate the quality of a denoised result. The MSE equation is defined as:

$$\mathrm{RMSE} = \sqrt{\frac{1}{N}\sum_{i=1}^{N}\left(\boldsymbol{x}_i - \widetilde{\boldsymbol{x}}_i\right)^2} \tag{18}$$

where $\boldsymbol{x}_i$ represents the desired output corresponding to the input, $\widetilde{\boldsymbol{x}}_i$ represents the value obtained by the denoising approach, and N represents the number of samples.

The MAE is defined as:

$$\mathrm{MAE} = \frac{1}{N}\sum_{i=1}^{N}\left|\boldsymbol{x}_i - \widetilde{\boldsymbol{x}}_i\right| \tag{19}$$

The lower the MSE and the MAE are, the better the quality of the denoising validity is.

The FCM algorithm is applied to partition original data and denoised data. The original parameter settings of FCM are preferred. In this paper, the clustering performance is evaluated in terms of the partition coefficient (PC) [38–40] and partition entropy (PE) [41,42]. The partition coefficient is calculated as follows:

$$PC = \frac{1}{n}\sum_{i=1}^{c}\sum_{k=1}^{n}\mu_{ik}^2 \tag{20}$$

The higher the PC is, the better the quality of the cluster validity is.

The partition entropy is defined by:

$$PE = -\frac{1}{n}\sum_{i=1}^{c}\sum_{k=1}^{n}\mu_{ik}\log_2\mu_{ik} \tag{21}$$

The lower the PE is, the better the quality of the cluster validity is.

The numerical experiments were performed on a computer with an Intel Core i7-10700 CPU at 3.8 GHz, 32 G RAM.

3.3. Data Denoising Results and Discussion

To examine the denoising performance of our approach with published approaches, we run experiments using three typical schemes, and compare the visual effects and objective evaluations. The wavelet basis function applied in HT denoising and ST denoising is Daubechies 6 (Db6) which comes from the Daubechies wavelet family. Since the wavelet transform has some limitations in the denoising of high-dimensional data, and VMD can only decompose one-dimensional data at one time, we analyze the penetration rate of dataset 1 and the rotation speed of the cutterhead of dataset 2. The penetration rate (V) and rotation speed of the cutterhead (RC) are both important reference attributes for TBM designers and operators in design, analysis, and control of TBM.

For visual comparison, the real values and denoised values with these models are provided in Figures 6 and 7. It can be seen that the HT method and ST method cannot eliminate outliers, which is not conducive to subsequent data analysis. We also tested other wavelet basis functions, and most of them failed to eliminate outliers. The ST method can eliminate noise more precisely than the HT method and avoid local jitter after denoising. Both the VMD method and the proposed method can reduce the influence of outliers on the dataset. Moreover, due to the proposed method considering the correlation between other attributes, it has a reasonable approximation to the outliers. Although the denoising data obtained by VMD looks smoother than other methods, some local trends have changed, as can be seen from two figures. The tunneling data are characterized by weak periodicity and the change in local trend is very unfavorable.

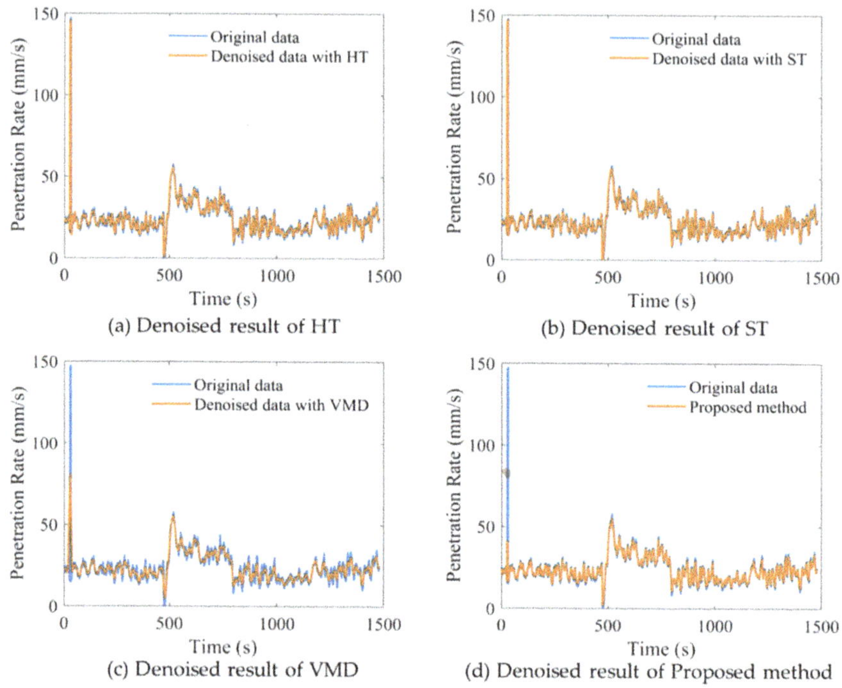

Figure 6. Comparisons between real and denoised data for dataset 1.

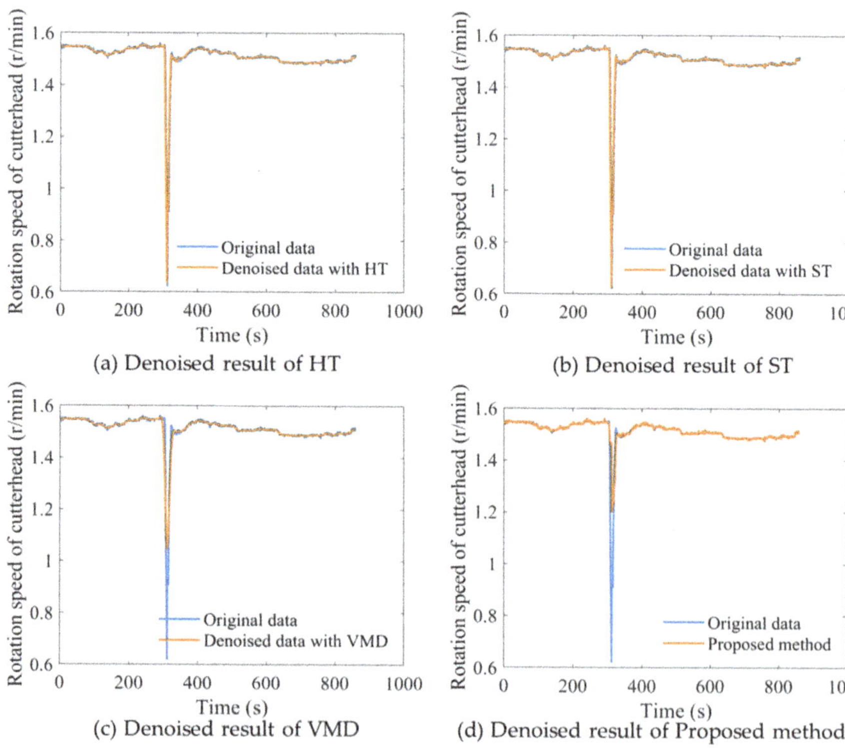

Figure 7. Comparisons between real and denoised data for dataset 2.

The quantitative results of the RMSE and MAE for different denoising results are listed in Table 3. The part of the sequence containing outliers was deleted to evaluate the denoising performance. Of course, this result does not mean that the performance of other denoising methods is not good, but they may not be suitable for data denoising of TBM data.

Table 3. Denoising performance comparison of different methods.

Method	Dataset 1		Dataset 2	
	RMSE	MAE	RMSE	MAE
HT	2.6603	3.7010	0.0129	0.0048
ST	2.4648	3.4039	0.0128	0.0047
VMD	3.2004	4.5471	0.0191	0.0064
Proposed	1.4159	2.0590	0.0178	0.0037

In fact, our method can denoise data for all attributes in the dataset. In order to further demonstrate the denoising performance of the proposed method, the real values and denoised values of other attributes are provided in Figure 8.

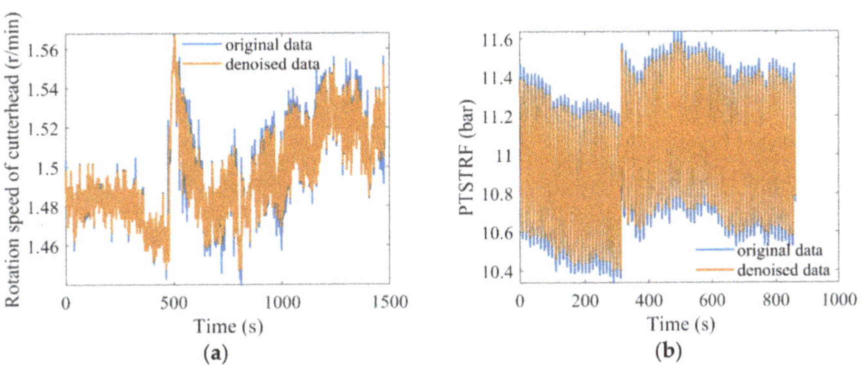

Figure 8. Denoising results of proposed method: (**a**) rotation speed of cutterhead and (**b**) pressure of tail skin system at top right front.

3.4. Data Clustering Results and Discussion

In real world applications, in-suit engineering data contain noise and outliers, and that are disturbances for clustering methods. As a well-known method, FCM is applied to cluster data in unsupervised learning. In this framework, the distance metric is utilized, but the existence of outliers has a bad effect on the division of clusters. The proposed denoising method can provide a reasonable approximation to outliers by considering the relationship between different attributes. In order to verify data denoising that can improve clustering accuracy, clustering experiments are carried out on the original data set and the denoised data set, respectively. Since the clear cluster number of the dataset was unknown, five groups of experiments with cluster numbers of 4, 5, 6, 7, 8 were carried out, respectively.

The clustering comparisons of TBM operational data before and after the proposed approach denoising is shown in Tables 4 and 5. It can be seen that the clustering performance of the denoising data is obviously better than that of the original data, which indicates the effectiveness of data denoising for helping the TBM data clustering task. For dataset 1, the mean PC and PE values of no-denoising data are 0.882 and 0.366. After data denoising using the proposed method, the mean PC value increases to 0.890, and the mean PE value drops to 0.341, respectively. For dataset 2, the mean PC value increased from 0.836 to 0.864, and the mean PE value decreased from 0.456 to 0.384. Moreover, it is observed that all cluster validity indexes achieve the best results when the cluster is 4.

Table 4. Clustering performance comparison of different datasets for Dataset 1.

Cluster	Original Data		Denoised Data	
	PC	PE	PC	PE
4	0.934	0.203	0.936	0.196
5	0.899	0.314	0.905	0.294
6	0.884	0.365	0.890	0.340
7	0.855	0.446	0.869	0.403
8	0.837	0.500	0.849	0.471

Table 5. Clustering performance comparison of different datasets for Dataset 2.

Cluster	Original Data		Denoised Data	
	PC	PE	PC	PE
4	0.852	0.404	0.881	0.331
5	0.843	0.427	0.866	0.370
6	0.847	0.426	0.874	0.357
7	0.845	0.446	0.877	0.354
8	0.793	0.577	0.824	0.509

4. Conclusions

In this paper, a new denoising approach was proposed for TBM in mixed ground conditions. According to the results, it can be concluded that the improved LRMR model is effective in eliminating noise and outliers of TBM, and can improve the accuracy of data clustering. In fact, there are various kinds of noise in TBM operation data during the tunneling process, and some attributes of TBM operational data are highly correlated. It novel to attempt to eliminate the noise between shield data by considering the correlation between parameters. Rank minimization is a common method for exploring correlation, and our research is constructed under the RPCA framework. To overcome the difficulty that the sparsity of noise cannot be fully guaranteed, the weighted nuclear norm is utilized to improve the ability of the model in sparse decomposition. Besides this, we employ a Forbenius norm to improve the stability of model solutions. The proposed LRMR model is a jointly nonconvex problem, we introduce the IALM technique and an alternating minimization strategy to solve it. The existence of outliers and noise will affect the clustering accuracy because most clustering methods use distance to measure the relationship between different clusters. By comparing the clustering performance of original data and denoised data, the clustering accuracy of the denoised data is obviously improved. However, there still exist some issues we have not discussed. In future, it would be an interesting and helpful task to simultaneously achieve noise suppression and accurately cluster in noisy TBM data clustering.

Author Contributions: Conceptualization, Y.W.; methodology, Y.W.; software, Y.W.; validation, Y.W.; formal analysis, Y.W. and Y.P.; investigation, X.S.; resources, X.S. and W.S.; data curation, Y.P.; writing—original draft preparation, Y.W.; writing—review and editing, Y.W. and X.S.; visualization, Y.W. and Y.P.; supervision, X.S. and W.S.; project administration, X.S.; funding acquisition, X.S. All authors have read and agreed to the published version of the manuscript.

Funding: This research was funded by the National Key R&D Program of China (Grant No. 2018YFB1702502) and the National Natural Science Foundation of China (Grant No. 52075068).

Institutional Review Board Statement: Not applicable.

Informed Consent Statement: Not applicable.

Data Availability Statement: Not applicable.

Conflicts of Interest: The authors declare no conflict of interest.

Abbreviations

Abbreviation	Attribute and Unit
RC	Rotation speed of cutterhead (r/min)
FP	Propelling pressure (bar)
FPA	Pressure of A group of hydraulic cylinders (bar)
FPB	Pressure of B group of hydraulic cylinders (bar)
FPC	Pressure of C group of hydraulic cylinders (bar)
FPD	Pressure of D group of hydraulic cylinders (bar)
PEB	Pressure of equipment bridge (bar)
PAS	Pressure of articulation system (bar)
PTSTRF	Pressure of tail skin system at top right front (bar)
PTSRF	Pressure of tail skin system at right front (bar)
PTSBRF	Pressure of tail skin system at bottom right front (bar)
PTSTLF	Pressure of tail skin system at top left front (bar)
PTSRB	Pressure of tail skin system at top right back (bar)
PTSBLF	Pressure of tail skin system at right front (bar)
PTSBRB	Pressure of tail skin system at bottom right back (bar)
PTSTLB	Pressure of tail skin system at top left back (bar)
PSCP	Pressure of screw conveyor pump (bar)
PCTL	Pressure of chamber at top left (bar)
PCT	Pressure of chamber at top (bar)
PCBL	Pressure of chamber at bottom left (bar)
PCTR	Pressure of chamber at top right (bar)
PB	Pressure of bentonite (bar)
GPTL	Grout pressure at top left (bar)
GPTR	Grout pressure at top right (bar)
GPBL	Grout pressure at bottom left (bar)
PTSLF	Pressure of tail skin system at left font (bar)
PTSBLF	Pressure of tail skin system at bottom left front (bar)
PTSLB	Pressure of tail skin system at left back (bar)
PTSBLB	Pressure of tail skin system at bottom left back (bar)
BPSS	Bentonite pressure of shield shell (bar)
PSCF	Pressure of screw conveyor at front (bar)
SA	Displacement of A group of thrust cylinders (mm)
SB	Displacement of B group of thrust cylinders (mm)
SC	Displacement of C group of thrust cylinders (mm)
SD	Displacement of D group of thrust cylinders (mm)
SATR	Displacement of articulated system at top right (mm)
SABR	Displacement of articulated system at bottom right (mm)
SATL	Displacement of articulated system at top left (mm)
SABL	Displacement of articulated system at bottom left (mm)
F	Thrust of cutterhead (kN)
V	Penetration rate (mm/s)

References

1. Pang, Y.; Shi, M.; Zhang, L.; Sun, W.; Song, X. A multivariate time series segmentation algorithm for analyzing the operating statuses of tunnel boring machines. *Knowl.-Based Syst.* **2022**, *242*, 108362. [CrossRef]
2. Song, X.; Shi, M.; Wu, J.; Sun, W. A new fuzzy c-means clustering-based time series segmentation approach and its application on tunnel boring machine analysis. *Mech. Syst. Signal. Process.* **2019**, *133*, 106279. [CrossRef]
3. Shi, M.; Zhang, T.; Zhang, L.; Sun, W.; Song, X. A fuzzy c-means algorithm based on the relationship among attributes of data and its application in tunnel boring machine. *Knowl.-Based Syst.* **2020**, *191*, 105229. [CrossRef]
4. Oropeza, V.; Sacchi, M. Simultaneous seismic data denoising and reconstruction via multichannel singular spectrum analysis. *Geophysics* **2011**, *76*, 25–32. [CrossRef]

5. Sendur, L.; Selesnick, I.W. Bivariate shrinkage functions for wavelet-based denoising exploiting interscale dependency. *IEEE Trans. Signal Process.* **2022**, *50*, 2744–2756. [CrossRef]
6. He, Q.; Wang, X.; Zhou, Q. Vibration Sensor Data Denoising Using a Time-Frequency Manifold for Machinery Fault Diagnosis. *Sensors* **2013**, *14*, 382–402. [CrossRef]
7. Zhang, M.; Gunturk, B.K. Multiresolution bilateral filtering for image denoising. *IEEE Trans. Image Process.* **2008**, *17*, 2324–2333. [CrossRef]
8. Zheng, Y.; Fu, H.; Au, O.K.C.; Tai, C.L. Bilateral normal filtering for mesh denoising. *IEEE Trans. Vis. Comput. Graph.* **2010**, *17*, 1521–1530. [CrossRef]
9. Zhang, S.; Zhang, H.; Liu, H. An improved EMD-based hard thresholding denoising algorithm. *Comput. Meas. Control* **2014**, *22*, 3659–3661.
10. Donoho, D.L. De-noising by soft-thresholding. *IEEE Trans. Inf. Theory* **2002**, *41*, 613–627. [CrossRef]
11. Huang, N.E.; Shen, Z.; Long, S.R.; Wu, M.C.; Snin, H.H.; Zheng, Q.; Yem, N.C.; Tung, C.C.; Liu, H.H. The empirical mode decomposition and the Hubert spectrum for nonlinear and non-stationary time series analysis. *Proc. R. Soc. A Math. Phys. Eng.* **1998**, *454*, 903–995. [CrossRef]
12. Rilling, G.; Flandrin, P.; Goncalves, P. On empirical mode decomposition and its algorithms. *IEEE-EURASIP Workshop Nonlinear Signal Image Process.* **2003**, *3*, 8–11.
13. Chen, T.; Ma, K.K.; Chen, L.H. Tri-state median filter for image denoising. *IEEE Trans. Image Process.* **1999**, *8*, 1834–1838. [CrossRef]
14. Simon, D. Kalman filtering. *Embed. Syst. Program.* **2001**, *14*, 72–79.
15. Kazubek, M. Wavelet domain image denoising by thresholding and Wiener filtering. *IEEE Signal Proc. Lett.* **2003**, *10*, 324–326. [CrossRef]
16. Zhang, Q.; Liu, Z.; Tan, J. Prediction of geological conditions for a tunnel boring machine using big operational data. *Autom. Constr.* **2019**, *100*, 73–83. [CrossRef]
17. Zhou, C.; Xu, H.; Ding, L.; Wei, L.; Zhou, Y. Dynamic prediction for attitude and position in shield tunneling: A deep learning method. *Autom. Constr.* **2019**, *105*, 102840. [CrossRef]
18. Zhang, S.; Li, Y. Seismic exploration desert noise suppression based on complete ensemble empirical mode decomposition with adaptive noise. *J. Appl. Geophys.* **2020**, *180*, 104055. [CrossRef]
19. Yaslan, Y.; Bican, B. Empirical mode decomposition based denoising method with support vector regression for time series prediction: A case study for electricity load forecasting. *Meas. J. Int. Meas. Confed.* **2017**, *103*, 52–61. [CrossRef]
20. Zhang, S.; Liu, K.; Liu, Q.; Zhang, C.; Zhang, Q.; Nan, Y. Tide variation monitoring based improved GNSS-MR by empirical mode decomposition. *Adv. Space Res.* **2019**, *63*, 3333–3345. [CrossRef]
21. Li, Y.; Xu, C.; Yi, L.; Fang, R. A data-driven approach for denoising GNSS position time series. *J. Geod.* **2018**, *92*, 905–922. [CrossRef]
22. Li, H.; Wang, Y.; Yang, Z.; Wang, R.; Li, X.; Tao, D. Discriminative dictionary learning-based multiple component decomposition for detail-preserving noisy image fusion. *IEEE Trans. Instrum. Meas.* **2019**, *69*, 1082–1102. [CrossRef]
23. Zhao, Y.; Yang, J. Hyperspectral image denoising via sparse representation and low-rank constraint. *IEEE Trans. Geosci. Remote Sens.* **2015**, *53*, 296–308. [CrossRef]
24. Gomes, P.R.B.; da Costa, J.P.C.L.; de Almeida, A.L.F.; De Sousa, R.T. Tensor-based multiple denoising via successive spatial smoothing, low-rank approximation and reconstruction for r-d sensor array processing. *Digit Signal Process.* **2019**, *89*, 1–7. [CrossRef]
25. Candes, E.; Li, X.; Ma, Y.; Wright, J. Robust principal component analysis? *J. ACM* **2011**, *58*, 1–37. [CrossRef]
26. Peng, Y.; Ganesh, A.; Wright, J. RASL: Robust alignment by sparse and low-rank decomposition for linearly correlated images. *IEEE Trans. Pattern Anal. Mach. Intell.* **2012**, *34*, 2233–2246. [CrossRef]
27. Bezdek, J.C. Objective function clustering. In *Pattern Recognition with Fuzzy Objective Function Algorithms*; Plenum Press: New York, NY, USA, 1981.
28. Lin, Z.; Chen, M.; Ma, Y. The augmented Lagrange multiplier method for exact recovery of corrupted low-rank matrices. Technical Report UILU-ENG-09-2215, UIUC. *arXiv* **2009**, arXiv:1009.5055.
29. Gu, S.; Zhang, L.; Zuo, W.; Feng, X. Weighted nuclear norm minimization with application to image denoising. In Proceedings of the 2014 IEEE Conference on Computer Vision and Pattern Recognition, Columbus, OH, USA, 23–28 June 2014; pp. 2862–2869.
30. Zou, H.; Zhang, H. On the adaptive elastic-net with a diverging number of parameters. *Ann. Stat.* **2009**, *37*, 1733–1751. [CrossRef]
31. Afonso, M.V.; Bioucas-Dias, J.M.; Figueriredo, M. Fast image recovery using variable splitting and constrained optimization. *IEEE Trans. Image Process.* **2010**, *19*, 2345–2356. [CrossRef] [PubMed]
32. Daubechies, I.; Defrise, M.; De, M. An iterative thresholding algorithm for linear inverse problems with a sparsity constraint. *Commun. Pure Appl. Math.* **2004**, *57*, 1413–1457. [CrossRef]
33. Beck, A.; Teboulle, M. A fast iterative shrinkage-thresholding algorithm for linear inverse problems. *SIAM J. Imaging Sci.* **2009**, *2*, 183–202. [CrossRef]
34. Zhuang, L.; Gao, H.; Lin, Z.; Ma, Y.; Zhang, X.; Yu, N. Non-negative low rank and sparse graph for semi-supervised learning. In Proceedings of the 2012 IEEE Conference on Computer Vision and Pattern Recognition, Providence, RI, USA, 16–21 June 2012; pp. 2328–2335.

35. Zhang, Z.; Ganesh, A.; Liang, X.; Yi, M. TILT: Transform-invariant low-rank textures. *Int. J. Comput. Vision* **2012**, *99*, 1–24. [CrossRef]
36. Cai, J.F.; Candès, E.J.; Shen, Z. A singular value thresh-olding algorithm for matrix completion. *SIAM J. Optimiz.* **2010**, *20*, 1956–1982. [CrossRef]
37. Dragomiretskiy, K.; Zosso, D. Variational Mode Decomposition. *IEEE Trans. Signal. Process.* **2013**, *62*, 531–544. [CrossRef]
38. Hathaway, R.J.; Bezdek, V. Optimization of clustering criteria by reformulation. *IEEE Trans. Fuzzy Syst.* **1995**, *3*, 241–245. [CrossRef]
39. Pratama, M.; Lu, J.; Zhang, G. Evolving type-2 fuzzy classifier. *IEEE Trans. Fuzzy Syst.* **2016**, *24*, 574–589. [CrossRef]
40. Trauwaert, E. On the meaning of dunns partition coefficient for fuzzy clusters. *Fuzzy Sets Syst.* **1988**, *25*, 217–242. [CrossRef]
41. Bezdek, J.C. Cluster validity with fuzzy sets. *J. Cybernet.* **1974**, *3*, 58–73. [CrossRef]
42. Wu, K.L.; Yang, M.S. A cluster validity index for fuzzy clustering. *Pattern Recogn. Lett.* **2005**, *26*, 1275–1291. [CrossRef]

energies

MDPI

Article

Design and Research on High Power Density Motor of Integrated Motor Drive System for Electric Vehicles

Shaopeng Wu *, Jinyang Zhou, Xinmiao Zhang and Jiaqiang Yu

Department of Electrical Engineering, Harbin Institute of Technology, Harbin 150080, China; 20s106146@stu.hit.edu.cn (J.Z.); 21s106171@stu.hit.edu.cn (X.Z.); 1180610621@stu.hit.edu.cn (J.Y.)
* Correspondence: wushaopeng@hit.edu.cn

Abstract: Although many PMSMs are used as the driving source for electric vehicle motor drive systems, there is still a gap compared with the power density index in the DOE roadmap. Considering that the motor occupies a large space in the motor drive system, it is of great significance for the system to increase the motor power density and thus reach the system power density index. This article starts with electrical machine basic design theory and finds the motor power density influence factors. Guided by the theory and considering motor driver influence, this article proposes a high power density motor for electric vehicle integrated motor drive system. The motor for the system is a five-phase interior permanent magnet synchronous motor (IPMSM) with a double-layer rotor structure and fractional slot distributed winding. Compared with Ver1.0 motor, Ver2.0 motor power density improves significantly. In order to prevent damage from excessive temperature, a temperature field solution model is established in this article to compare the cooling effect and pressure loss of the spiral, dial, and axial water jackets. The temperature is checked at motor main operating conditions using an optimal cooling structure. Finally, the power density of the designed Ver2.0 motor reaches 3.12 kW/kg in mass and 15.19 kW/L in volume.

Keywords: electric vehicle; high power density motor design; interior permanent magnet synchronous motor; axial water jacket

Citation: Wu, S.; Zhou, J.; Zhang, X.; Yu, J. Design and Research on High Power Density Motor of Integrated Motor Drive System for Electric Vehicles. *Energies* **2022**, *15*, 3542. https://doi.org/10.3390/en15103542

Academic Editor: Sheldon Williamson

Received: 9 April 2022
Accepted: 10 May 2022
Published: 12 May 2022

1. Introduction

Compared with traditional vehicles, electric vehicles are powered by a motor drive system (MDS), which consists of a motor controller, inverter, and motor. In traditional motor drive systems, in order to achieve high efficiency, high speed, and overload performance of the drive system, more attention is paid to the performance of the drive system than the motor size, resulting in the system's low-power density and large space [1].

The electrical and electronics technical team roadmap issued by the U.S. Department of Energy (DOE) proposes high power density, high efficiency, and low-cost requirements for MDSs [2]. In the roadmap, the motor peak volume power density should be greater than 50 kW/L, while the mass power density should be greater than 5.7 kW/kg. Finally, MDS power density should be greater than 33 kW/L in 2025.

An integrated motor drive system (IMDS) is an integrated system of a motor and its driver [3]. According to the positional relationship between the motor and its driver, it can be divided into radial mount and axial mount [4]. After integration, MDS power density can be improved while the size can be reduced [5,6]. Because of the above advantages, the IMDS is the hotspot of the current research on motor drive systems for electric vehicles.

The IMDS designed by UQM Technologies Inc adopts a radial mount, and a brushless DC motor was used inside the IMDS, with peak power of 75 kW. However, motor mass power density is only 0.83 kW/kg [7]. Tesla's Model S axially mounts the motor and motor driver, but the total volume of the two is still large, resulting in low system power density [8]. Daimler AG has designed a ring motor driver, which is mounted on the stator axial surface to reduce the volume of the IMDS, but the 45 kW IMDS's size of motor is so large that the system power density was dragged down seriously. WEMPEC at the University of Wisconsin applies the five-phase motor and modular motor driver to IMDS, which makes the system more compact. Through the compact structure, IMDS power density and reliability are improved effectively. The only disadvantage of the motor drive system is the power of only 6 kW, which may not be suitable for high-power electric vehicles [9,10]. Reference [4] compared the characteristics of the radial mount and axial mount. From the outer diameter aspect, the axial-mount motor drive system is smaller than the radial-mount, which is more conducive to achieving a high power density index. However, from the heat-dissipation aspect, a radial mount can take better advantage of the motor's good radial heat dissipation and can manage system temperature better.

In the field of motor drive control, many control strategies can improve the performance of motor drive systems and bring changes to IMDS. Reference [11] proposed an advanced DTC strategy for a three-phase PMSM drive system to improve the torque performance, which is beneficial for the high performance of MDSs. Moreover, an advanced SVPWM technique for a motor driver combined with a model predictive control improves the motor THD performance and torque performance [12].

The main driver motor is the power component in the electric vehicle. Reference [13] summarizes the electric vehicle's main driver motor power density from 2004 to the present, including Prius, Camry, and BMW as typical cases. According to comparison, it is found that most motors' mass power density is less than 3 kW/kg, while most motors' volume power density is less than 10 kW/L. According to the table, Prius 2017 and BMW i3 are the most optimal case of motor power density.

Prius 2004 main driver motor adopts V shape rotor structure with a 48/8 slot-pole combination, the maximum motor speed is 6000 rpm, and power density is 3.3 kW/L [14]. On this basis, Prius 2010 main driver motor adopts the ∇ shape rotor structure, maximum speed is increased to 13,500 rpm, and power density is increased to 4.8 kW/L [15]. Prius 2017 main driver motor optimizes the ∇ shape rotor structure to make it more compact, reducing rotor outer diameter to make the motor endure a higher speed, and digging holes in rotor low-flux density region to reduce motor weight. After the redesign, motor power density reaches 1.7 kW/kg in mass and 5.7 kW/L in volume, which realizes the high power density optimization [16]. BMW i3 main driver motor adopts a 72/12 slot/pole combination with a double-bar shape rotor structure; the maximum output torque reaches 250 Nm, the peak power reaches 125 kW, and the motor power density reaches 3 kW/kg in mass and 9.1 kW/L in volume [17]. From the development trend of the above examples, it can be found that increasing the motor speed and optimizing the rotor topology is critical for high power density motor design.

As the power core of the MDS, the motor affects the system's inherent performance. At the same time, the motor occupies the motor drive system's vast majority of weight and volume, which are the main factor affecting the power density of the MDS. However, in the above MDS design example, more attention is paid to the compact design of the motor driver rather than the high power density design of the motor. As a result, the motor drive system cannot achieve high power density even with a compact motor driver. In view of the current research status of electric vehicles' main driver motor, there is still a gap compared with the DOE roadmap. In order to improve MDS power density and narrow the gap, the research and design of high power density motors for electric vehicles are crucial.

In this paper, a 60 kW five-phase IPMSM for electric vehicles is designed to meet the motor's high power density requirements. Through theoretical derivation, the factors affecting the motor power density are found. According to the theoretical guidance and the Ver1.0 motor index [18], Ver2.0 motor electromagnetic design and optimization considering the stress field are carried out. Then, a horizontal comparison was made with the Ver1.0 motor. Finally, to ensure that the Ver2.0 motor does not exceed the allowable temperature, the motor temperature-field solution model is established, and a water jacket structure is selected. Combined with the fluid–solid coupling theory, the temperature field is solved and analyzed under the motor's main operating conditions.

2. Design Method for High Power Density Motor

An electric motor is a strongly coupled nonlinear system, and its performance is affected by factors such as the phase number, slot/pole combination, and the rotor structure. For the design of HPDM, the key factors affecting motor performance should be found from the electrical machine basic design theory and combined with the torque performance and field weakening performance characteristics of electric vehicles to analyze and design HPDM.

2.1. High Power Density Motor Theory

The high power density of the motor means that the motor can still maintain a large output power when the size of the motor is reduced. The following formula gives the relationship between output power and speed:

$$P_w = 2\pi n \cdot T_e \tag{1}$$

In the formula, P_w is motor output power in kW, n is the motor speed in rpm, and T_e is motor output torque in Nm.

The above formula expresses that the motor power is proportional to the torque. If torque density expression can be reached, it can derive power density with the same or similar rotational speeds. While maintaining the output torque, increasing the motor speed is helpful in improving the output power, according to electrical machine basic design theory [19].

$$\frac{D_i^2 l_{ef} n}{P_w} = \frac{6.1}{\alpha_p K_w A B_\delta} \tag{2}$$

where D_i is the diameter of the armature in mm, l_{ef} is the effective axial length in mm, α_p is the pole arc coefficient, K_w is the winding coefficient, B_δ is the air gap flux density peak value in T, A is s linear current density of the stator in A/m, and A can be written as s follow expression:

$$A = \frac{2I_s N m}{\pi D_i} \tag{3}$$

By converting the formula, a further variant of the formula can be obtained. If we take $D_i{}^2 l_{ef}$ on the left side of Equation (2) as the motor volume, we can obtain power density by dividing "motor volume" by the denominator P_w. On the right side of Equation (2), use current density J_c and bare copper area in slot S_{copper} to replace stator linear current density A.

$$\rho_e = \frac{\alpha_p K_w 2mN J_c S_{copper} B_\delta n}{6.1\pi D_i} \tag{4}$$

where ρ_e is motor power density in kW/kg, m is phase number, N is the number of conductors in the slot, and I_s is the stator phase current RMS value in A.

According to Equation (4), increasing the motor speed, phase number, winding coefficient, and pure copper area in the slot is beneficial to increasing the machine power density. The detailed design will be completed in the next section.

2.2. Phase Number of High Power Density Motor

Traditional electric vehicles' main driver motor always uses a three-phase drive motor. Three-phase motors have relatively mature technology and low control difficulty compared with polyphase motors. However, a three-phase motor cannot operate under fault conditions and has poor fault tolerance ability. In order to increase electric vehicles' operation safety, improve the machine power density, and reduce the burden of line voltage, it is essential to select a suitable motor phase number. Reference [20] compared the torque performance of three-phase, dual three-phase, four-phase, five-phase, and six-phase motors under normal operating conditions and fault operating conditions. It is found that five-phase motor and six-phase motor have almost the best torque performance under fault operating conditions. However, the six-phase motor driver has many bridge arms and is complicated to control. Therefore, the five-phase motor is selected to design an HPDM in this paper.

According to Equation (4), increasing motor phase number is beneficial in designing a high power density motor. However, farther than this, increasing phase number can also help to reduce phase voltage. The line voltage peak value is usually limited by DC bus voltage, and too high phase voltage will lead to winding insulation breakdown. Adopting a five-phase motor can reduce phase voltage by a larger angle between adjacent current vectors, which is 144° in a five-phase and 120° in a three-phase motor. If the phase voltage of the three-phase motor is represented by U_p, the corresponding phase voltage in the five-phase motor can be expressed as 0.9 U_p. The phase voltage in a five-phase motor decreases by 10% compared with the phase voltage in a three-phase motor, and it greatly reduces winding insulation breakdown possibility and increases motor fault tolerance capability.

Apart from this, a five-phase motor can reduce harmonic composite magnetic motive force compared with the three-phase motor. After derivation, a three-phase motor has $6\,k \pm 1$ ($k = 0, 1, 2, \ldots$) order harmonic composite magnetic motive force, while a five-phase motor only has $10\,k \pm 1$ ($k = 0, 1, 2, \ldots$) order harmonic composite magnetic motive force (MMF). From harmonic order, the five-phase motor has less order than the three-phase motor, and it will cause lower torque ripple and core loss.

To summarize the above analysis, using five-phase winding can not only benefit the motor power density but also help alleviate the motor phase voltage, reduce torque ripple and motor loss, and increase motor fault-tolerance performance.

2.3. Slot/Pole Combination

Inductance reflects the ability of the current to create a magnetic field, which affects the electromagnetic performance of the motor. The relationship between the electromagnetic torque and d-axis and q-axis inductance is as follows [21]:

$$T_e = \frac{m}{2} p \left[\psi_f i_q + (L_q - L_d) i_d i_q \right] \tag{5}$$

In the equation, p is the machine pole pairs number, Ψ_f is permanent magnet flux linkage, L_d and L_q are the d-axis inductance and q-axis inductance, respectively, in H, where d-axis is defined as the permanent magnet magnetizing direction, and the q-axis is defined as 90° ahead of d-axis; i_d and i_q are the d-axis current component and q-axis current component, respectively, in A.

Compared with the surface-mounted permanent magnet synchronous machine (SPMSM), IPMSM has reluctance torque because of the difference in the L_d and L_q according to Equation (5). When the motor speed increases and reaches the voltage limit, the increase in i_d causes the angle between the phase current vector and q-axis to change and become larger to reduce the motor magnetic field. The motor enters the flux-weakening state to allow motor speed raising under the voltage limit. Figure 1 shows the change curve of the motor torque component with the angle between the phase current vector and the q-axis.

As the angle increases, the proportion of reluctance torque gradually increases, which proves that the flux-weakening process is dominated by reluctance torque. Because of the reluctance torque, IPMSM can achieve higher speed than SPMSM, and on account of the excellent high-speed characteristics, IPMSM is more suitable for an electric vehicle.

L_d and L_q affect the torque capability and flux-weakening capability of the motor. According to the derivation, the expression of L_d and L_q is obtained:

$$\begin{cases} L_d = \frac{m(K_w N)^2 Z \sin(\theta)}{\pi^2 p^2} \left(\lambda_0 + \frac{1}{2}\lambda_2 \right) l_{ef} \\ L_q = \frac{m(K_w N)^2 Z \cos(\theta)}{\pi^2 p^2} \left(\lambda_0 - \frac{1}{2}\lambda_2 \right) l_{ef} \end{cases} \tag{6}$$

where Z is the number of stator slots, and λ_0 and λ_2 are the fundamental and secondary components of the machine's permeance, respectively.

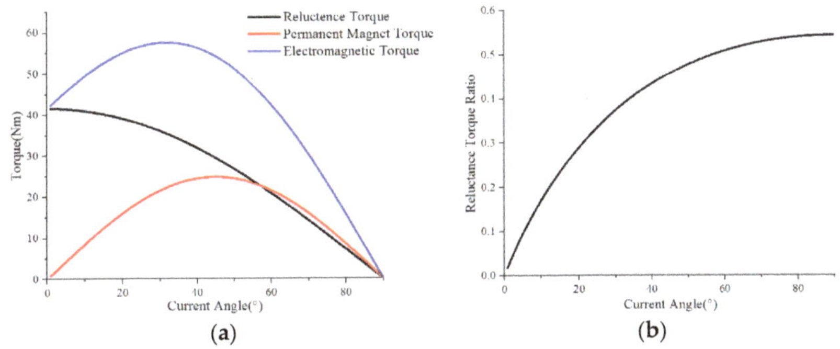

Figure 1. Torque varies with the current angle: (**a**) torque component; (**b**) reluctance torque ratio.

L_d and L_q are influenced by motor slot/pole combinations. In order to improve the torque performance and flux-weakening performance of the five-phase IPMSM, the slot/pole combination needs to be selected. According to Equation (4), a high winding coefficient can improve the motor power density. The highest winding coefficient under different slot/pole combinations are shown in Table 1, which includes fractional slot concentrated winding (FSCW), fractional slot distributed winding (FSDW), and integer slot distributed winding (ISWD). For comparison purposes, refer to the designed motor index before (Ver1.0) three main operating conditions specifications. Three main operating conditions for electric vehicles are shown in Table 2.

Table 1. Winding coefficient under different slot/pole combination.

Slot/Pole	6	8	12
10	0.951 (y = 2)	0.951 (y = 1)	0.951 (y = 1)
20	0.975 (y = 3)	0.951 (y = 2)	0.951 (y = 2)
30	1.000 (y = 5)	0.980 (y = 4)	0.951 (y = 3)
40	0.982 (y = 7)	1.000 (y = 1)	0.975 (y = 3)

y is winding pitch, y = 1: concentrated winding; y > 1: distributed winding.

Table 2. Motor three main operating conditions.

Parameter	Base Speed	High Speed	Overload
Power	60 kW	60 kW	110 kW
Speed	6000 rpm	15,000 rpm	5000 rpm
Torque	>95.5 Nm	>38.2 Nm	>200 Nm
Operating time	60 min	3 min	3 min

Seven different slot/pole combinations, including FSCW, FSDW, and ISDW, are selected from Table 1 and shown in Figure 2.

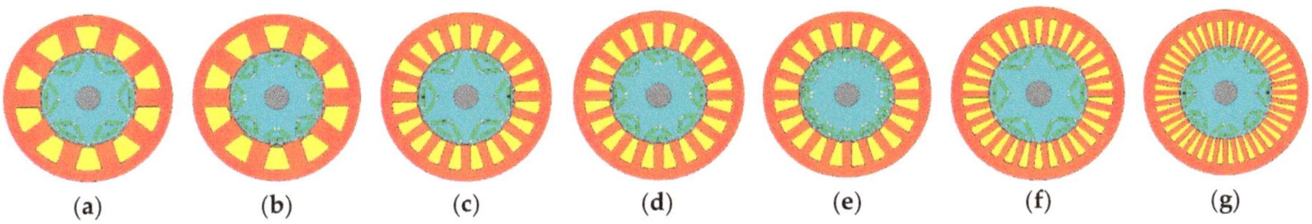

Figure 2. Selected Seven Different Slot/Pole Combinations: (**a**) 10s6p. (**b**) 10s8p. (**c**) 20s6p. (**d**) 20s8p. (**e**) 20s12p. (**f**) 30s6p. (**g**) 40s8p.

To ensure the stator outer diameter, each phase conductor number and PM mass is the same, and the torque and inductance performance of selected seven slot/pole combinations under three operating conditions are solved; the results are summarized in Figures 3 and 4.

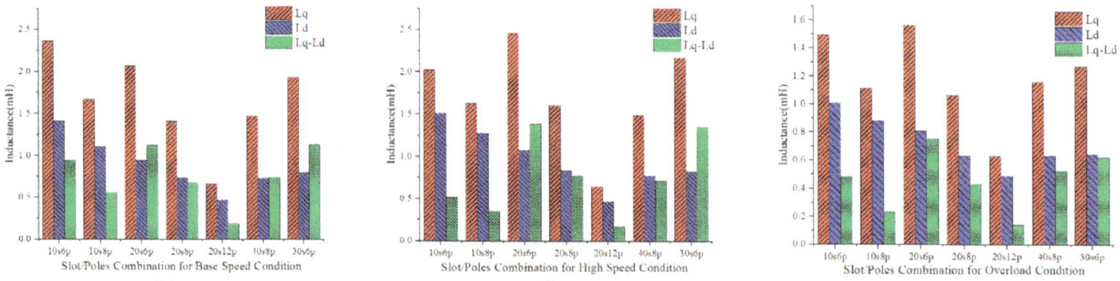

Figure 3. Inductance characteristics in different slot/pole combinations: (**a**) base speed condition, (**b**) high-speed condition, and (**c**) overload speed condition.

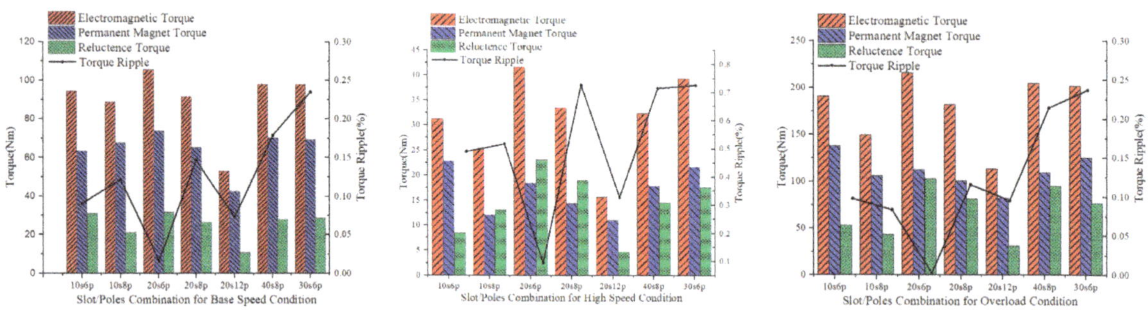

Figure 4. Torque characteristics in different slot/pole combinations: (**a**) base speed condition, (**b**) high-speed condition, and (**c**) overload speed condition.

Through the comparative analysis, it is found that when the pole pair number is the same, the difference in value between the L_d and L_q becomes larger as the slot number increases, and the torque performance under high-speed conditions is better at the same time; when the slot number is the same, the torque performance under high-speed condition becomes worse as the pole pair number increases. Compared with the integer slot winding, the fractional slot has a smaller torque ripple. It can be concluded that a motor that adopts a fractional slot can obtain better torque performance with a low torque ripple. Compared with the concentrated winding, the distributed winding has a larger difference in inductance, causing a larger reluctance torque. Although most motors choose FSCW or ISDW in basic design, they often benefit from the traditional three-phase motor but not the five-phase motor design. In order to make the five-phase motor with more choices of high winding coefficient slot/pole combination, FSDW can be seen as one suitable choice. Five-phase motor with an FSDW owns less end-winding length and torque ripple than the ISDW and owns a higher winding coefficient compared with a five-phase motor with FSCW. Considering both power density improvement and torque performance, 20s6p slot/pole combination FSDW is selected.

2.4. *IPMSM Permanent Magnet Layer Number*

The permanent magnet layer of the rotor affects the air gap-flux density and electromagnetic characteristics. It is significant to find how the PM layer number influences the motor electromagnetic characteristics and how to choose the PM layer number to obtain high power density. Figure 5 shows three rotor structures with bar shape, V shape, and ∇ shape, respectively, wherein the bar shape and V shape are single-layer rotor structures, and the ∇ shape is a double-layer rotor structure. Figure 6 shows the comparison of the no-load air gap flux density and its harmonic content under three different rotor structures.

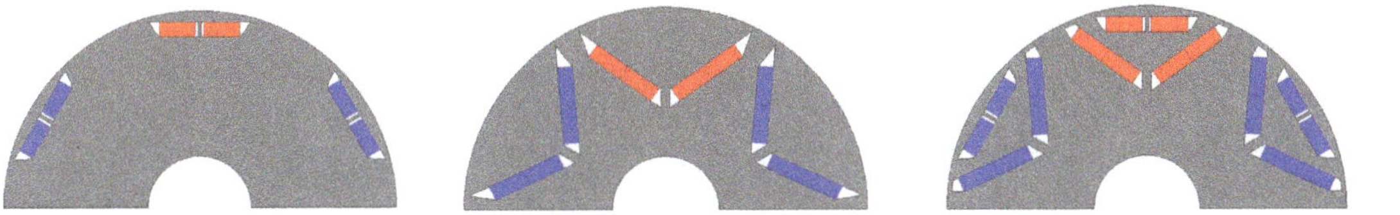

(a) (b) (c)

Figure 5. IPMSM Rotor Structures: (**a**) Bar Shape. (**b**) V Shape. (**c**) ∇ Shape.

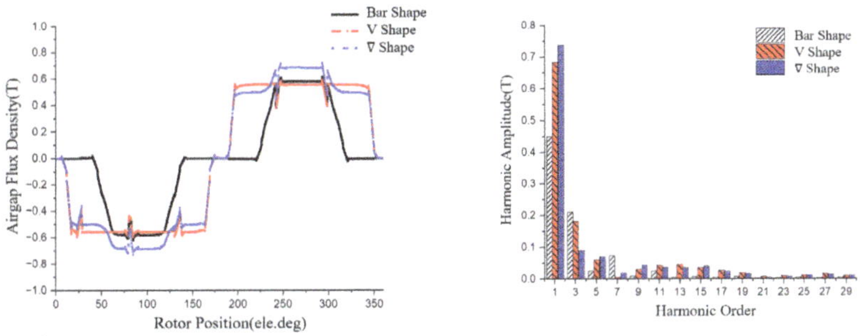

(a) (b)

Figure 6. IPMSM rotor structures air gap flux density performance: (**a**) air gap flux density waveform, (**b**) harmonic content.

Compared with the single-layer rotor structure, the double-layer rotor structure's air gap flux density peak value is larger, which, according to Equation (4), can improve motor power density. From the view of air gap flux density harmonic content, the fundamental content is higher than single-layer, while the other harmonics content is less than single-layer. If we define harmonic distortion rate as the ratio of all harmonics RMS, which excludes fundamental order, to the fundamental RMS, then the harmonic distortion rate of zero corresponds to the sinusoidal degree equal to 1. It can be clearly seen that the double-layer rotor structure has a better sinusoidal degree than the single-layer rotor structure.

By observing the double-layer ∇ shape rotor structure cloud map of the no-load magnetic field distribution in Figure 7, it is found that the nether layer PM pushes the upper layer magnetic field lines upward, resulting in lower magnetic flux leakage at the end of the upper layer PM. Increasing the air gap flux density peak value in the middle part of the double-layer rotor structure makes the no-load air gap flux density approach the sine waves, resulting in an improved sinusoidal degree of the air gap flux density. According to Equation (4), selecting the double-layer rotor structure contributes to motor high power density design.

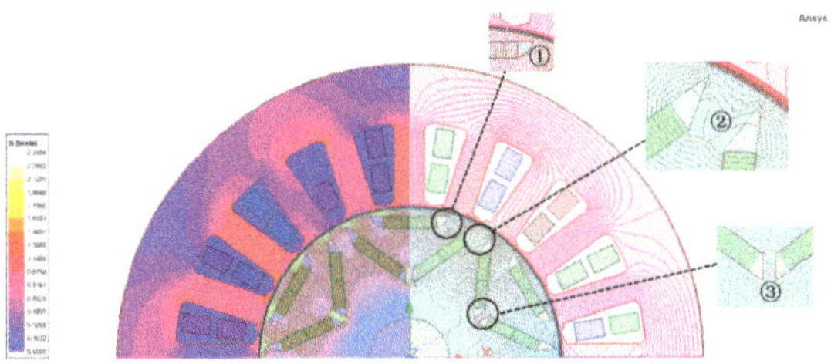

Figure 7. No-load magnetic field distribution cloud map in ∇ shape rotor structure.

3. EM Analysis of Five-Phase CSDW High Power Density Motor

In the integrated motor drive system designed previously, called Ver1.0 type, it was found that an IMDS has a large volume and the motor inside has a low motor power density, which is still a distance from the DOE roadmap. In order to reflect high power density design method rationality and correctly, all indicators remain the same as in Ver1.0 type [18]. According to these indicators, the electromagnetic analysis and optimization of the Ver2.0 motor were carried out and compared with the Ver1.0 motor. The indicators for both Ver1.0 and Ver 2.0 types are shown in Table 3, which include motor indicators and part of motor driver indicators.

Table 3. Main indicators of HPDM.

Parameter	Value
Rate power	60 kW
Peak power	110 kW
Base speed	6000 rpm
Maximum speed	15,000 rpm
Peak Torque	>200 Nm
Torque Ripple	<5%
Current density	<12 A/mm^2
DC bus voltage	650 V
Peak current	<200 A
Stator outer diameter	≤240 mm
Volume power density	>10 kW/L
Mass power density	>3 kW/kg

What needs illustration is that the peak current is affected by the selected switching device. Conventional IGBT devices cannot stand the extremely adverse working environment, but with the development of a wide bandgap semiconductor technology, SiC devices can adapt to high-power, high-temperature, and high-frequency applications. Considering the high power of IMD, the SiC switching device was selected, and the 1200 V and 300 A SiC half-bridge module of CREE series CAS300M12BM2 was selected as the switching device in the driver. The voltage and current indicators in Table 3 have taken the margins into account according to the chosen switching device.

The IMDS topology is shown in Figure 8a, which is the same structure as the Ver1.0 type. Through the analysis in Section 2, it is determined that the motor structure is a five-phase 20s6p double-layer rotor structure IPMSM, which can improve motor power density and ensure the torque and flux-weakening performance for an electric vehicle. The motor topology is shown in Figure 8b.

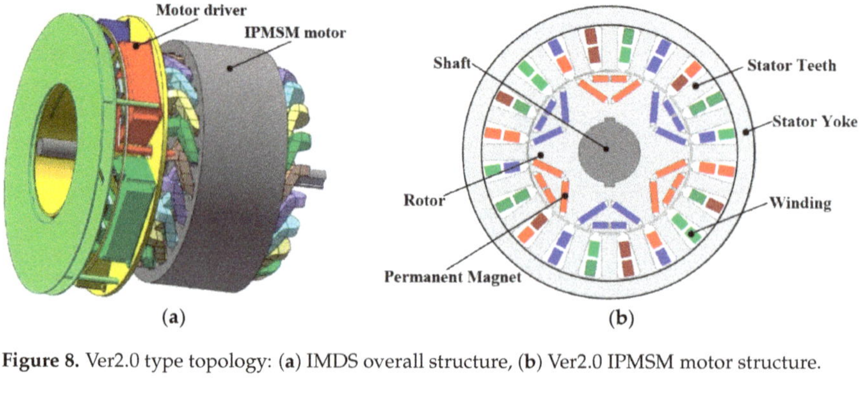

Figure 8. Ver2.0 type topology: (**a**) IMDS overall structure, (**b**) Ver2.0 IPMSM motor structure.

3.1. Magnetic Bridge

The ∇-type rotor inner structure's connection to the outer relies only on two tangential magnetic bridges. The rotor stress is concentrated under high-speed conditions to cause rotor damage. In order to reduce the rotor stress at high speeds, radial magnetic bridges should be added between the permanent magnets in the same layer. There are three magnetic bridges that the motor can freely set, which are respectively recorded as magnetic bridges 1, 2, and 3, as shown in Figure 9. The effect of different magnetic bridge combinations on motor stress and electromagnetic performance will be discussed below.

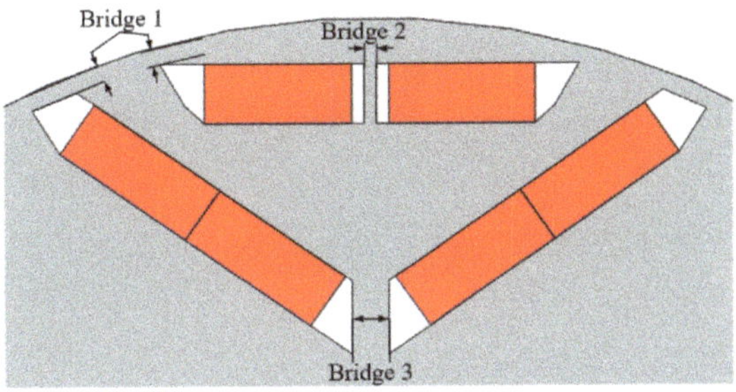

Figure 9. Rotor magnetic bridge schematic diagram.

Figure 10 shows the rotor stress results for different magnetic bridge combinations under Maximum speed conditions. It is found that when there is only bridge 1, or when adopting bridge 1 and bridge 2 combination, the rotor stress exceeds the yield strength of the 10JNEX900 silicon steel sheet (570 MPa) [22]. When adopting the bridge 1 and bridge 3 combination, the rotor's maximum stress is significantly reduced compared with the above two combinations. It can be concluded that the radial magnetic bridge 3 has an obvious improvement effect on the rotor intensity under maximum speed conditions.

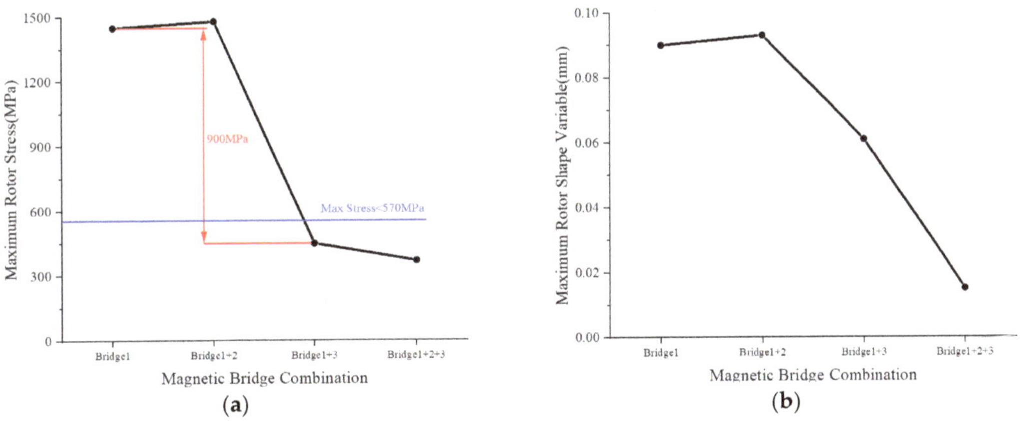

Figure 10. Rotor stress with different magnetic bridge combinations at maximum speed conditions: (**a**) rotor maximum von Mises stress, (**b**) rotor maximum deformation quantity.

Although widening a magnetic bridge helps to reduce rotor stress, an excessively wide magnetic bridge will lead to serious leakage flux of the permanent magnet ending, which will influence torque performance and motor power density. Figures 11 and 12 show the motor torque performance with different magnetic bridge 1 width and different magnetic bridge 3 width, respectively.

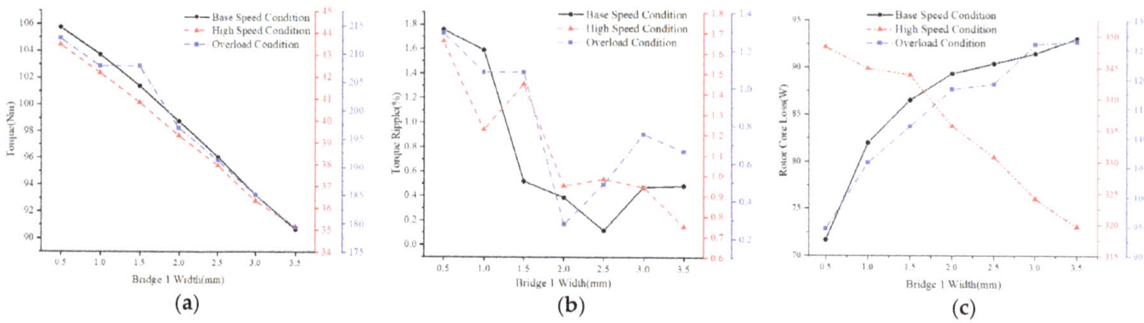

Figure 11. Motor electromagnetic characteristics with different bridge 1 width: (**a**) output torque, (**b**) torque ripple; (**c**) rotor core loss.

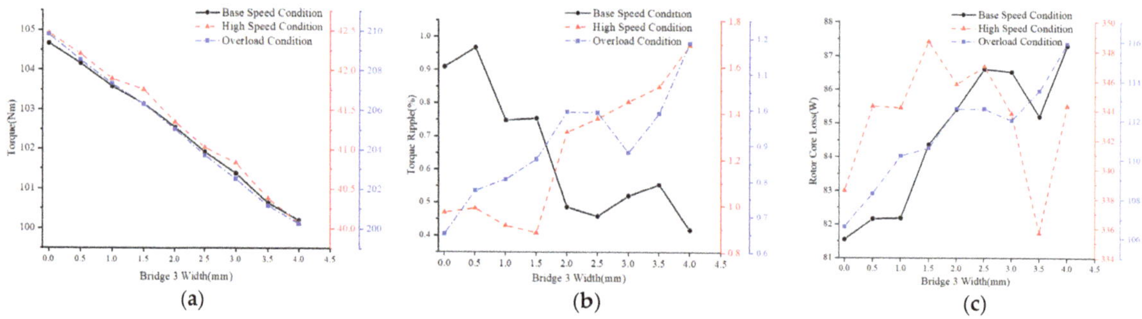

Figure 12. Motor electromagnetic characteristics with different bridge 3 width: (**a**) output torque, (**b**) torque ripple, (**c**) rotor core loss.

Magnetic bridge 1 is a tangential magnetic bridge, which mainly affects motor electromagnetic characteristics. A narrow magnetic bridge 1 will cause serious saturation at the magnetic bridge, leading to a rotor flux leakage increase. More permanent magnet flux leakage leads to permanent magnet utilization reduction, affecting motor torque performance and a larger rotor core loss under saturated conditions. Compared with the magnetic bridge 1, magnetic bridge 3 is a radial magnetic bridge. Although an excessively wide magnetic bridge can effectively relieve rotor stress, it also causes serious flux leakage onto the lower layer permanent magnet's terminal, leading to a motor output torque decrease. Torque ripple and rotor core loss under high-speed conditions and overload conditions will increase when the magnetic bridge is too wide. However, its relative, the tangential magnetic bridge has less influence on output torque.

In general, the radial magnetic bridge mainly affects the mechanical stress of the rotor, and the tangential magnetic bridge mainly affects the electromagnetic characteristics.

3.2. Stator Structure Optimization

The stator structure parameters are shown in Figure 13, which are mainly divided into the slot size and the stator yoke size. In terms of improving the power density, according to Equation (4), increasing slot pure copper area is beneficial to motor high power density. However, too large a slot area will lead to teeth narrowing and serious saturation. Therefore, the torque performance and motor loss should be comprehensively considered in optimizing stator structure.

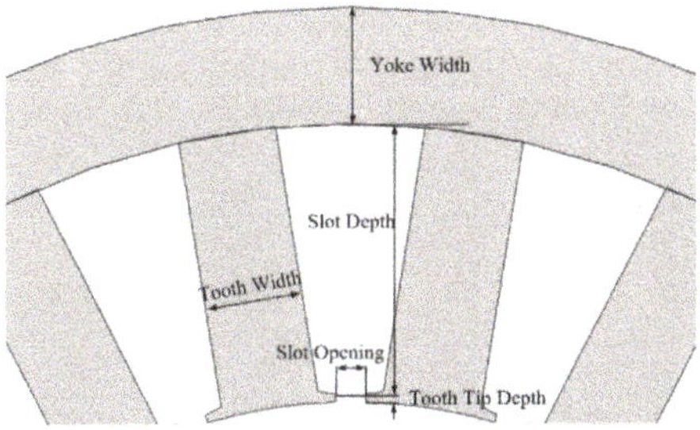

Figure 13. Stator structure schematic diagram.

Saturation can be defined as the following: stator core can be called an unsaturated state when the magnetic flux density is in the linear region of the B–H curve, and a saturated state is when the magnetic flux density is higher than the maximum magnetic density in the linear region. In the Ver2.0 motor, to alleviate excessive rotor loss at high speed and maintain motor torque performance, JFE super iron 10JNEX900, which has a high saturation magnetic flux density, low iron loss, and high permeability, is selected. The saturated magnetic flux intensity corresponding to 10JNEX900 is 1.8 T.

In order to find the optimal stator structure for the 60 kW electric vehicle high power density motor Ver2.0, stator tooth width was changed under the condition that the fixed stator outer diameter, each phase parallel branch current, and the slot copper space factor are the same. Under different stator tooth widths, the electromagnetic characteristic change curves at three operating conditions are shown in Figure 14.

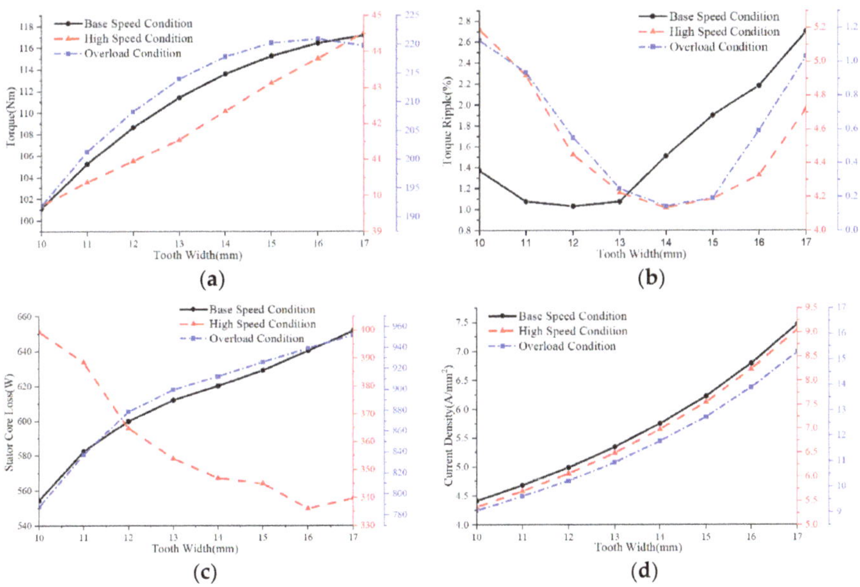

Figure 14. Electromagnetic characteristics vary with different tooth width: (**a**) output torque, (**b**) torque ripple, (**c**) stator core loss, (**d**) current density.

When the tooth width increases, the output torque becomes larger, and the torque ripple increases first and then decreases. Motor torque ripple at a high-speed condition is much larger than in the other two operating conditions. When the tooth width becomes wider, the tooth saturation can be effectively alleviated, and the output torque increases correspondingly. However, the magnetic circuit of the stator runs through the stator tooth and stator yoke. When the tooth width exceeds the yoke width, the yoke saturation degree deepens, and the torque ripple increases, so the torque ripple shows a trend of decreasing first and then increasing. Figure 15 shows the stator teeth and yoke saturation in stator teeth width of 11 mm and 20 mm, respectively.

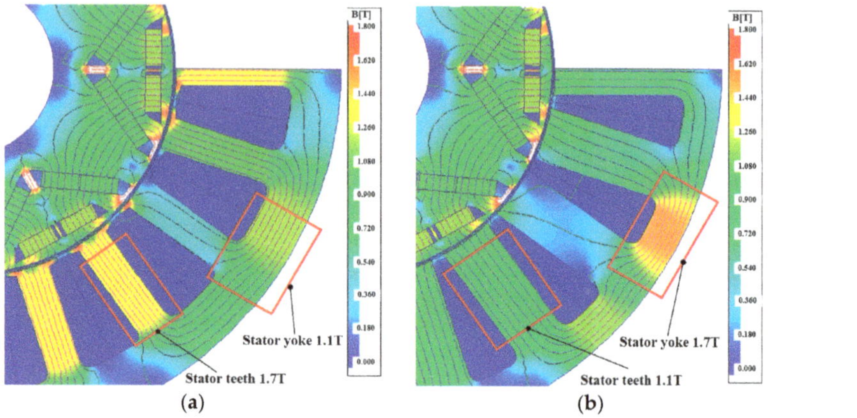

Figure 15. Stator flux density distribution in different stator tooth widths: (**a**) tooth width of 11 mm, and (**b**) tooth width of 20 mm.

According to Figure 14c, it can be seen that, at the base speed condition and the overload condition, the stator core loss increases with the tooth width widening, while at a high-speed condition, the stator core loss decreases with the tooth width widening. To find the reason for the two types of stator loss, the stator tooth loss and stator yoke loss should be calculated separately according to the following expression:

$$p_{\text{yoke}} = p_{10/50} B_{\text{yoke}}^2 \left(\frac{f}{50} \right)^{1.3} G_{\text{yoke}} \tag{7}$$

$$p_{\text{tooth}} = p_{10/50} B_{\text{tooth}}^2 \left(\frac{f}{50} \right)^{1.3} G_{\text{tooth}} \tag{8}$$

In the formula, p_{yoke} and p_{tooth} are the core loss of the yoke and teeth, respectively, B_{yoke} and B_{tooth} are the average flux density of the yoke and tooth, respectively, in T, and G_{yoke} and G_{tooth} are the core weight of the yoke and the tooth, respectively, in kg.

The stator core loss at base speed condition generally shows an upward trend, in which the stator yoke core loss gradually increases with the tooth width, and the stator tooth core loss gradually decreases with the tooth width. When the tooth width becomes wider, the stator tooth saturation degree decreases, so the loss of the stator tooth decreases, as can be seen in Figure 16a.

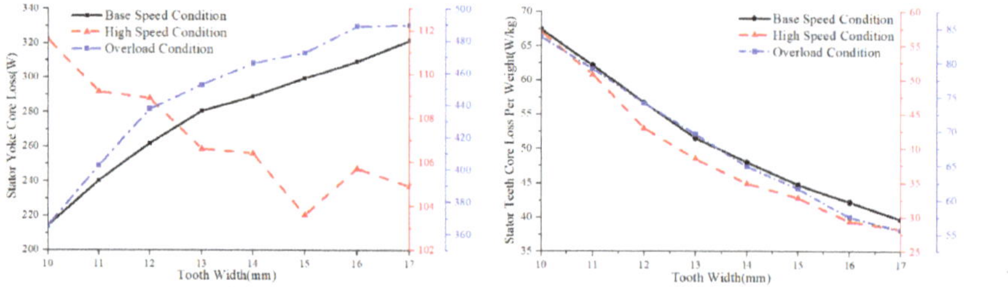

Figure 16. Stator core loss under different tooth widths: (**a**) stator yoke core loss, (**b**) stator teeth core loss per weight.

While stator tooth width increases, stator tooth allows more magnetic flux lines to travel, but it also requires that the stator yoke travels more magnetic flux lines. As described earlier, stator yoke width should be kept constant, so that magnetic flux lines may be borne under a too wide a stator tooth, as shown in Figure 15b, which will cause the saturation of the stator yoke and increase the stator yoke core loss. At the overload condition, the winding current density is about twice the density measured at the base speed condition. Even if the tooth width is widened, the loss of the yoke is still saturated, and the stator total core loss shows an upward trend. The motor in the high-speed condition is in the flux-weakening state, the motor magnetic field is further weakened, and both tooth flux density and yoke flux density drop significantly, so the total stator core loss shows a downward trend.

On the whole, considering the motor torque performance and motor loss, 13 mm tooth width is selected for the Ver2.0 motor.

3.3. Motor Electromagnetic Characteristic under Different Modulation Algorithms

Compared with a three-phase motor, a five-phase motor has higher reliability and a degree of control freedom. Reliability is mainly reflected in that when one phase or two phases have faults such as winding short circuit fault or winding open fault, the remaining phases can maintain the original operating condition. As for the degree of control, in a five-phase motor, it is reflected in the voltage vector space. In a three-phase motor, there is only one length between any two phases, while in the five-phase motor, there are two lengths. Therefore, expect a zero-coordinate system, the five-phase motor has the fundamental coordinate system and third harmonic coordinate system, which is called the fundamental α_1–β_1 coordinate system, and third harmonic α_3–β_3 coordinate system in this article.

Figure 17 shows the conventional five-phase half-bridge inverter topological structure. It can be seen that the five-phase inverter has 2^5 switching states, corresponding to 32 space voltage vectors. These 32 voltage vectors can be divided into 10 long vectors, 10 medium length vectors, 10 small vectors, and 2 zero vectors according to the vector length in the fundamental α_1–β_1 coordinate system. Moreover, these vectors in the fundamental α_1–β_1 coordinate system vector distribution and the third harmonic α_3–β_3 coordinate system vector distribution are shown in Figure 18, respectively.

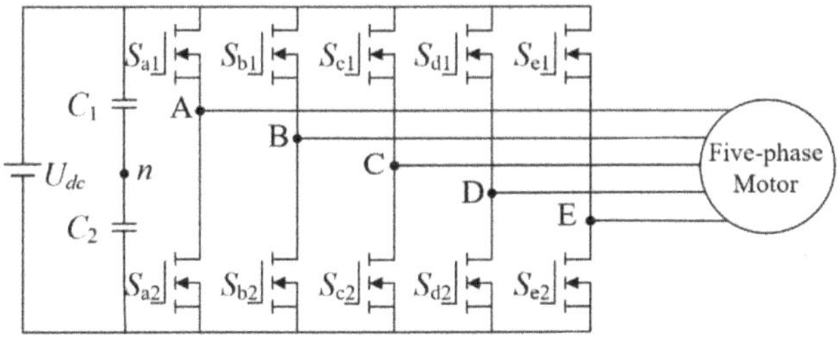

Figure 17. The topology of the five-phase half-bridge inverter.

Similar to the three-phase motor SVPWM modulation algorithm for using two adjacent voltage vectors to obtain the motor working voltage vector, the five-phase motor can also synthesize motor rotating vector using two adjacent long vectors in the fundamental coordinate system, which are usually called near two vectors SVPWM (NTV-SVPWM) [23]. However, when adopting the NTV-SVPWM modulation algorithm, the third harmonic will be generated in the third harmonic coordinate system, resulting in excessive voltage harmonics of the motor. At the same time, using NTV-SVPWM does not take advantage of the five-phase motor's high freedom control, causing bad motor electromagnetic characteristics such as high torque ripple, high line voltage, and high loss.

To solve the above problems, near four vectors SVPWM (NFV-SVPWM) modulation algorithm is proposed. NFV-SVPWM modulation algorithm uses two adjacent long vectors and two medium length vectors to synthesize the working voltage vector. NFV-SVPWM can adjust medium length vector length and long vector length by action time. In a fundamental coordinate system, the four vectors combined into a working voltage vector ensure the motor's normal operation as the NTV-SVPWM. However, in the third-harmonic coordinate system, the synthetic vectors multiplied by the vector length and action time can be arranged into a parallelogram so that the third harmonic can be eliminated in NFV-SVPWM.

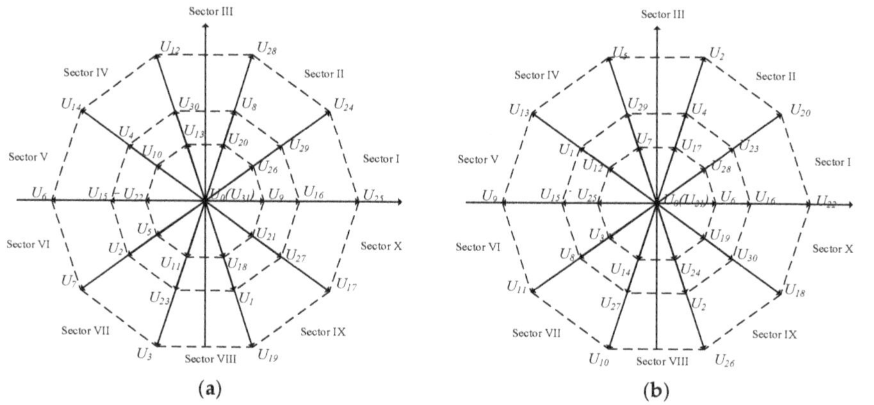

Figure 18. Space voltage vector distribution in the fundamental and third harmonic coordinate system. (**a**) Fundamental α_1–β_1 coordinate system vector distribution. (**b**) third harmonic α_3–β_3 coordinate system vector distribution.

Figure 19 shows the maximum torque per ampere (MTPA) control of the five-phase PMSM block diagram. The NTV-SVPWM and NFV-SVPWM modulation algorithms are respectively applied, and the corresponding current excitation results obtained are shown in Figure 20 for the power device operating at a 15 kHz switching frequency.

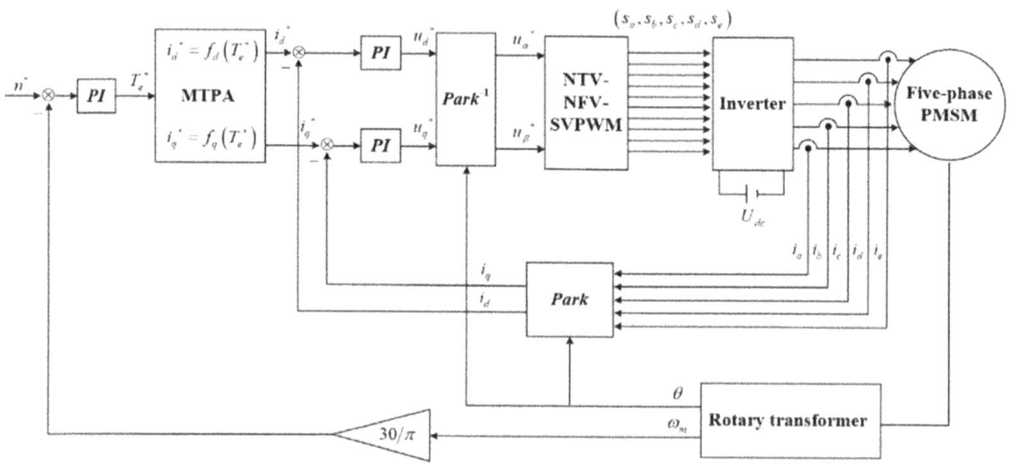

Figure 19. MPTA control of the five-phase PMSM block diagram.

Different modulation algorithms can influence the motor's electromagnetic characteristics. Figure 21 shows motor torque at the base speed condition using Figure 20 current excitation. In addition, motor loss in different modulation algorithms is summarized in Table 4.

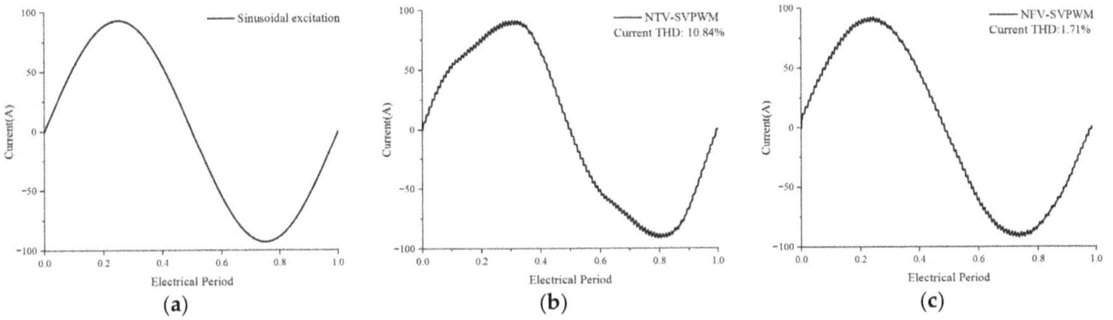

Figure 20. Current excitation under different modulation algorithm: (**a**) ideal sinusoidal, (**b**) NTV-SVPWM, and (**c**) NFV-SVPWM.

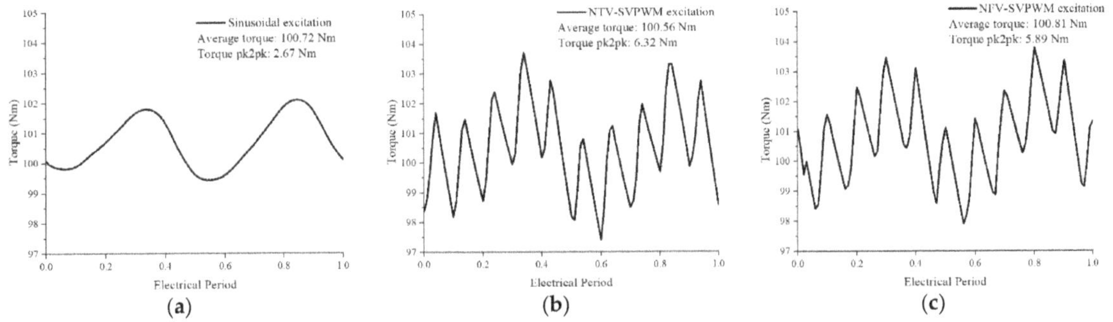

Figure 21. Motor torque under different modulation algorithms (**a**) ideal sinusoidal, (**b**) NTV-SVPWM, and (**c**) NFV-SVPWM.

Table 4. Motor loss under different modulation algorithms.

Loss	Ideal Sinusoid	NTV-SVPWM	NFV-SVPWM
Stator core loss	492.36 W	1320.9 W	1230.4 W
Rotor core loss	29.24 W	66.8 W	62 W
Magnet loss	12.58 W	40.25 W	33.67 W
Copper loss	332.45 W	321.1 W	317.84 W

It can be seen from Figure 21 that, compared with the NTV-SVPWM modulation algorithm, the NFV-SVPWM modulation algorithm has the smallest torque ripple and an increase in core loss, and magnet loss caused by the NTV-SVPWM modulation algorithm's high current harmonics. Large torque ripple influences electric vehicle's comfort level, while high loss affects the IMDS temperature distribution, increasing the difficulty in IMDS thermal management. Based on electromagnetic characteristics, NFV-SVPWM is selected as the driver modulation algorithm.

When the inverter switching frequency is 15 kHz, the core loss in the stator is almost twice that of the motor under ideal sinusoidal excitation, which shows that the inverter can influence the motor's electromagnetic characteristic. Compared with ideal sinusoidal excitation, the current obtained by the modulation algorithm contains harmonics, which are generated by the comparison between modulating wave and carrier wave. When the modulation wave frequency is low, the current harmonic content is large; when the carrier frequency is high, the current harmonic content is small. The switching frequency affects the modulation frequency; a lower switching frequency leads to serious current distortion because of which the motor cannot perform with the electromagnetic characteristics of the ideal design. In order to reduce the driver-side influence on the motor, the switching frequency should be appropriately selected, and the motor's electromagnetic characteristics under different switching frequencies should be studied so as to achieve great motor electromagnetic characteristics. Table 5 shows the motor loss distribution of each part when the switching frequency changes from 3 kHz to 45 kHz.

Table 5. Motor loss under different switching frequencies using NFV-SVPWM.

Switching Frequency	Current THD	Stator Core Loss	Rotor Core Loss	Magnet Loss
Ideal sinusoid	0	492.4 W	29.2 W	12.6 W
3 kHz	8.06%	754.5 W	80.5 W	201.9 W
9 kHz	2.88%	675.5 W	68.5 W	75.9 W
15 kHz	1.71%	615.4 W	62 W	33.7 W
30 kHz	1.54%	568.2 W	46.8 W	23.5 W
45 kHz	1.37%	522.9 W	31.6 W	13.2 W

As can be seen from Table 5, with the increase in switching frequency, the distortion rate of the motor current waveform gradually decreases, and the motor loss component gradually approaches that of the sinusoidal excitation. It can be concluded that the increase in switching frequency is beneficial in reducing motor loss, especially the decrease in eddy current loss of permanent magnet and stator core loss.

However, when the switching frequency is too high, the power device switching loss will also increase. Although the motor driver adopts SiC devices, which can effectively reduce the power device switching loss, when the switching frequency is too high, power device switching loss will still be large. Therefore, the switching frequency of the power tube is selected as 30 kHz.

3.4. Ver2.0 Electromagnetic Characteristic

Through the selection of motor structure, slot/pole combination, IPMSM rotor structure, rotor magnetic bridge position, and optimization analysis of the stator and rotor, the final IPMSM topology structure is shown in Figure 22, where Figure 22a is Ver1.0 motor topology, and Figure 22b is Ver2.0 motor topology.

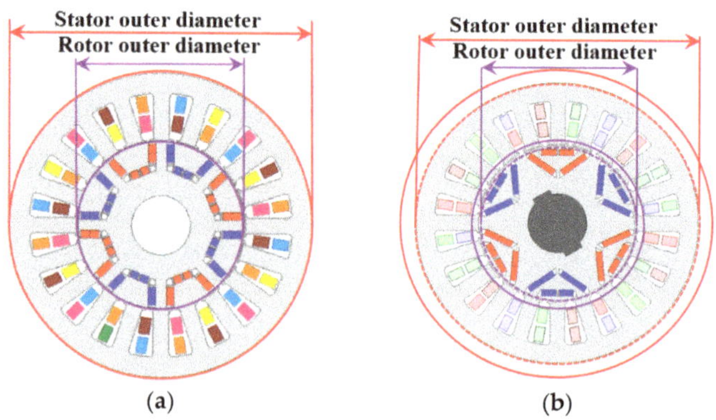

Figure 22. Motor in IMDS topology: (**a**) Ver1.0 motor topology, (**b**) Ver2.0 motor topology.

The Red and purple lines in Figure 22a,b represent the stator outer diameter and the rotor outer diameter, respectively, and the solid-line curve and dash-line curve represent Ver1.0 motor size and Ver2.0 motor size, respectively. It can be concluded that, through the design method of the high power density motor in this article, the stator and rotor diameters are significantly reduced in the Ver2.0 motor compared with the Ver1.0 motor.

The electric vehicle motor designed in this article has three operating conditions which are called the base speed operating condition, high-speed condition, and an overload condition. Base speed is the motor's long time operation condition, which requires high efficiency and reliability. Overload is an instantaneous condition, which requires the motor to have high torque capacity, but high torque usually means large current density, large copper loss, and serious distorted iron loss, leading to motor electromagnetic characteristic reduction. High speed is also an instantaneous condition, which requires the motor to have a certain torque capacity at high speeds. In high-speed conditions, the current frequency is so high that the motor core loss increases heavily, affecting the motor's life and safety. Considering that the motor loses too much under high-speed and overload conditions, which brings high motor temperature problems, besides adopting segmented PM to reduce magnet eddy current loss, motor material needs to be selected emphatically. The materials selected for Ver2.0 are expressed in Table 6. The Ver2.0 motor selected 10JNEX900 as the stator and rotor core because of the material's high saturation magnetic flux density, low iron loss, and high permeability advantages, as stated above. In addition, the N38UH was selected for Ver2.0 as the magnet material on account of its high magnetic energy product and excellent demagnetization curve linearity at a high temperature of 150 °C.

Table 6. Ver2.0 motor's space dimensions.

Component	Material	Mass	Volume
Stator	10JNEX900	14.01 kg	1.87 L
Rotor	10JNEX900	6.17 kg	0.83 L
Winding	Copper	10.14 kg	1.14 L
Magnet	N38UH	1.49 kg	0.19 L
Shaft	20CrMnTi	2.88 kg	0.39 L
Overall	-	35.2 kg	4.42 L

The Ver2.0 motor's spatial dimensions of each part are shown in Table 6. The Ver2.0 motor's main size and structure after optimization can be seen in Table 7. The Ver2.0 driving performance is shown in Table 8.

Table 7. Ver2.0 motor's main size and structure.

Parameters	Ver1.0	Ver2.0
Stator outer diameter	260 mm	240 mm
Rotor inner diameter	145 mm	133 mm
Air gap length	1 mm	1 mm
Stator teeth width	14 mm	13 mm
Magnet segment	3	9
Axial length	90 mm	90 mm
Slot/pole combination	20/8	20/6
Rotor structure	U	∇

Table 8. Ver2.0 motor and Ver1.0 motor electromagnetic characteristics.

Parameters	Base Speed		High Speed		Overload	
	Ver1.0	Ver2.0	Ver1.0	Ver2.0	Ver1.0	Ver2.0
Speed	6000 rpm		15,000 rpm		5000 rpm	
Output torque	101 Nm	105 Nm	47 Nm	48 Nm	199 Nm	209 Nm
Torque ripple	3.9%	3.5%	19.9%	4.64%	12.54%	1.16%
Line voltage amplitude	626 V	635 V	641 V	596 V	639 V	569 V
Line voltage THD	22.9%	9.2%	75.4%	56.2%	46.5%	12.9%
Current density	5.3 A/mm^2		10.9 A/mm^2		6.5 A/mm^2	
Efficiency	95.6%	96.9%	90.6%	90.4%	94.6%	95.5%

Comparing the Ver1.0 motor and Ver2.0 motor designed by the above methods and processes, the motors have smaller torque ripple because of the larger output torque, which is mainly attributed to the double-layer rotor structure and suitable slot/pole combination to improve the PM utilization rate and reduce the MMF harmonic content. In addition, while improving the motor torque performance, line voltage THD is greatly reduced, and the line voltage pressure under high speed and overload conditions can be relieved.

The Ver1.0 motor's peak mass power density is 2.94 kW/kg, and the peak volume power density is 12.9 kW/L. The Ver2.0 motor, which is designed by the high power density method in this article, has a peak mass power density of 3.12 kW/kg and a peak volume power density of 15.19 kW/L. It is worth noting that when calculating the volume power density, the volume is selected as the cylinder volume composed of the total axial length, which includes winding end axial length, stator axial length, and the stator outer diameter. It can be seen that compared with the Ver1.0 motor, the Ver2.0 motor power density is higher.

According to the torque performance, the efficiency, DC bus voltage pressure, and the motor power density verify that the motor designed by the above method has better performance.

4. Analysis of Temperature Field of HPDM

HPDM requires a motor with a small volume and large output power, but too high output power will lead to large copper loss, core loss, and windage loss. However, HPDM has a small space/size for heat dissipation. If the motor temperature field is not analyzed, the temperature will become the limit of the HPDM. At higher temperatures, there will be hidden dangers of irreversible demagnetization of the permanent magnet and winding insulation inside the motor, resulting in the internal failure and reduced reliability of the motor.

4.1. Temperature Field Solution Model

Conventional cooling methods can be divided into active cooling and passive cooling. Passive cooling mainly relies on natural convection to dissipate heat and does not require additional cooling circuits. Active cooling mainly relies on forced convection of fluids and requires additional cooling circuits. The vehicle IMDS has high power, high speed, and a large loss of each component. Passive cooling cannot effectively dissipate heat from the vehicle's high-power machine. Therefore, this paper uses water-cooled active cooling with strong heat dissipation capacity to dissipate heat.

Using the traditional steady-state thermal circuit method to calculate the temperature rise cannot output the accurate 3D temperature field distribution of the motor. Therefore, this section calculates the temperature field of the motor based on the numerical finite element method. The temperature field solution model includes a water jacket, front and back end caps, stator and rotor cores, windings, magnets, shafts, and bearings. The temperature field solution model of the established 60 kW HPDM is shown in Figure 23.

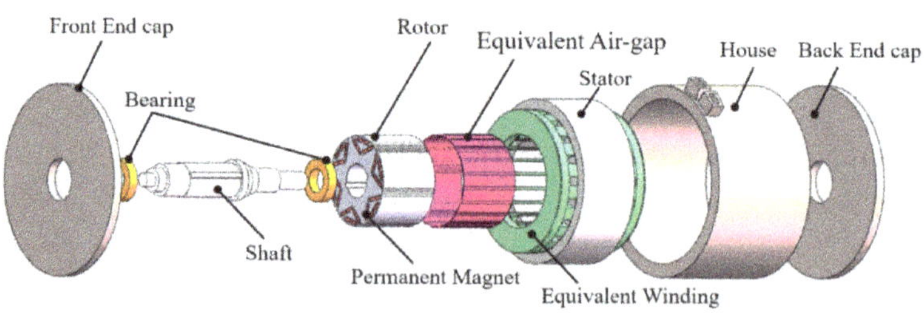

Figure 23. A 60 kW high power density machine temperature field solution model.

It is worth mentioning that, in order to reduce the model complexity, the rotary transformer installed at the front end of the bearing for position detection is not drawn in the temperature field solution model. Additionally, for the same reason of reducing the simulation difficulty and obtaining more accurate thermal simulation results, the temperature field solution model is equivalent to the winding and the stator–rotor air gap. The winding replaces all bare copper wires in the slot with a copper winding, which is placed in the center of the slot; the other parts in the slot, except the copper winding, are represented by equivalent insulation, and the heat transfer coefficient is treated equivalently according to the quality.

4.2. Water Jacket Structure

The water jacket provides radial heat dissipation for the motor and takes away most of the heat. However, the water jacket has various structures, and the heat dissipation capacity and pressure loss of different water jackets are different for HPDM because of the motor's large power and small heat dissipation area. Therefore, the most effective heat dissipation capacity should be taken as the key consideration factor in the selection of water jackets. This paper compares the three water jacket structures: spiral type water jacket structure I, dial type water jacket structure II, and axial water jacket structure III, as shown in Figure 24.

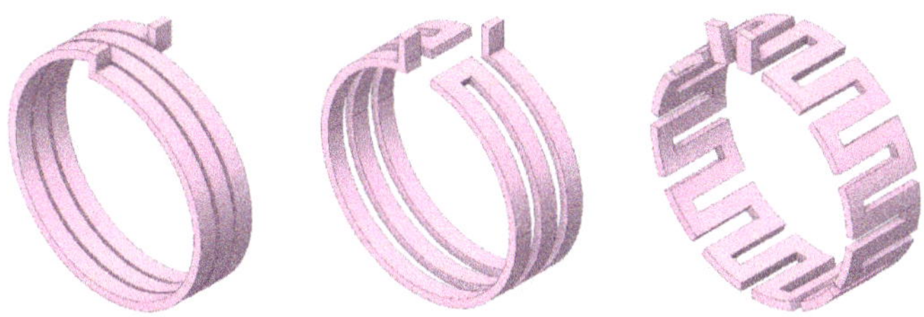

(a) (b) (c)

Figure 24. Different cooling water jacket structures (**a**) spiral, (**b**) dial, and (**c**) axial.

To ensure that different water jackets have the same axial length and 6 L/min fluid flow, a simplified motor temperature field solution model including water jackets, end caps, stator core, and equivalent winding is used. The heat source density of the stator core and the winding heat source density are set as the base speed condition corresponding to the heat sources 329,196 W/m^3 and 544,613 W/m^3. At the initial temperature of the water flow of 65 °C, the flow rate, pressure loss, and temperature comparison of each component of the three structures are shown in Figure 25.

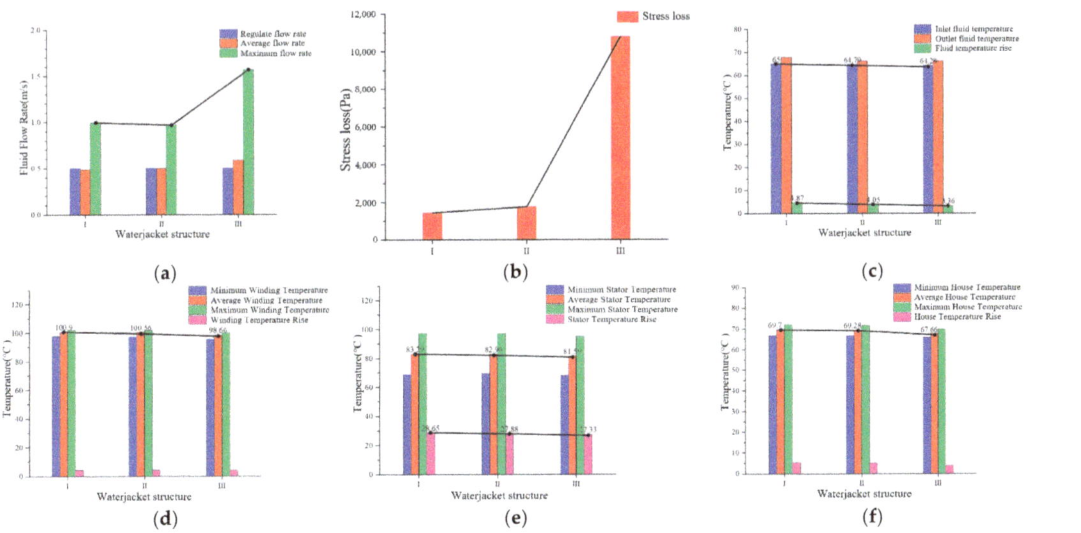

Figure 25. Comparison of three water jacket structures: (**a**) fluid flow rate, (**b**) stress loss, (**c**) fluid temperature, (**d**) winding temperature, (**e**) stator temperature, and (**f**) water jacket temperature.

By comparing the flow velocity, pressure loss, and temperature results of the three structures in Figure 25, it is found that the axial water jacket has the strongest heat dissipation capacity, the dial water jacket is the second, and the spiral water jacket structure is the worst. However, from the perspective of pressure loss, it can be seen that the spiral water jacket has the smallest pressure loss because of almost no turning heads, followed by the dial water jacket structure. The axial water jacket structure has the largest number of turning heads and the largest pressure loss. The dial channel structure is the best choice for comprehensive heat dissipation capacity and pressure loss. However, for HPDM, the base speed condition is not the most serious internal heating condition. Considering the extreme conditions of high speed and overload, the axial water jacket structure III had the best heat dissipation capacity. So, the axial water jacket structure should be selected for better thermal management inside the Ver2.0 motor.

4.3. Temperature Solution of Main Operating Conditions

The main heating components of the HPDM designed in this paper are stator core, rotor core, winding, and the permanent magnet. Under the three main operating conditions, the heat generation rate of the heating element is obtained by dividing the loss of each heating element by the volume of the heating element. The heat generation rate of each element is shown in Table 9.

Table 9. Motor heat generation rate under main operating condition.

Component	Base Speed	High Speed	Overload
Stator yoke	268,248 W/m^3	110,862 W/m^3	451,831 W/m^3
Stator tooth	491,935 W/m^3	278,326 W/m^3	430,712 W/m^3
Winding	543,908 W/m^3	799,036 W/m^3	2,275,536 W/m^3
Rotor	165,534 W/m^3	450,413 W/m^3	145,721 W/m^3
Permanent magnet	15,772 W/m^3	46,481 W/m^3	37,487 W/m^3

Using the established HPDM temperature field solution model, the temperature field is solved for the base speed condition, high-speed condition, and an overload condition by assigning relevant boundary conditions, material properties, and heat generation rate of heat-generating components. Among them, the base speed condition is a long-term working point, and its steady-state temperature is observed. The high-speed condition and the overload condition are short-time working conditions, and the temperature distribution of each part of the motor is observed under the operation for 180 s. The initial temperature of the water flow is given as 65 °C, the temperature section diagrams under three working conditions are shown in Figure 26, and the temperature change curves of each component under transient conditions are shown in Figure 27.

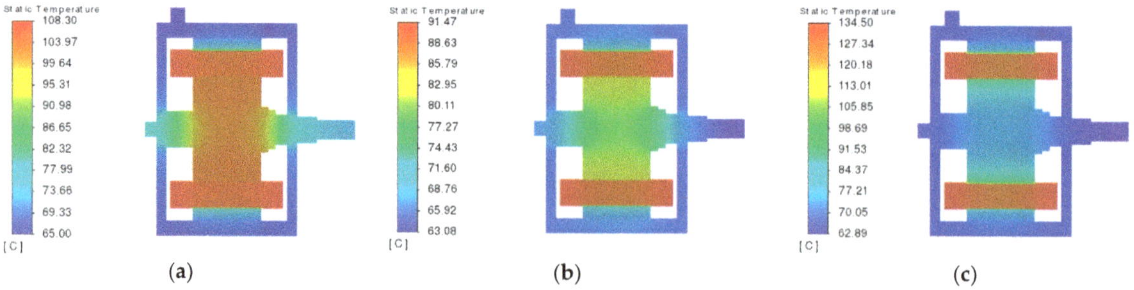

Figure 26. Temperature field solution results under three main operation conditions: (**a**) base speed conditions, (**b**) high-speed conditions, and (**c**) overload conditions.

From the temperature field solution results, it can be found that under the three main operating conditions of the base speed, high speed, and overload, the windings are the most serious parts of heat generation, and the highest temperature occurs at the end of the winding at the overload point, which is 135 °C. The temperature rises to 70 °C, and the maximum temperature rise do not exceed 155 °C and 105 °C for Class F insulation. The PMs in both the steady-state and the transient-state do not exceed the maximum allowable temperature of the material N38UH, which is 150 °C. The rotary transformer selected for Ver2.0 is TAMAGAWA TS2225N903E102; its allowable temperature, according to the manual, is −40 °C to 150 °C. In this article, the motor's cooling system solutions' rotary transformer can operate normally. It can be seen that the axial water jacket structure can effectively manage the heat of the designed 60 kW high power density motor.

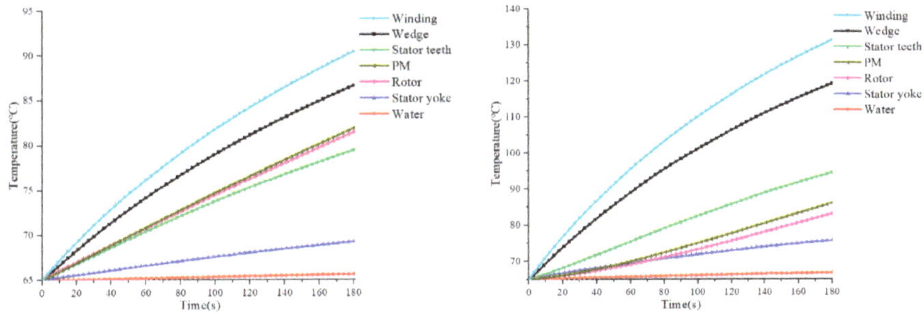

Figure 27. Temperature change curve under transient operating conditions: (**a**) high-speed conditions, (**b**) overload Conditions.

5. Conclusions

This paper researched the design method of the HPDM for electric vehicles IMDS. Through the method, a five-phase high power density motor with a rated power of 60 kW, a maximum speed of 15,000 r/min, and a peak power density of 15.19 kW/L has been designed. Compared with the Ver1.0 motor designed before, the Ver2.0 motor power density is higher. In addition, the temperature field is checked for the three main operating conditions of base speed, high speed, and overload. The following conclusions are drawn:

(1) The motor power density is related to the phase number, winding utilization, copper space factor, air gap flux density, and motor rotational speed. When designing a high power density motor, the above factors can be weighed;

(2) Increasing IPMSM rotor PM layer number can make air gap flux density step up, increase air gap flux density amplitude, and improve the sinusoidal degree at the same time, which is beneficial to improving the motor torque performance. For a five-phase motor, FSDW has less end winding length and torque ripple than an ISDW and has a higher winding coefficient compared with a five-phase motor with FSCW. FSDW is suitable for the five-phase motor with the high power density and electromagnetic characteristics in some cases;

(3) Driver modulation algorithm affects motor electromagnetic characteristics. Comparing the NTV-SVPWM modulation algorithm with the NFV-SVPWM modulation algorithm, NFV-SVPWM can eliminate harmonics in the third harmonic plane by taking advantage of the five-phase motor's two vector planes and obtain higher sinusoidal output excitation. Through the comparative analysis of the output current excitations with different switching frequencies, it is found that a high switching frequency is more conducive to improving motor electromagnetic characteristics and reducing motor loss. However, the switching frequency should be selected considering power device switching loss;

(4) For the three water jackets—the spiral, dial, and axial—the spiral structure has the smallest pressure loss and the worst heat dissipation capacity. The axial structure has the largest pressure loss and the strongest heat dissipation capacity.

Author Contributions: S.W., orientation, question answering, editing of the paper and providing study conditions; J.Z., the fourth section research, and paper writing; X.Z., the second and third section study; J.Y., editing the equations and drawing figures. All authors have read and agreed to the published version of the manuscript.

Funding: This work was supported by the key project of the National Key Research and Development Program—Intergovernmental Cooperation in International Science and Technology Innovation under Grant 2019YFE0123500, part of the Power Electronics Science and Education Development Program of Delta Group under Grant DREM2019004.

Data Availability Statement: Not applicable.

Conflicts of Interest: The authors declare no conflict of interest.

Abbreviations

The following abbreviations are used in this manuscript:

MDS	Motor drive system
IMDS	Integrated motor drive system
THD	Total harmonic distortion
HPDM	High power density motor
SPMSM	Surface-mounted permanent magnet synchronous machine
IPMSM	Interior permanent magnet synchronous machine
PM	Permanent magnet
RMS	Root mean square
MMF	Magnetic motive force
FSCW	Fractional slot concentrated winding
FSDW	Fractional slot distributed winding
ISDW	Integral slot distributed winding
MTPA	Maximum torque per ampere

References

1. Jahns, T.M.; Dai, H. The past, present, and future of power electronics integration technology in motor drives. *CPSS Trans. Power Electron. Appl.* **2017**, *2*, 197–216. [CrossRef]
2. Available online: https://www.energy.gov/sites/default/files/2017/11/f39/EETT%20Roadmap%2010-27-17.pdf (accessed on 9 May 2022).
3. Abebe, R.; Vakil, G.; Calzo, G.L.; Cox, T.; Lambert, S.; Johnson, M.; Gerada, C.; Mecrow, B. Integrated motor drives: State of the art and future trends. *IET Electr. Power Appl.* **2016**, *10*, 757–771. [CrossRef]
4. Martin, M.; Schimanek, E.; Billmann, M. Towards an integrated drive for hybrid traction. In Proceedings of the CPES Power Electronic Conference, Blacksburg, VA, USA, 17–19 April 2005.
5. Deng, X.; Lambert, S.; Mecrow, B.; Mohamed, M.A.S. Design Consideration of a High-Speed Integrated Permanent Magnet Machine and its Drive System. In Proceedings of the 2018 XIII International Conference on Electrical Machines (ICEM), Alexandroupoli, Greece, 3–6 September 2018; pp. 1465–1470. [CrossRef]
6. Lee, W.; Li, S.; Han, D.; Sarlioglu, B.; Minav, T.A.; Pietola, M. A Review of Integrated Motor Drive and Wide-Bandgap Power Electronics for High-Performance Electro-Hydrostatic Actuators. *IEEE Trans. Transp. Electrif.* **2018**, *4*, 684–693. [CrossRef]
7. Gieras, J.F.; Bianchi, N. Electric Motors for Light Traction. *EPE J.* **2004**, *1*, 12–23. [CrossRef]
8. Jahns, T.M.; Sarlioglu, B. The Incredible Shrinking Motor Drive: Accelerating the Transition to Integrated Motor Drives. *IEEE Power Electron. Mag.* **2020**, *7*, 18–27. [CrossRef]
9. Ugur, M.; Keysan, O. Design of a GaN Based Integrated Modular Motor Drive. In Proceedings of the 2018 XIII International Conference on Electrical Machines (ICEM), Alexandroupoli, Greece, 3–6 September 2018; pp. 1471–1477. [CrossRef]
10. Wang, J.; Li, Y.; Han, Y. Evaluation and design for an integrated modular motor drive (IMMD) with GaN devices. In Proceedings of the 2013 IEEE Energy Conversion Congress and Exposition, Denver, CO, USA, 15–19 September 2013; pp. 4318–4325. [CrossRef]
11. Nasr, A.; Gu, C.; Wang, X.; Buticchi, G.; Bozhko, S.; Gerada, C. Torque-Performance Improvement for Direct Torque-Controlled PMSM Drives Based on Duty-Ratio Regulation. *IEEE Trans. Power Electron.* **2022**, *37*, 749–760. [CrossRef]
12. Korpe, U.U.; Gokdag, M.; Koc, M.; Gulbudak, O. Modulated Model Predictive Control of Permanent Magnet Synchronous Motors with Improved Steady-State Performance. In Proceedings of the 2021 3rd Global Power, Energy and Communication Conference (GPECOM), Antalya, Turkey, 5–8 October 2021; pp. 67–72. [CrossRef]
13. Chowdhury, S.; Gurpinar, E.; Su, G.-J.; Raminosoa, T.; Burress, T.A.; Ozpineci, B. Enabling Technologies for Compact Integrated Electric Drives for Automotive Traction Applications. In Proceedings of the 2019 IEEE Transportation Electrification Conference and Expo (ITEC), Detroit, MI, USA, 19–21 June 2019; pp. 1–8. [CrossRef]

14. Hsu, J.S. *Report on Toyota Prius Motor Thermal Management*; ORNL: Oak Ridge, TN, USA, 2005; pp. 5–25. [CrossRef]
15. Burress, T.A.; Campbell, S.L.; Coomer, C.; Ayers, C.W.; Wereszczak, A.A.; Cunningham, J.P.; Marlino, L.D.; Seiber, L.E.; Lin, H.T. *Evaluation of the 2010 Toyota Prius Hybrid Synergy Drive System*; Oak Ridge National Laboratory (ORNL): Oak Ridge, TN, USA, 2011; pp. 44–69. [CrossRef]
16. Burress, T. Benchmarking of Competitive Technologies. Available online: https://www.energy.gov/sites/prod/files/2014/03/f10/ape006_burress_2011_o.pdf (accessed on 12 December 2018).
17. Li, Y.; Yang, H.; Lin, H.; Fang, S.; Wang, W. A Novel Magnet-Axis-Shifted Hybrid Permanent Magnet Machine for Electric Vehicle Applications. *Energies* **2019**, *12*, 641. [CrossRef]
18. Wu, S.; Tian, C.; Zhao, W.; Zhou, J.; Zhang, X. Design and Analysis of an Integrated Modular Motor Drive for More Electric Aircraft. *IEEE Trans. Transp. Electrif.* **2020**, *6*, 1412–1420. [CrossRef]
19. Pyrhönen, J.; Jokinen, T.; Hrabovcová, V. *Design of Rotating Electrical Machines*; John Whiley and Sons, Inc.: Hoboken, NJ, USA, 2008.
20. Villani, M.; Tursini, M.; Fabri, G.; Castellini, L. High reliability permanent magnet brushless motor drive for aircraft application. *IEEE Trans. Ind. Electron.* **2012**, *59*, 2073–2081. [CrossRef]
21. Sanabria-Walter, C.; Polinder, H.; Ferreira, J.A. High-Torque-Density High-Efficiency Flux-Switching PM Machine for Aerospace Applications. *IEEE J. Emerg. Sel. Top. Power Electron.* **2013**, *1*, 327–336. [CrossRef]
22. Ou, J.; Liu, Y.; Breining, P.; Gietzelt, T.; Wunsch, T.; Doppelbauer, M. Experimental Characterization and Feasibility Study on High Mechanical Strength Electrical Steels for High-Speed Motors Application. *IEEE Trans. Ind. Appl.* **2021**, *57*, 284–293. [CrossRef]
23. Zong, Z.L.; Wang, K.; Zhang, J.Y. Control strategy of five-phase PMSM utilizing third harmonic current to improve output torque. In Proceedings of the 2017 Chinese Automation Congress (CAC), Jinan, China, 20–22 October 2017; pp. 6112–6117. [CrossRef]

energies

MDPI

Article

Design and Analysis of a 30 kW, 30,000 r/min High-Speed Permanent Magnet Motor for Compressor Application

Zhenning Qi [1], Yue Zhang [1,*], Siyang Yu [2] and Zhenyao Xu [2]

1 School of Electrical Engineering, Shandong University, Jinan 250061, China; zhenningqi@163.com
2 School of Electrical Engineering, Shenyang University of Technology, Shenyang 110870, China; yusy@sut.edu.cn (S.Y.); xuzy@sut.edu.cn (Z.X.)
* Correspondence: yzhang35@sdu.edu.cn

Abstract: In this paper, the design and analysis of a 30 kW, 30,000 r/min high-speed permanent magnet motor (HSPMM) for compressor application are provided. In order to provide a reasonable electromagnetic design scheme, the electromagnetic performances of the HSPMM under different structures are analyzed and compared by the finite element method (FEM). The thermal performances and cooling system of the HSPMM are, respectively, analyzed and designed by computational fluid dynamics (CFD). Finally, the HSPMM's rotor strength is studied by both FEM and analytical methods, and the influencing factors of which are also researched in this paper.

Keywords: high-speed permanent magnet motor; FEM; electromagnetic design; CFD; thermal analysis; analytical method; rotor strength

Citation: Qi, Z.; Zhang, Y.; Yu, S.; Xu, Z. Design and Analysis of a 30 kW, 30,000 r/min High-Speed Permanent Magnet Motor for Compressor Application. *Energies* **2022**, *15*, 3923. https://doi.org/10.3390/en15113923

Academic Editor: Anna Richelli

Received: 20 April 2022
Accepted: 21 May 2022
Published: 26 May 2022

1. Introduction

High-speed permanent magnet motors (HSPMMs) have the advantages of compact structure, high power density, and high efficiency [1–3]. Compared with the traditional method of driving high-speed loads by using a constant speed motor and gearbox, it is more effective to reduce the volume of the driving system, increase power density and improve operation efficiency by using HSPMM directly, so that HSPMM has a promising future and a wide range of applications such as high-speed compressors, flywheel energy storage, and so on. However, the performances of HSPMM in multiple physical fields also need to be further investigated because of their high frequency, high speed, and high loss density [4]. Therefore, a 30 kW, 30,000 r/min HSPMM for compressor application is designed and analyzed in multi-physics fields in this paper.

Firstly, this paper provides analyses of the influencing factors of the HSPMM's electromagnetic performances such as the motor main dimensions, the number of poles, the number of stator slots, and the magnetization method of PM.

Due to the high power density and compact size, the heat dissipation capacity of HSPMM is limited, which may result in the demagnetization of PM. Therefore, the thermal analysis and cooling system design of the HSPMM are particularly necessary.

Considering the PM materials are fragile in tensile, it is necessary to install a sleeve outside the rotor for mechanical protection. This paper analyzes the rotor stress distribution of the HSPMM by both analytical and FEM methods. The influencing factors such as sleeve thickness and interference are also analyzed in this paper.

2. Electromagnetic Analysis

The electromagnetic performances of HSPMMs with different structures are presented and compared by FEM in this chapter [5,6]. As for material selection, NdFeB is selected as the PM material due to its high remanence and coercivity, and a silicon steel sheet is selected as the core material.

2.1. Main Dimensions of Motor

The stator inner diameter D and the motor active length L_{ef} which are called the main dimensions of the motor play an important role in motor design. The two main dimensions can be calculated by the following formulas:

$$D^2 L_{ef} = \frac{6.1}{\alpha'_p K_{nm} K_{dp} A B_\delta} \cdot \frac{P'}{n} \tag{1}$$

$$\lambda = \frac{L_{ef}}{\tau} \tag{2}$$

$$\tau = \frac{\pi D}{2P} \tag{3}$$

where P' is the apparent power, n is the motor speed, $\alpha_P{}'$ is the equivalent pole arc coefficient, K_{nm} is the air gap flux factor, K_{dp} is the winding factor, A is the line load, B_δ is the air gap flux density, λ is the main dimension ratio, τ is the pole distance, and P is the number of pole pairs.

The estimated values of each parameter in the above formula are shown in Table 1. Therefore, it can be roughly obtained that the D and the L_{ef} are 45 mm and 105 mm, respectively, which still needs to be further adjusted according to the subsequent analysis results.

Table 1. Calculation parameters of main dimensions of motor.

Parameters	Value	Parameters	Value
P' (kW)	27	A (A/m)	50,000
n (r/min)	30,000	B_δ (T)	0.7
$\alpha_P{}'$	0.7	λ	3
K_{nm}	1.11	P	2
K_{dp}	0.92		

Then, the air gap thickness and the outer diameter of the rotor can be set to 1.5 mm and 42 mm, which also need to be further optimized.

2.2. Number of Poles

The number of poles plays an important role in HSPMMs' electromagnetic performances. In order to limit the operating frequency of HSPMMs and suppress the losses caused by high-frequency current and high-frequency magnetic flux density, the number of poles of HSPMMs should better be set as two or four. The structural parameters and electromagnetic performances of two HSPMMs with two poles and four poles, respectively, are analyzed and compared in this paper. The two HSPMMs are designed with the same power (30 kW), same speed (30,000 r/min), and the same stator slots number (12).

The two HSPMMs' structural parameters are presented and compared in Table 2. It can be found the two poles motor has a lower frequency (500 Hz) than the four poles one (1000 Hz), which is beneficial to reduce stator iron loss and winding copper loss. However, the two poles motor has a larger size and lower power density than the four poles motor.

The two motors' back EMF and air gap magnetic flux density with FFT analysis results under no-load conditions are compared and shown in Figures 1 and 2, respectively.

Table 2. Structural parameters of 2 and 4 poles HSPMMs.

Parameters	2 Poles	4 Poles
Power (kW)	30	30
Speed (r/min)	30,000	30,000
Frequency (Hz)	500	1000
Stator outer diameter (mm)	118	103
Stator inner diameter (mm)	45	45
Slot number	12	12
Core length (mm)	130	105
Rotor outer diameter (mm)	42.2	42.2
Number of winding turns	7	8
Winding pitch	5	2

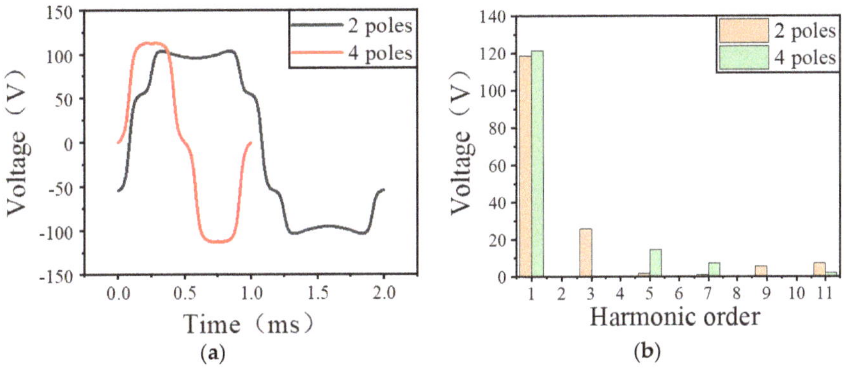

Figure 1. Waveforms and FFT analysis results of back EMF for 2 poles and 4 poles HSPMMs. (**a**) Waveforms; (**b**) FFT analysis results.

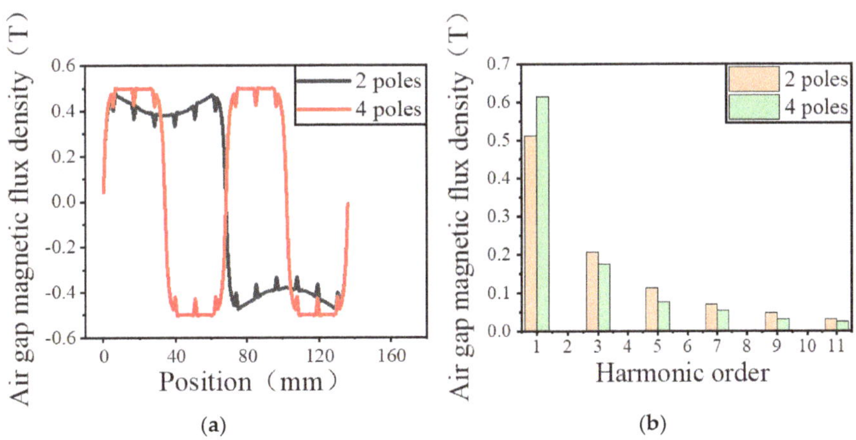

Figure 2. Waveforms and FFT analysis results of air gap magnetic flux density for 2 poles and 4 poles HSPMMs. (**a**) Waveforms; (**b**) FFT analysis results.

Figure 1 shows the two motors' back EMF waveforms with FFT analysis results. It can be found that the four poles motor has a larger fundamental component, the fifth harmonic component, and the seventh harmonic component, while other harmonic components are smaller than the two poles motor, especially the third harmonic component. In general, the back EMF of the four poles motor has lower harmonics, and the waveform of which is closer to the sine wave.

The two motors' air gap magnetic flux density waveforms with FFT analysis results are presented and compared in Figure 2. Each harmonic component of the air gap magnetic flux density of the four poles motor is lower than the two poles motor except for the fundamental component, indicating the four poles structure has better performance in reducing harmonics. Additionally, adopting the four poles structure for HSPMMs can also improve the utilization of PMs due to their higher fundamental component of air gap magnetic flux density.

The two motors' performances under full-load condition are also compared in this paper and shown in Figure 3 and Table 3. Due to the lower frequency, the two poles HSPMM has lower stator iron loss and winding copper loss than the four poles. Although the four poles HSPMM has lower eddy current loss and wind friction loss because of their smaller size and lower harmonics, the efficiency of the two poles HSPMM is still higher than the four poles HSPMM. Additionally, it can be found from Figure 3b and Table 3 that the two motors have the same average torque under full-load conditions, while the four poles motor has a larger torque ripple (13.84%) than the two poles (5.32%).

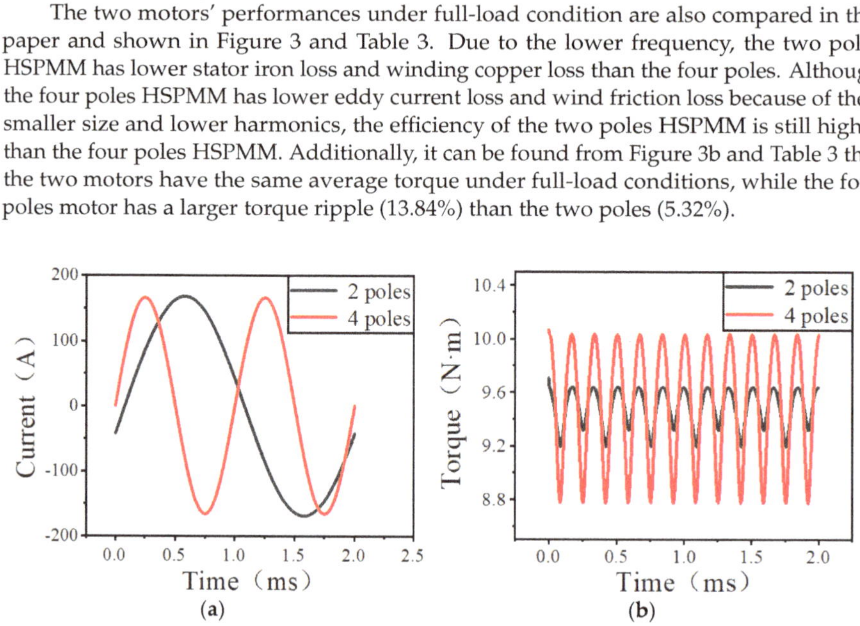

Figure 3. The comparisons of current and torque waveforms of the HSPMM under full-load conditions. (**a**) Current waveforms; (**b**) torque waveforms.

Table 3. Performances of 2 and 4 poles HSPMMs under full-load condition.

Parameters	2 Poles	4 Poles
RMS current (A)	117.10	117.28
Torque (N·m)	9.48	9.49
Torque ripple (%)	5.32	13.84
Stator iron loss (W)	321.43	558.68
Winding copper loss (W)	257.22	264.09
Rotor eddy current loss (W)	218.19	210.48
Wind friction loss (W)	72.53	58.58
Efficiency (%)	97.10	96.36

Finally, this paper chooses to adopt the four poles structure due to its lower harmonics, smaller size, higher power density, and lower rotor eddy current loss.

2.3. Number of Slots

As shown in Figure 4, there are three normally used stator structures of HSPMMs including slotless structure, less slot structure, and multi-slot structure. Considering the advantages of higher air-gap flux density, higher utilization of PMs, and lower harmonic amplitude, there is no doubt that the multi-slot structure has been the first choice in the design of HSPMM.

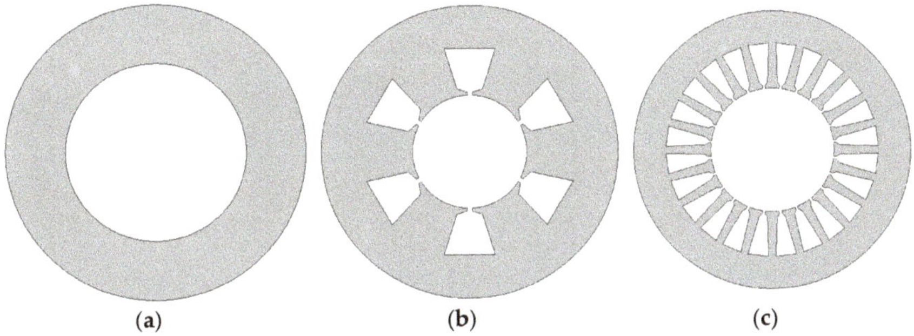

Figure 4. Stator structures of HSPMMs. (**a**) Slotless structure; (**b**) less slot structure; (**c**) multi slot structure.

In this paper, 3 HSPMMs with 12 slots, 18 slots, and 24 slots, respectively, are designed and compared to figure out the influence of the number of slots on HSPMMs. The three HSPMMs are designed with the same power, speed, frequency, number of poles, and size for a fair comparison.

Figure 5 shows the three HSPMMs' back EMF waveforms with FFT analysis results. It can be found that the 24 slots HSPMM has the largest fundamental component. Except for the third harmonic component, the other order harmonic components of the 24 slots HSPMM are lower than other motors, indicating that the multi-slot structure can help reduce harmonic components of back EMF.

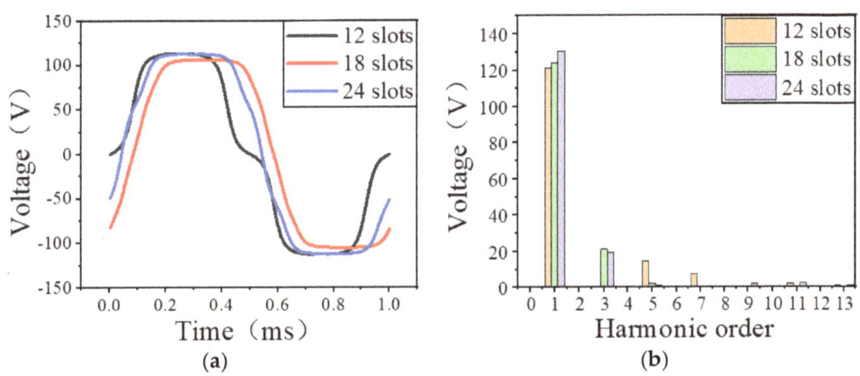

Figure 5. Waveforms and FFT analysis results of back EMF for 12 slots, 18 slots, and 24 slots HSPMMs. (**a**) Waveforms; (**b**) FFT analysis results.

The three HSPMMs' air gap magnetic flux density waveforms with FFT analysis under no-load conditions are presented in Figure 6. Due to the same rotor structure, the air gap magnetic flux density waveforms and FFT analysis results of the three motors are very similar to each other.

The performances of the three HSPMMs under full-load conditions are also compared and presented in Table 4. With the increase in the number of slots, the RMS current and torque ripple of HSPMMs is decreasing gradually. The three motors have nearly equal stator iron loss and wind friction loss since they have the same frequency and rotor size. The 18 slots motor has the lowest winding copper loss, though the RMS current of which is larger than the 24 slot one. The eddy current loss of 24 slots motor is much lower than other motors, which further proves that increasing the number of slots has good performance in the reduction of the harmonic components and the eddy current loss of HSPMMs.

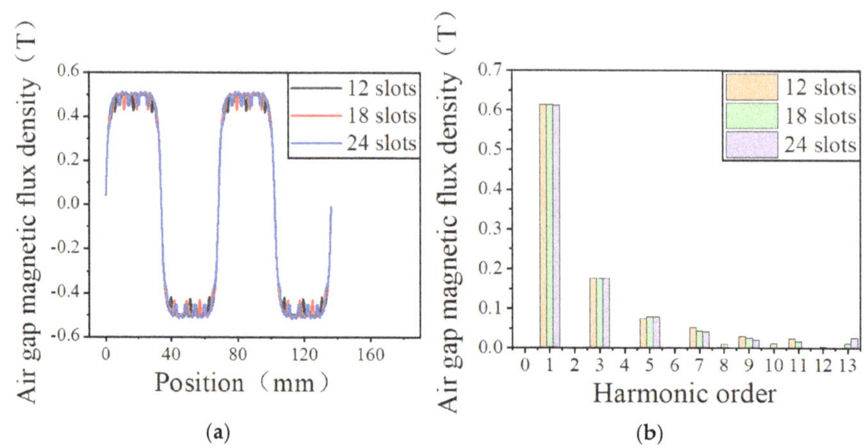

Figure 6. Waveforms and FFT analysis results of air gap magnetic flux density for 12 slots, 18 slots, and 24 slots HSPMMs. (**a**) Waveforms; (**b**) FFT analysis results.

Table 4. Performances of 12 slots, 18 slots and 24 slots HSPMMs under full-load condition.

Parameters	12 Slots	18 Slots	24 Slots
RMS current (A)	117.28	114.61	108.59
Torque (N·m)	9.49	9.54	9.54
Torque ripple (%)	13.84	5.05	2.95
Stator iron loss (W)	558.68	554.87	558.81
Winding copper loss (W)	264.09	225.73	244.81
Rotor eddy current loss (W)	210.48	47.15	21.22
Wind friction loss (W)	58.58	58.58	58.58
Efficiency (%)	96.36	97.05	97.06

In summary, the 24 slots structure is a more reasonable choice for HSPMMs.

2.4. PM Magnetization Methods

The PM magnetization methods of HSPMMs mainly include radial magnetization, parallel magnetization, and Halbach magnetization, as shown in Figure 7. In order to figure out the influence of PM magnetization methods on the electromagnetic performances of HSPMMs, three HSPMMs with different PM magnetization methods are designed and compared in this paper. The three HSPMMs have the same power, speed, and frequency.

Figure 8 shows the magnetic flux density and line distributions of the three HSPMMs. According to Figure 8, It can be found that the color of the magnetic flux density cloud diagram in Figure 8c is generally lighter, indicating that the magnetic flux density of the motor with the Halbach magnetization method is lower than the other motors.

The three HSPMMs' back-EMF waveforms with FFT analysis results are presented in Figure 9. Then the RMS value of the total back EMF of the three motors calculated by FEM are 91.31 V (radial), 84.96 V (parallel), and 79.63 V (Halbach), respectively, indicating the HSPMM with the Halbach magnetization method has the smallest RMS value of total back EMF. Additionally, the back EMF waveform of the HSPMM with the Halbach magnetization method is closest to a sine wave due to its lower harmonic components.

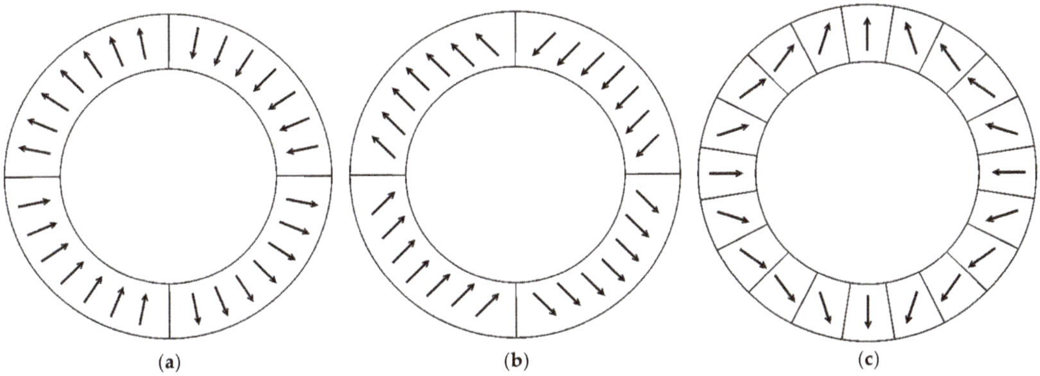

Figure 7. PM magnetization methods. (**a**) Radial magnetization; (**b**) parallel magnetization; (**c**) Halbach magnetization.

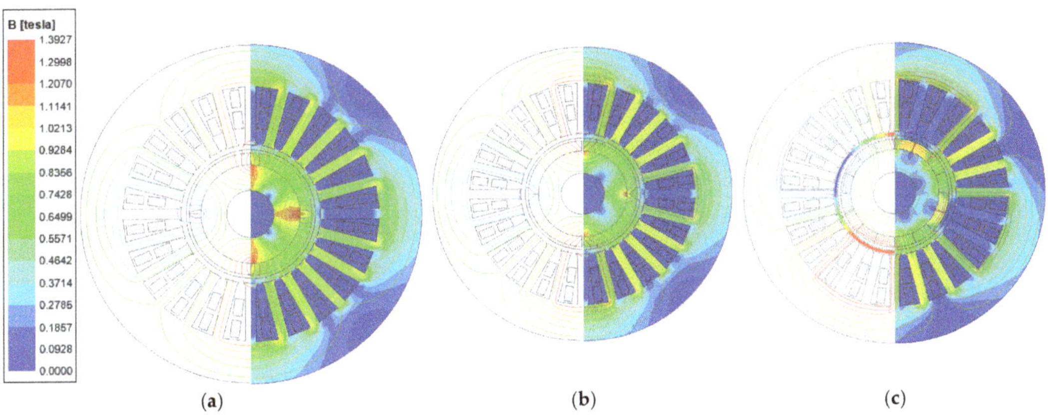

Figure 8. Magnetic flux density and lines distributions of HSPMMs with different magnetization methods under no-load condition. (**a**) Radial magnetization; (**b**) parallel magnetization; (**c**) Halbach magnetization.

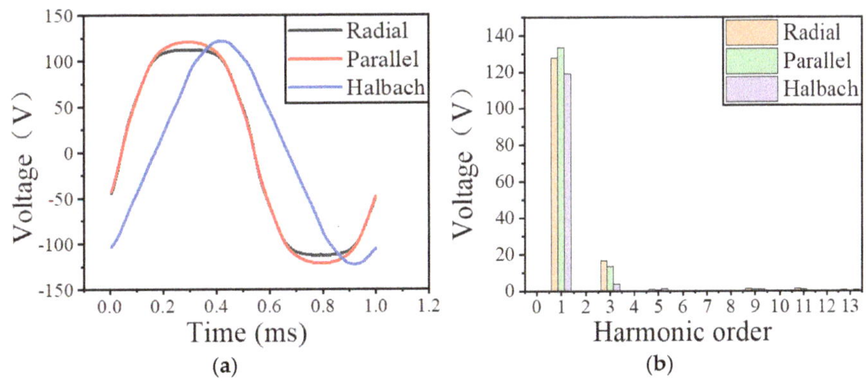

Figure 9. Waveforms and FFT analysis results of back EMF for HSPMMs with different magnetization methods. (**a**) Waveforms; (**b**) FFT analysis results.

The air gap magnetic flux density waveforms with FFT analysis results for the three HSPMMs are presented in Figure 10. It is learned that each harmonic component of air gap magnetic flux of the HSPMM with the Halbach magnetization method is lower than other motors except for the fifth harmonic component.

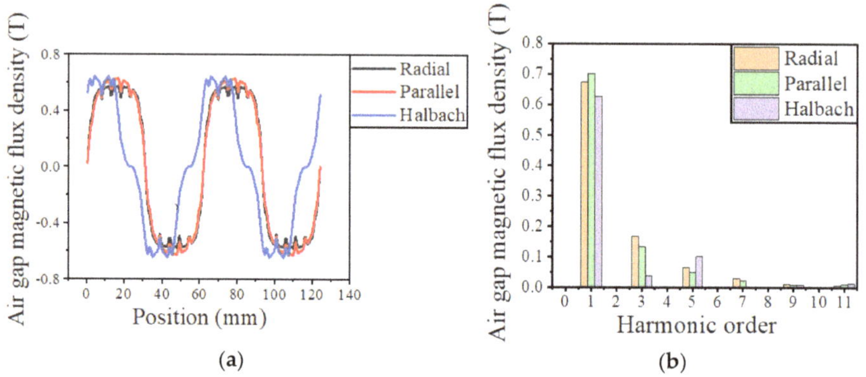

Figure 10. Waveforms and FFT analysis results of air gap magnetic flux density for HSPMMs with different magnetization methods. (**a**) Waveforms; (**b**) FFT analysis results.

The three HSPMMs' performances under full-load conditions are presented in Table 5. It can be learned that the HSPMM with PM parallel magnetized has the lowest current, the lowest torque ripple, the lowest rotor eddy current loss, and the highest efficiency.

Table 5. Performances of the three HSPMMs with different PM magnetization methods under full-load condition.

Parameters	Radial	Parallel	Halbach
RMS current (A)	110.89	106.15	119.02
Torque (N·m)	9.55	9.56	9.55
Torque ripple (%)	2.07	1.42	1.99
Stator iron loss (W)	511.09	478.63	437.87
Winding copper loss (W)	303.32	273.81	378.37
Rotor eddy current loss (W)	11.84	10.23	11.37
Wind friction loss (W)	33.16	33.16	33.16
Efficiency (%)	97.22	97.42	97.21

2.5. Final Electromagnetic Design Scheme

Based on the previous analysis, the electromagnetic design scheme of a 30 kW, 30,000 r/min HSPMM is finally completed in this paper. Figure 11 shows the current and torque waveforms of the HSPMM under full-load conditions, and the designed motor's structure and electromagnetic parameters are presented in detail in Table 6.

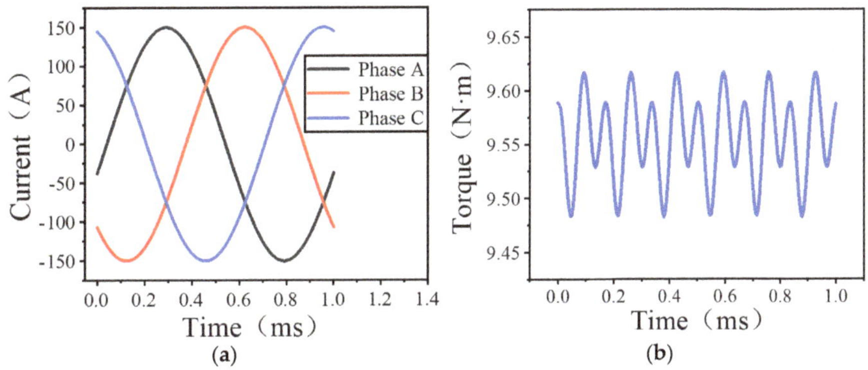

Figure 11. The current and torque waveforms of the HSPMM under full-load conditions. (**a**) Current waveform; (**b**) torque waveform.

Table 6. Structure and electromagnetic parameters of the 30 kW, 30,000 r/min HSPMM.

Parameters	Value	Parameters	Value
Power (kW)	30	Pole number	4
Speed (r/min)	30000	Slot number	24
Frequency (Hz)	1000	Stator outer diameter (mm)	100
RMS current (A)	106.15	Stator inner diameter (mm)	41
Torque (N·m)	9.56	Core length (mm)	103
Torque ripple (%)	1.42	Air gap thickness (mm)	1.5
Stator iron loss (W)	478.63	Rotor outer diameter (mm)	38
Winding copper loss (W)	273.81	Sleeve thickness (mm)	1
Rotor eddy current loss (W)	10.23	PM material	NdFeB
Wind friction loss (W)	33.16	PM thickness (mm)	3.4
Efficiency (%)	97.42	Magnetization direction	Parallel

3. Thermal Analysis

HSPMMs have limited heat dissipation capability due to their compact size and high loss density so a reasonable cooling system design is particularly necessary for HSPMMs. In this paper, CFD is selected to provide the thermal analysis for HSPMMs and study the cooling effects of different water-cooling methods [7,8].

3.1. Cooling Channel Structure

The structure of the three cooling channels is presented in Figure 12. Then, the comparative analysis of the thermal performances of HSPMMs with different cooling channel structures is performed in this paper. The three cooling channels are designed with the same inner surface area (13,229 mm^2), thickness (3 mm), and volume (40,774 mm^3) for comparison.

The calculation results of the heat transfer coefficient for the three HSPMMs are presented in Figure 13. It can be found the heat transfer coefficient of the spiral structure channel is higher than the other structures, indicating the spiral structure channel has better performance in heat dissipation. Therefore, as shown in Table 7, the maximum temperature of each component of the HSPMM with a spiral structure channel is the lowest of the three HSPMMs due to its higher heat transfer coefficient.

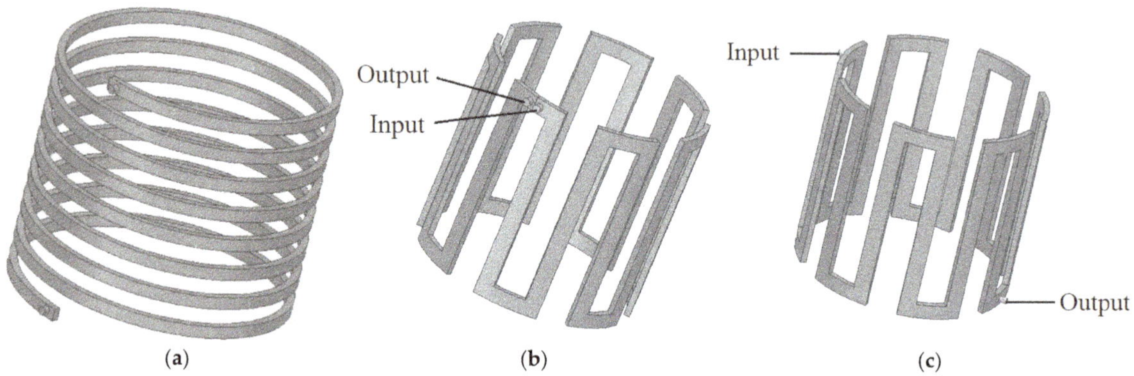

Figure 12. Cooling channel structures. (**a**) Spiral structure; (**b**) axial structure 1; (**c**) axial structure 2.

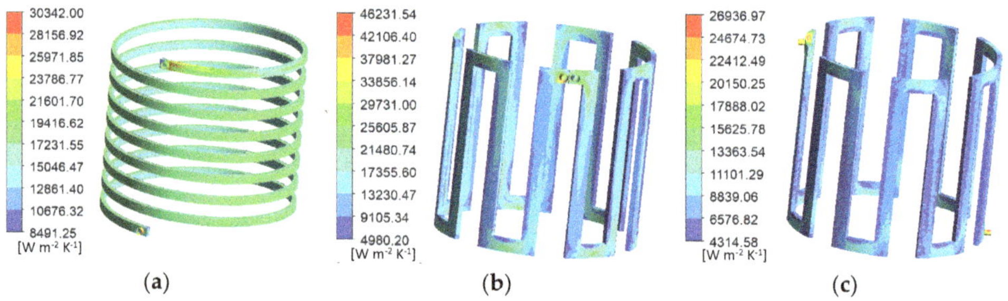

Figure 13. Heat transfer coefficient distribution of the three cooling channel structures. (**a**) Spiral structure; (**b**) axial structure 1; (**c**) axial structure 2.

Table 7. The maximum temperature of each part of motors with different cooling channel structures.

Parameters	Spiral	Axial 1	Axial 2
Cooling water (°C)	36.49	38.67	44.53
Stator core (°C)	66.08	66.63	68.80
Stator winding (°C)	69.96	70.84	73.99
Sleeve (°C)	124.89	125.01	131.91
PM (°C)	124.81	124.92	131.79

Figure 14 shows the pressure distribution of the three cooling channel structures. The axial channel 1 has the largest inlet pressure due to its larger channel angle and flow resistance. Correspondingly, the spiral channel has a smaller channel angle and flow resistance which contribute to its lower inlet pressure. In axial channel 2, the cooling water flows along two paths on both sides of the channel so that the actual flow distance is only half of the spiral channel and the axial channel 1. Therefore, the inlet pressure of the axial channel 2 is lower than other channels though it has a larger channel angle and flow resistance, too.

3.2. Number of Cooling Channel Turns

Considering the spiral cooling channel has a better cooling effect than other channel Structures, this paper provides further analysis of the number of spiral cooling channel turns.

The thermal analysis results of HSPMMs with a different number of cooling channel turns are shown in Figure 15. It can be found that with the rise of the number of cooling channel turns, the maximum temperature of cooling water and motor gradually reduced so that increasing the number of cooling channel turns is helpful to enhance the cooling effect.

However, when the number of channel turns is larger than eight, each motor component's maximum temperature hardly changed anymore, and even the cooling water maximum temperature rose slightly. Meanwhile, the eight-turn cooling channel has the largest heat transfer coefficient. Therefore, the number of cooling channel turns should be designed more reasonably with the considerations of both cooling effect and manufacturing difficulty.

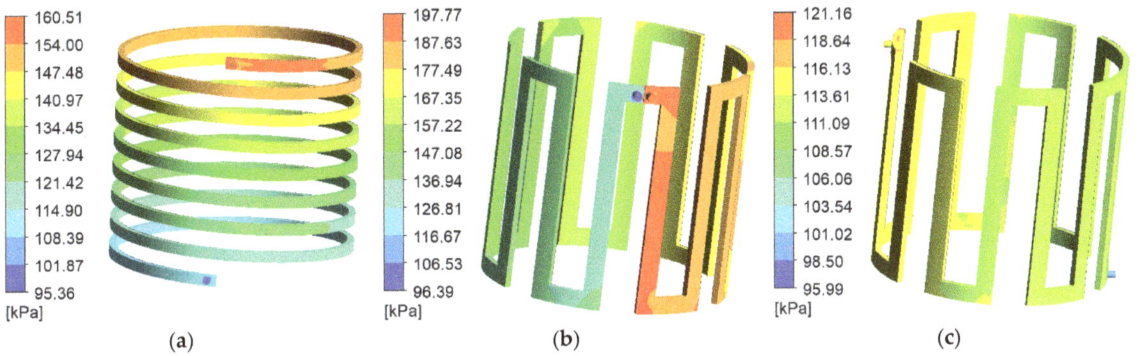

Figure 14. Pressure distribution of the three cooling channel structures. (**a**) Spiral structure; (**b**) axial structure 1; (**c**) axial structure 2.

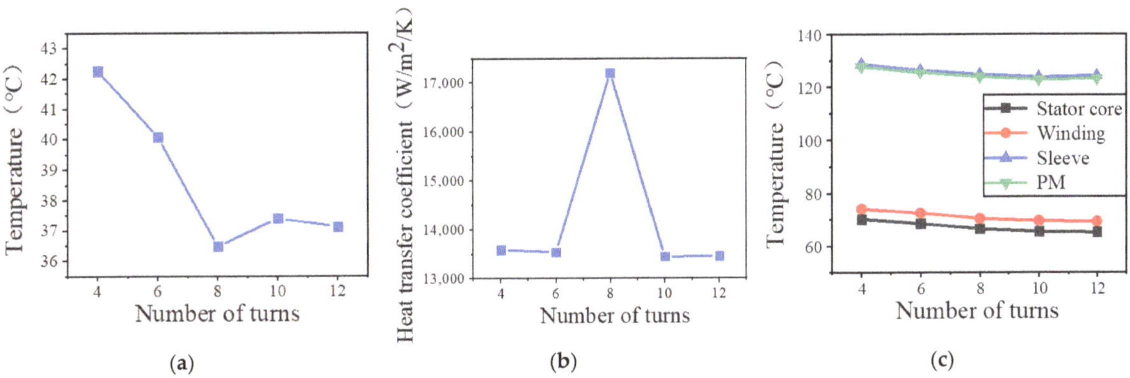

Figure 15. Thermal analysis results of HSPMMs with different number of cooling channel turns. (**a**) Cooling water maximum temperature; (**b**) Heat transfer coefficient; (**c**) Motor maximum temperature.

3.3. Cooling Water Flow Rate

The flow rate of cooling water also plays an important role in motor heat dissipation. Figure 16 shows the thermal analysis results of an HSPMM at different cooling water flow rates. It is learned that with the rise of the cooling water flow rate, the maximum temperature of the cooling water and the motor gradually decreased while the heat transfer coefficient of the cooling channel increased, indicating that increasing the cooling water flow rate can effectively enhance the cooling effect. However, when the flow rate is up to 3 m/s, the maximum temperature of the motor and the cooling water changes more slowly, indicating that continuing to increase the flow rate has few influences on the cooling effect, though the heat transfer coefficient still keeps increasing. Additionally, a high flow rate will also lead to an increase in inlet pressure and water pump power consumption.

Therefore, it is more reasonable to set the cooling water flow rate to 3 m/s.

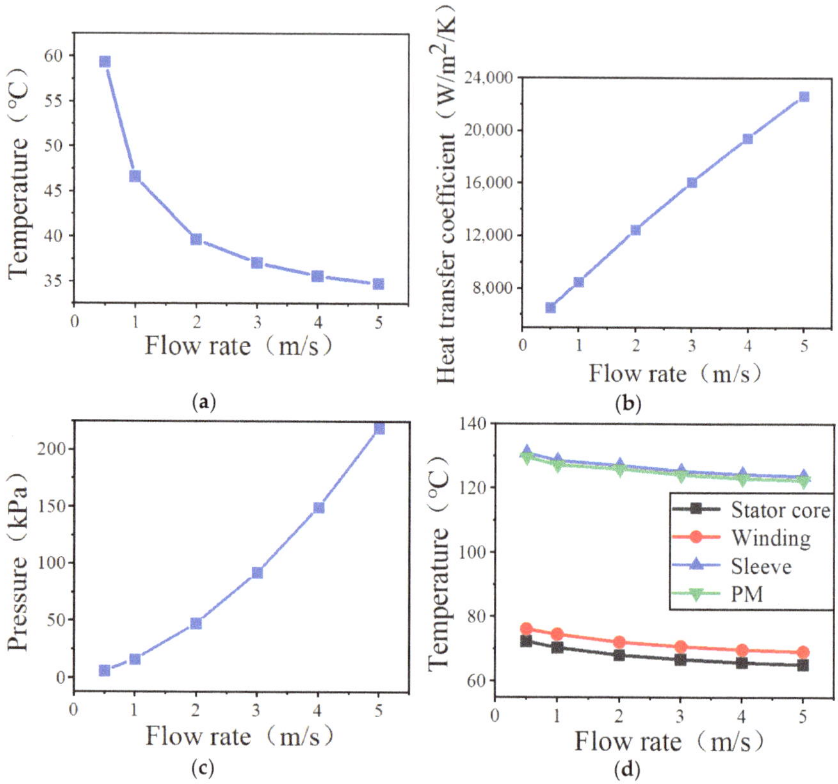

Figure 16. Thermal analysis results of HSPMMs at different cooling water flow rates. (**a**) Cooling water maximum temperature; (**b**) heat transfer coefficient; (**c**) inlet pressure; (**d**) motor maximum temperature.

3.4. Temperature Calculation Result

Finally, an eight turns spiral water channel is selected to protect the designed motor from overheating. The water flow rate is set as 3 m/s. The motor's temperature distribution calculated by CFD is shown in Figure 17. It can be found that the maximum temperature of the PM is 81.74 °C, which is much lower than the NdFeB's maximum permitted working temperature (180 °C), indicating the HSPMM designed in this paper has lower rotor loss density and the designed cooling method has better heat dissipation effect.

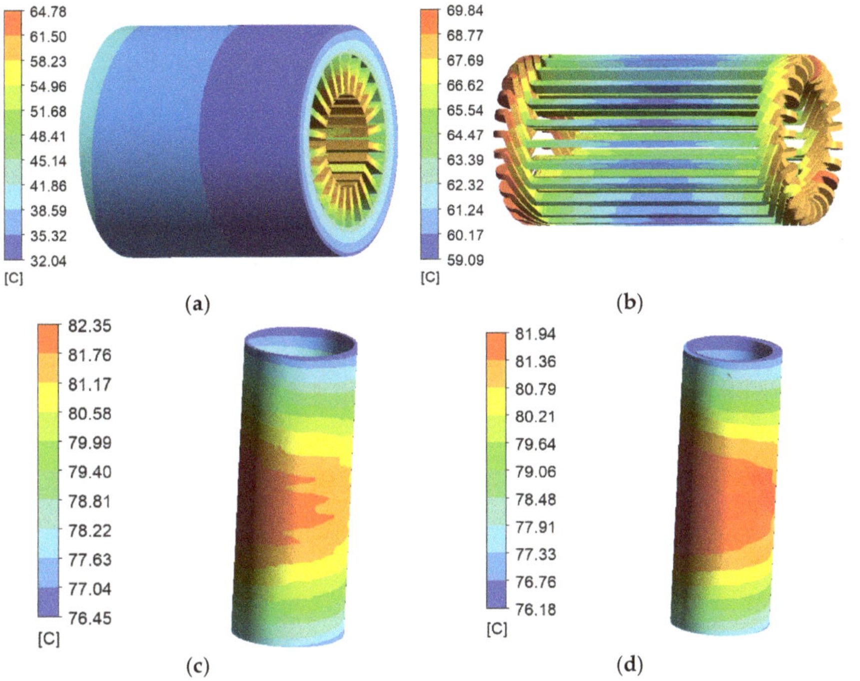

Figure 17. Temperature distribution of the designed motor. (**a**) Stator; (**b**) winding; (**c**) sleeve; (**d**) PM.

4. Stress Analysis

This paper provides the stress analysis for the sleeve–PM–core–shaft structure rotor of HSPMM by both the analytic method and FEM method [9]. The stainless sleeve with interference is adopted to protect the PM due to its better heat dissipation and strength. The stress analysis model of that structure rotor is shown in Figure 18.

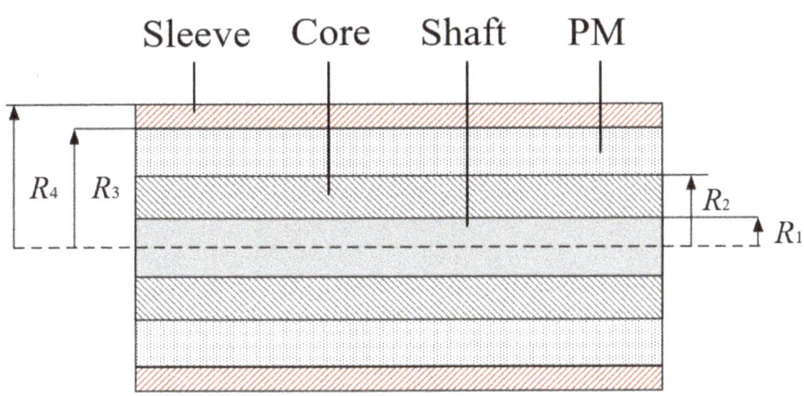

Figure 18. Stress analysis model of the sleeve–PM–core–shaft structure rotor.

4.1. Analytic Method

The sleeve–PM–core–shaft structure rotor can be equivalent to four thick-walled cylinder models, and the adjacent models are closely fitted. The pressure on the inner and outer surfaces of each cylinder model can be considered uniformly distributed. All the relevant properties of the cylinder model materials are isotropic. Then, a single-cylinder model is analyzed at first.

According to the material mechanics theory and geometric relationship, the strain-displacement relationship of the cylinder model can be described by [10]:

$$\begin{cases} \varepsilon_r = \frac{du}{dr} \\ \varepsilon_\theta = \frac{du}{dr} \end{cases} \tag{4}$$

where the ε_r is the cylinder radial strain, ε_θ is the cylinder tangential strain, u is the cylinder displacement.

According to the Hooke's law, the stress–strain relationship can be described by:

$$\begin{cases} \varepsilon_r = \frac{\sigma_r - \mu\sigma_\theta}{E} + \alpha\Delta T \\ \varepsilon_\theta = \frac{\sigma_\theta - \mu\sigma_r}{E} + \alpha\Delta T \end{cases} \tag{5}$$

where the σ_r is the cylinder radial stress, σ_θ is the cylinder tangential stress, μ is the cylinder Poisson's ratio, E is the cylinder young's modulus, α is the cylinder thermal expansion coefficient, ΔT is the cylinder temperature rise.

The force balance equation of the cylinder in the rotating condition can be described by:

$$\frac{d\sigma_r}{dr} + \frac{\sigma_r - \sigma_\theta}{r} + \rho\omega^2 r = 0 \tag{6}$$

where the ρ is the cylinder density, ω is the cylinder rotational angular velocity.

According to Equations (4)–(6), the u, σ_r, and σ_θ of a single-cylinder model can be obtained as follows:

$$u = Mr + \frac{N}{r} + \frac{(\mu^2 - 1)\rho\omega^2 r^3}{8E} \tag{7}$$

$$\begin{cases} \sigma_r = \frac{ME}{1-\mu} - \frac{NE}{(1+\mu)r^2} - \frac{(3+\mu)\rho\omega^2 r^2}{8} - \frac{E\alpha\Delta T}{1-\mu} \\ \sigma_\theta = \frac{ME}{1-\mu} + \frac{NE}{(1+\mu)r^2} - \frac{(1+3\mu)\rho\omega^2 r^2}{8} - \frac{E\alpha\Delta T}{1-\mu} \end{cases} \tag{8}$$

where the M and N is the calculation coefficient.

According to Equations (7) and (8), the calculation equations for the displacement and stress of the sleeve–PM–core–shaft structure rotor can be obtained as follows:

$$u_a = M_a r + \frac{N_a}{r} + \frac{({\mu_a}^2 - 1)\rho_a\omega^2 r^3}{8E_a} \tag{9}$$

$$\begin{cases} \sigma_{ra} = \frac{M_a E_a}{1-\mu_a} - \frac{N_a E_a}{(1+\mu_a)r^2} - \frac{(3+\mu_a)\rho_a\omega^2 r^2}{8} - \frac{E_a\alpha_a\Delta T_a}{1-\mu_a} \\ \sigma_{\theta a} = \frac{M_a E_a}{1-\mu_a} + \frac{N_a E_a}{(1+\mu_a)r^2} - \frac{(1+3\mu_a)\rho_a\omega^2 r^2}{8} - \frac{E_a\alpha_a\Delta T_a}{1-\mu_a} \end{cases} \tag{10}$$

where the a = 1, 2, 3, and 4, corresponding to the equations of the shaft, core, PM, and sleeve, respectively. Considering that the displacement at the center of the shaft is 0, so that the N_1 is 0.

The stress and displacement boundary conditions for the sleeve–PM–core–shaft structure rotor are as follows:

$$\begin{cases} u_1(r = R_1) = u_2(r = R_1) \\ u_2(r = R_2) = u_3(r = R_2) \\ u_4(r = R_3) - u_3(r = R_3) = \delta \\ \sigma_{r1}(r = R_1) = \sigma_{r2}(r = R_1) \\ \sigma_{r2}(r = R_2) = \sigma_{r3}(r = R_2) \\ \sigma_{r3}(r = R_3) = \sigma_{r4}(r = R_3) \\ \sigma_{r4}(r = R_4) = 0 \end{cases} \tag{11}$$

where the δ is the interference between sleeve and PM.

Finally, the calculation coefficients in Equations (9) and (10) and the complete stress analytical equations of the sleeve–PM–core–shaft structure rotor can be obtained by Equation (11).

4.2. FEM Verification

In order to verify the accuracy of the proposed analytic method, this paper adopts both the analytic method and FEM to calculate the stress distributions of a sleeve–PM–core–shaft structure rotor for comparison.

The stress distributions calculated by the two methods are shown in Figure 19. It can be found that there are certain acceptable errors in both the PM radial stress and tangential stress calculated by the two methods. Meanwhile, the analytical result of sleeve equivalent stress agrees well with the FEM one and the calculation error rate between the two methods is only 0.36%. Therefore, it can be considered that the analytic method provided by this paper can accurately calculate the stress distribution of the sleeve–PM–core–shaft structure rotor.

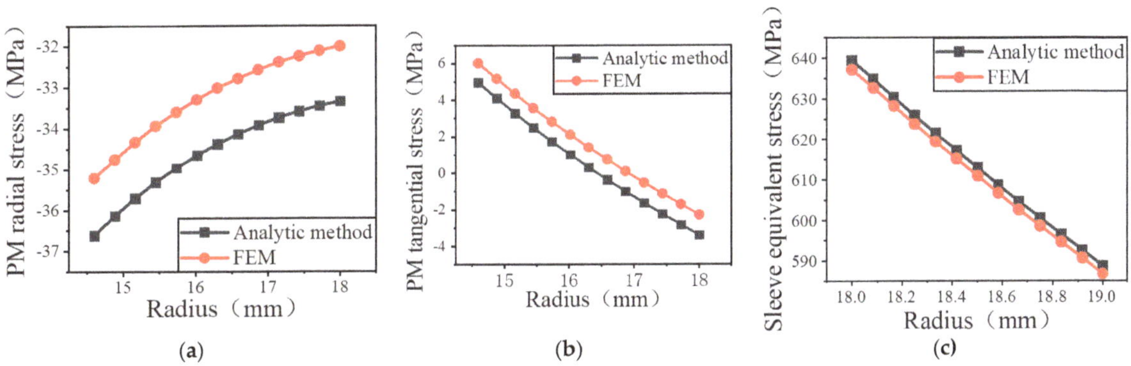

Figure 19. Stress distribution of the sleeve–PM–core–shaft structure rotor. (**a**) PM radial stress; (**b**) PM tangential stress; (**c**) sleeve equivalent stress.

4.3. Influencing Factors

Both the sleeve thickness and the interference have significant effects on the rotor stress distribution. The maximum stress of the rotor variations with sleeve thicknesses and interferences is presented in Figure 20. It can be found that increasing the interference is more conducive to reducing the maximum tensile stress of the PM in both radial and tangential directions, while the maximum equivalent stress of the sleeve will increase accordingly. Additionally, the larger thickness of the sleeve also has the advantage of reducing the tensile stress of the PM, but it has little effect on the reduction of sleeve equivalent stress.

4.4. Final Rotor Mechanical Protection Scheme

According to the above analysis, this paper finally adopts a stainless steel sleeve of 1 mm to mechanically protect the rotor of the HSPMM, and the interference between the sleeve and the PM is set to 0.05 mm. Then, the designed motor's rotor stress distribution is calculated by FEM and the result of which is shown in Figure 21. It is learned that the PM bears compressive stress in both radial and tangential directions, while the PM material has a strong tolerance to compressive stress. Meanwhile, the maximum equivalent stress of the sleeve is 537.19 MPa, which is lower than the permitted value of stainless steel (1100 MPa). Therefore, the final rotor mechanical protection scheme designed in this paper is reliable.

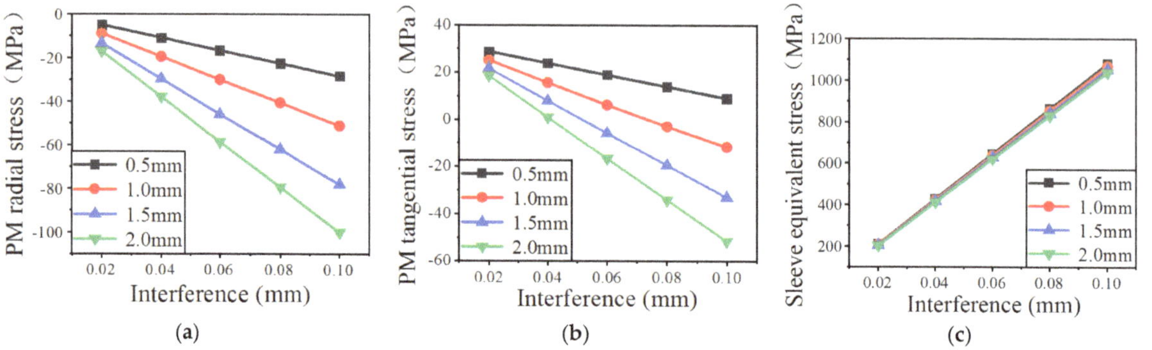

Figure 20. The maximum stress of the rotor variations with sleeve thickness and interference. (**a**) PM maximum radial stress; (**b**) PM maximum tangential stress; (**c**) sleeve maximum equivalent stress.

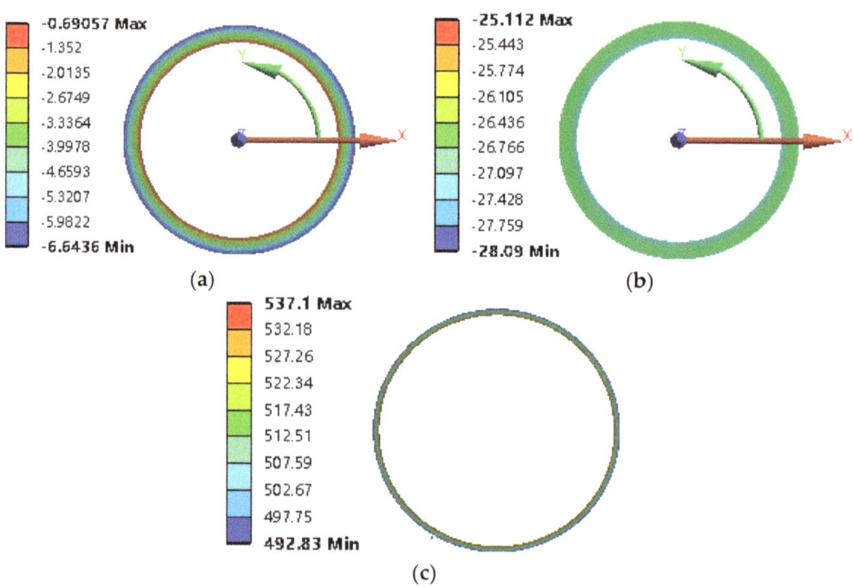

Figure 21. The designed motor's rotor stress distribution. (**a**) PM radial stress; (**b**) PM tangential stress; (**c**) sleeve equivalent stress.

5. Conclusions

This paper provides the design and analysis of a 30 kW, 30,000 r/min high-speed permanent magnet motor (HSPMM) for compressor application. Firstly, the effects of the number of poles, the number of slots, and the PM magnetization methods on the electromagnetic performance of the motor are studied. It is found the four poles, and twenty-four slots motor with PM parallel magnetized has better electromagnetic performance. Then, the CFD is adopted to provide thermal analysis for HSPMMs with different cooling channel structures and different channel turn numbers, and it is found the eight turns spiral water channel with a flow rate of 3 m/s is more beneficial for heat dissipation. The analytic method of the sleeve–PM–core–shaft structure rotor stress distribution is proposed in this paper. This paper provides a comparative analysis between the analytic method and the FEM of the rotor stress, proving the analytic method is reliable. Finally, the influencing factors of rotor stress distribution are analyzed in this paper. According to the analysis results, this paper adopts a stainless steel sleeve of 1 mm to mechanically protect the rotor of the HSPMM, and the interference between the sleeve and the PM is set to 0.05 mm. The rotor stress FEM calculation results of the designed motor show the fact that the rotor protection scheme adopted in this paper is reliable.

Author Contributions: Conceptualization, Z.Q. and Y.Z.; methodology, Z.Q. and Y.Z.; software, Y.Z., S.Y. and Z.X.; validation, Y.Z., S.Y. and Z.X.; formal analysis, Y.Z.; investigation, Z.Q. and Y.Z.; resources, Y.Z.; data curation, Z.Q.; writing—original draft preparation, Z.Q.; writing—review and editing, Z.Q. and Y.Z.; visualization, Z.Q. and Y.Z.; supervision, Y.Z.; project administration, Y.Z.; funding acquisition, Y.Z. All authors have read and agreed to the published version of the manuscript.

Funding: This research was funded by the Ministry of Science and Technology under the National Key R&D Program of China (Grant No. 2021YFE0108600). And this research is also funded by the National Natural Science Foundation of China under grant 52077121, and the National Natural Science Foundation of China Key International Cooperation project under grant 51920105011. This research is also funded by the Natural Science Foundation of Liaoning province under grant 2021-KF-24-01.

Institutional Review Board Statement: Not applicable.

Informed Consent Statement: Not applicable.

Data Availability Statement: Not applicable.

Conflicts of Interest: The authors declare no conflict of interest.

References

1. Gerada, D.; Mebarki, A.; Brown, N.L.; Gerada, C.; Cavagnino, A.; Boglietti, A. High-speed Electrical Machines: Technologies, Trends, and Developments. *IEEE Trans. Ind. Electron.* **2014**, *61*, 2946–2959. [CrossRef]
2. Bianchi, N.; Bolognani, S.; Luise, F. Potentials and Limits of High-Speed PM Motors. *IEEE Trans. Ind. Appl.* **2004**, *40*, 1570–1578. [CrossRef]
3. Li, S.; Li, Y.; Choi, W.; Sarlioglu, B. High-Speed Electric Machines: Challenges and Design Considerations. *IEEE Trans. Transp. Electrif.* **2016**, *2*, 2–13. [CrossRef]
4. Zhang, F.; Dai, R.; Liu, G.; Cui, T. Design of HSIPMM Based on Multi-physics Fields. *IET Electr. Power Appl.* **2018**, *12*, 1098–1103. [CrossRef]
5. Jing, L.; Tang, W.; Wang, T.; Ben, T.; Qu, R. Performance Analysis of Magnetically Geared Permanent Magnet Brushless Motor for Hybrid Electric Vehicles. *IEEE Trans. Transp. Electrif.* **2022**, *8*, 2874–2883. [CrossRef]
6. Jing, L.; Pan, Y.; Wang, T.; Qu, R.; Cheng, P.T. Transient Analysis and Verification of a Magnetic Gear Integrated Permanent Magnet Brushless Machine with Halbach Arrays. *IEEE J. Emerg. Sel. Top. Power Electron.* **2022**, *10*, 1881–1890. [CrossRef]
7. Dong, J.; Huang, Y.; Jin, L.; Lin, H.; Yang, H. Thermal Optimization of a High-Speed Permanent Magnet Motor. *IEEE Trans. Magn.* **2014**, *50*, 7018504. [CrossRef]
8. Tong, W.; Sun, R.; Zhang, C.; Wu, S.; Tang, R. Loss and Thermal Analysis of a High-Speed Surface-Mounted PMSM with Amorphous Metal Stator Core and Titanium Alloy Rotor Sleeve. *IEEE Trans. Magn.* **2019**, *55*, 8102104. [CrossRef]
9. Du, G.; Xu, W.; Zhu, J.; Huang, N. Rotor Stress Analysis for High-Speed Permanent Magnet Machines Considering Assembly Gap and Temperature Gradient. *IEEE Trans. Energy Convers.* **2019**, *34*, 2276–2285. [CrossRef]
10. Shen, J.; Qin, X.; Yao, L.; Wang, Y. Rotor Strength Analysis and Retaining Sleeve Design for High-speed PM Machines. *Proc. CSEE* **2022**, *42*, 2334–2346.

energies

MDPI

Article

Insulation Degradation Analysis Due to Thermo-Mechanical Stress in Deep-Sea Oil-Filled Motors †

Jian Zhang [1], Rui Wang [1], Youtong Fang [1] and Yuan Lin [2,*]

1 College of Electrical Engineering, Zhejiang University, Hangzhou 310027, China; jian_zhang_zju@zju.edu.cn (J.Z.); 22010150@zju.edu.cn (R.W.); youtong@zju.edu.cn (Y.F.)
2 Institute of Ocean Engineering and Technology, Ocean College, Zhejiang University, Zhoushan 316021, China
* Correspondence: samylin@zju.edu.cn; Tel.: +86-177-0653-8206
† This paper is an extended version of our paper published in IEEE 4th Student Conference on Electric Machines and Systems (SCEMS), Huzhou, China, 1–3 December 2021.

Abstract: With the wide application of motors in deep sea exploration, deep-sea motors require a higher power density and a longer lifetime. Motor lifetime mainly depends on the thermo-mechanical stress (TMS) load on its stator insulation. Unlike normal motors, deep-sea motors are usually filled with oil to compensate for the high pressure generated by seawater, which leads to high additional viscous drag loss. This, combined with the high pressure, will greatly change the TMS distribution and further influence motor insulation lifetime. Thus, the insulation degradation analysis of deep-sea oil-filled (DSOF) motors due to TMS has become important. This paper presents a TMS analytical model of DSOF motor insulation, considering the joint effects of high pressure and motor temperature. The CFD method is adopted to perform motor thermal analysis, considering temperature effects on viscous drag loss. The FEA method is adopted for thermo-mechanical analysis and to verify the analytical model accuracy. Rainflow counting and the Miner fatigue method are adopted to evaluate motor lifetime. Results show that compared with motors working in normal environments, TMS on DSOF motor insulation can be reduced by up to 59.5% due to high pressure and the insulation lifetime can be increased by up to 16.02%. Therefore, this research can provide references for high power density DSOF motor design.

Keywords: viscous loss; oil-filled motor; thermo-mechanical stress; CFD analysis; insulation degradation

Citation: Zhang, J.; Wang, R.; Fang, Y.; Lin, Y. Insulation Degradation Analysis Due to Thermo-Mechanical Stress in Deep-Sea Oil-Filled Motors. *Energies* **2022**, *15*, 3963. https://doi.org/10.3390/en15113963

Academic Editor: Issouf Fofana

Received: 31 March 2022
Accepted: 24 May 2022
Published: 27 May 2022

1. Introduction

Stator insulation systems (SISs) play a key role in affecting motor service life. The differences in thermal expansion coefficients among insulation components, including ground insulation, copper wires, coatings and epoxy fillings in stators, will induce thermo-mechanical stress (TMS) in the SIS. TMS is considered to be a critical factor causing insulation degradation [1]. Deep-sea motors usually adopt oil-filled methods to balance the seawater pressure, which brings two challenges to TMS assessment compared to motors running under normal conditions. One is that the oil filled in the air gap will lead to high additional temperature-dependent viscous drag loss and changing motor heat dissipation conditions, which will greatly influence the motor thermal distribution. The other is that the motor SIS is subjected to high pressure caused by seawater, which combined with the thermal distribution will change the TMS distribution greatly. In order to analyze the insulation degradation of a DSOF motor due to the TMS, these two challenges have to be considered.

In recent years, many researchers have addressed the thermal analysis issue of motors working on land, but few studies have focused on DSOF motors. The motor loss and heat dissipation conditions in DSOF motors are quite different from normal motors. The viscous loss induced by oil may account for more than 50% of the total loss [2]. Additionally, the thermal conductivities and heat transfer coefficients are tightly related to the fluid and temperature fields of DSOF motors. Therefore, numerical analyses such as the finite element method (FEM) and computational fluid dynamics (CFD) method should be adopted for the motor thermal analysis instead of analytical calculation. The advantage of the CFD method in calculating viscous drag loss compared with analytic methods has been demonstrated on a motor test rig [3]. The necessity of coupled field analysis by taking the viscous loss component into consideration in DSOF motor design has been illustrated in [4]. Reference [5] constructs a three-dimensional fluid-thermal coupled model of an oil-filled motor to obtain the heat transfer coefficients, fluid flow characteristics and temperature distribution. However, most of the published literature ignores the influence of temperature on the gap oil viscosity and the viscous loss. This paper focuses on this issue by using the viscosity user-defined function (UDF) to improve the accuracy of the DSOF motor temperature field calculation, based on which TMS analysis and lifetime assessment of the DSOF motor are performed.

TMS has been extensively studied in various industrial fields. The mismatch of materials' thermal expansion coefficients can induce TMS in electronic assemblies not only by temperature cycling during normal operation, but also due to the high temperatures experienced during fabrication, shipping and storage [1]. References [6,7] present a degradation model investigating the thermo-mechanical fatigue in insulated gate bipolar transistor modules and apply it to degradation estimation as well as to the accurate lifetime assessment of power modules. References [8,9] analyze the failure of a ceramic ball grid array package used for an FPGA chip caused by TMS and both analytical modeling and numerical simulations are performed to analyze the TMS, which is validated to be the main factor causing FPGA chip failure. In the paper [10], fatigue life evaluation for multiple epoxy laminate composites considering the stress versus cycle life has been investigated. Reference [11] focuses on the thermo-mechanical fatigue of silicon and molybdenum (SiMo) alloyed ductile iron by performing FEA and testing the stresses of the specimen. In IEEE Standard-56, TMS is regarded as one of the aging factors of motor insulation [12]. As for electric motors, researchers have derived the TMS analytical equation of a single wire in epoxy-impregnated traction motor stator windings and established 3D FEA models to demonstrate it [13]. Reference [14] analyzes cooling methods to reduce the mechanical stress, which will cause fatigue and the degradation of motor insulation, induced by the thermal loading on motor windings impregnated with epoxy, and builds an experimental set-up to validate the numerical analysis results. Some researchers have performed thermo-structural simulations on a segmented stator winding geometry numerically to quantify the TMS induced in the motor windings due to high temperature [5]. However, the work on insulation degradation analysis caused by TMS in electric motors is far from sufficient, especially for motors working in the deep sea. For DSOF motors, the SIS bears a high pressure varying with the operating depth, which can reach more than 100 MPa for the working depth of 10,000 m. The high pressure will change the TMS distribution and further affect the motor lifetime. This paper seeks to address the insulation TMS distribution issue considering a high-pressure environment based on the accurate thermal calculation results of DSOF motors.

This paper will take the brushless DC (BLDC) motor as the target due to its outstanding performance and wide application in deep-sea equipment. The main contributions of this paper are as follows:

1. Considering the deep-sea high-pressure effects and structural characteristics of the DSOF motor SIS, an analytical model of TMS for stator insulation in DSOF motors is proposed. The effects of slot dimension and copper fill factor on the TMS are investigated, and the analytical model shows that TMS in DSOF motor stator insulation is closely related to the external seawater pressure, the motor temperature rises and the motor structure dimensions.

2. To obtain the DSOF motor temperature rises for the further TMS analysis, a 3D CFD model of a DSOF motor is established considering the sensitivity of oil viscosity to temperature. Two UDFs are compiled for the temperature field calculation. In addition, the influence of viscosity on the heat transfer coefficient of the motor is analyzed, and the importance of the viscous loss and viscosity UDFs in the temperature field in respect of the accurate calculation of the DSOF motor is demonstrated.

3. Based on the calculation results of the motor temperature distribution, the analytical TMS model is used to obtain the stator insulation TMS variation trend with the seawater pressure. It is found that the seawater pressure can decrease the TMS to 59.5% at most. Then, based on the DSOF motor model, numerical analysis is carried out to obtain the coupled TMS through thermal-mechanical simulation, which further demonstrates the accuracy of the proposed analytical model.

4. The TMS spectra of dangerous points on the stator insulation are synthesized. By counting the rainflow of the TMS spectra, combining the S-N curve of stator insulation materials and using Miner fatigue theory, the lifetime of the DSOF motor is predicted. The research can afford a new perspective for high power density DSOF motor design. Figure 1 shows the flow chart of fatigue lifetime evaluation of DSOF motors.

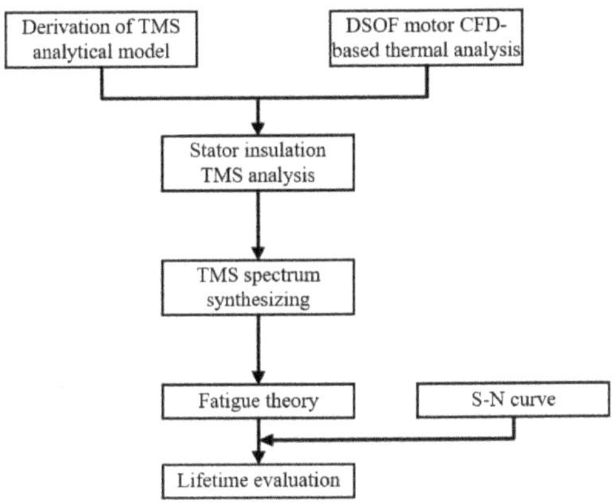

Figure 1. Flow chart of DSOF motor fatigue lifetime prediction.

This paper is organized as follows. Section 2 derives the analytical model of TMS in DSOF motor SISs by considering deep-sea high pressure, motor temperature rises and structure dimensions. Section 3 focuses on the temperature field calculation of DSOF motors based on fluid-thermal coupling analysis. In Section 4, the temperature results obtained by CFD analysis are applied to further calculate the TMS. Thermal-structure simulations are conducted to verify the accuracy of the proposed TMS analytical model. In addition, the motor insulation degradation and lifetime prediction are investigated based on the TMS analysis results. Conclusions are given in Section 5.

2. Analytical Model

Figure 1 shows the flow chart of fatigue lifetime evaluation of DSOF motors. The derivation of the TMS analytical model is performed first. For a DSOF motor, the SIS is adhered to stator laminations and compressed copper wires. The polymer coating and copper have similar thermal expansion coefficients, which are quite different from those of the stator core and epoxy impregnation. Thus, high TMS will be introduced into the stator insulation when the motor temperature rises. Moreover, the high density of the copper wires and irregular shape of the stator slots will make derivation of the TMS analytical model very complex. In order to facilitate the task while ensuring the model accuracy, the derivation is implemented by a combination of preliminary derivation and post-modification. That is, we disregard the effect of stator slots and windings in the preliminary derivation and investigate them in the post-modification.

2.1. Preliminary Derivation of TMS Analytical Model

The stator model ignoring the stator slots and windings is shown in Figure 2, where the outer layer is the stator core layer and the inner layer is the impregnation insulation layer. The preliminary aim is to obtain the TMS analytical model of the inner layer. In a deep-sea high-pressure environment, the inner and outer surfaces of the model are subjected to the same pressure due to the oil-filled method.

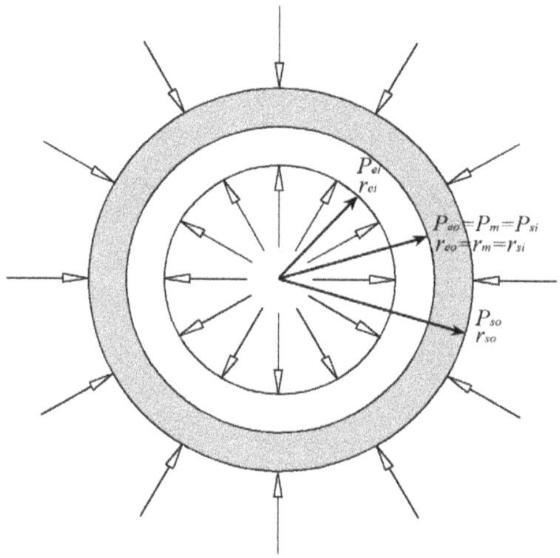

Figure 2. Cross-section of simplified DSOF motor stator model subjected to both internal and external pressures.

Since the geometry of motor stator, the operating temperature distribution and the pressure constraints are all symmetrical about the central axis, the derivation of the TMS analytical model can be performed based on a cylindrical coordinate system. According to the strain–stress theory, strain-displacement correlations and Lamé equations for spatially axisymmetric hollow cylinders [15–17], the stress components of the inner layer can be obtained provided the pressure in the common boundary is known. The two layers have equal radial displacements at the common boundary. Accordingly, the outer surface pressure of the insulation layer can be determined. As shown in Figure 2, P_{si} and P_{so} are the internal and external pressures of the stator core layer, respectively. r_{si} and r_{so} are the stator core layer inner and outer radii. As for the insulation layer, P_{ei} and P_{eo} are its internal and external pressures, respectively. r_{ei} and r_{eo} are its inner and outer radii, respectively. The pressure at the common surface is P_m, which is equal to P_{eo} and P_{si}. The radius of the common boundary is r_m, which is equal to r_{eo} and r_{si}. By means of Lamé equations, the stresses at the outer surface of the insulation layer and the stresses at the inner surface of the stator core layer are obtained, which can be further substituted into the strain–stress equation. Then, the radial displacements of the insulation layer and stator core layer at the common boundary, u_{eo} and u_{si}, can be obtained by means of strain-displacement correlation as follows:

$$u_{eo} = \frac{r_{eo}}{E_e}\left[\frac{2P_{ei}-P_{eo}\left(R_e^2+1\right)}{R_e^2-1} - \mu_e\left(\frac{P_{ei}-R_e^2P_{eo}}{R_e^2-1} - P_{eo}\right)\right] + r_{eo}\alpha_e T \quad (1)$$

$$u_{si} = \frac{r_{si}}{E_s}\left[\frac{P_{si}\left(R_s^2+1\right)-2R_s^2P_{so}}{R_s^2-1} - \mu_s\left(\frac{P_{si}-R_s^2P_{so}}{R_s^2-1} - P_{si}\right)\right] + r_{si}\alpha_s T \quad (2)$$

where

$$R_e = \frac{r_{eo}}{r_{ei}}$$

$$R_s = \frac{r_{so}}{r_{si}}$$

R_e denotes the ratio of the insulation layer outer radius to its inner radius. R_s denotes the ratio of the stator layer outer radius to its inner radius. E_e is Young's modulus, μ_e is Poisson's ratio and α_e is the coefficient of thermal expansion of the insulation layer. E_s, μ_s and α_s denote the same properties of the stator core layer. T denotes model temperature rise. Considering a seawater pressure P equal to P_{ei} and P_{so}, and with Equation (1) equal to Equation (2), the pressure P_m can be expressed as

$$P_m = \frac{V_{io}P - V_nT}{V_m} \tag{3}$$

where

$$V_{io} = \frac{2-\mu_e}{R_e^2-1} + \frac{R_s^2(2-\mu_s)}{R_s^2-1} \cdot \frac{E_e}{E_s}$$

$$V_m = \frac{(1-2\mu_e)R_e^2+\mu_e+1}{R_e^2-1} + \frac{(\mu_s+1)R_s^2-2\mu_s+1}{R_s^2-1} \cdot \frac{E_e}{E_s}$$

$$V_n = E_e(\alpha_s - \alpha_e)$$

V_{io}, V_m and V_n are the coefficients related to the motor dimensions and material properties. Substituting Equation (3) and insulation layer parameters into the Lamé equations then using the von Mises theory, the average TMS on the stator insulation layer, σ_a, can be expressed as

$$\sigma_a = \varsigma(r) \cdot \sqrt{\frac{[(V_m - V_{io})P + V_nT]^2}{V_m^2}} \tag{4}$$

where

$$\varsigma(r) = 2.25\frac{r_{eo}^2}{(r_{eo}-r_{ei})(R_e^2-1)}\int_{r_{ei}}^{r_{eo}} \frac{1}{r^2}\mathrm{d}r$$

$\varsigma(r)$ is a function of the stator insulation layer radius. The preliminary TMS analytical model notes that there is a minimum value of average TMS on the SIS as the deep-sea pressure increases. The TMS will first decrease to a minimum value, and then increase with the deep-sea pressure rising.

2.2. Post-Modification

Subsequently, the influence of the stator slot dimension and copper fill factor on the TMS of the motor SIS are explored using Ansys software to refine the analytical model.

2.2.1. The Effect of Slot Dimension

For investigating the influence of slot dimension on the TMS, three stator models A_1, A_2 and A_3 with stator teeth widths of 9, 13.4 and 16 mm are analyzed, respectively. The results are presented in Figure 3. The black line is the preliminary analytical model solution, while the others are from the simulation results. The minimum TMS points of the three model simulation results show no difference from the analytical solution, but the slopes are inconsistent and the difference will become larger when the stator slots become narrower. Consequently, the slot shape factor (SSF), s, denoting the ratio of the stator slot width to the stator yoke height is introduced to weigh the slot shape. Interestingly, the SSF only affects the changing rate of the TMS, not the minimum value. Therefore, the slope-modified coefficient Q_s is introduced to refine the preliminary derivation model.

To further quantify the influence of SSF, another six FEA models A_4 to A_9 with different SSFs are constructed and simulated. The correlation of the modified coefficient Q_s with s is shown in Figure 4. Remarkably, the slope-modified coefficient Q_s will increase to a stable value of 0.9123 as the slot shape factor grows.

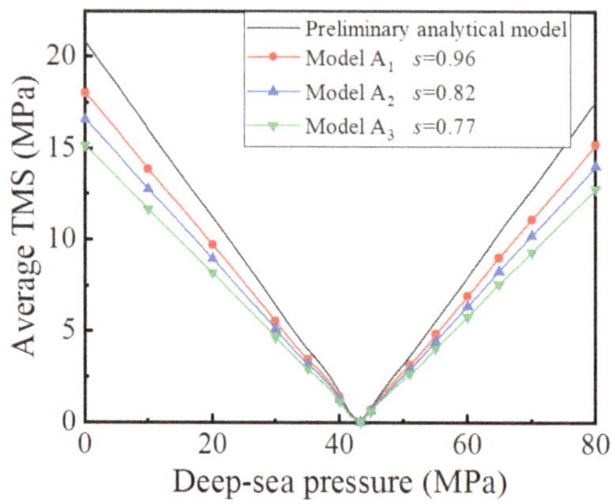

Figure 3. Average TMS of different slot shape factor variations with deep-sea pressure.

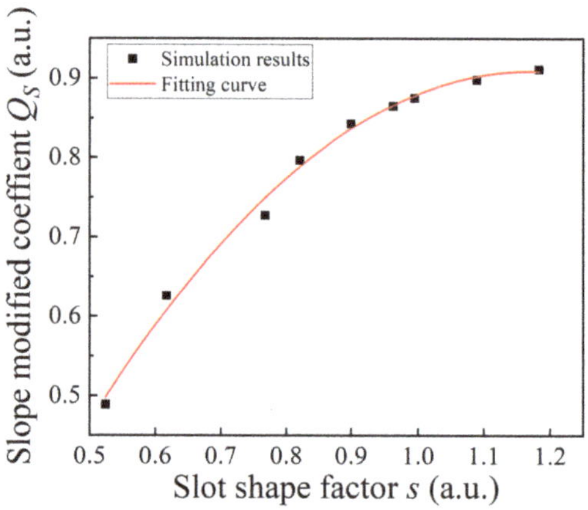

Figure 4. Slope-modified coefficient versus slot shape factor.

2.2.2. The Effect of Copper Fill Factor

For investigating the effects of the influence of the copper wire fill factor on the average TMS of the SIS, FEA simulations are performed on three stator FEA models with same SSF of 0.96 and different copper fill factors (CFFs), f, of 0, 0.3 and 0.4 are simulated. The results are presented in Figure 5. The black line still represents the analytical solution of the preliminary model, while others are from the simulation results. The three model simulation results of deep-sea pressure corresponding to the minimum TMS agree with the analytical solution, while the curve slope shows some inconsistencies. When the CFF becomes larger, the TMS changing rate will decrease, and the decrease rate will become smaller. Meanwhile, the lowest point moves upwards, and the slope closer to the minimum is gentler. Therefore, another slope-modified coefficient Q_f is introduced to improve the preliminary derivation model.

To further explore the influence of CFF, five FEA models with different CFFs are investigated. The variation in the slope-modified coefficient Q_f with f is shown in Figure 6. Notably, the slope-modified coefficient Q_f will decrease to a stable value of 0.7913 as the CFF grows.

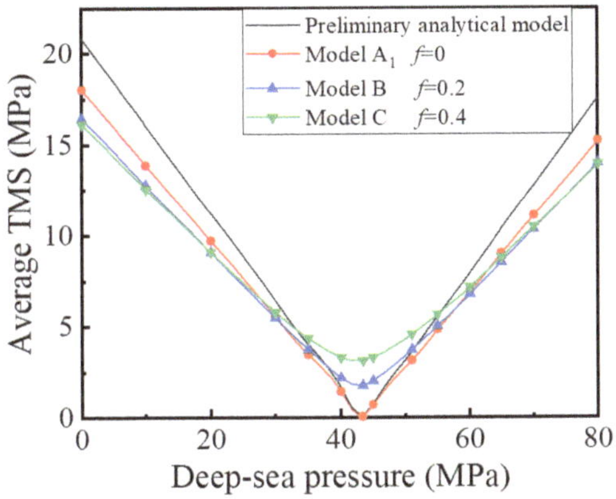

Figure 5. Average TMS of different copper fill factor variations with deep-sea pressure.

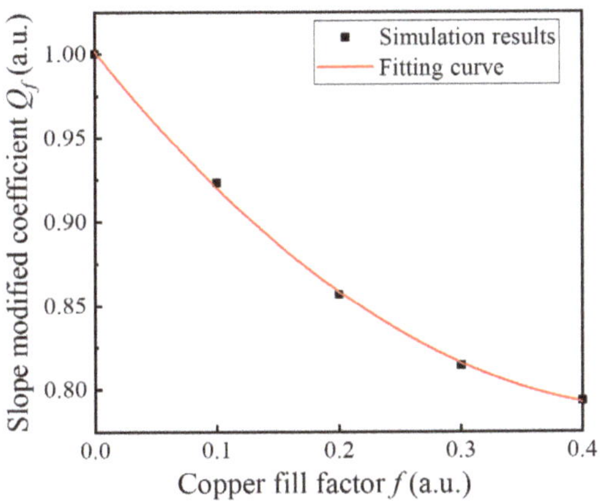

Figure 6. Slope-modified coefficient versus copper fill factor.

Eventually, the modified TMS analytical model can be expressed as

$$\sigma_a = Q_s Q_f \varsigma(r) \cdot \sqrt{\frac{[(V_m - V_{io})P + V_n T]^2}{V_m^2}} \quad (5)$$

In Equation (5), the average TMS for the DSOF motor SIS is not only related to the material properties and dimensions of the motor, but also to the temperature change and the ambient pressure. When the motor material, dimensions and temperature rise are fixed, the TMS on the insulation will first decline to a minimum value and then increase as the DSOF motor work depth increases. By means of the analytical model, the external pressure corresponding to the minimum TMS based on the DSOF motor temperature rise can be predicted.

3. Temperature Distribution of DSOF Motors Based on Fluid-Thermal Coupling Analysis

DSOF motor thermal analysis needs to focus on heat sources and heat dissipation. For heat sources, except copper loss, iron loss and eddy current loss, additional viscous loss generated by oil filled in the air gap has to be considered. For heat dissipation, direct contact with oil and seawater makes the heat removal of DSOF motors quite different from normal ones. Thus, the key to the thermal analysis of DSOF motors is the accurate calculation of viscous loss and heat transfer coefficients. Viscous loss is closely related to rotation speed and oil viscosity. For an oil-filled motor, the inner and outer radii of the air gap are r_1 and r_2, respectively, and according to the boundary conditions of the gap oil and fluid momentum equation, the mechanical loss consumed by the force of oil acting on the rotor surface can be determined. This mechanical loss, namely viscous loss, is defined as Equation (6) [18,19]:

$$P_{\text{oil}} = \frac{4\pi\mu\omega^2 L{r_2}^2{r_1}^2}{\delta(r_2 + r_1)} \tag{6}$$

where μ denotes fluid dynamic viscosity (Pa · s), ω denotes rotor angular velocity, L is rotor axial length and δ is the air gap width. It can be noted that viscous loss has a linear relation with dynamic viscosity for a motor with determined dimensions and rotational speed. Obviously, viscosity has a critical impact on viscous loss.

3.1. Viscous Drag Loss at Different Temperature

Viscosity has a close relation with fluid temperature. The distance between fluid molecules will expand and the fluid viscosity will decrease significantly as the temperature increases [20,21]. Hence, viscous loss calculation needs to take the temperature into account. The experience formula describing the viscosity variations with temperature, namely the Poisson formula, is shown in Equation (7) [22,23]:

$$\mu_t = v\rho = \mu_0 e^{-\lambda(t-t_0)} \tag{7}$$

where μ_t denotes the dynamic viscosity at temperature t, μ_0 denotes the dynamic viscosity at temperature t_0, and λ is the viscosity-temperature coefficient (VTC) of the liquid, reflecting the viscosity decreasing rate. Clearly, the viscosity has an e-exponential correlation with the temperature. Unlike no-oil motors, the empirical value of VTC 0.035 cannot be directly adopted for DSOF motors [24]. It is necessary to introduce CFD numerical calculation methods. A 24-slot and 8-pole DSOF motor (Sanao Electrical, Shanghai, China) with 2.62 KW rated power and 5000 rpm operating speed was adopted as the research subject. Its 3D CFD model was established as presented in Figure 7. Figure 8 shows the meshing details, comprising a total of 3,533,129 meshing cells.

Figure 7. 3D CFD model of the DSOF motor.

According to the CFD fluid field model of the motor, oil viscosities at different temperatures are obtained and the corresponding viscous losses are determined. For subsequent motor temperature analysis, the temperature and corresponding viscous loss are fitted and compiled into a UDF, namely a viscous loss UDF. The relation between the viscous loss and temperatures is presented as Equation (8), where P_{oil} is viscous loss and T is the temperature of the gap oil.

$$P_{\text{oil}} = 3.576 \times 10^{-5}T^4 - 1.064 \times 10^{-2}T^3 + 1.154 \times 10T^2 - 54.356T + 1050.62 \tag{8}$$

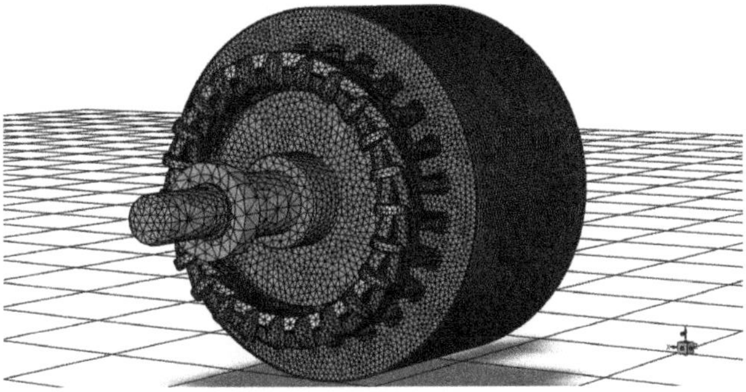

Figure 8. Meshing results of the DSOF motor.

3.2. The Need for a Viscous Loss UDF and Viscosity UDF

Viscosity changing with temperature not only affects the heat source (viscous loss) but also the heat dissipation of the DSOF motor. Therefore, the function of viscosity changing with temperature should be compiled into a viscosity UDF and adopted in the temperature field numerical calculation. By fitting with a polynomial, the relation between the oil viscosity and temperature is expressed as follows:

$$\begin{aligned}\mu = &-1.201 \times 10^{-10}T^5 + 5.387 \times 10^{-8}T^4 - 9.284 \times 10^{-6}T^3 + \\ &7.735 \times 10^{-4}T^2 - 3.162 \times 10^{-2}T + 0.533\end{aligned} \quad (9)$$

where μ is the oil viscosity. Then, the viscosity UDF is compiled and added to the CFD analysis. After fully investigating the heat source and the dissipation condition, a thermal analysis of the DSOF motor is performed. CFD analysis both with and without the UDFs is carried out for investigating the influence of the UDFs. The temperature results of each motor component are compared in Table 1. The initial temperature is set to 293 K.

Table 1. Thermal analysis with and without viscous loss and viscosity UDFs.

Component	Temperature (K)					
	Minimum		Maximum		Average	
	No UDFs	With UDFs	No UDFs	With UDFs	No UDFs	With UDFs
Windings	331.7	325.1	332.7	326.1	332.2	325.6
Rotor	332.1	325.2	333.8	326.4	333.0	325.8
Magnets	333.9	326.5	335.2	327.6	334.7	327.2
Shaft	327.2	321.2	331.9	325.0	330.8	324.1
Stator	326.0	320.3	331.2	324.6	328.3	322.2

It is noted that motor component temperatures with viscous UDF and viscosity UDF are lower than those without the UDFs. Temperature rises of the rotor, permanent magnets and stator of the DSOF motor have decreased by 17.96%, 18.06% and 17.17%, respectively, indicating that viscosity indeed affects the heat transfer coefficient and heat source of the DSOF motor, and further demonstrating the necessity of viscous and viscosity UDFs in the accurate thermal calculation of DSOF motors.

3.3. Temperature Distribution of DSOF Motor

The temperature distribution of the whole DSOF motor is shown in Figure 9. The motor maximum temperature is 327 K and minimum temperature is 321 K. It is noted that the temperature rises of all motor components are almost the same.

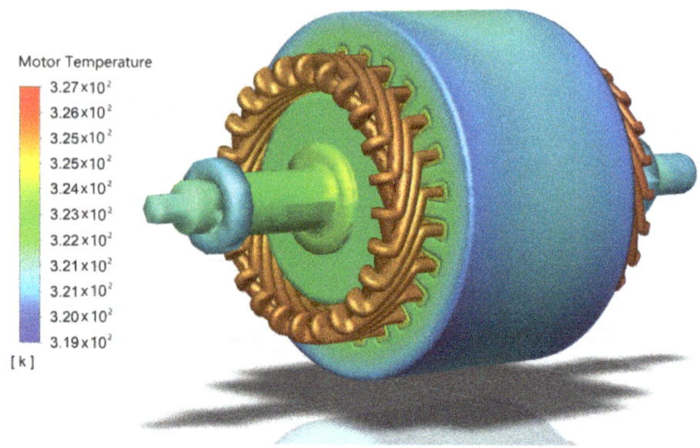

Figure 9. DSOF motor temperature distribution.

Figure 10 shows cross-section temperature contour details of the DSOF motor. Obviously, for the stator components, the maximum temperature difference is only 4 K. Therefore, the temperature distribution of the SIS can be regarded as uniform and spatially axisymmetric. The thermal analysis results will subsequently be used for the following TMS analysis of the DSOF motor.

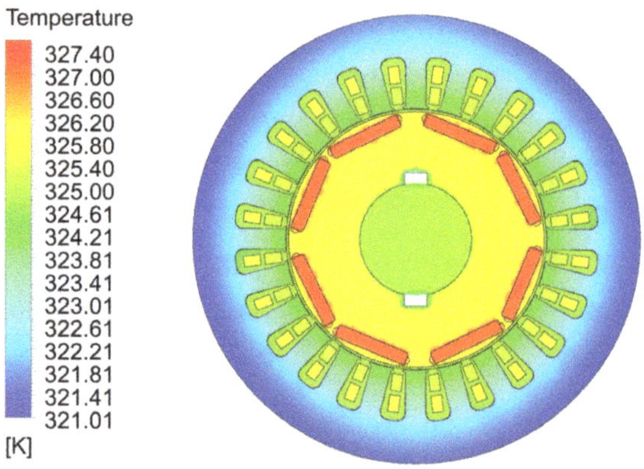

Figure 10. DSOF motor cross-section temperature field contour.

4. Thermo-Mechanical Simulation and Lifetime Analysis of the DSOF Motor

4.1. Thermo-Mechanical Simulation

Before analyzing the TMS of the DSOF motor by means of the analytical model proposed in Section 2, thermal-mechanical simulations are conducted to verify the model accuracy. The relevant material properties are listed in Table 2.

Table 2. Relevant material properties of the motor stator.

Properties	Electrical Sheet	Polymer Coating	Epoxy Impregnation
Young's modulus (MPa)	2×10^5	7.4×10^3	3.5×10^3
Poisson's ratio	0.29	0.42	0.34
Coefficient of thermal expansion (1/K)	1.2×10^{-5}	1.6×10^{-5}	5×10^{-5}
Thermal conductivity (W/m·K)	28	0.2	0.21
Heat capacity (J/kg·K)	440	1090	400
Density (kg/m^3)	7650	1530	1180

According to previous thermal processing results and the proposed TMS analytical model, the optimal working environment pressure for the DSOF motor is 20 MPa, where the TMS on the stator insulation system is the smallest. Therefore, thermal-mechanical simulations of the motor under normal pressure and 20 MPa are compared for further investigation. Figure 11 presents the simulation results.

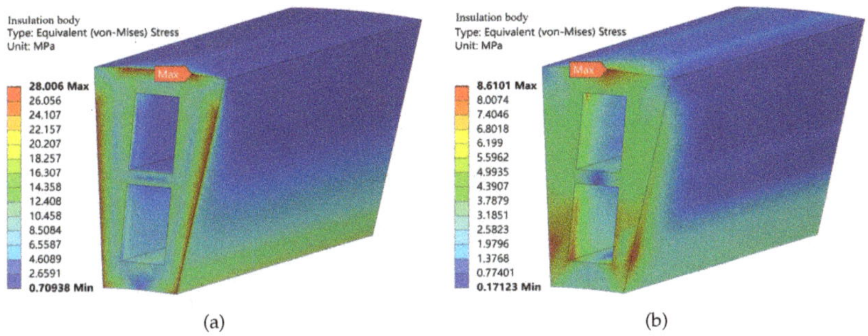

Figure 11. TMS in the DSOF motor stator insulation under seawater pressure. (**a**) Normal pressure. (**b**) Deep-sea pressure 20 MPa.

It is noted that the TMS at the stator insulation component is decreased when the seawater pressure changes from normal pressure to 20 MPa. For the stator insulation body, the most critical regions, where the maximum TMS is located, are on the top boundaries of the free ends. One is 8.61 MPa under the external pressure of 20 MPa, decreased by 69.25% compared to the other under normal pressure. Additionally, the average TMS in the SIS of the DSOF motor is further researched, and its variations with seawater pressure and temperature rise are illustrated in Figure 12. Obviously, the average TMS can be affected by both temperature rise and environment pressure, and the varying trends are consistent with the analytical model.

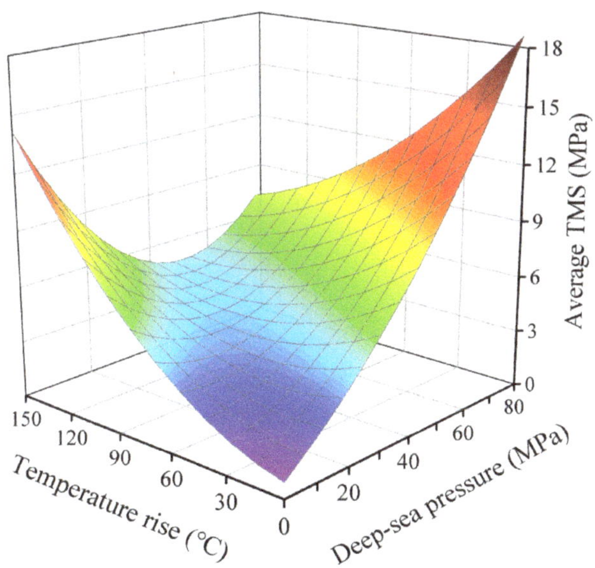

Figure 12. Average TMS in the SIS of the DSOF motor variation with seawater pressure and temperature rise.

Analytical solutions and simulation results of the external pressure variation corresponding to the minimum TMS with DSOF motor temperature rise are illustrated in Figure 13, where the analytical results agree well with the numerical ones and the maximum error rate is 4%, proving the accuracy of the TMS analytical model in this study.

For the targeted DSOF motor, the TMS in the motor SIS influenced by the deep-sea water pressure is then obtained as shown in Figure 14.

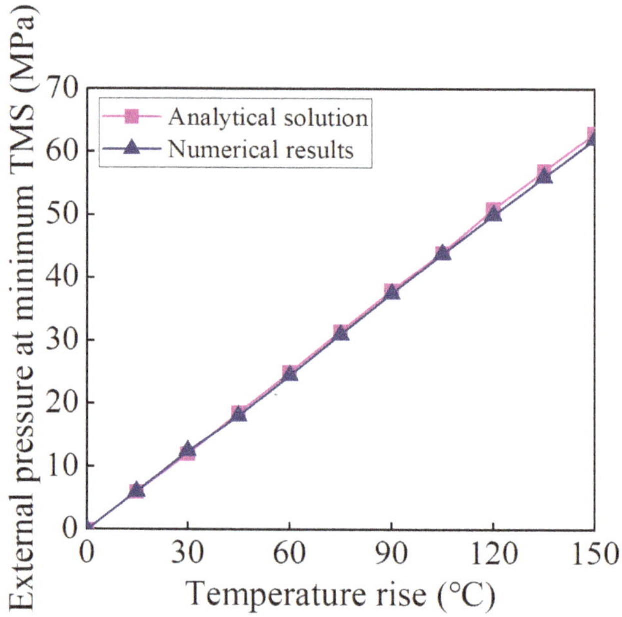

Figure 13. Comparison of analytical solution and numerical results of external pressure at the minimum TMS.

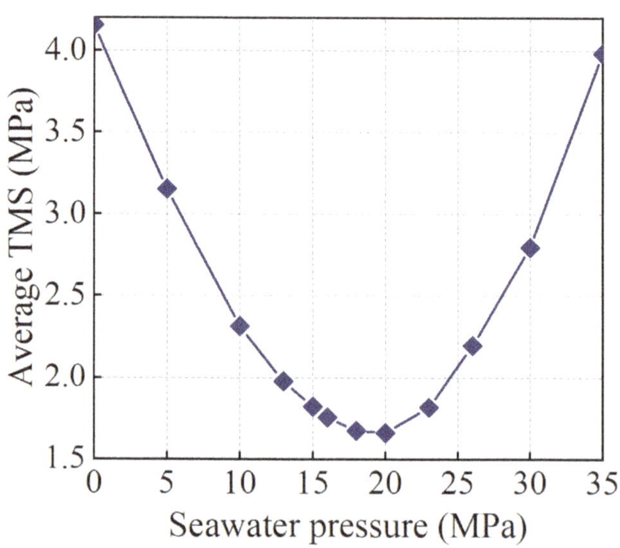

Figure 14. TMS in the SIS of the DSOF motor variation with seawater pressure based on the thermal results.

As working depth becomes larger, the average TMS in the DSOF motor insulation system will first decrease from 4.2 MPa to 1.7 MPa and then increase, and the stress can be reduced by up to 59.5%. These results indicate that the seawater pressure can ease stator insulation degradation and extend the DSOF motor service life. For further investigating the influence of the deep-sea working environment on the fatigue process of the DSOF motor, the insulation degradation analysis is then carried out.

4.2. Insulation Degradation Analysis

For replicating a typical DSOF motor operating cycle, an actual-use heating rate value of 0.02 °C/s and temperature change range of 5 °C relative to steady-state temperature are adopted [25]. Thus, for transient thermal loads of the targeted motor, the copper heating rate and thermal cycle are set to 0.02 °C/s and 350 s. Figure 15 shows the motor stator cycling temperature distribution results. Accordingly, the TMS spectra of the most dangerous point, where the maximum TMS is located, of the stator insulation are analyzed under normal pressure and 20 MPa, as shown in Figure 16. The stress spectra can be used for motor stator insulation fatigue lifetime prediction.

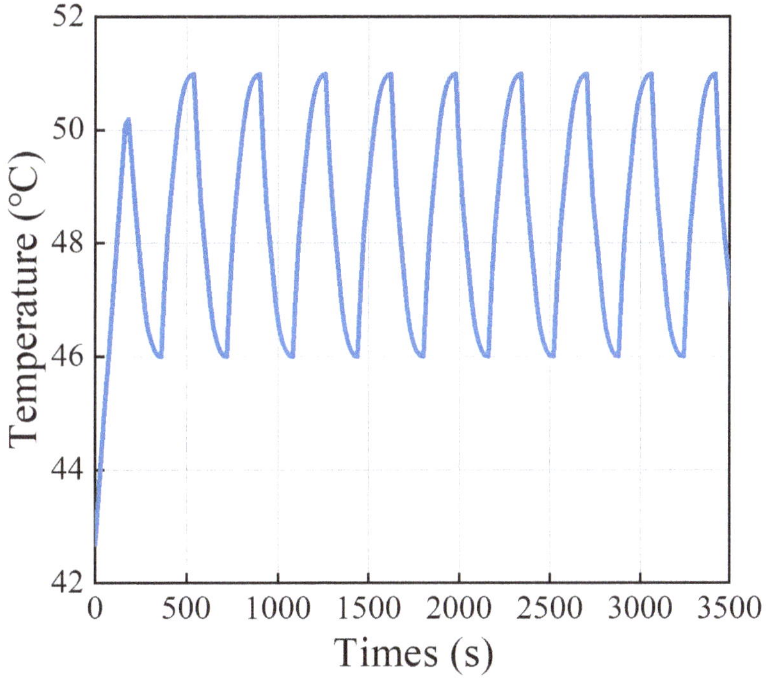

Figure 15. Motor stator temperature.

The mean amplitude and the number of TMS cycles are calculated by a rainflow cycle counter. The results are shown in Figure 17, where (a) represents the TMS results under normal pressure and (b) illustrates the TMS results under 20 MPa pressure. Obviously, there are 10 typical cycles in each case.

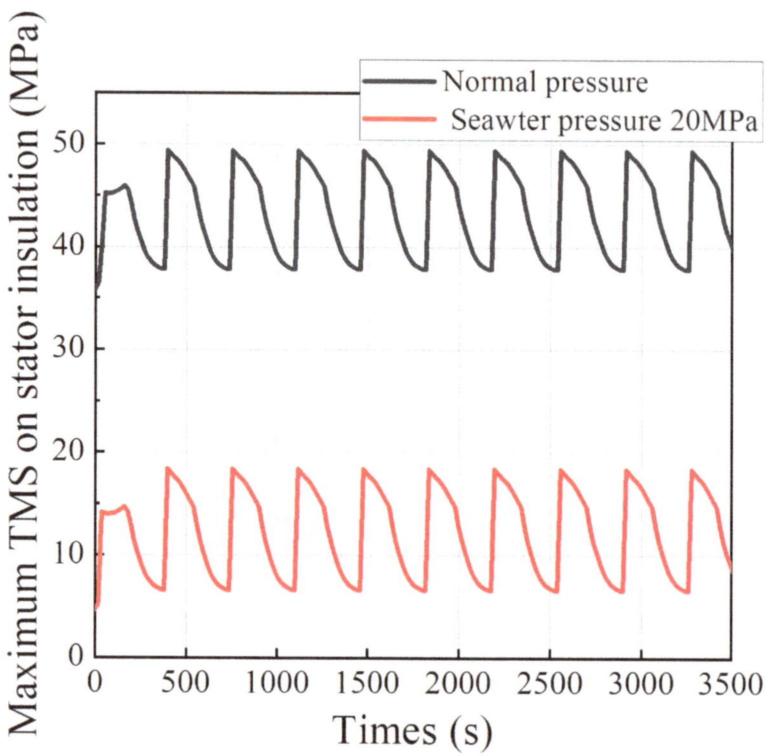

Figure 16. TMS spectra of stator insulation under normal pressure and 20 MPa.

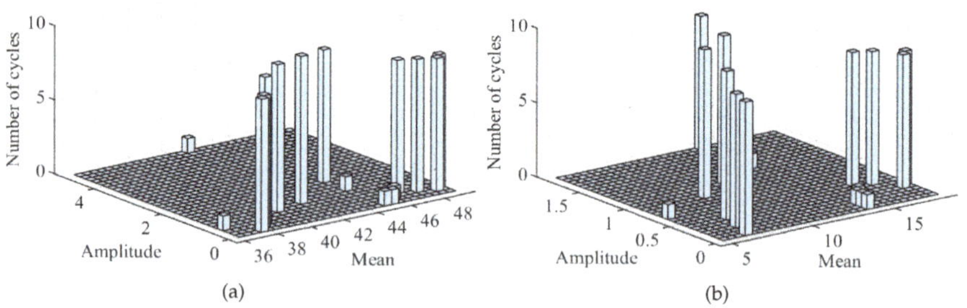

Figure 17. Results from rainflow cycle counter of the maximum TMS of stator insulation. (**a**) Normal pressure. (**b**) Deep-sea pressure 20 MPa.

By applying the Goodman equation to the rainflow counting results of the TMS spectra, the equivalent zero-mean stress amplitudes are obtained [26]. Then, combining the S-N curve of the insulation material and Miner fatigue theory, the cumulative damage of the dangerous point in the stator insulation is obtained as 6.08×10^{-5} under normal pressure and 5.24×10^{-5} under 20 MPa. It is assumed that when fatigue damage reaches 1, fatigue failure occurs. Thus, the number of cycles to the insulation fatigue failure is 16,436 under normal pressure and 19,069 under 20 MPa. As presented in Figure 16, the cycle time is 0.97 h. By multiplying the number of cycles to failure, the stator insulation lifetimes are 15,943 h and 18,497 h under normal pressure and 20 MPa, respectively. It is noted that compared with motors working in normal conditions, the lifetime of a motor running in a deep-sea environment will increase by 16.02% at most, which indicates that deep-sea high-pressure can slow down the degradation of stator insulation under TMS.

5. Conclusions

This paper has made an effort to analyze the insulation degradation for deep-sea oil-filled (DSOF) motors due to TMS. This was done by first proposing a TMS analytical model of DSOF motor stator insulation, considering the two key factors; namely, deep-sea pressure and motor temperature rise. Then, CFD analysis was performed by investigating the thermal effects on oil viscosity and viscous drag loss to obtain the motor temperature rise. Based on the temperature rise, thermo-mechanical simulations on a DSOF motor were performed to quantify the TMS and verify the proposed TMS analytical model. According to the TMS analysis results, the DSOF motor lifetime was finally evaluated and compared for two different working pressure conditions. The thermo-mechanical simulation results were compared with the analytical solutions obtaining a 4% deviation rate, which demonstrates the accuracy of the proposed TMS analytical model. Results show that the TMS in the DSOF motor insulation system will first decrease then increase as the DSOF motor working depth becomes larger. Compared to the normal pressure working conditions, the TMS on the motor stator insulation can be reduced by 59.5% at most under seawater pressure, and DSOF motor insulation lifetime can be increased by 16.02% at most, which indicates that the seawater pressure can ease stator insulation degradation and extend the DSOF motor service life. This research can provide references for high power density DSOF motor design. In the future, the relevant accelerated aging experiments for a DSOF motor will be carried out to further verify the proposed model. The high power density DSOF motor design will be conducted based on this research.

Author Contributions: Conceptualization and methodology, J.Z.; software and writing—original draft preparation, R.W.; supervision, Y.F.; investigation, Y.L. All authors have read and agreed to the published version of the manuscript.

Funding: This research was funded by the National Natural Science Foundation of China under Grant No. 51977192, the Ningbo Science and Technology Innovation 2025 Major Project (No. 20211ZDYF020218), and the Zhejiang Provincial Ten-Thousand-Talent Plan under Grant No. 2019R52003.

Institutional Review Board Statement: Not applicable.

Informed Consent Statement: Not applicable.

Data Availability Statement: Not applicable.

Conflicts of Interest: The authors declare no conflicts of interest.

References

1. Silwal, B.; Sergeant, P. Thermally induced mechanical stress in the stator windings of electrical machines. *Energies* **2018**, *11*, 2113. [CrossRef]
2. Osada, M.; Zhang, G.; Carneiro, F.; Muromachi, M.; Yoshimoto, S.; Yamamoto, A. Effect of insulating liquid viscosity on output force of a high-power synchronous electrostatic motor. *Int. J. Mech. Sci.* **2022**, *220*, 107149. [CrossRef]
3. Bell, S.B.; Blair, A.C.; Wagner, L.N.; Zou, V.; Buendia, A.R.; Ashrafzadeh, F. Survey of computational fluid dynamics software for rotational purposes. In Proceedings of the ASME 2019 International Mechanical Engineering Congress and Exposition, Salt Lake City, UT, USA, 11–14 November 2019; Volume 59421, p. V005T07A014.
4. Min, F.; Hai, Y.; Zhaoyang, Y.; Wenliang, W.; Chunxu, M. Coupling calculation of 3d whole domain steady flow and temperature field for underwater oil-filled brushless dc motors. In Proceedings of the 2019 22nd International Conference on Electrical Machines and Systems (ICEMS), Harbin, China, 11–14 August 2019; pp. 1–6.
5. Torres, R.A.; Dai, H.; Jahns, T.M.; Sarlioglu, B.; Lee, W. Cooling design of integrated motor drives using analytical thermal model, finite element analysis, and computational fluid dynamics. In Proceedings of the 2021 IEEE Applied Power Electronics Conference and Exposition (APEC), Phoenix, AZ, USA, 14–17 June 2021; p. 1509.
6. Czerny, B.; Khatibi, G.; Liedtke, M.; Nicolics, J. Thermomechanical reliability investigation of insulated gate bipolar transistor module. In Proceedings of the 2018 41st International Spring Seminar on Electronics Technology (ISSE), Zlatibor, Serbia, 16–20 May 2018.

7. Chen, Y.; Li, B.; Wang, X.; Yan, Y.; Wang, Y.; Qi, F. Investigation of heat transfer and thermal stresses of novel thermal management system integrated with vapour chamber for igbt power module. *Therm. Sci. Eng. Prog.* **2019**, *10*, 73–81. [CrossRef]
8. Shao, S.; Niu, Y.; Wang, J.; Liu, R.; Park, S.; Lee, H.; Yip, L.; Refai-Ahmed, G. Design guideline on board-level thermomechanical reliability of 2.5 d package. *Microelectron. Reliab.* **2020**, *111*, 113701. [CrossRef]
9. Reddy, V.V.; Ume, I.C.; Williamson, J.; Sitaraman, S.K. Evaluation of the quality of bga solder balls in fcbga packages subjected to thermal cycling reliability test using laser ultrasonic inspection technique. *IEEE Trans. Compon. Packag. Manuf. Technol.* **2021**, *11*, 589–597. [CrossRef]
10. Panduranga, R.; Shivakumar, K. Mode-ii total fatigue life model for unidirectional im7/8552 carbon/epoxy composite laminate. *Int. J. Fatigue* **2017**, *94*, 97–109. [CrossRef]
11. Lekakh, S.N.; Buchely, M.; O'Malley, R.; Godlewski, L.; Li, M. Thermo-cycling fatigue of simo ductile iron using a modified thermo-mechanical test. *Int. J. Fatigue* **2021**, *148*, 106218. [CrossRef]
12. *IEEE Std 56–2016*; IEEE Guide for Insulation Maintenance of Electric Machines-Redline. IEEE: New York, NY, USA, 2016; pp. 1–233.
13. Huang, Z.; Reinap, A.; Alaküla, M. Degradation and fatigue of epoxy impregnated traction motors due to thermal and thermal induced mechanical stress-part i: Thermal mechanical simulation of single wire due to evenly distributed temperature. In Proceedings of the 8th International Conference on Power Electronics, Machines and Drives PEMD, Glasgow, UK, 19–21 April 2016.
14. Silwal, B.; Mohamed, A.H.; Nonneman, J.; Paepe, M.D.; Sergeant, P. Assessment of different cooling techniques for reduced mechanical stress in the windings of electrical machines. *Energies* **2019**, *12*, 1967. [CrossRef]
15. Yeo, W.; Purbolaksono, J.; Aliabadi, M.; Ramesh, S.; Liew, H. Exact solution for stresses/displacements in a multilayered hollow cylinder under thermo-mechanical loading. *Int. J. Press. Vessel. Pip.* **2017**, *151*, 45–53. [CrossRef]
16. Evci, C.; Gülgeç, M. Functionally graded hollow cylinder under pressure and thermal loading: Effect of material parameters on stress and temperature distributions. *Int. J. Eng. Sci.* **2018**, *123*, 92–108. [CrossRef]
17. Salifu, S.; Desai, D.; Kok, S.; Ogunbiyi, O. Thermo-mechanical stress simulation of unconstrained region of straight x20 steam pipe. *Procedia Manuf.* **2019**, *35*, 1330–1336. [CrossRef]
18. Li, J. Loss and Temperautre Field Analysis for Deep-Sea Brushless dc Motor. Ph.D. Thesis, Harbin Institute of Technology, Harbin, China, 2011.
19. Asami, F.; Miyatake, M.; Yoshimoto, S.; Tanaka, E.; Yamauchi, T. A method of reducing windage power loss of a high-speed motor using a viscous vacuum pump. *Precis. Eng.* **2017**, *48*, 60–66. [CrossRef]
20. Kelkar, M.S.; Maginn, E.J. Effect of temperature and water content on the shear viscosity of the ionic liquid 1-ethyl-3-methylimidazolium bis (trifluoromethanesulfonyl) imide as studied by atomistic simulations. *J. Phys. Chem. B* **2007**, *111*, 4867–4876. [CrossRef] [PubMed]
21. Miller, R.; Griffiths, P.; Hussain, Z.; Garrett, S.J. On the stability of a heated rotating-disk boundary layer in a temperature-dependent viscosity fluid. *Phys. Fluids* **2020**, *32*, 024105. [CrossRef]
22. Chen, W.; Fan, C. Non-newtonian fluid machanics. *Chin. J. Nat.* **1985**, *8*, 243–247.
23. Ramos, E.A.; Treviño, C.; Lizardi, J.J.; Méndez, F. Non-isothermal effects in the slippage condition and absolute viscosity for an electroosmotic flow. *Eur. J. Mech.-B/Fluids* **2022**, *93*, 29–41. [CrossRef]
24. Kacem, R.; Alzamel, N.; Ouerfelli, N. Sensitivity of viscosity arrhenius parameters to polarity of liquids. *Russ. J. Phys. Chem. A* **2017**, *91*, 1654–1659. [CrossRef]
25. Huang, Z.; Reinap, A.; Alaküla, M. Degradation and fatigue of epoxy impregnated traction motors due to thermal and thermal induced mechanical stress-part ii: Thermal mechanical simulation of multiple wires due to evenly and unevenly distributed temperature. In Proceedings of the 8th International Conference on Power Electronics, Machines and Drives PEMD, Glasgow, UK, 19–21 April 2016.
26. Griffo, A.; Tsyokhla, I.; Wang, J. Lifetime of machines undergoing thermal cycling stress. In Proceedings of the 2019 IEEE Energy Conversion Congress and Exposition (ECCE), Baltimore, MD, USA, 29 September–3 October 2019; pp. 3831–3836.

energies

MDPI

Article

Comprehensive Comparison of a High-Speed Permanent Magnet Synchronous Motor Considering Rotor Length–Diameter Ratio

Wentao Gao, Yufeng Zhang, Guanghui Du *, Tao Pu and Niumei Li

School of Electrical and Control Engineering, Xi'an University of Science and Technology, Xi'an 710054, China; gaowentao0704@163.com (W.G.); xkdzhangyufeng@xust.edu.cn (Y.Z.); pt18482163557@163.com (T.P.); liniumei@stu.xust.edu.cn (N.L.)

* Correspondence: duguanghui1104@163.com

Abstract: For high-speed permanent magnet machines (HSPMMs), many design schemes of rotor length–diameter ratios can satisfy the constraints of multiple physical fields during the motor design period. The rotor length–diameter ratio greatly impacts the comprehensive performances of multiple physical fields. However, these analyses are missing in the existing literature. Therefore, this paper focuses on the influence of the rotor length–diameter ratio on comprehensive performances. Firstly, finite element models (FEM) of multiple physical fields are built by ANSYS Workbench platform and Motor-CAD software. Then, the comprehensive performances of multiple physical fields are comparatively analyzed. Finally, the designed HSPMM is implemented, based on one prototype of 60 kW, 30,000 rpm to verify the results of comparative analysis. Based on the comparative analysis above, the influent laws of rotor length–diameter ratios on comprehensive performances of multiple physical fields are discussed and summarized, which can be used as a reference for the rotor structural design of HSPMMs.

Keywords: high-speed permanent magnet machines; comprehensive characteristics; comparative; rotor length–diameter ratio

Citation: Gao, W.; Zhang, Y.; Du, G.; Pu, T.; Li, N. Comprehensive Comparison of a High-Speed Permanent Magnet Synchronous Motor Considering Rotor Length–Diameter Ratio. *Energies* **2022**, *15*, 5256. https://doi.org/10.3390/en15145256

Academic Editors: Mario Marchesoni and Antonio Cano-Ortega

Received: 8 June 2022
Accepted: 18 July 2022
Published: 20 July 2022

1. Introduction

In recent years, with the development of electronic devices, high-speed permanent magnet machines (HSPMMs) access industry and life more and more widely, because of their excellent electromagnetic performance and simple control characteristics [1–3]. However, when the HSPMMs rotate at high speed, various problems occur, which hinder the development of HSPMMs, and solving these problems has become a research focus and hot spot for researchers [4–6]. In the design of HSPMMs, the key structural parameters and performances must meet the constraints of multiple physical fields, which include electromagnetic, mechanical, and thermal physical fields [7,8]. However, many rotor shapes can satisfy the constraints of multiple physical fields, from short and thick to long and thin with rotor shape, which leads to the uncertain selection of the rotor shape in the design period. Therefore, the influence of the rotor length–diameter ratio on the comprehensive performances is worthy of discussion and research. In this paper, the influence of rotor length–diameter ratio on comprehensive performances is studied and discussed, in order to obtain better comprehensive performances for HSPMMs.

In the existing literatures, the comprehensive performances of HSPMMs are comparatively investigated through key performance parameters. For electromagnetic and loss properties, some key electromagnetic parameters are comparatively analyzed, such as line back-EMF at no load, radial air-gap flux density distribution, stator iron-core loss, permanent magnet loss, air-friction loss, etc. [9,10]. For rotor stress characteristics, the rotor stresses are calculated, including radial and tangential stresses of permanent magnets, and von-Mises stress distributions of sleeve [11]. For rotor dynamics, the rotor mode and critical speed are simulated [12]. For the thermal behavior, the cooling system and temperature distribution are investigated according to the designed parameters and calculated loss characteristics [13]. It can be seen that the design of HSPMMs must meet the needs of comprehensive performances of multiple physical fields. For HSPMMs, many rotor shapes can satisfy all the needs of comprehensive performances, but different rotor length–diameter ratios have great influence on the comprehensive performances of multiple physical fields. A shorter and thicker rotor shape can effectively reduce the motor temperature rises and improve the rotor dynamics' reliability, but the permanent magnet stress and the sleeve stress are increased due to the poor centrifugal force of the larger rotor diameter [14,15]. A longer and slender rotor shape can effectively improve the rotor stress, while the rotor dynamics and thermal characteristics may not be satisfied due to the longer rotor core length or the smaller rotor diameter [16]. It is clear from the above description that the rotor length–diameter ratio can affect motor performances, including electromagnetic, mechanical, and thermal characteristics. Hence, the influence of rotor length–diameter ratio on comprehensive performances cannot be ignored. In the existing literatures, the influence of rotor length–diameter ratio on motor performance mainly focuses on the analysis of single physical fields and the influence of the rotor length–diameter ratio on the comprehensive performance of multiple physical fields is not clear. To obtain an accurate relationship between the rotor length–diameter ratio and comprehensive performances, comparative analysis of multiple physical fields (such as electromagnetic, mechanical, and thermal) is necessary. The rotor length–diameter ratio should be investigated based on a comparative analysis of comprehensive performances, which include electromagnetic and loss characteristics, mechanical strength, rotor dynamics, and thermal performance.

Based on the above analysis, the rotor length–diameter ratio has a great influence on the comprehensive performances of multiple physical fields. However, the effect has rarely been focused on in the existing literature, which leads to the uncertain selection of the rotor length–diameter ratio in the design period of high-speed permanent magnet motors. Therefore, in this paper, the influence of the rotor length–diameter ratio on comprehensive performances is discussed and comparatively investigated. Firstly, finite element models of multiple physical fields are built. Then the influent laws of rotor length–diameter ratio on comprehensive performances are summarized. Finally, a prototype is manufactured and the comprehensive performances are experimentally measured. The conclusions obtained can be used as a reference for the rotor structural design of HSPMMs. The initial electromagnetic design scheme is introduced in Section 2. In Section 3, the process of comparative analysis is determined and three cases with representative rotor length–diameter ratios are selected. In Section 4, the comprehensive performances of these three cases are comparatively analyzed in detail. In Section 5, the designed HSPMM are manufactured and experiments are conducted. Finally, the conclusions summarized are drawn in Section 6, and some key simulation results and the influence of the rotor length–diameter ratio on comprehensive performances are summarized, which can be used as a reference for rotor structural design of HSPMMs.

2. HSPMM Structure and Main Performance

In the design of a high-speed permanent magnet motor, electromagnetic performance, losses, temperature distribution, rotor stress, and rotor dynamics should be considered comprehensively. It is necessary to analyze the comprehensive characteristics of the motor, and then the designed motor scheme can meet all needs. In the design process, it is important to determine the mechanical structure of the motor firstly, because the different mechanical structures will directly affect other performance of the designed motor [17].

In this paper, a high-speed permanent magnet machine for 60 kW at 30,000 rpm is designed and the main parameters of the motor can be obtained, as shown in Table 1. A surface-mounted permanent magnet rotor is designed. To ensure the reliability of rotor permanent magnets (PMs) rotating at high speed, a carbon fiber sleeve of 5 mm is used to improve the rotor stress, and the tensile strength of carbon fiber materials is 1400 MPa. The stator core is composed of 0.35 mm cold-rolled silicon of low-loss steel.

Table 1. Main parameters of the designed motor for HSPMM.

Parameters	Values
Output power (kW)	60
Rated rotating speed (rpm)	30,000
Rated load voltage (V)	380
Rated power factor	0.98
Stator slots	24
Carbon fiber sleeve thickness (mm)	5

3. Initial Comprehensive Characteristics Comparative Analysis for HSPMM

3.1. Comprehensive Comparative Analysis Process for HSPMM

To analyze the influence of main structural parameters on the comprehensive characteristics, a process is presented, as shown in Figure 1.

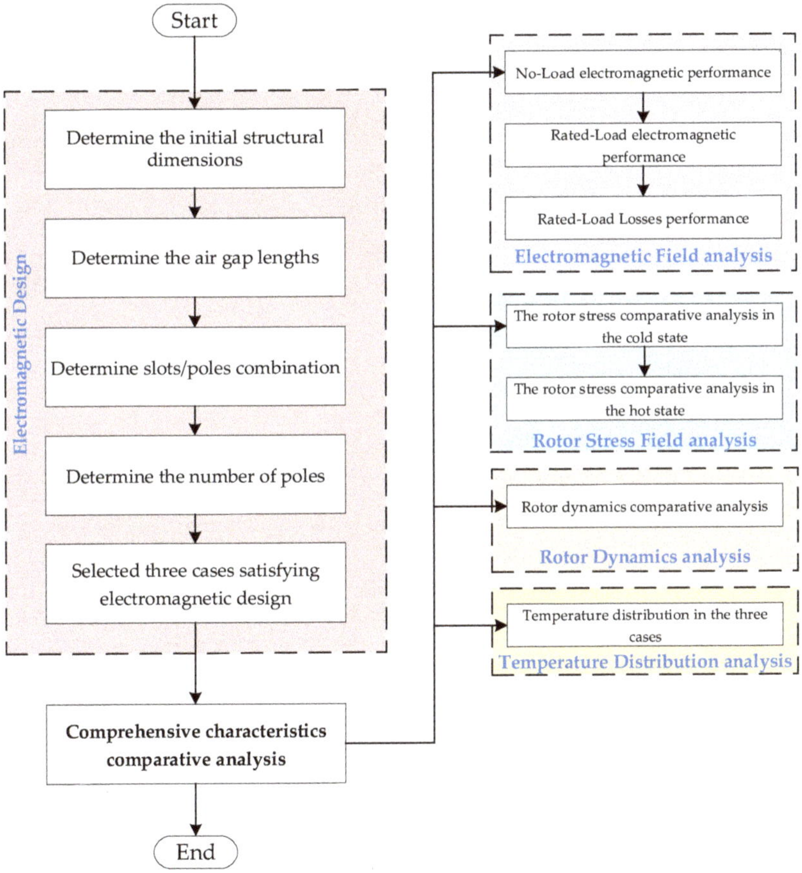

Figure 1. Comprehensive characteristics comparative analysis and design process.

Step 1: According to the basic performance parameters and rotor diameter constraints, the initial structural dimensions are determined. Step 2: The electromagnetic characteristics of different air-gap lengths are compared and analyzed, and the air-gap length is determined. Step 3: The influence of slots/poles combination on the electromagnetic characteristics is analyzed. Step 4: The number of poles is determined through the influence of the variable number of poles on electromagnetic performance. Step 5: Three cases satisfying electromagnetic design are selected, mainly considering the difference in rotor shapes. Step 6: The electromagnetic and loss characteristics are comparatively analyzed under three cases. Step 7: The rotor stresses are comparatively analyzed under three cases. Step 8: The rotor dynamics are comparatively analyzed under three cases. Step 9: The temperature distribution is comparatively analyzed under three cases.

3.2. *Multiple Physic Fields Constraints*

The rotor dimensions of HSPMMs are limited by the electromagnetic, mechanical, and temperature characteristics. For the electromagnetic performance, the rotor dimensions must satisfy the electromagnetic load. For the mechanism, the rotor length–diameter ratio must be within a specific range, so that the rotor will not be damaged by centrifugal force and resonance. For temperature rises, the heat dissipation design of the motor should take into account the selected cooling system and working mode, so that the temperature distribution of the motor is within the constraints. Three aspects restrict each other and determine the rotor dimensions of HSPMMs. In this paper, the following basic constraints are determined according to the actual requirements.

1. Geometric constraints: interference fit $\delta_s = 0.15$ mm, the thickness of the sleeve $h_{sleeve} = 5$ mm, the thickness of the magnets 7 mm $\leq h_{PM} \leq$ 8.5 mm, and the stator outer diameter $D_{is} = 157$ mm.
2. Electromagnetic constraints: the amplitude of the line to line Back-EMF at no-load is between 500 V and 540 V, and the air-gap magnetic flux density is between 0.4 T and 0.6 T. The output power in the rated load is $P_{out} \geq 60$ kW, the thermal load required $AJ \leq 200$ A^2/mm^3(determined by experience).
3. Strength constraints: the tensile strength of permanent magnet should be less than 80 MPa, and the tensile strength of carbon fiber sleeve should be less than 1400 MPa. Safety margins should be considered in the engineering, so the tensile strength limitation of the permanent magnet at rated speed is 64 MPa, and the tensile strength limitation of the sleeve at rated speed is 1100 MPa.
4. Critical speed constraints: the rated speed of the rotor is required to be less than 0.7 times the first-order critical speed.
5. Thermal constraints: the limited maximum working temperature of the HSPMM is 130 °C, and stator housing water cooling system is adopted.

3.3. *Influence Analysis of Main Parameters on Electromagnetic Performance*

3.3.1. Air-Gap Length

For different air-gap lengths, the electromagnetic performance of the motor is analyzed, such as the air-gap lengths of 0.5, 1.5, 2.0, and 3.0 mm. In this analysis, rated output power is maintained, and the thickness of the permanent magnet is increased with the increase of air-gap length. The permanent magnet thickness and electromagnetic characteristics are shown in Table 2.

Table 2. Electromagnetic characteristics under different air-gap lengths.

Case	1	2	3	4
Air-gap length (mm)	0.5	1.5	2.0	3.0
Magnet thickness (mm)	6.5	8	8.8	13
No-load line back-EMF (V)	500.93	497.70	490.46	488.13
RMS current (A)	94.35	95.25	95.13	94.21
Cogging torque (%)	2.70	2.20	1.93	1.67
Efficiency (%)	97.09	97.05	97.01	96.94

The radial air-gap flux density distribution with the mechanical angle under no load and rated load is shown in Figure 2. The amplitude of the radial air-gap flux density is mainly affected by the air-gap length, and the impact of the air-gap length on harmonic order and the distribution of the radial air-gap flux density of the motor is little.

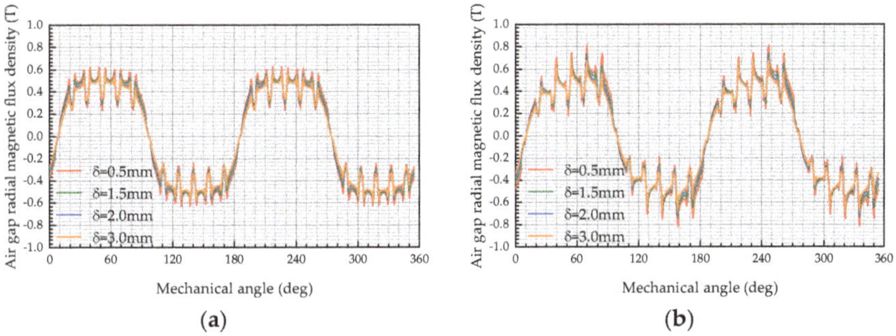

Figure 2. Radial air−gap flux density under no load and rated load: (**a**) no-load; (**b**) rated-load.

The rotor eddy-current losses under no load and rated load with different air-gap lengths are shown in Figure 3. It can be shown that the air-gap length is smaller and the rotor eddy-current loss is larger because a smaller air-gap length will cause more current harmonics to enter the rotor, resulting in more rotor eddy-current losses. As the air-gap length increases, more current harmonics will be shielded, so the rotor eddy-current losses gradually decrease. When using a smaller air-gap length, the dosage of magnets is smaller, while the rotor eddy-current loss is larger. When using a larger air-gap length, the dosage of magnets is larger, and the rotor eddy-current loss is smaller.

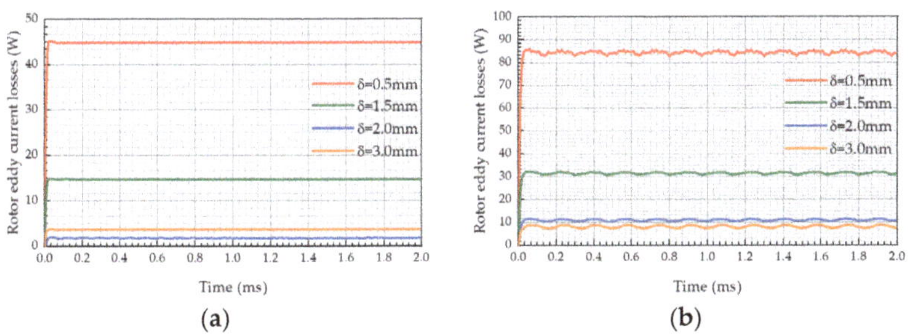

Figure 3. Rotor eddy-current losses on the magnets under no load and rated load: (**a**) no-load; (**b**) rated load.

Therefore, the air-gap length is a compromise between rotor eddy-current losses and electromagnetic performance. Although the rotor eddy-current loss is smallest under the 3.0 mm air-gap length, electromagnetic performance cannot be guaranteed, especially for the no-load back-EMF. The rotor eddy-current losses of the 0.5 mm and 1.5 mm air-gap lengths are larger, which ultimately lead to a higher motor temperature rise. According to the level of influence of the air-gap length on rotor eddy-current loss in Figure 4, rotor eddy-current loss decreases gradually with the increase of air-gap length. Electromagnetic performance and the rotor eddy-current loss are comprehensively considered and the air-gap length of 2 mm is selected as a compromise solution.

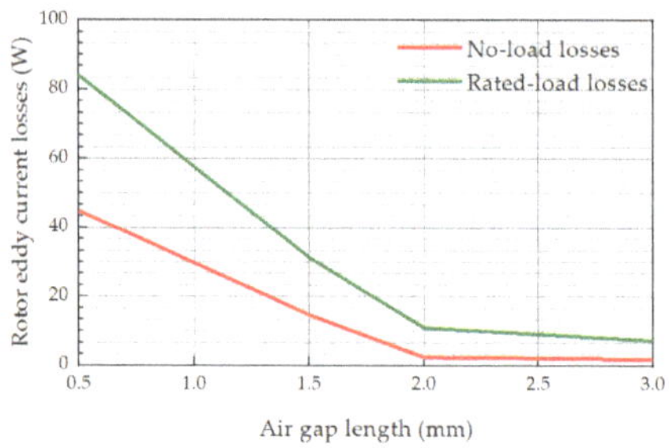

Figure 4. Effect of air-gap length on rotor eddy-current losses under no load and rated load.

3.3.2. Slots/Poles Combination

The slots/poles combination will affect the winding factor and the cogging torque period. Therefore, the impact of slots/poles combination on electromagnetic characteristics is not ignored.

The winding factor, cogging torque period, and load power frequency with different pole numbers are discussed in Table 3. Because of the same minimum common multiple of the poles and slots, the cogging torque period is the same. However, the stator winding factors of two-pole and four-pole motors are 0.949, and 0.933, respectively, indicating the high stator windings utilization ratios of two-pole and four-pole motors. When the power frequency is high, it will greatly increase the stator iron-core loss and rotor eddy-current loss. Therefore, in the HSPMMs, a large power-supply frequency should not be used. In conclusion, this paper adopts 24 slots and two poles or four poles in the designed motor.

Table 3. Electromagnetic performance comparative analysis under different slots/poles combination.

Slots/Poles	Slot per Pole per Phase	Pole Pitch/Pitch	Winding Factor	Cogging Torque Period	Load Frequency
24/2	4	12/11	0.949	24	500 Hz
24/4	2	6/5	0.933	24	1000 Hz
24/6	4/3	4/3	0.885	24	1500 Hz
24/8	1	3/1	0.5	24	2000 Hz

3.3.3. Number of Poles

The number of poles plays a major role in HSPMMs performance. Electromagnetic characteristics of two-pole and four-pole motors under no load and rated load are analyzed. In Figure 5, for the two-pole motor, the magnetic flux density of stator yoke is greater than 1.8 T, which makes the stator yoke saturated under the same stator outer diameter of 157 mm. Therefore, the stator outer diameter of the two-pole motor should be expanded.

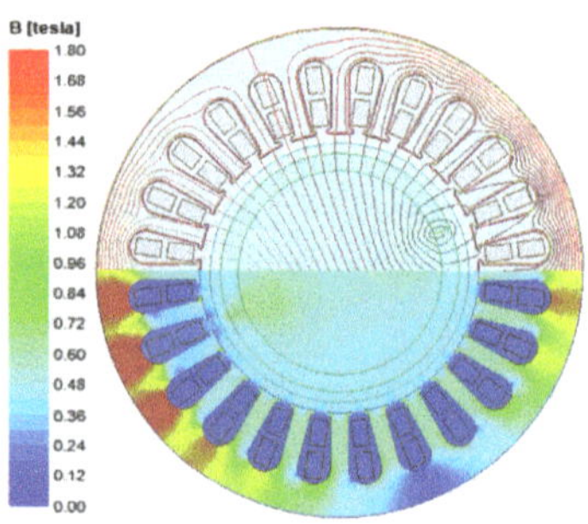

Figure 5. Magnetic flux density and lines distributions of two-pole motor with the stator outer diameter 157 mm.

The stator outer diameter of the two-pole motor must be expanded to the stator yoke non-saturation. The main structural parameters of the two-pole motor and the four-pole motor are compared, as shown in Table 4. The stator outer diameter of the two-pole motor is extended by 15% over the four-pole motor, ensuring basic electromagnetic characteristics.

Table 4. Structural parameters of two-pole and four-pole motors for HSPMM.

Parameters	Two-Pole	Four-Pole
Power (kW)	60	60
Speed (rpm)	30,000	30,000
Frequency (Hz)	500	1000
Slot number	24	24
Stator outer diameter (mm)	181	157
Stator inner diameter (mm)	89	89
Rotor outer diameter (mm)	86	86
Effective core length	110	110

Electromagnetic characteristics of two-pole and four-pole motors are shown in Table 5. Although the four-pole motor has almost double stator iron-core loss density compared to the two-pole motor, the stator iron-core loss of the four-pole motor has almost 3/2 times that of the two-pole motor, as they have different iron volumes.

Table 5. Electromagnetic characteristics comparison of two-pole motor and four-pole motor for HSPMM under rated load.

Parameters	Two-Pole	Four-Pole
RMS current (A)	95.38	95.25
Line back-EMF (V)	497.83	497.70
Torque (Nm)	19.10	19.11
Stator-teeth flux density (T)	1.04	1.04
Stator-yoke flux density (T)	1.31	1.31
Thermal load (A^2/mm^3)	168.8	162.07
Power density (kW/kg)	2.66	4.10
Materials of stator	DW310-35	DW310-35
Iron loss density (W/kg)	20.28	41.85
Stator-core loss (W)	457.581	611.792

The two-pole and four-pole motor are compared and analyzed from the stator yoke magnetic flux density distribution, rotor eddy current, winding end length, radial air-gap flux density distribution, and line-to-line back-EMF.

In Figure 6, the maximum values of flux densities are basically the same for the two-pole and the four-pole motors, which are about 1.38 T in the stator yoke. However, the stator outer diameter is 157 mm for the four-pole motor and 181 mm for the two-pole motor, thus the utilization of the stator-yoke material is higher for the four-pole motor. The four-pole motor allows the smaller dimensions and an improved material utilization ratio.

Rotor eddy-current loss in permanent magnets is part of the cause of motor temperature rise. Rotor eddy-current density distribution on permanent magnets of the two-pole and the four-pole motors is shown in Figure 7. Obviously, the rotor eddy current of the two-pole motor is much higher than that of the four-pole motor. According to the 2D-FEA calculation, the rotor eddy-current loss of the four-pole motor is 31.52 W, which is lower than the 235.56 W of the two-pole rotor. The four-pole motor is formed by an arrangement of four magnets, and the two-pole motor consists of two magnets.

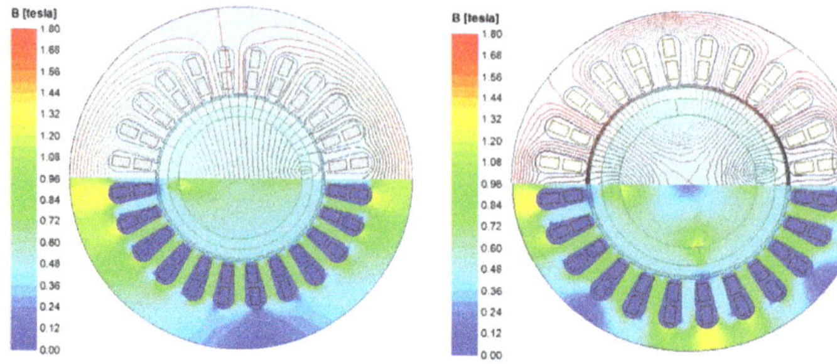

(**a**) (**b**)

Figure 6. Magnetic flux density and lines distributions of two-pole and four−pole motors at no-load: (**a**) two-pole motor; (**b**) four-pole motor.

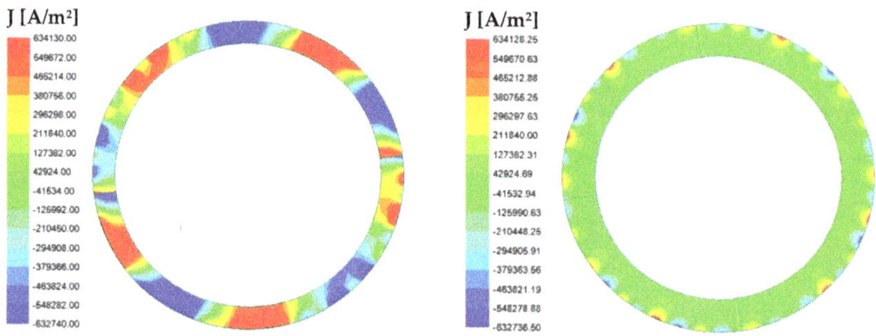

(**a**) (**b**)

Figure 7. The rotor magnet eddy-current density distribution of two-pole and four-pole motors under rated load: (**a**) two-pole motor; (**b**) four-pole motor.

The winding end length of the two-pole motor is longer than that of the four-pole motor, because the stator winding coils of the two-pole motor have a larger pitch than that of the four-pole motor. The stator weights and coils' half-turn lengths of two-pole and four-pole motors are compared in Table 6. According to the comparison in Table 6, the coils' half-turn length of the four-pole motor is much shorter than that of the two-pole motor, and the weight and material consumption are much smaller than that of the two-pole motor. The radial air-gap flux density distribution with the mechanical angle and Fourier transforms results for the two-pole and the four-pole motors are shown in Figures 8 and 9.

Table 6. Structure parameter value comparison of stator winding coils for HSPMM.

Parameters	Two-Pole Motor	Four-Pole Motor
Coils half-turn length (mm)	299.492	197.258
Total winding weight (kg)	22.5634	14.6193
Armature core steel consumption (kg)	24.1621	17.4648

The fundamental amplitude of radial air-gap flux density of thefour4-pole motor is better, and the odd harmonic amplitude is lower than that of the two-pole motor. The amplitudes of the 11th and 13th harmonics of the four-pole motor are larger than that of the two-pole motor, because the two-pole motor adopts a shorter pitch winding, which weakens the 11th and 13th harmonics.

At rated load, the radial air-gap flux density distribution of two-pole and four-pole motor is almost the same at no load. However, at rated load, the fundamental amplitude of radial air-gap flux density of the four-pole motor is much greater than that of the two-pole motor, which can provide a higher air-gap flux density to ensure the motor generates sufficient electromagnetic torque.

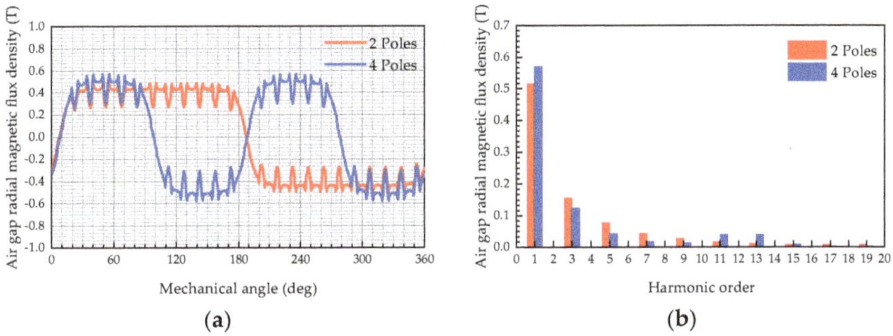

Figure 8. Radial air-gap flux density of two-pole and four−pole motors under no load: (**a**) radial air-gap flux density with mechanical angle; (**b**) Fourier transforms.

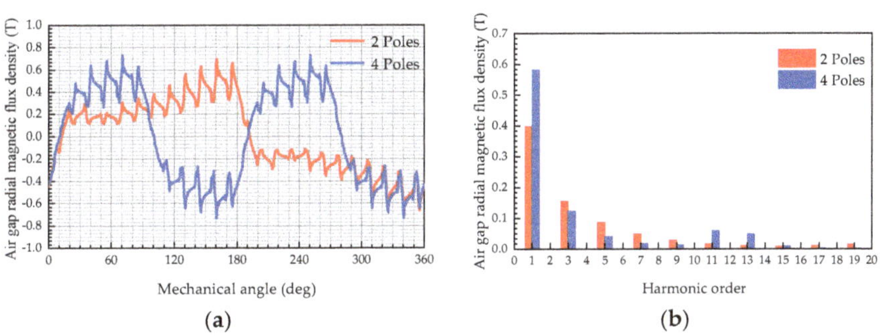

Figure 9. Radial air-gap flux density of two-pole and four-pole motor under rated load: (**a**) radial air-gap flux density with mechanical angle; (**b**) Fourier transforms.

The line-to-line back-EMF and the Fourier transforms results of two-pole and four-pole motors at no-load are shown in Figure 10. It can be found that the line-to-line back-EMF waveform of the two-pole motor is the flat-top wave, because the fifth and seventh harmonics amplitudes of the line-to-line back-EMF of the two-pole motor is higher, and the fundamental amplitude of the two-pole motor is slightly greater than that of the four-pole motor.

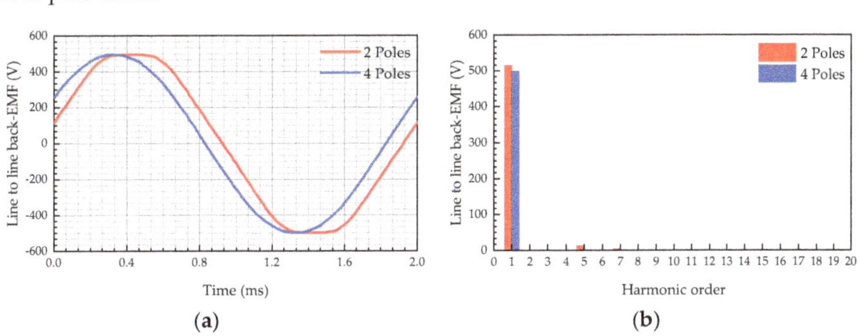

Figure 10. Line to line back−EMF of two-pole and fours-pole motor at no load: (**a**) line to line back-EMF waveform; (**b**) Fourier transform.

Based on the above analysis, for electromagnetic performance between two-pole and four-pole motors, the following results can be obtained.

- For a two-pole motor, due to the stator core's magnetic saturation, the motor dimension, coils length, and the total weight are larger than those for a four-pole motor. It is reasonable to consider a four-pole motor for the strict requirements on motor size applications.
- For the stator-core loss, a four-pole motor is larger than that of a two-pole motor, and the main reason is the lower load frequency for the four-pole motor.
- For the rotor eddy-current density, rotor eddy-current loss of the four-pole motor is 31.52 W, which is much lower than the 235.56 W of the two-pole rotor.
- For the radial air-gap flux density, the fundamental amplitude of a two-pole motor is lower than that of a four-pole motor, and the high-order harmonic amplitudes of a two-pole motor are greater than that of a four-pole motor.
- For the line-to-line back-EMF, the waveform of a four-pole motor is more sinusoidal. The fifth and seventh harmonic amplitudes of the line-to-line back-EMF of a two-pole motor are much larger than those of a four-pole motor.

3.3.4. Permanent Magnetic Materials

At present, permanent magnetic materials of HSPMMs are SmCo permanent magnet and NdFeB permanent magnet. Permanent magnetic materials usually have low mechanical strength and high thermal sensitivity. The permanent magnetic material SmCo has better temperature stability, while NdFeB has the low bending strength, low tensile strength, and compressive strength. Compared to SmCo permanent magnet, the permanent magnetic material NdFeB has poorer temperature stability, but the better magnetic property, better mechanical properties, and the lower price. The performances of the two permanent magnetic materials are shown in Table 7. To ensure the mechanical strength of the rotor, permanent magnet NdFeB is used.

Table 7. Performance of permanent magnet materials for HSPMM.

Parameters	NdFeB	SmCo
Residual flux density	$\leq$1.47 T	0.85 T $\sim$1.15 T
Coercive force	$\leq$1200 kA/m	$\leq$800 kA/m
(BH)max	$\leq$398 kJ/m^3	$\leq$258.6 kJ/m^3
Density (kg/m^3)	7400	8300
Thermal conductivity $(\mathrm{W}/(\mathrm{m}\cdot\mathrm{K}))$	8.9	11
Poisson ratio	0.27	0.24
Working temperature	$\leq$150 °C	$\leq$300 °C
Coefficient of the remanent	$-(0.095 \sim 0.15)\%/\mathrm{K}$	$-(0.03 \sim 0.09)\%/\mathrm{K}$
Tensile strength (MPa)	80	35

3.4. Comprehensive Comparative Analysis Schemes for Rotor Shape

For the requirements of electromagnetic performance, some designers design the rotor short and thick, and some designers design the rotor long and thin. The length–diameter ratio of the rotor is defined as $\lambda = L_{ef}/D_r$.

There are few studies in the literature on the influence of the length–diameter ratio on the multiple physical characteristics of the motor. In this paper, under the constraints of the multiple characteristics, three design cases are selected for comprehensive comparative analysis. Case 1 has the rotor shape of short and thick, and the length–diameter ratio is smallest. Case 3 has the rotor shape of long and thin, and the length–diameter ratio is biggest. Case 2 is an intermediate state between the rotor shapes of Case 1 and Case 3. The structural parameters of the rotor shape for the three cases are shown in Table 8.

Table 8. Structural parameters of the rotor shape for HSPMM under three cases.

Parameters	Case 1	Case 2	Case 3
Effective core length (mm)	55	110	200
Air-gap length (mm)	2	2	2
Stator outer diameter (mm)	203.6	165.4	131.1
Stator inner diameter (mm)	114	92.6	73.4
Rotor outer diameter (mm)	110	88.6	69.4
Length–diameter ratio	0.5	1.24	2.88
Sleeve thickness (mm)	5	5	5
Slots/poles combination	24/4	24/4	24/4

4. Comprehensive Comparison for Rotor Shape

4.1. Electromagnetic and Losses Characteristics Analysis

The two-dimensional finite element model (2D-FEM) is established for the HSPMM with three rotor shapes. The magnetic flux density distribution and flux line distribution are calculated for the rated load condition, as shown in Figure 11. In Case 1, the magnetic flux density is the highest, which is about 1.72 T. The magnetic flux density of the motor in Case 2 is the lowest, which is about 1.38 T. For Case 3, the magnetic flux density is about 1.506 T. If the magnetic flux density is too high, the iron core would be saturated, and the iron-core loss and temperature rise increased.

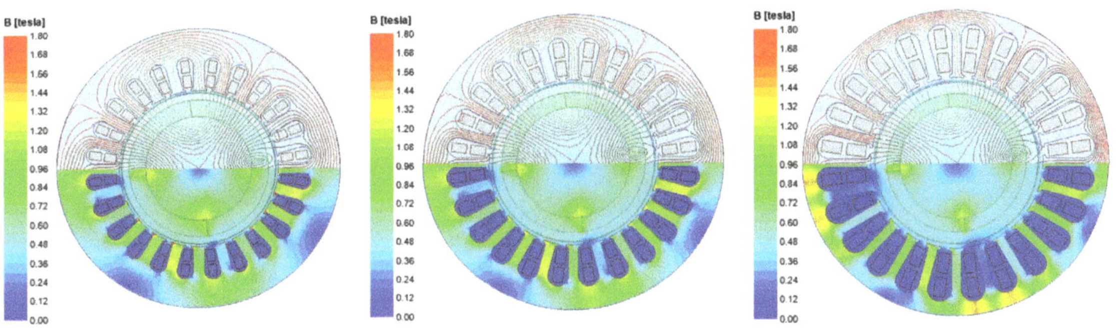

Figure 11. Magnetic flux density and lines distributions under different cases at rated load: (**a**) Case 1; (**b**) Case 2; (**c**) Case 3.

The line-to-line back-EMF and its Fourier transform results in three cases under no-load conditions are shown in Figure 12. In Case 1, the fundamental amplitude is slightly lower, and the 5th and 11th harmonics are slightly higher. Case 2 and Case 3 are the same.

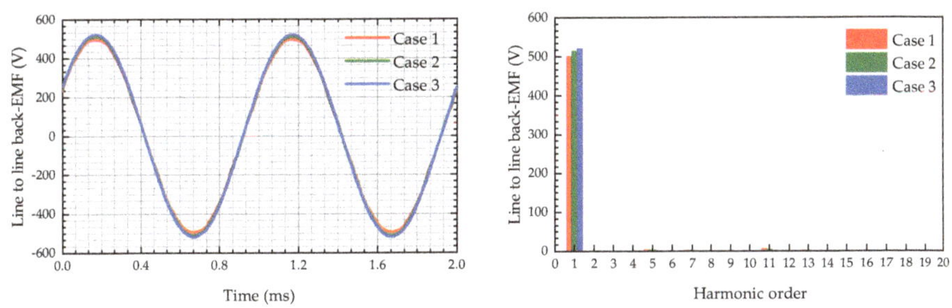

Figure 12. The line to line back-EMF and its Fourier transforms under the three cases at no load: (**a**) line to line back-EMF waveform; (**b**) Fourier transform results.

The radial air-gap flux density with mechanical angle waveform and Fourier transforms results are compared and analyzed for the three cases under the no load and rated load, as shown in Figure 13. The fundamental amplitude of Case 1 is the biggest, while the corresponding high-order harmonics amplitude is also highest. The fundamental amplitude of Case 2 is lower than Case 1, while the amplitude of the high-order harmonics is also smaller than Case 1. In Case 3, the fundamental amplitude is about 0.46 T, which is smallest and is not conducive to the electromagnetic performance of the motor.

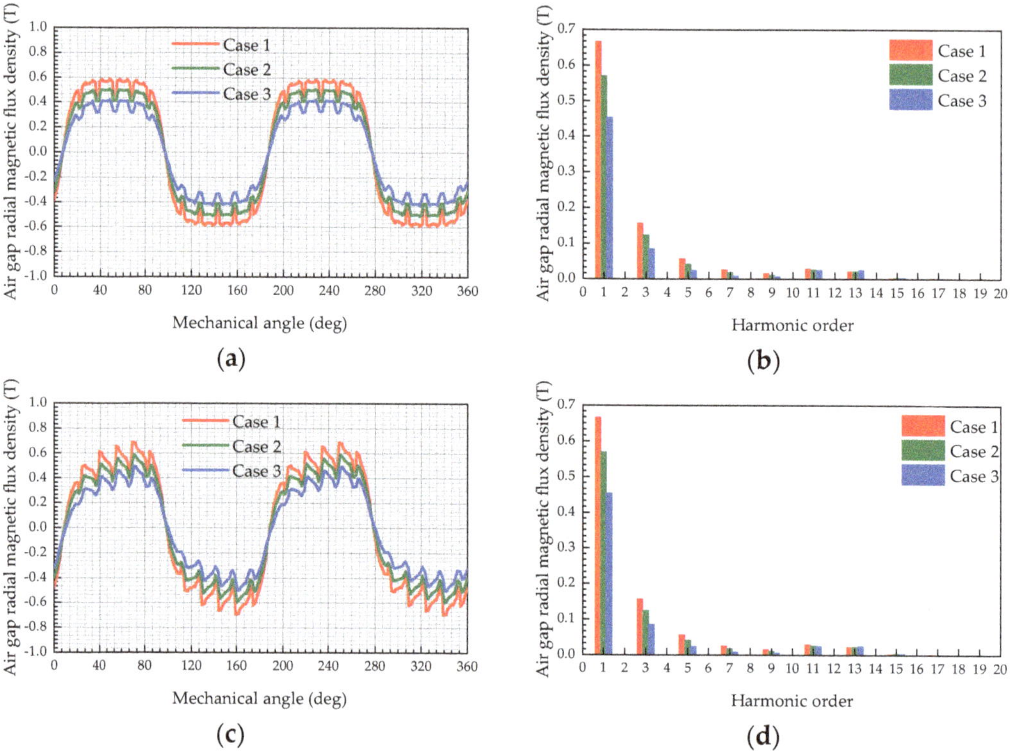

Figure 13. Radial air-gap flux density and Fourier transforms under three cases at no load and rated load: (**a**) radial air-gap flux density at no load; (**b**) Fourier transforms results at no load; (**c**) radial air-gap flux density at rated load; (**d**) Fourier transforms results at rated load.

4.2. Losses Comparative Analysis

4.2.1. Rotor Eddy-Current Loss Comparative Analysis

The rotor eddy-current density in the three rotor shapes at rated load is shown in Figure 14. It can be found that the rotor eddy-current density on the permanent magnets in Case 3 is the smallest, and the rotor eddy-current density on the permanent magnets in Case 1 is the largest. Rotor eddy-current losses of the permanent magnets are calculated, as shown in Figure 15. The rotor eddy-current loss in Case 1 is the highest, which is about 121.4 W, indicating that the rotor temperature rises in Case 1 would be serious. Case 3 has smallest rotor eddy-current loss, which is about 5.7 W. Obviously, as the rotor length–diameter ratio grows, the rotor eddy-current loss decreases.

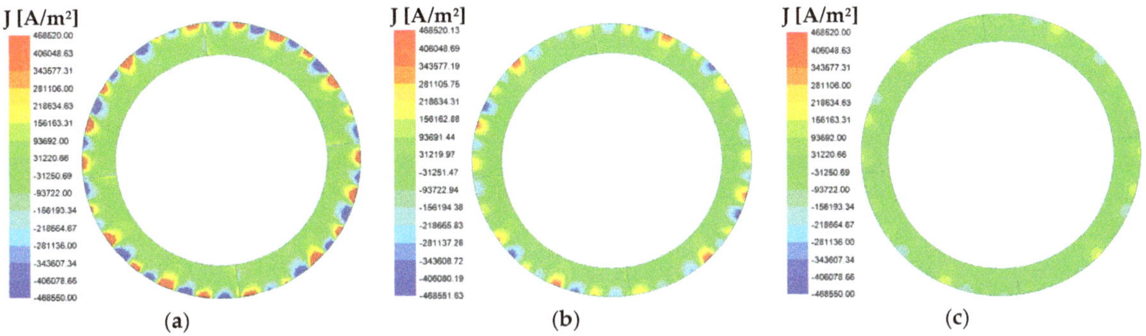

Figure 14. Rotor eddy-current density in three cases at rated load: (**a**) Case 1; (**b**) Case 2; (**c**) Case 3.

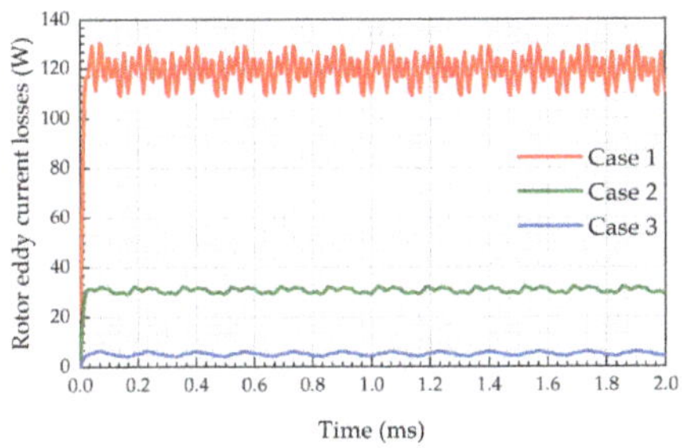

Figure 15. Eddy-current losses in the three cases at rated load.

4.2.2. Stator Iron-Core Loss Comparative Analysis

The stator iron-core losses calculation results under the three cases are shown in Figure 16. According to Berttoti's discrete model of iron-core losses, the iron-core losses are related to the amplitude and frequency of magnetic flux density in the stator core. From the magnetic flux density distribution in Figure 11, the magnetic flux density amplitude of the stator yoke in Case 3 is the highest, so the iron-core loss is the largest.

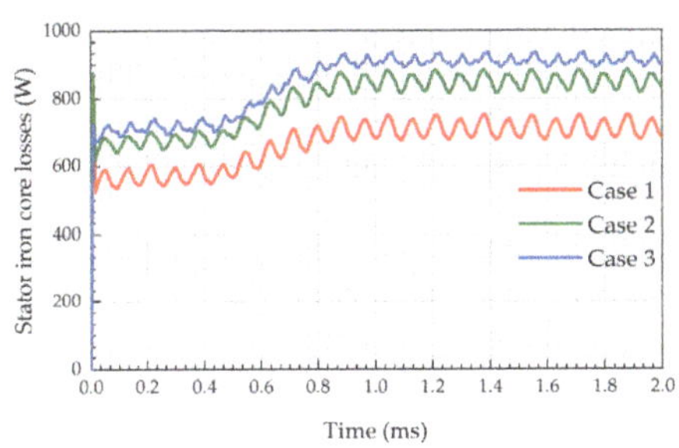

Figure 16. Stator iron-core losses in the three cases at rated load.

4.2.3. Air-Friction Losses Comparative Analysis

The rotor surface linear velocity of HSPMMs is higher, air-friction loss is larger than that of the low-speed motor. Air-friction loss of HSPMMs has a large influence on the design of the rotor shape structure and motor temperature rise. Air-friction loss calculation values of the three cases are shown in Figure 17. It can be found that the bigger the rotor diameter, the larger the air-friction loss at rated speed.

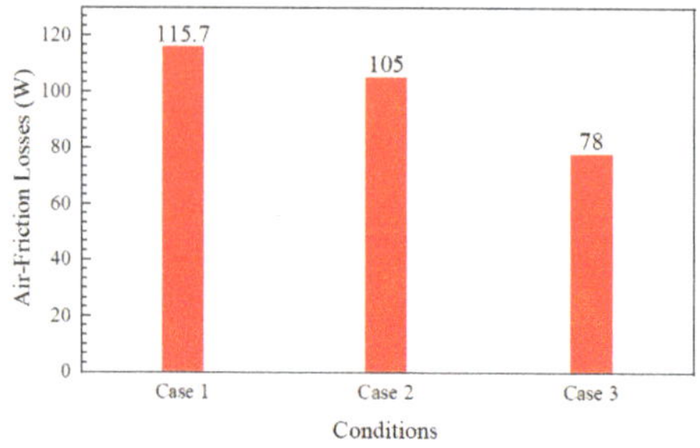

Figure 17. Air-friction losses in the three cases at rated speed.

4.3. Rotor Stress Characteristics Analysis

4.3.1. Rotor Stress Comparative Analysis at a Cold State

In this paper, 3D-FEM are established for the stress of permanent magnets and sleeve under three cases. The radial and tangential stress of permanent magnets and the Von-Mises stress of the sleeve at a cold state are obtained. Figure 18 shows the Von-Mises stress distribution diagram of the sleeve. It can be found that the Von-Mises stress of the sleeve decreases with the decrease in the rotor diameter at a cold state, and the results are consistent with the theoretical analysis.

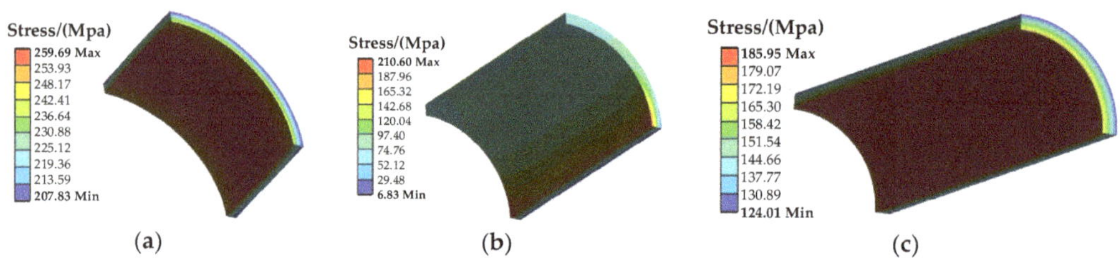

Figure 18. Sleeve Von-Mises stress distribution under the three cases at rated speed at a cold state: (**a**) Case 1; (**b**) Case 2; (**c**) Case 3.

Figure 19 indicates the tangential stress distribution of permanent magnets. It can be found that the tangential stress of permanent magnets in Case 1 is the highest at 40.246 MPa in a cold state. The tangential stress of permanent magnets in Case 3 is the lowest at 8.74 MPa in a cold state. The tangential stress of permanent magnets is mainly related to the interference fit. When the interference fit increases, the assembly of permanent magnets becomes closer, and its tangential stress decreases correspondingly. However, excessive interference fit would increase the radial pressure of permanent magnets and increase the technology assembly and manufacturing difficulty.

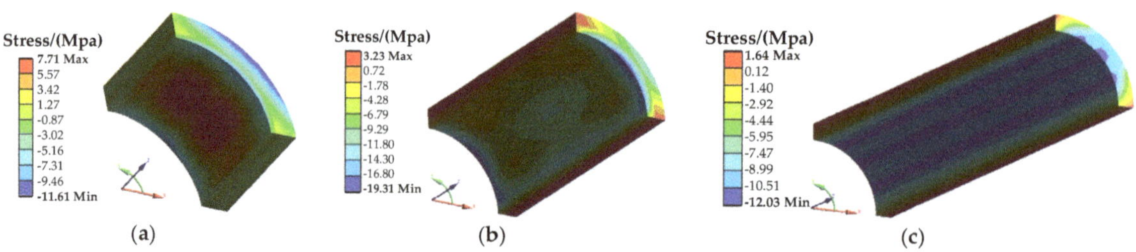

Figure 19. Permanent magnets tangential stress distribution under the three cases at rated speed at a cold state: (**a**) Case 1; (**b**) Case 2; (**c**) Case 3.

The radial stress diagram of the permanent magnets under the three cases is shown in Figure 20. The radial stress of permanent magnets indicates the centrifugal force of the permanent magnets when the rotor is rotating at the rated speed. It can be found that the stress at the periphery of the permanent magnets is minimal, due to the radial pressure exerted by the sleeve on the permanent magnets and the centrifugal force of the rotor canceling each other out.

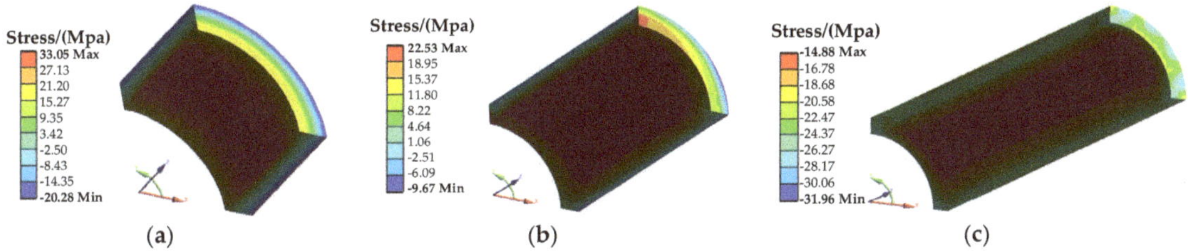

Figure 20. Permanent magnets radial stress distribution under the three cases at rated speed at a cold state: (**a**) Case 1; (**b**) Case 2; (**c**) Case 3.

4.3.2. Rotor Stress Comparative Analysis at a Hot State

For the three cases, the Von-Mises stresses of sleeve, tangential stress, and radial stress of permanent magnets are comparatively analyzed at a hot state. The Von-Mises stress distribution of the sleeve is shown in Figure 21, the tangential stress distribution of permanent magnets is shown in Figure 22, and the radial stress distribution of permanent magnets is shown in Figure 23. At the hot state, it can be found that the radial and tangential stress of magnets increase with the temperature rise, while the carbon fiber sleeve Von-Mises stress did not change, indicating that the thermal stability of the permanent magnets is poorer, and the performance of the permanent magnets is strongly influenced by temperature rise. According to the stress calculation results at a hot state, the tangential stress of permanent magnets under Case 1 is 87.5 MPa, which has exceeded the thermal constraints. Therefore, the rotor shape of Case 1 cannot meet the constraints of rotor stress.

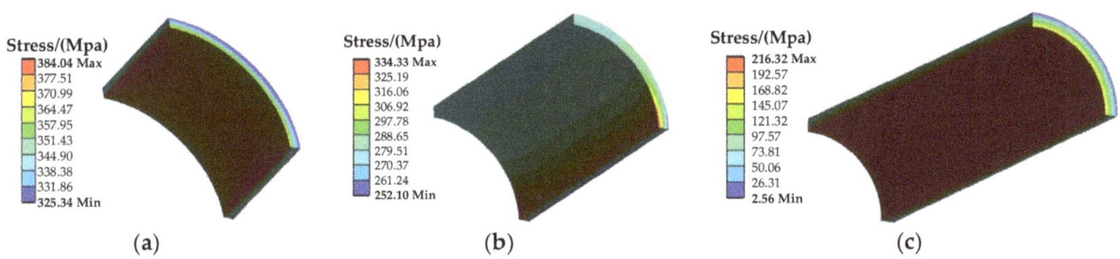

Figure 21. Sleeve Von−Mises stress distribution under the three cases at rated speed in a hot state: (**a**) Case 1; (**b**) Case 2; (**c**) Case 3.

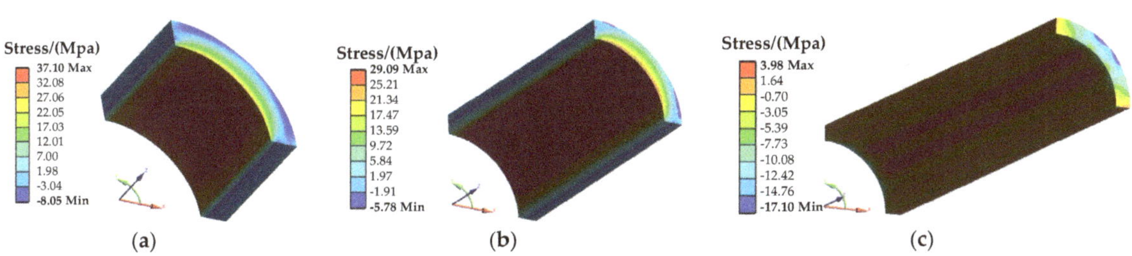

Figure 22. Permanent magnets' tangential stress distribution under the three cases at rated speed at a hot state: (**a**) Case 1; (**b**) Case 2; (**c**) Case 3.

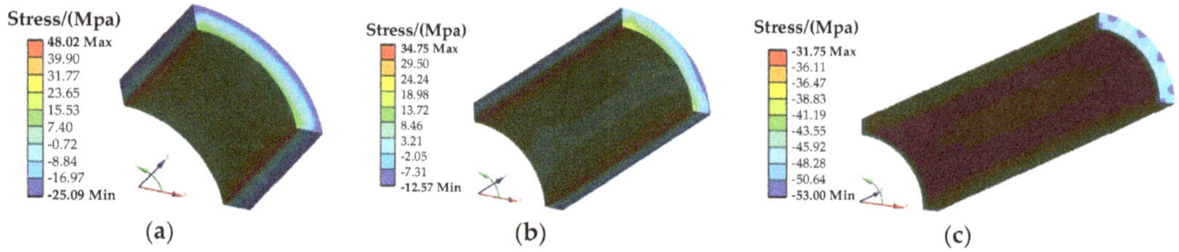

Figure 23. Permanent magnets' radial stress distribution under the three cases at rated speed at a hot state: (**a**) Case 1; (**b**) Case 2; (**c**) Case 3.

From the above rotor stresses analysis under the cold and hot state, it can be found that as the rotor length-diameter ratio increases, the rotor outer diameter gradually decreases, thus the rotor stress gradually decreases, indicating that the mechanical reliability of the rotor gradually increases.

4.4. Rotor Dynamics Characteristics Analysis

When the high rotor speed of HSPMMs reaches its critical speed, the rotor would resonate and damage the rotor. To avoid resonance, the rated speed of the rotor should be less than 0.7 times the first-order critical speed.

The first-order critical speed of the rotor is related to the diameter and length of the rotor. For case 1, $D_r/L_{ef}^2 = 2.5 \times 10^{-2}$, for case 2, $D_r/L_{ef}^2 = 5.12$, and for case 3, $D_r/L_{ef}^2 = 1.2 \times 10^{-3}$.

The supporting stiffness of the rotor is set as 3.5×10^4 N/mm. Modes of oscillation are obtained by the 3D-FEM calculation. Figure 24 shows the first order and second order modes of the rotor for Case 1, Case 2, and Case 3. It is found that the first critical speed of the rotor is 43,800 rpm for Case 1, 168,220 rpm for Case 2, and 44,663 rpm for Case 3, which indicates that the rotor dynamics meet the stability requirements for the three cases, and smaller and larger length–diameter ratios may cause the rotor to resonate more easily.

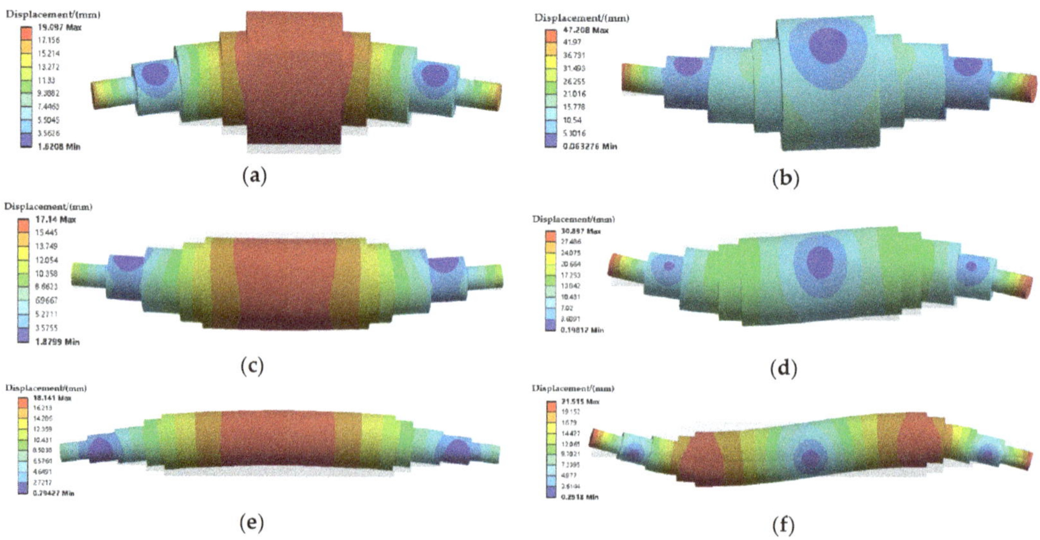

Figure 24. Rotor dynamic characteristics under the three cases: (**a**) first-order mode for Case 1; (**b**) second-order mode for Case 1; (**c**) first-order mode for Case 2; (**d**) second-order mode for Case 2; (**e**) first-order mode for Case 3; (**f**) second-order mode for Case 3.

4.5. Thermal Characteristics Analysis for Three Cases

In this paper, the housing water cooling system is used for heat dissipation design. The cooling system must ensure the safe operation and lifetime of the designed motor. The maximum tolerable temperature of the permanent magnets is limited to less than 130 °C to increase the reliability. Furthermore, the maximum winding temperature is set to 130 °C to increase the service life of the HSPMM.

To meet the temperature constraints, a housing water cooling system is designed for the motor. The water duct cools the motor through the spiral waterway on the stator housing. The distribution of spiral ducts is shown in Figure 25. In this design, the cooling water temperature and the ambient temperature are both 30 °C, and the water flow rate is set to 2 m^3/h.

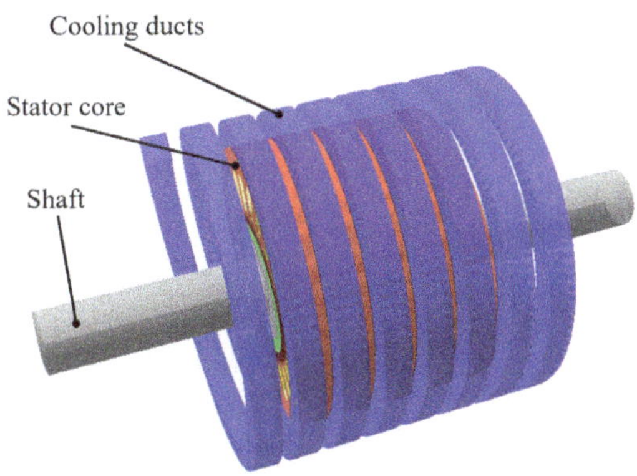

Figure 25. Spiral water ducts for designed motor stator housing cooling system.

The temperature distribution and comparison of the stator windings, stator cores, and permanent magnets are shown in Figures 26 and 27. For the three cases, the maximum temperatures of the winding and rotor are lower than the limiting temperature values. However, the temperature of permanent magnets of Case 1 is slightly lower than that in the other two cases. When the motor temperature requirements are extremely strict and the cooling system is also limited, the smaller rotor length–diameter ratio design can reduce the permanent magnet temperature slightly. In Figure 27, it can be found that as the rotor length–diameter ratio increases, the temperature of stator reduces gradually, and the temperature of rotor increases.

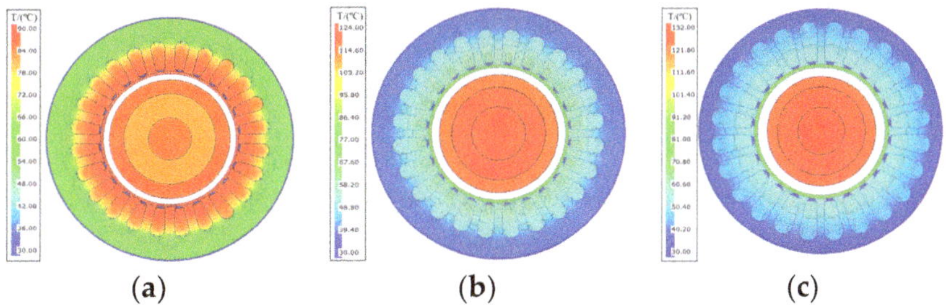

Figure 26. The temperature distribution under the three cases at rated speed: (**a**) Case 1; (**b**) Case 2; (**c**) Case 3.

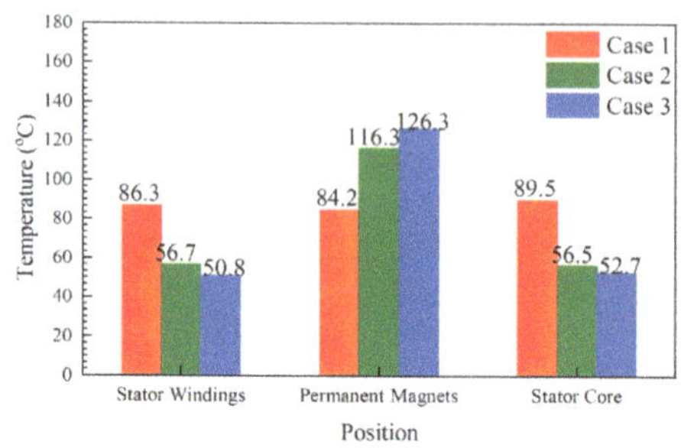

Figure 27. The temperatures on the different positions under the three cases at rated speed.

5. Prototype and Experimental Tests

Based on the above analysis, Case 2 can satisfy the comprehensive performance constraints best, so Case 2 is selected as the final design parameters of the 60 kW at 30,000 rpm HSPMM. Then, a prototype is manufactured, as shown in Figure 28. Electromagnetic and thermal tests are also measured, including the power, back-EMF, and winding temperature, as shown in Table 9.

Figure 28. Prototype.

Table 9. Structure parameters of the rotor shape for HSPMM under the three cases.

Parameters	Measurement	Calculation
Power (kW)	60	60
Back-EMF (V)	520	523
Stator winding temperature (°C)	60.9	56.7

The line-to-line back-EMF of no-load operation is measured with the prototype driven by another motor. The measured peak-to-peak value of line-to-line back-EMF at 30,000 rpm is 520 V, which is nearly the same as the designed voltage. The stator windings temperature at rated load is measured by the Pt100 resistance temperature detectors installed in the stator slots. It can be found that the temperature of the measured windings is 60.9 °C, which is close to the FEM calculation results of 56.7 °C. Obviously, for the designed HSPMM, the measured value of electromagnetic performance and temperature characteristics are close to the calculated results, to satisfy the electromagnetic and thermal constraints.

In addition, the prototype is shown to work at the rated speed for a long period. There is no damage to the rotor, which indicates that the design of the HSPMM is also reasonable and that the designed motor satisfied all the comprehensive performance constraints in the actual industry.

Overall, the experimental prototype tests show that the designed motor meets all physical field constraints, including electromagnetic, mechanical, and thermal characteristics, and as this paper proposed, the influent law of rotor length–diameter ratio on comprehensive characteristics is beneficial for the HSPMMs.

6. Conclusions

For the high-speed permanent magnet machines design, many rotor shape schemes can satisfy the multiple physical fields' comprehensive performances. A variety of rotor length–diameter ratios are available, from stubby to slender. However, different rotor length–diameter ratios impact on multiple physical fields' performances. The effect of rotor length–diameter ratio on multiple physical fields' comprehensive performances have rarely been focused on in the existing literature, which leads to an uncertain selection of the rotor length–diameter ratio in the design period of high-speed permanent magnet motors.

In this paper, finite element models of electromagnetic properties, rotor stresses, rotor dynamics, and temperature are built by the corresponding finite element analysis software, and the multiple physical fields performances are analyzed numerically and comprehensively comparatively analyzed through the finite element models.

The following key simulation results can be drawn from a comprehensive comparative analysis.

- When keeping the same stator outer diameters for the two-pole motor and four-pole motor, the stator yoke of the two-pole motor is magnetically saturated. The stator outer diameter of the two-pole motor is enlarged by 15% to avoid the magnetic flux density of stator-core saturation. The stator coils half-turn length of the four-pole motor is 197.258 mm, which is much shorter than the 299.492 mm of the two-pole motor. Rotor eddy-current loss of the four-pole motor is 31.52 W, which is much lower than the 235.56 W of the two-pole rotor. Compared to the two-pole motor, the four-pole motor has the higher power density.
- With the smallest rotor length–diameter ratio, Case 1 has the largest rotor eddy-current loss (121.4 W), lowest core loss (720.8 W), and largest air-friction loss (115.7 W). With the largest rotor length–diameter ratio, Case 3 has the smallest rotor eddy-current loss (5.7 W), highest core loss (923.3 W), and smallest air-friction loss (78.0 W).
- For the three design cases, Case 1 has the smallest rotor length–diameter ratio and the largest rotor outer diameter, and thus the sleeve stress and permanent magnet tangential and radial stresses are both the largest regardless of operating in cold and hot states, which are 259 MPa, 7.71 MPa, and 33.051 MPa for the cold state, and 384 MPa, 37.1 MPa, and 48.02 MPa for the hot state. Case 3 has the largest rotor length–diameter ratio and the smallest rotor outer diameter, and thus has the smallest sleeve stress and permanent magnet tangential and radial stresses in both the cold state and the hot state, which are 185.95 MPa, 1.64 MPa, and 14.881 MPa for the cold state, and 216.32 MPa, 3.98 MPa, and 31.745 MPa for the hot state.
- The influence of the rotor length–diameter ratio on rotor dynamics performance is compared and investigated. The first critical speed of the rotor is 43,800 rpm for Case 1, 68,220 rpm for Case 2, and 44,663 rpm for Case 3.
- The temperatures of the stator windings for the three cases are 86.3 °C, 56.7 °C, and 50.8 °C, respectively. The temperatures of permanent magnets for the three cases are 84.2 °C, 116.3 °C, and 126.3 °C, respectively.

In this paper, the analysis results show different rotor length–diameter ratios greatly influence multiple physical fields' comprehensive performances, and these influent laws are crucial to the design of high-speed permanent magnet motors. Thus, in the design period of high-speed permanent magnet machines, the influence of rotor length–diameter ratio on multiple physical fields' comprehensive performances should be considered and investigated, in order to obtain better multiple physical fields' comprehensive performances.

In order to verify the theoretical analysis, based on the comprehensive characteristics comparative analysis results, one 60 kW at 30,000 rpm HSPMM is designed. Multiple physical fields' comprehensive comparative analysis results are tested through prototype experiments. The RMS value of the no-load line back-EMF during rated operation is

520 V, which is very close to the calculated result 523 V and the stator winding stabilized temperature is 60.9 °C, which agrees with the analysis result 56.7 °C. The analytical method is confirmed by the experimental results of the prototype.

In the existing literature, for high-speed permanent magnet motor design, the influence of rotor length–diameter ratios on multiple comprehensive physical performances is not usually considered, so the rotor shape can be designed in various forms, such as a short and thick rotor or long and thin rotor. However, different comprehensive physical properties are produced by different rotor shapes, which are very important for the design of high-speed motors. In this paper, the influence of rotor length–diameter ratio on the comprehensive performance of multiple physical fields is considered and investigated, which includes electromagnetic characteristics, losses properties, rotor stresses distributions, rotor dynamics, and thermal behavior. Through comprehensive comparative analysis of different rotor length–diameter ratios, the following conclusions can be obtained, which have not been proposed in the previous literatures.

- For the electromagnetic characteristics, as the rotor length–diameter ratio goes up, the fundamental amplitude of the no-load line back-EMF increases gradually and the fundamental and high-order harmonical amplitudes of the no-load and rated-load radial air-gap flux density decreases.
- For the loss properties, as the rotor length–diameter ratio grows, the rotor eddy-current loss decreases, the stator-core loss increases, and the air-friction loss decreases.
- For the rotor stresses' distributions, as the rotor length–diameter ratio increases, the rotor outer diameter gradually decreases, thus the rotor stress gradually decreases, indicating that the mechanical reliability of the rotor gradually increases.
- For the rotor dynamics, smaller and larger length–diameter ratios may cause the rotor to resonate more easily.
- For the thermal behavior, as the rotor length–diameter ratio increases, the temperature of stator reduces gradually, and the temperature of rotor increases.

Based on the analysis above, it can be concluded that the influence of the rotor length–diameter ratio on multiple comprehensive physical performances cannot be ignored and different rotor length–diameter ratios greatly impact on the multiple comprehensive physical performances of HSPMMs. The conclusions obtained in this paper can be used as a reference for rotor structural design of HSPMMs.

In this paper, the destructive experiments usually performed to test rotor stresses could not be directly tested on to the rotor at high-speed rotation, due to the limitation of the test equipment. Thus, the reliability of the rotor is indirectly verified through the long-term operation of the HSPMMs.

In the future, to provide more valuable conclusions, the rotor length–diameter ratio should be further optimized by adopting an optimization analysis method, based on a coupled multi-physics performance, which could achieve a better multi-physics performance of the HSPMMs.

Author Contributions: Conceptualization, G.D. and Y.Z.; data curation, W.G. and G.D.; formal analysis, W.G.; funding acquisition, G.D.; investigation, W.G. and G.D.; methodology, W.G. and G.D.; project administration, G.D.; resources, G.D.; software, W.G.; supervision, Y.Z.; validation, G.D., Y.Z., T.P. and N.L.; visualization, W.G.; writing–original draft, W.G.; writing–review and editing, W.G., G.D., T.P. and N.L. All authors have read and agreed to the published version of the manuscript.

Funding: This research was funded by National Nature Science Foundation of China, grant number 52177056.

Institutional Review Board Statement: Not applicable.

Informed Consent Statement: Not applicable.

Data Availability Statement: Not applicable.

Conflicts of Interest: The authors declare no conflict of interest.

Nomenclature

δ_s	Interference fit.
h_{sleeve}	The thickness of the sleeve.
h_{PM}	The thickness of the permanent magnet.
D_{is}	Stator outer diameter.
P_{out}	Output power at the rated load.
AJ	Thermal load.
δ	Air-gap length.
B	Magnetic flux density.
J	Current density.
λ	Rotor length–diameter ratio.
L_{ef}	Effective core length.
D_r	Rotor outer diameter.

References

1. Du, G.H.; Zhou, Q.X.; Liu, S.L.; Huang, N.; Chen, X. Multiphysics design and multiobjective optimization for high-speed permanent magnet machines. *IEEE Trans. Transp. Electrif.* **2020**, *6*, 1084–1092. [CrossRef]
2. Du, G.H.; Xu, W.; Zhu, J.G.; Huang, N. Power loss and thermal analysis for high-power high-speed permanent magnet machines. *IEEE Trans. Ind. Electron.* **2020**, *67*, 2722–2733. [CrossRef]
3. Du, G.H.; Huang, N. Multiphysics analysis of high-speed permanent magnet generators for waste heat application. *IET Electr. Power Appl.* **2020**, *14*, 937–942. [CrossRef]
4. Jacek, F.G.; Saari, J.H. Performance calculation for a high-speed solid-rotor induction motor. *IEEE Trans. Ind. Electron.* **2012**, *59*, 2689–2700.
5. Flyur, R.I.; Viacheslav, Y.V.; Valentina, V.A. Design aspects of a high-speed electric machine series. In Proceedings of the 2020 International Conference on Industrial Engineering, Applications and Manufacturing (ICIEAM), Sochi, Russia, 18–22 May 2020.
6. Nikita, U.; Alexander, S.; Cheol, H.P.; Ahn, J.H.; Janne, H.; Juha, P. Design aspects of high-speed electrical machines with active magnetic bearings for compressor applications. *IEEE Trans. Ind. Electron.* **2017**, *64*, 8427–8436.
7. Chen, Y.Y.; Ding, Y.; Zhuang, J.H.; Zhu, X.Y. Multi-objective optimization design and multi-physics analysis a double-stator permanent-magnet doubly salient machine. *Energies* **2018**, *11*, 2130. [CrossRef]
8. Wang, T.Y.; Wang, F.X.; Bai, H.R.; Xing, J.Q. Optimization design of rotor structure for high speed permanent magnet machines. In Proceedings of the 2007 International Conference on Electrical Machines and Systems (ICEMS), Seoul, Korea, 8–11 October 2007.
9. Du, G.H.; Huang, N.; He, H.C.; Lei, G.; Zhu, J.G. Parameter design for a high-speed permanent magnet machine under multiphysics constraints. *IEEE Trans. Energy Convers.* **2020**, *35*, 2025–2035. [CrossRef]
10. Zhang, F.G.; Du, G.H.; Wang, T.Y.; Wang, F.X.; Cao, W.P.; James, L.K. Electromagnetic design and loss calculations of a 1.12-MW high-speed permanent-magnet motor for compressor applications. *IEEE Trans. Energy Convers.* **2016**, *31*, 132–140. [CrossRef]
11. Chen, L.L.; Zhu, C.S.; Zhong, Z.X.; Liu, B.; Wan, A.P. Rotor strength analysis for high-speed segmented surface-mounted permanent magnet synchronous machines. *IET Electr. Power Appl.* **2018**, *12*, 979–990. [CrossRef]
12. Gong, C.; Li, S.F.; Thomas, G.H. Rotor dynamic analysis of ultra-high speed switched reluctance machines over 1 million rpm. In Proceedings of the 2018 IEEE Energy Conversion Congress and Exposition (ECCE), Portland, OR, USA, 23–27 September 2018.
13. Du, G.H.; Huang, N.; Zhao, Y.Y.; Lei, G.; Zhu, J.G. Comprehensive sensitivity analysis and multiphysics optimization of the rotor for a high speed permanent magnet machine. *IEEE Trans. Energy Convers.* **2021**, *36*, 358–367. [CrossRef]
14. Fang, H.Y.; Li, D.; Qu, R.H.; Li, J.; Wang, C.; Song, B. Rotor design and eddy-current loss suppression for high-speed machines with a solid-PM rotor. *IEEE Trans. Ind. Appl.* **2019**, *55*, 448–457. [CrossRef]
15. Xu, H.; Geng, H.P.; Lin, H.; Qi, Y.H.; Yin, X.L. Rotor design and analysis of a high speed permanent magnet synchronous motor for cryogenic centrifugal pump. In Proceedings of the 2019 IEEE International Conference on Mechatronics and Automation (ICMA), Tianjin, China, 4–7 August 2019.
16. Fang, H.Y.; Qu, R.H.; Li, J.; Zheng, P.; Fan, X.G. Rotor design for high-speed high-power permanent-magnet synchronous machines. *IEEE Trans. Ind. Appl.* **2017**, *53*, 3411–3419. [CrossRef]
17. Du, G.H.; Xu, W.; Zhu, J.G.; Huang, N. Rotor stress analysis for high-speed permanent magnet machines considering assembly gap and temperature gradient. *IEEE Trans. Energy Convers.* **2019**, *34*, 2276–2285. [CrossRef]

energies

Article

A Novel Model-Free Predictive Control for T-Type Three-Level Grid-Tied Inverters

Zheng Yin [1], Cungang Hu [1,*], Kui Luo [2], Tao Rui [3,*], Zhuangzhuang Feng [1], Geye Lu [4] and Pinjia Zhang [1,4,*]

1 School of Electrical Engineering and Automation, Anhui University, Hefei 230601, China
2 State Key Laboratory of Operation and Control of Renewable Energy & Storage Systems, China Electric Power Research Institute, Beijing 100192, China
3 School of Internet, Anhui University, Hefei 230601, China
4 Department of Electrical Engineering, Tsinghua University, Beijing 100190, China
* Correspondence: hcg@ahu.edu.cn (C.H.); 19057@ahu.edu.cn (T.R.); pinjia.zhang@mail.tsinghua.edu.cn (P.Z.)

Abstract: The model-free predictive control (MFPC) scheme is an effective scheme to enhance the parameter robustness of model predictive control. However, the MFPC scheme can be affected by the current gradient updating frequency. This paper proposes an improved MFPC scheme for a T-type three-level inverter. First, a novel current gradient updating method is designed to estimate all current gradients per control period, which uses the current gradient relationship between different voltage vectors and eliminates the effect of current gradients updating stagnation. Moreover, a sector judgment method based on the current gradient is proposed. Redundant small vectors are accurately judged and the computational burden is greatly reduced. Finally, simulation and experimental comparisons on a T-type three-level inverter verify the effectiveness of the proposed MFPC scheme.

Keywords: model-free predictive control; T-type three-level inverter; current gradient; stagnation effect; sector judgment

Citation: Yin, Z.; Hu, C.; Luo, K.; Rui, T.; Feng, Z.; Lu, G.; Zhang, P. A Novel Model-Free Predictive Control for T-Type Three-Level Grid-Tied Inverters. *Energies* **2022**, *15*, 6557. https://doi.org/10.3390/en15186557

Academic Editor: Adam Smoliński

Received: 28 July 2022
Accepted: 6 September 2022
Published: 8 September 2022

1. Introduction

In recent years, grid-tied inverters, including two-level grid-tied inverters and multi-level grid-tied inverters, have been widely applied in various new energy generation systems, such as wind power generation systems [1] and photovoltaic power generation systems [2]. As the crucial interface between renewable energy generation systems and the grid, the grid-tied inverters are needed to meet the various requirements with better steady-state performance, faster dynamic performance, higher reliability, and so on [3,4]. Although two-level inverters are commonly used because of their flexible control and reliable operation, they also have some problems such as high harmonics. Compared to two-level inverters, multi-level inverters have many advantages. Among them, the T-type three-level inverter has been studied by many scholars because of its advantages of low output harmonic, low switching loss, and high efficiency [5,6].

With the increasing application of T-type three-level inverters, various control schemes have also been developed. In particular, model predictive control (MPC) schemes have been studied in T-type three-level inverters, with the advantages of high-speed response, comprehensive control schemes, and multi-objective control ability [7–9]. However, the control performance of MPC schemes mainly depends upon the accuracy of model parameters [10]. Many methods have been designed to improve the parameter robustness of MPC. System parameters are identified online to reduce parameter errors [11–13]. The disturbances generated by inaccurate parameters are calculated and compensated in the prediction [14,15]. In these studies, it is essential to tune parameters in different observers, which increases the calculation burden in traditional processors. Moreover, an MPC based on sum of squares optimization has been developed in [16], which improves output performance by formulating the largest possible region of attraction for the considered equilibrium point and guaranteeing the stability of the Lyapunov function. Recently, model-free predictive control (MFPC) has been reported to overcome this challenge by providing robust and accurate state predictions without a complete model of the system.

Various MFPC schemes have been reported based on the ultra-local model [17,18], the auto-regressive with exogenous input (ARX) model [19], and the look-up table (LUT) [20–31]. The model-free predictive control (MFPC) scheme based on LUT was firstly introduced in [20] for permanent magnet synchronous motor (PMSM) drives, which is a computationally light yet very effective scheme. This scheme uses the current gradient of each voltage vector to predict currents. However, the current gradients of remaining voltage vectors keep the old values, which is considered as the current gradient updating stagnation effect. In the most serious case, the long-term stagnation effect can affect the stability of the control system.

To improve the updating frequency of current gradients, an MFPC scheme based on minimum updating frequency is proposed to update each gradient within the specified time [21]. The implementation process is that, if a voltage vector is not used in the past predefined frequency, it will be forcibly used in the next period and its current gradient will be updated. However, this updating method [21] affects the control performance since the non-optimal voltage vector is used frequently. In [22,23], the remaining current gradients are estimated according to the updated current gradients of the used voltage vector in the past three times. This method is effective only when the voltage vectors of the past three consecutive periods are different from each other; otherwise, the update can stagnate. In [24], the current gradients of all unused voltage vectors are estimated by using the current gradient relationship between different voltage vectors. Nevertheless, the updating can be affected when the applied vectors are the same for two consecutive periods. In [25–29], multi-vector MFPC is studied, in which two or three current gradients of the two or three applied voltage vectors are updated by using multiple adaptive sampling points. Although the updated number of current gradients increases with the increase in the number of applied voltage vectors, the stagnation effect still exists. The MFPC scheme is used in the three-level inverter-fed interior PMSM system [30], and all the current gradients are divided into seven categories. Each type of current gradient is updated according to the amplitude relationship of the corresponding voltage vector. However, among the 27 current gradients corresponding to the 27 voltage vectors of the three-level inverter system, up to six current gradients can be updated. In [31], the extended adjacent state scheme is used to reduce the possible number of remaining gradients, then the measured current gradient is used to update the remaining current gradients.

The stagnation effect becomes more obvious as the number of voltage vectors increases when the MFPC scheme is used in a T-type three-level grid-tied inverter system. To enhance the parameter robustness of predictive control, eliminate the current gradient updating stagnation effect, and improve the efficiency of voltage vector optimization, this paper proposes an improved MFPC scheme. The main contributions are shown below.

(1) A simplified look-up table (LUT) is designed. The redundant vector formed by two or three voltage vectors is considered as one corresponding current gradient. Hence, the number of LUTs decreased from 27 to 19.
(2) A novel current gradient updating method is proposed to eliminate the stagnation effect caused by the conventional updating method. The current gradient relationship between different voltage vectors is derived, so all the current gradients can be estimated in each period.
(3) A sector judgment method based on the current gradient is proposed. The proposed judgment method avoids using mathematical models to calculate the reference voltage. Hence, the number of candidate voltage vectors is reduced from 27 to 3, and the calculation speed is greatly improved.

The contents of this paper are organized as follows. In Section 2, the topology, vectors, and conventional MPC scheme are reviewed. In Section 3, the proposed scheme is introduced in detail. In Section 4, the proposed MFPC scheme is validated. In Section 5, the conclusion is given.

2. Conventional MPC Scheme

2.1. Topology and Voltage Vectors

The topology of the T-type three-level grid-tied inverter studied in this paper is shown in Figure 1, where u_{dc} is the dc-link voltage, and C_1 and C_2 are the upper and lower dc-link capacitors, respectively. O is the neutral point of dc-link capacitors and the zero-potential reference point. i_{C1} and i_{C2} are the current of C_1 and C_2, respectively. i_g is the output current. It can be seen from Figure 1 that each T-type three-level inverter bridge arm contains four switching tubes, which are S_1~S_4. There are three switching modes, including P, O, and N modes, based on the different switching combinations. All bridge arms share the neutral point of dc-link capacitors.

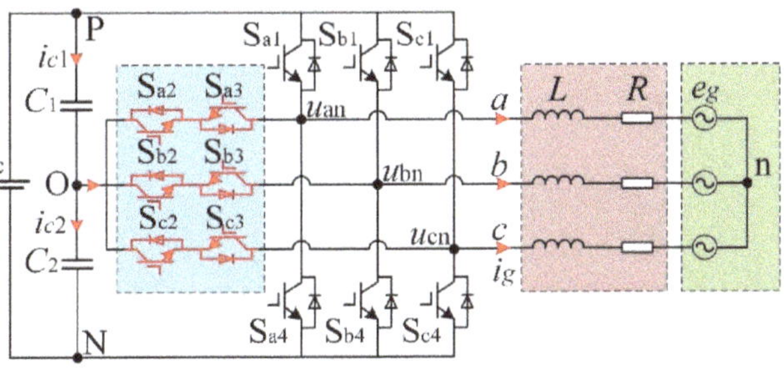

Figure 1. Topology of the T-type three-level grid-tied inverter.

There are 27 voltage vectors that can be generated by a three-phase three-level inverter, as shown in Figure 2. The specific positions in Figure 2 corresponding to multiple space vectors are called redundant vectors, whose output characteristics are equivalent. According to the amplitude of the voltage vector, the 27 switching states can be divided into a zero vector (PPP, OOO, NNN), small vector (POO, PPO, OPO, OPP, OOP, ONN, OON, NON, NOO, NNO, ONO), medium vector (PON, OPN, NPO, NOP, ONP, PNO), and large vector (PNN, PPN, NPN, NPP, NNP, PNP).

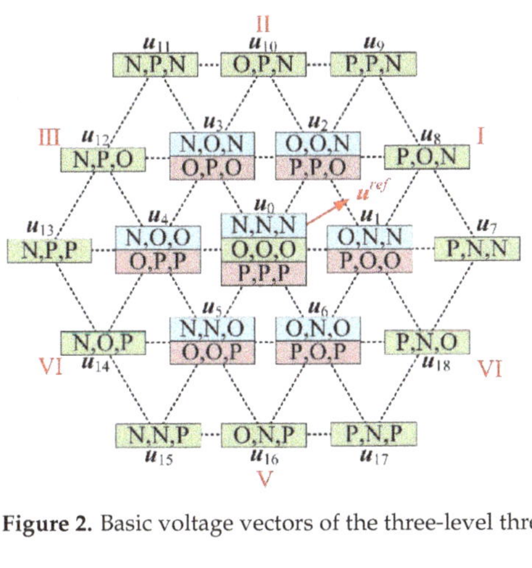

Figure 2. Basic voltage vectors of the three-level three-phase inverter.

2.2. Conventional Predictive Model

The principle of the conventional MPC scheme has been studied in many papers. Firstly, the mathematical model based on the LR-filtered three-level inverter connecting to grids in the α-β coordinate system can be expressed as

$$L\frac{di_g}{dt} = u_x - Ri_g - e_g \tag{1}$$

where $u_x = [u_{x\alpha}, u_{x\beta}]^T$, $i_g = [i_{g\alpha}, i_{g\beta}]^T$, and $e_g = [e_{g\alpha}, e_{g\beta}]^T$ represent the output voltage vectors of the inverter, output current vectors, and grid voltage vectors, respectively. L represents the filter inductance and R represents the stray resistance.

The discrete model of the inverter can be expressed as

$$i_g(k+1) = i_g(k) + \frac{T_s}{L}\left[u_x(k) - Ri_g(k) - e_g(k)\right] \tag{2}$$

where T_s is the control period.

The voltage vector selected in the last control period can be used to calculate the prediction current $i_g(k+1)$ by (2). Then, all the voltage vectors shown in Figure 2 can be further used to calculate prediction currents $i_{gx}(k+2)$ by (3).

$$i_{gx}(k+2) = \left(1 - \frac{RT_s}{L}\right)i_g(k+1) + \frac{T_s}{L}\left[u_x(k+1) - e_g(k+1)\right] \tag{3}$$

Finally, to evaluate the control performance of each voltage vector, a cost function shown in (4) is defined. The voltage vector that minimizes the cost function is used as the optimal vector and applied to the next control period.

$$G_x = \left(i_{ref}(k+2) - i_{gx}(k+2)\right)^2 \tag{4}$$

where $i_{ref}(k+2)$ are the reference currents.

In a three-level inverter control system, in addition to the output current, the neutral point voltage needs to be controlled. To avoid the design of weighting factor values when multiple control targets exist, MPC schemes without weighting factors have been proposed. In the MPC scheme without weighting factors, the P-type small vectors (POO, PPO, OPO, OPP, OOP) are considered as candidate vectors when the capacitor voltage $u_{c1} \geq u_{c2}$, and the N-type small vectors (ONN, OON, NON, NOO, NNO, ONO) are considered as candidate vectors when the capacitor voltage $u_{c1} < u_{c2}$. The control block diagram of MPC is shown in Figure 3.

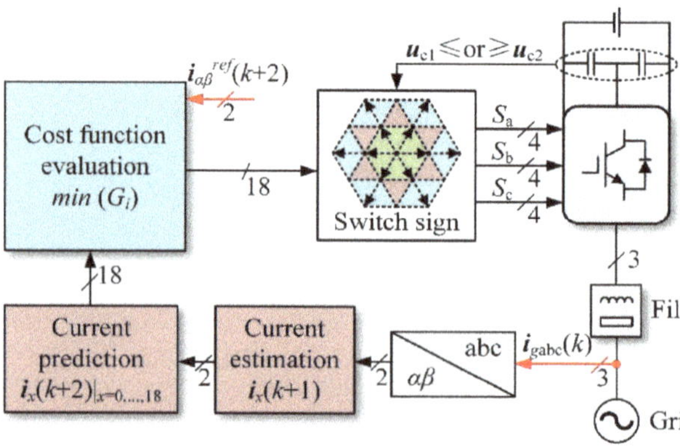

Figure 3. Control block diagram of the conventional MPC scheme.

However, the prediction currents are affected when model parameters are mismatched with the controller parameters. The prediction currents with parameter errors can be expressed as

$$i_{gerr}(k+1) = i_g(k) + \frac{T_s}{L+\Delta L}\left[u_x(k) - (R+\Delta R)i_g(k) - e_g(k)\right] \tag{5}$$

where ΔL represents the inductance error between its actual value and its controller value, and ΔR represents the resistance error between its actual value and its controller value. Hence, it is important to enhance the parameter robustness of predictive control.

3. The Proposed MFPC Scheme

To eliminate the dependence on model parameters, an improved MFPC scheme based on a novel current gradient updating method is proposed in this section, including the basic principle of MFPC, the current gradient updating stagnation effect analysis, the proposed current gradient updating method, the sector judgment method, and the implementation steps.

3.1. Basic Principle of MFPC

In the MFPC scheme, the mathematical model (1) can be rewritten as [17]

$$\begin{aligned}\Delta i_x(k-1) &= \tfrac{T_s}{L}\left(u_x(k-1) - e_g(k-1) - Ri_g(k-1)\right) \\ &= i_g(k) - i_g(k-1)\end{aligned} \tag{6}$$

where $\Delta i_x(k-1) = [\Delta i_{x\alpha}(k-1), \Delta i_{x\beta}(k-1)]^{\mathrm{T}}$ are the current gradients of the applied voltage vector $u_x(k-1)$. Although the switching states of redundant vectors are different, their coordinate components are the same. The multiple current gradients corresponding to redundant vectors are considered to be consistent. For example, voltage vector u_1 corresponds to two switching states of ONN and POO, so the current gradient corresponding to the two states is uniformly defined as Δi_1. Hence, nineteen current gradients Δi_x are stored in the look-up table (LUT) for current predictions, as shown in Table 1.

Table 1. The LUT of the T-type three-level inverter.

Voltage Vector	u_0	u_1	u_2	u_3	u_4	u_5	u_6	u_7	u_8	u_9	u_{10}	u_{11}	u_{12}	u_{13}	u_{14}	u_{15}	u_{16}	u_{17}	u_{18}
Current Gradient	Δi_0	Δi_1	Δi_2	Δi_3	Δi_4	Δi_5	Δi_6	Δi_7	Δi_8	Δi_9	Δi_{10}	Δi_{11}	Δi_{12}	Δi_{13}	Δi_{14}	Δi_{15}	Δi_{16}	Δi_{17}	Δi_{18}

Then, the prediction currents at the $(k+1)$th instant and the $(k+2)$th instant can be calculated based on (7) and (8), respectively.

$$i_{gx}(k+1) = i_g(k) + \Delta i_x(k) \tag{7}$$

$$i_{gx}(k+2) = i_{gx}(k+1) + \Delta i_x(k+1) \tag{8}$$

Finally, the optimal voltage vector u_{op} selected by cost function (4) is applied in the next control period.

3.2. Current Gradient Updating Stagnation Effect Analysis

It is obvious that the current prediction of MFPC is based on the current gradients which directly affect the accuracy of current prediction and the result of optimal voltage vector selection. The current gradients of applied voltage vectors are calculated based on (6) in the conventional MFPC scheme [20], however, the updating of the remaining current gradients stagnates. When the MFPC scheme is used in a three-level inverter system, the stagnation effect can be more obvious with the increasing number of voltage vectors, which not only affects the current performance but also affects the control of the neutral point voltage.

To reduce the updating stagnation effect, various current gradient updating methods have been reported in [21–24,30]; however, the updating stagnation is still existing. Hence, it is necessary to design an updating method to totally eliminate the stagnation effect.

3.3. The Proposed Current Gradient Updating Method

In the proposed updating method, sampling points are set before switching states in each control period to avoid the current spikes. Based on (6), the relationship between the current gradient of remaining voltage vector Δi_y and the current gradient of applied voltage vector Δi_x at $(k-1)$th instant and $(k-2)$th instant can be expressed as (9) and (10), respectively.

$$\Delta i_y(k-1) - \Delta i_x(k-1) = \frac{T_s}{L}(u_y(k-1) - u_x(k-1)) \tag{9}$$

$$\Delta i_y(k-2) - \Delta i_x(k-2) = \frac{T_s}{L}(u_y(k-2) - u_x(k-2)) \tag{10}$$

The filter inductance L can be considered as a constant when T_s is short enough. Then, T_s/L of (9) and (10) can be eliminated by the division between (9) and (10), and the relationship between the current gradient in the two control periods can be written as

$$\Delta i_y(k-1) = \frac{u_y(k-1) - u_x(k-1)}{u_y(k-2) - u_x(k-2)}(\Delta i_y(k-2) - \Delta i_x(k-2)) + \Delta i_x(k-1) \tag{11}$$

However, (11) cannot be used to update the current gradient when $u_y(k-2) - u_x(k-2) = 0$. In an α-axis coordinate system, seven situations may cause the stagnation effect. They are $u_{8\alpha} = u_{18\alpha}$, $u_{1\alpha} = u_{9\alpha} = u_{17\alpha}$, $u_{2\alpha} = u_{6\alpha}$, $u_{0\alpha} = u_{10\alpha} = u_{16\alpha}$, $u_{3\alpha} = u_{5\alpha}$, $u_{4\alpha} = u_{11\alpha} = u_{15\alpha}$, and $u_{12\alpha} = u_{14\alpha}$, respectively. In a β-axis coordinate system, five situations may cause the stagnation effect. They are $u_{9\beta} = u_{10\beta} = u_{11\beta}$, $u_{2\beta} = u_{3\beta} = u_{8\beta} = u_{12\beta}$, $u_{0\beta} = u_{1\beta} = u_{4\beta} = u_{7\beta} = u_{13\beta}$, $u_{5\beta} = u_{6\beta} = u_{14\beta} = u_{18\beta}$, and $u_{15\beta} = u_{16\beta} = u_{17\beta}$, respectively. To eliminate the stagnation effect, (9) under the α-axis and β-axis is derived as (12) and (13), respectively. For example, when the applied voltage vector is u_2, $\Delta i_{2\alpha\beta}$ can be obtained based on (6). Then, in addition to $\Delta i_{6\alpha}$ (calculated based on (12)), other α-axis current gradients can be calculated based on (11). Moreover, in addition to $\Delta i_{3\beta}$, $\Delta i_{8\beta}$, and $\Delta i_{12\beta}$ (calculated based on (13)), other β-axis current gradients can be calculated based on (11).

$$\Delta i_{y\alpha}(k-1) = \Delta i_{x\alpha}(k-1) \tag{12}$$

$$\Delta i_{y\beta}(k-1) = \Delta i_{x\beta}(k-1) \tag{13}$$

As a result, the stagnation effect is eliminated and the nineteen current gradients can be updated in each control period, as shown in Figure 4.

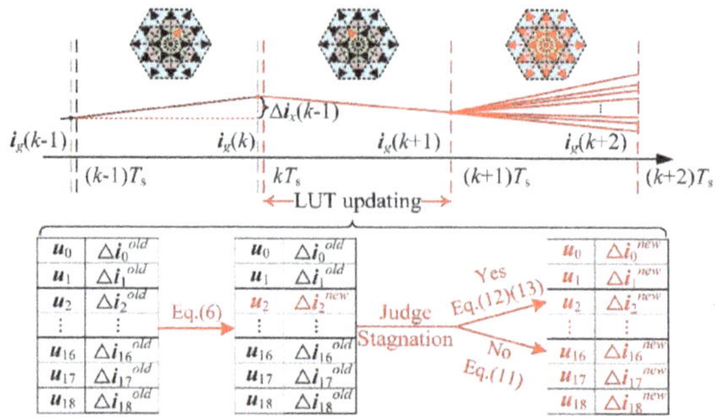

Figure 4. Proposed current gradient updating method.

3.4. Proposed Sector Judgment Method

Conventionally, in order to find the optimal voltage vector through the cost function (4), 27 voltage vectors need to be traversed. When the redundancy state of the zero vector is ignored and the small vectors are selected as P-type small vectors or N-type small vectors according to the dc-link capacitor voltage difference, there are still 19 voltage vectors to be traversed. Since the calculation of reference voltage depends on the mathematical model, the sector judgment method based on reference voltage may cause sector judgment deviation when the model parameters are mismatched. Therefore, a sector judgment method based on the current gradient is proposed in this section.

Figure 5 shows the 19 voltage vectors and their corresponding 19 current gradients. The three-level inverter space vector coordinate system is divided into 6 large sectors and 24 small sectors. The position of the large sector where the reference current is located is judged according to the prediction currents of the medium vectors. For example, if the reference current is in large sector I, the cost of u_8 calculated by (4) is the smallest compared with the cost of u_{10}, u_{12}, u_{14}, u_{16}, and u_{18}.

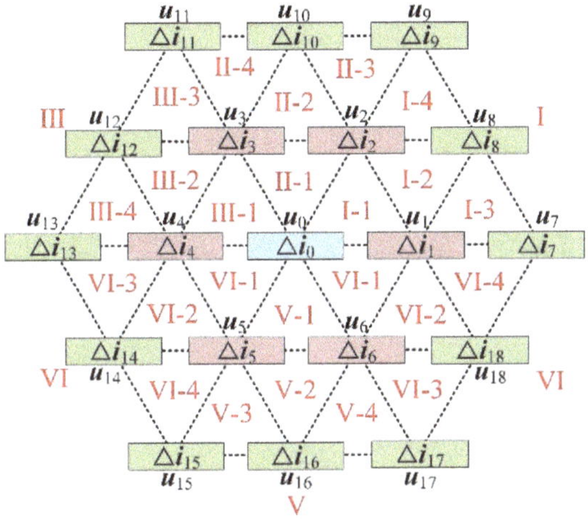

Figure 5. Big and small sectors of the three-level three-phase inverter.

After judging the large sector, it is necessary to judge the small sector. First, the prediction currents can be calculated by (8). Then, the weighted error square ε^2 of the four small sectors can be calculated by (14) and selecting the smallest ε^2 corresponding sector as the target sector. Finally, there are only three voltage vectors as candidate vectors.

$$\varepsilon^2 = \sum_{x=1}^{3} \left(i_{ref}(k+2) - i_{gx}(k+2) \right)^2 \tag{14}$$

3.5. Implementation Steps

Figure 6 shows the control diagram of the proposed MFPC scheme. The proposed MFPC scheme is realized to eliminate parameter dependence, the novel current gradient updating method is realized to eliminate the stagnation effect, and the sector judgment method is also realized to further reduce the computational burden over the conventional MFPC scheme. The detailed implementation steps are as follows.

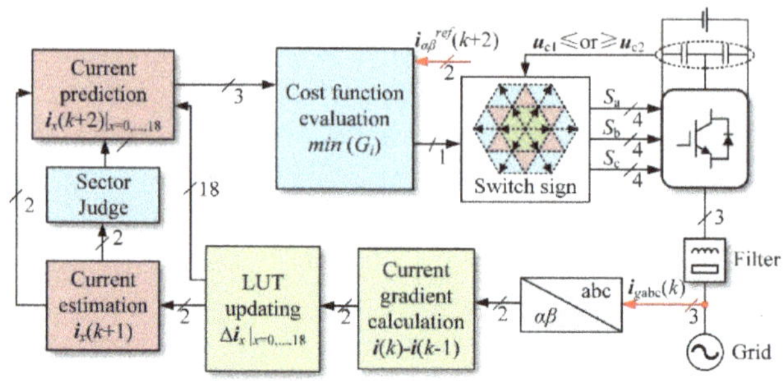

Figure 6. Control block diagram of the proposed MFPC scheme.

Step I: Sample the dc-link capacitor voltage $u_{c1}(k)$ and $u_{c2}(k)$, select P-type basic voltage vectors or N-type basic voltage vectors based on $u_{c1}(k) - u_{c2}(k) \leq$ or >0.
Step II: Sample the output current $i_g(k)$ and calculate current gradient $\Delta i_x(k-1)$ by (6).
Step III: Update the remaining current gradients without stagnation effect by (11), and update the remaining current gradients with stagnation effect by (12) and (13).
Step IV: Calculate prediction current $i_{gx}(k+1)$ and $i_{gx}(k+2)$ by (7) and (8), respectively.
Step V: Judge large sectors by (4), and judge small sectors by (14).
Step VI: Evaluate the cost of the three voltage vectors by (4) and select the optimal vector.

4. Simulation and Experimental Evaluation

In order to verify the effectiveness of the proposed MFPC scheme, simulations in MATLAB/Simulink and experiments under the conventional MPC scheme, conventional MFPC schemes, and the proposed MFPC scheme are carried out. Figure 7 depicts the T-type three-level three-phase grid-connected inverter experimental setup. The main control chip of the inverter is DSP28335. The sampling frequency is set as 20 kHz. The inverter parameters are listed in Table 2.

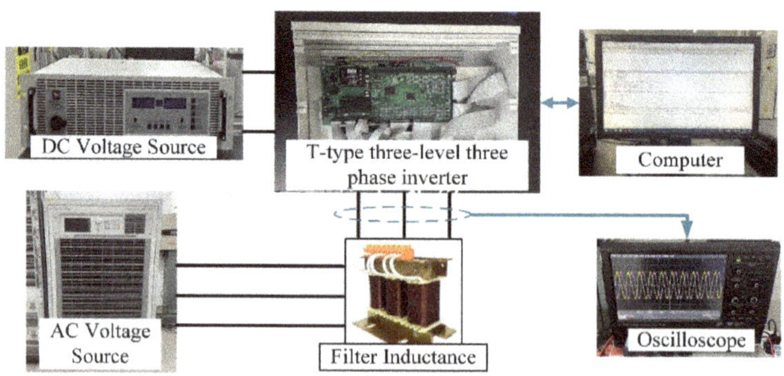

Figure 7. T-type three-level three-phase grid-connected inverter experimental platform.

Table 2. System and control parameters.

Parameters	Symbol	Values
DC-link voltage	u_{dc}	300 V
Peak of grid phase voltage	e	150 V
Grid angular frequency	ω_g	314.16 rad/s
Control time	T_s	50 μs
Parasitic resistance	R	0.05 Ω
Filter inductance	L_0	10 mH

4.1. Impact of Current Gradient Updating Stagnation

The proposed current gradient updating method is compared with the updating method in [20] and the updating method in [30]. Figure 8a illustrates the α-axis and β-axis current gradients when the updating method in [20] is used. In this current gradient updating method, there is a notable stagnation effect in both the current gradients Δi_α and Δi_β because the current gradients of the applied voltage vector can be updated for the LUT; however, the remaining current gradients keep the old values. Compared to the two-level inverter system [20], the stagnation effect becomes more obvious with the increasing number of voltage vectors when the updating method [20] is used in the three-level inverter system. As shown in Figure 8b, although the updating frequency is improved, the stagnation effect still exists in the α-axis and β-axis current gradients when the updating method in [30] is used. When the updating method is converted to the proposed updating method, it can be seen that there is no stagnation effect existing in the current gradient waveforms, as shown in Figure 8b, which shows the effectiveness of the proposed updating method.

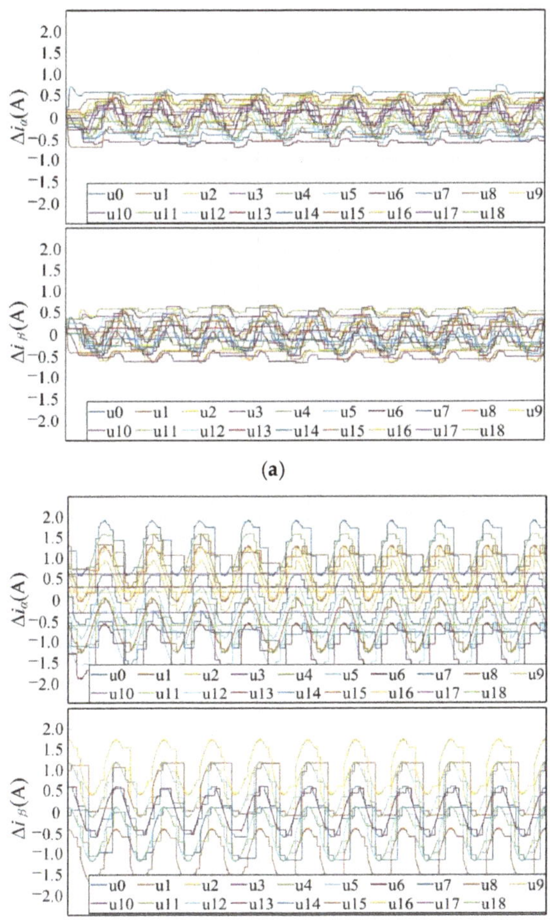

(a)

(b)

Figure 8. *Cont.*

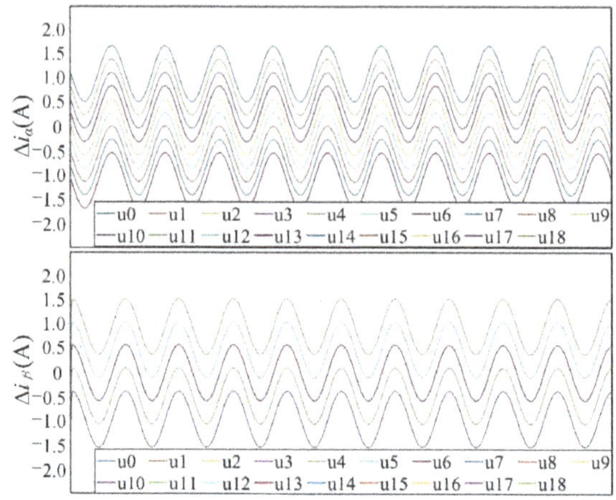

(c)

Figure 8. Current gradient comparisons under different updating methods (i_{ref} = 5A). (**a**) Updating method in [20]. (**b**) Updating method in [30]. (**c**) The proposed updating method.

4.2. Steady-State Experimental Evaluation

Figure 9 shows the steady-state performance comparisons of the conventional MPC scheme, conventional MFPC scheme in [20], conventional MFPC scheme in [30], and the proposed MFPC scheme when model parameters match controller parameters. Based on the steady-state experimental waveforms, it can be seen that both the conventional MPC scheme, the conventional MFPC scheme in [30], and the proposed MFPC scheme can achieve tracking the reference currents and balancing the neutral point voltages. However, the neutral point voltage of the conventional MFPC scheme in [20] is affected by the current gradient stagnation effect. The harmonic spectra of the current are drawn by MATLAB/Simulink with the experimental data obtained from the oscilloscope, as shown in Figure 9.

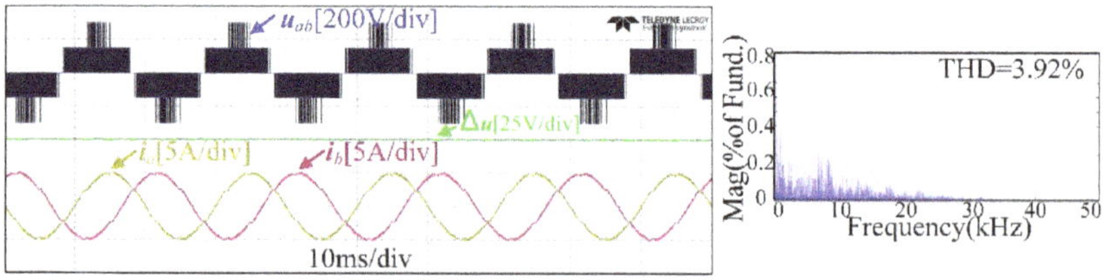

(a)

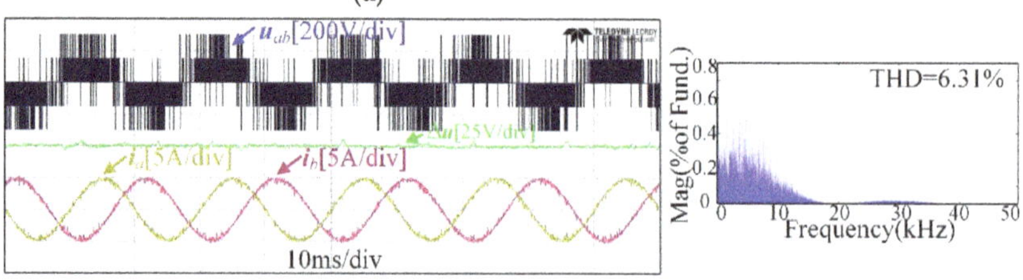

(b)

Figure 9. *Cont.*

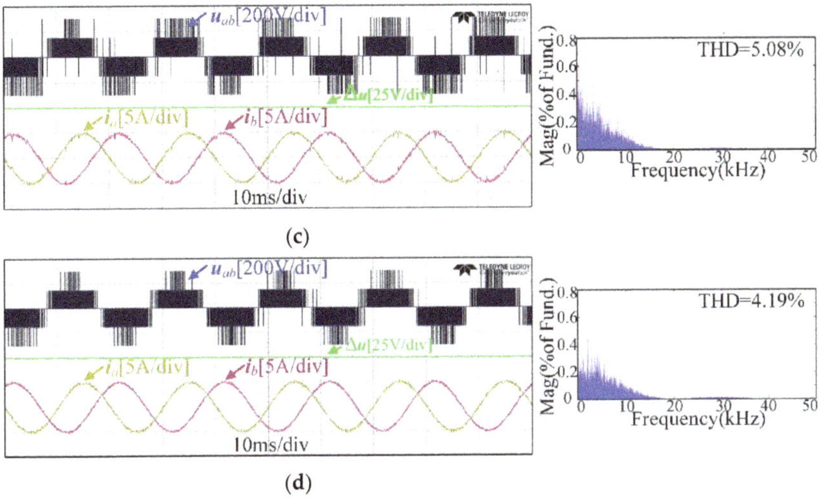

Figure 9. Experimental waveform comparisons (i_{ref} = 5A). (**a**) Waveforms under the conventional MPC scheme. (**b**) Waveforms under the conventional MFPC scheme in [20]. (**c**) Waveforms under the conventional MFPC scheme in [30]. (**d**) Waveforms under the proposed MFPC scheme.

For the conventional MPC scheme, the total harmonic distortion (THD) of phase-a output current is 3.92%. For the conventional MFPC scheme in [20], the THD of phase-a output current changes from 3.92% to 6.31% because the current performance is affected by the stagnation effect and the sampling noise effect. For the conventional MFPC scheme in [30], the THD of phase-a output current changes from 6.31% to 5.08% because the stagnation effect is improved, but it cannot be eliminated. Moreover, the current spikes caused by the stagnation effect can be observed simultaneously in Figure 9b,c. Compared with the conventional MFPC scheme in [30], the THD of phase-a output current reduces from 5.08% to 4.19% because the stagnation effect is eliminated, as shown in Figure 9d. However, the current performance of the proposed scheme is worse than that of the conventional MPC scheme because the current performance of the MFPC scheme is affected by the sampling noise.

To further verify the effectiveness of the proposed MFPC scheme, other experimental comparisons under different reference currents are carried out. The THDs of phase-a current under four control schemes are shown in Figure 10. From Figure 10, the THDs of the four control schemes are reduced with the increased reference currents. The proposed MFPC still has a similar current THD to the MPC and has a lower current THD than the MFPC in [20] and the MFPC in [30].

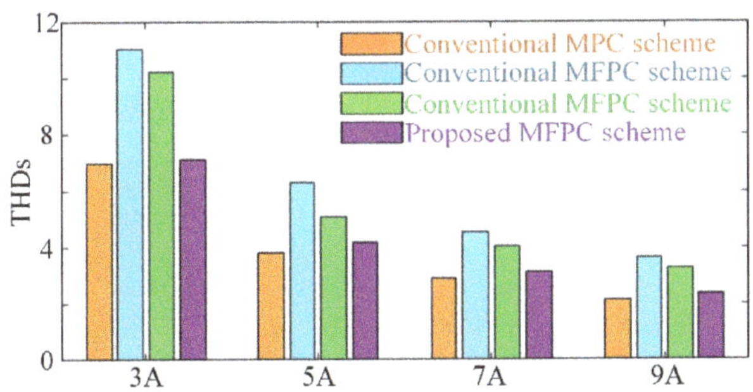

Figure 10. The comparison of phase-a current THDs under different reference currents.

4.3. Experimental Evaluation under Mismatched Model Parameters

To verify the effectiveness of the proposed MFPC scheme in enhancing parameter robustness, the control performance of the proposed scheme and the conventional MPC scheme are compared in this section when the filter inductance as the model parameter is taken as different actual values from the controller values L_0. It can be seen that compared with the conventional MPC scheme, the proposed scheme reduces the THD from 8.98% to 6.25% when the actual filter inductance is set as 0.05 H (i.e., $0.5L_0$) as shown in Figure 11. Additionally, the proposed scheme reduces the THD from 2.78% to 1.87% when the actual filter inductance is set as 0.2 H (i.e., $2L_0$), as shown in Figure 12. Under the above two experimental conditions, the current prediction errors of the proposed scheme are lower than that of the conventional MPC scheme. The current waveforms, the THDs, and the current prediction errors verify that the proposed scheme has stronger parameter robustness than the conventional MPC scheme.

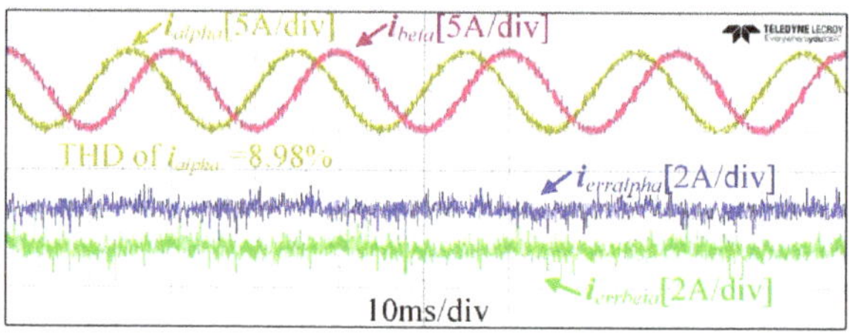

(**a**)

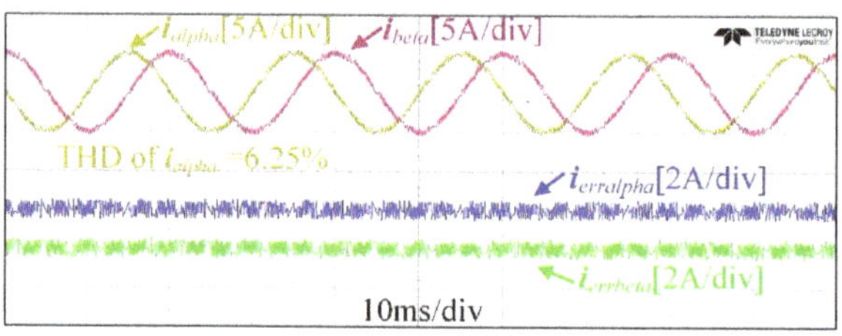

(**b**)

Figure 11. Experimental waveforms comparisons when $L/L_0 = 0.5$ ($i_{ref} = 5A$). (**a**) Waveforms under the conventional MPC scheme. (**b**) Waveforms under the proposed MFPC scheme.

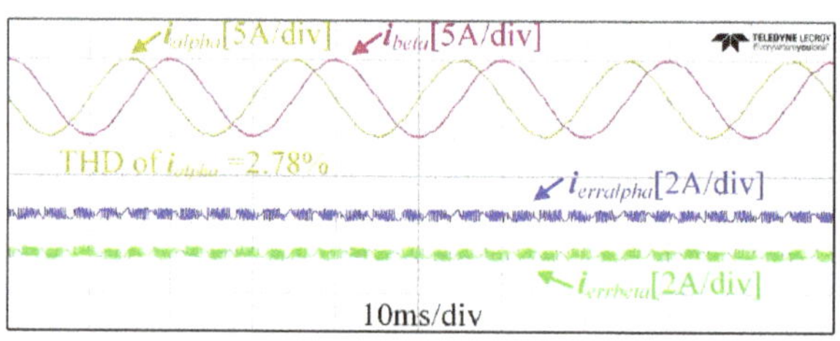

(**a**)

Figure 12. *Cont.*

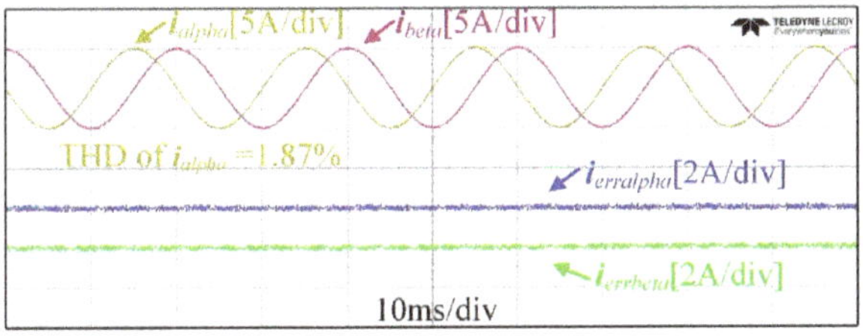

(b)

Figure 12. Experimental waveforms comparisons when $L/L_0 = 2$ (i_{ref} = 5A). (**a**) Waveforms under the conventional MPC scheme. (**b**) Waveforms under the proposed MFPC scheme.

4.4. Performance Comparison of All Schemes

In order to further demonstrate the advantages of the proposed MFPC, various performances of all schemes are compared and summarized in Table 3, including THDs, computational burden, parameter robustness, and stagnation effect. The proposed MFPC has a similar THD to MPC when model parameters are accurate and have better robustness against mismatched parameters than MPC. Though 18 current gradients in the proposed MFPC should be updated per control period, the speed of optimization is effectively reduced by the proposed sector judgment method. Hence, the proposed MFPC has a similar computational burden to MPC. For the MFPC in [20,30], although they cannot be affected by the mismatched parameters, their THDs are higher than those of the proposed MFPC because of the stagnation effect. For the MFPC in [20], it has the smallest computational burden because there is no current gradient updating exists, which leads to the largest THD. For MFPC in [30], it has the largest computational burden because there is all current gradient is updated and candidate voltage vectors are not reduced.

Table 3. Performance comparison of different control schemes.

Schemes	THDs	Computational Burden	Parameter Robustness	Stagnation Elimination
Conventional MPC	3.92%	35.9 μs	No	-
Conventional MFPC in [20]	6.31%	31.8 μs	Yes	No
Conventional MFPC in [30]	5.08%	41.2 μs	Yes	No
Proposed MFPC	4.19%	36.2 μs	Yes	Yes

5. Conclusions

This paper proposes an improved MFPC scheme based on the novel current gradient updating method and the effective sector judgment method for a T-type three-level grid-tied inverter. The proposed scheme has the following salient features: First, compared to conventional MPC schemes without a weighting factor, it totally eliminates the effect of mismatched model parameters and enhances the parameter robustness, especially when the actual parameters are less than the control parameters. Then, compared to conventional MPC schemes using a weighting factor, it avoids designing the weight factor and still realizes the good control of current and neutral point voltage. Moreover, compared to conventional MFPC schemes, it eliminates the current gradient stagnation effect by the designed updating method. The THDs of the proposed scheme decrease by up to 2.12% and the current spikes caused by the long updating stagnation are avoided. Finally, the discrimination accuracy of redundant small vectors is improved by using the proposed sector judgment method based on the current gradient, and the speed of optimization is increased by 31.5%.

Author Contributions: Conceptualization, Z.Y. and T.R.; methodology, Z.Y. and Z.F.; validation, C.H. and T.R.; formal analysis, P.Z., K.L. and Z.Y.; writing—original draft preparation, Z.Y.; writing—review and editing, Z.Y., T.R., G.L. and C.H.; supervision, T.R. and C.H.; project administration, C.H. All authors have read and agreed to the published version of the manuscript.

Funding: This work is supported in part by the Outstanding youth project of Anhui Natural Science (2108085J24) and in part by the Major science and technology projects in Anhui Province (202103a05020019).

Conflicts of Interest: The authors declare that there is no conflict of interest.

References

1. Zhang, Z.; Xu, H.; Xue, M.; Chen, Z.; Sun, T.; Kennel, R.; Hackl, C.M. Predictive control with novel virtual-flux estimation for back-to-back power converters. *IEEE Trans. Ind. Electron.* **2015**, *62*, 2823–2834. [CrossRef]
2. Zhao, T.; Chen, D. A power adaptive control strategy for further extending the operation range of single-phase cascaded H-bridge multilevel PV inverter. *IEEE Trans. Ind. Electron.* **2022**, *69*, 1509–1520. [CrossRef]
3. Adefarati, T.; Bansal, R.C. Integration of renewable distributed generators into the distribution system: A review. *IET Renew. Power Gener.* **2016**, *10*, 873–884. [CrossRef]
4. Freddy, T.K.S.; Rahim, N.A.; Hew, W.P.; Che, H.S. Comparison and analysis of single-phase transformerless grid-connected PV inverters. *IEEE Trans. Power Electron.* **2014**, *29*, 5358–5369. [CrossRef]
5. Schweizer, M.; Kolar, J.W. Design and implementation of a highly efficient three-level t-type converter for low-voltage application. *IEEE Trans. Power Electron.* **2013**, *28*, 1504–1511. [CrossRef]
6. Schweizer, M.; Lizama, I.; Friedli, T.; Kolar, J.W. Comparison of the chip area usage of 2-level and 3-level voltage source converter topologies. In Proceedings of the IECON 2010-36th Annual Conference on IEEE Industrial Electronics Society, Glendale, AZ, USA, 7–10 November 2010; pp. 391–396.
7. Karamanakos, P.; Geyer, T. Model Predictive Torque and Flux Control Minimizing Current Distortions. *IEEE Trans. Power Electron.* **2019**, *34*, 2007–2012. [CrossRef]
8. Yan, L.; Wang, F.; Dou, M.; Zhang, Z.; Kennel, R.; Rodríguez, J. Active Disturbance-Rejection-Based Speed Control in Model Predictive Control for Induction Machines. *IEEE Trans. Ind. Electron.* **2020**, *67*, 2574–2584. [CrossRef]
9. Guo, L.; Jin, N.; Gan, C.; Luo, K. Hybrid Voltage Vector Preselection-Based Model Predictive Control for Two-Level Voltage Source Inverters to Reduce the Common-Mode Voltage. *IEEE Trans. Ind. Electron.* **2020**, *67*, 4680–4691. [CrossRef]
10. Young, H.A.; Perez, M.A.; Rodriguez, J. Analysis of Finite-Control-Set Model Predictive Current Control with Model Parameter Mismatch in a Three-Phase Inverter. *IEEE Trans. Ind. Electron.* **2016**, *63*, 3100–3107. [CrossRef]
11. Zhang, X.; Zhang, L.; Zhang, Y. Model Predictive Current Control for PMSM Drives with Parameter Robustness Improvement. *IEEE Trans. Power Electron.* **2019**, *34*, 1645–1657. [CrossRef]
12. Zhang, Y.; Li, B.; Liu, J. Online Inductance Identification of a PWM Rectifier under Unbalanced and Distorted Grid Voltages. *IEEE Trans. Ind. Appl.* **2020**, *56*, 3879–3888. [CrossRef]
13. Niu, S.; Luo, Y.; Fu, W.; Zhang, X. Robust Model Predictive Control for a Three-Phase PMSM Motor with Improved Control Precision. *IEEE Trans. Ind. Electron.* **2021**, *68*, 838–849. [CrossRef]
14. Yan, L.; Dou, M.; Hua, Z. Disturbance Compensation-Based Model Predictive Flux Control of SPMSM with Optimal Duty Cycle. *IEEE J. Emerg. Sel. Top. Power Electron.* **2019**, *7*, 1872–1882. [CrossRef]
15. Wang, J.; Wang, F.; Wang, G.; Li, S.; Yu, L. Generalized Proportional Integral Observer Based Robust Finite Control Set Predictive Current Control for Induction Motor Systems with Time-Varying Disturbances. *IEEE Trans. Ind. Inf.* **2018**, *14*, 4159–4168. [CrossRef]
16. Zoghdar-Moghadam-Shahrekohne, B.; Pozzi, A.; Raimondo, D.M. SOS-based Stability Region Enlargement of Bilinear Power Converters through Model Predictive Control. In Proceedings of the 2021 29th Mediterranean Conference on Control and Automation (MED), Bari, Italy, 22–25 June 2021; pp. 847–854.
17. Zhang, Y.; Jiang, T.; Jiao, J. Model-Free Predictive Current Control of a DFIG Using an Ultra-Local Model for Grid Synchronization and Power Regulation. *IEEE Tran. Energy Convers.* **2020**, *35*, 2269–2280. [CrossRef]
18. Jin, N.; Chen, M.; Guo, L.; Li, Y.; Chen, Y. Double-Vector Model-Free Predictive Control Method for Voltage Source Inverter With Visualization Analysis. *IEEE Trans. Ind. Electron.* **2022**, *69*, 10066–10078. [CrossRef]
19. Rodríguez, J.; Heydari, R.; Rafiee, Z.; Young, H.A.; Flores-Bahamonde, F.; Shahparasti, M. Model-Free Predictive Current Control of a Voltage Source Inverter. *IEEE Access* **2020**, *8*, 211104–211114. [CrossRef]
20. Lin, C.; Liu, T.; Yu, J.; Fu, L.; Hsiao, C. Model-Free Predictive Current Control for Interior Permanent-Magnet Synchronous Motor Drives Based on Current Difference Detection Technique. *IEEE Trans. Ind. Electron.* **2014**, *61*, 667–681. [CrossRef]
21. Lin, C.; Yu, J.; Lai, Y.; Yu, H. Improved Model-Free Predictive Current Control for Synchronous Reluctance Motor Drives. *IEEE Trans. Ind. Electron.* **2016**, *63*, 3942–3953. [CrossRef]

22. da Rù, D.; Polato, M.; Bolognani, S. Model-free predictive current control for a SynRM drive based on an effective update of measured current responses. In Proceedings of the 2017 IEEE International Symposium on Predictive Control of Electrical Drives and Power Electronics (PRECEDE), Pilsen, Czech Republic, 4–6 September 2017; pp. 119–124.
23. Carlet, P.G.; Tinazzi, F.; Bolognani, S.; Zigliotto, M. An Effective Model-Free Predictive Current Control for Synchronous Reluctance Motor Drives. *IEEE Trans. Ind. Appl.* **2019**, *55*, 3781–3790. [CrossRef]
24. Ma, C.; Li, H.; Yao, X.; Zhang, Z.; de Belie, F. An Improved Model-Free Predictive Current Control with Advanced Current Gradient Updating Mechanism. *IEEE Trans. Ind. Electron.* **2021**, *68*, 11968–11979. [CrossRef]
25. Agustin, C.A.; Yu, J.-T.; Lin, C.-K.; Jai, J.; Lai, Y.-S. Triple-Voltage-Vector Model-Free Predictive Current Control for Four-Switch Three-Phase Inverter-Fed SPMSM Based on Discrete-Space-Vector Modulation. *IEEE Access* **2021**, *9*, 60352–60363. [CrossRef]
26. Agustin, C.A.; Yu, J.-T.; Cheng, Y.-S.; Lin, C.-K.; Huang, H.-Q.; Lai, Y.-S. Model-Free Predictive Current Control for SynRM Drives Based on Optimized Modulation of Triple-Voltage-Vector. *IEEE Access* **2021**, *9*, 130472–130483. [CrossRef]
27. Lin, C.-K.; Yu, J.-T.; Agustin, C.A.; Li, N.-Y. A Two-Staged Optimization Approach to Modulated Model-Free Predictive Current Control for RL-Connected Three-Phase Two-Level Four-Leg Inverters. *IEEE Access* **2021**, *9*, 147537–147548. [CrossRef]
28. Lin, C.-K.; Agustin, C.A.; Yu, J.-T.; Cheng, Y.-S.; Chen, F.-M.; Lai, Y.-S. A Modulated Model-Free Predictive Current Control for Four-Switch Three-Phase Inverter-Fed SynRM Drive Systems. *IEEE Access* **2021**, *9*, 162984–162995. [CrossRef]
29. Agustin, C.A.; Yu, J.-T.; Cheng, Y.-S.; Lin, C.-K.; Yi, Y.-W. A Synchronized Current Difference Updating Technique for Model-Free Predictive Current Control of PMSM Drives. *IEEE Access* **2021**, *9*, 63306–63318. [CrossRef]
30. Yu, F.; Zhou, C.; Liu, X.; Zhu, C. Model-Free Predictive Current Control for Three-Level Inverter-Fed IPMSM With an Improved Current Difference Updating Technique. *IEEE Trans. Energy Convers.* **2021**, *36*, 3334–3343. [CrossRef]
31. Ipoum-Ngome, P.G.; Mon-Nzongo, D.L.; Flesch, R.C.C.; Song-Manguelle, J.; Wang, M.; Jin, T. Model-Free Predictive Current Control for Multilevel Voltage Source Inverters. *IEEE Trans. Ind. Electron.* **2021**, *68*, 9984–9997. [CrossRef]

energies

MDPI

Article

A General Equivalent Modeling Method for DFIG Wind Farms Based on Data-Driven Modeling

Qianlong Zhu, Jun Tao, Tianbai Deng * and Mingxing Zhu

The College of Electrical Engineering and Automation, Anhui University, Hefei 230601, China
* Correspondence: dengtianbai@ahu.edu.cn; Tel.: +86-0551-6386-1412

Abstract: To enhance the stable performance of wind farm (WF) equivalent models in uncertain operating scenarios, a model-data-driven equivalent modeling method for doubly-fed induction generator (DFIG)-based WFs is proposed. Firstly, the aggregation-based WF equivalent models and the equivalent methods for aggregated parameters are analyzed and compared. Two mechanism models are selected from the perspective of practicality and complementarity of simulation accuracy. Secondly, the simulation parameters are set through two sampling methods to construct a training database. Next, the whole fault process is divided into five phases, and the weight coefficient optimization model is established according to the data-driven idea to achieve the adaptive configuration of the weight. Finally, the electromechanical transient simulations of the power systems with a DFIG-based WF is carried out by using the MATLAB/Simulink platform. Compared with the detailed WF model, the simulation time of the WF equivalent proposed in this paper can be significantly reduced by about 87%, and simulation results show that the proposed method can effectively improve the adaptability of the WF equivalent model in different wind scenarios and voltage dips.

Keywords: DFIG-based wind farm; general equivalent modeling; mechanism model; data-driven; weight coefficient

1. Introduction

Increasing the penetration of large-scale wind farms (WFs), wind power has become an important power source in power systems [1,2]. Wind power has randomness and volatility, which brings deep changes in the operation mechanism [3–5]. To support rapid development and reduce the operation risks, simulation technology is increasingly indispensable to reflect the behavior of the actual power systems and lead scientific construction of new-type electric power systems. Large-scale WFs may consist of hundreds or even thousands of wind turbines (WTs), which could significantly enlarge the size of A model and then cause the "curse of dimensionality" [6]. Therefore, the equivalent model, on the basis of reasonable reduction from the detailed model, is essential to be developed.

Currently, the aggregation-based method, which was originally used in synchronous generators, is widely applied to model a large-scale WF in the literature [7,8]. The aggregated model can be divided into a single-machine equivalent model (SEM) and a multi-machine equivalent model (MEM) [9]. The single-machine equivalent method, which aggregates the whole WF into one equivalent WT, requires a small amount of calculation, but it is hard to represent dynamic behaviors of a whole WF due to the distribution of collector cables and wind speed differences across the WF. The multi-machine equivalent method, which separates WTs of a WF into several clusters and aggregates each cluster into one equivalent WT, generally represents WF characteristics better and has wider applications in practice [10]. The MEM includes two steps: (1) identify the group of WTs with similar dynamic characteristics, and (2) obtain the aggregated parameters of the WTs, transformers, and collector cables [11].

During the past few decades, to obtain a better performance on WT clustering, wind speed [12,13], rotor speed [14], the power characteristic curve [15,16], crowbar action [17,18],

Citation: Zhu, Q.; Tao, J.; Deng, T.; Zhu, M. A General Equivalent Modeling Method for DFIG Wind Farms Based on Data-Driven Modeling. *Energies* **2022**, *15*, 7205. https://doi.org/10.3390/en15197205

Academic Editor: Surender Reddy Salkuti

Received: 6 September 2022
Accepted: 26 September 2022
Published: 30 September 2022

chopper action [19] and other quantities have been fully investigated to be selected as the clustering indicators. Wind speed is regarded as the primary clustering indicator. Crowbar or chopper action is considered more suitable for grouping WTs when simulating low voltage ride through characteristics. However, the selection of these clustering indicators is generally based on specific time spots, and limitations universally exist in these studies. For example, WT characteristics dynamically change along with the time that passes, and the time spot-based clustering results are suitable in limited situations. In [20], WTs connected to the same feeder were aggregated into one equivalent WT, but this approach, might introduce large equivalent errors when wind speeds of WTs within a feeder have large deviations. Recently, a WF equivalent modeling method based on feature influence factors and improved back propagation (BP) neuron networks algorithm was proposed, which improves the efficiency and accuracy when grouping large amounts of WTs [21]. The clustering result is still calculated at a specific steady state, and keeps changing due to the unpredictability of environmental scenarios. Therefore, the MEM based on the operation point is stochastic, and both structures and parameters of the equivalent model are random. Conclusions under different scenarios will also be different.

As for the uncertainty problem of the MEM, several studies propose a probabilistic clustering concept for aggregate modeling of WFs. In [22], a weighted graph representing relationship of the power of WTs was used to build the Markov chain in order to estimate the probability that WTs belong to the same cluster. In [23], a probabilistic equivalent model was constructed considering the probability distribution characteristics of wind speed and wind direction in an actual DFIG-based WF. On this basis, the probability distribution characteristics of the fault type is taken [24]. Moreover, the paper merges clustering results with insignificant differences through the significance test of the Fisher discriminant, and improves the generality of the probabilistic WF equivalent model. In [25], historical meteorological data were utilized to investigate the probability distribution of key equivalent parameters, such as capacity, wind speed and electrical impedance to the point of common coupling. In [26], an equivalent model for mixed WFs based on BP neural networks was constructed. However, it is difficult for this data-driven modeling approach to achieve the desired simulation accuracy.

Accordingly, this paper proposes an innovative general equivalent modeling method for DFIG-based WFs. The main contributions are listed as follows:

(1) The method of combining model-driven and data-driven is introduced into the research on general equivalent modeling of DFIG-based WFs. For uncertain scenarios, the established model has a wide range of adaptability.
(2) Meaningful insights into how to select mechanism models are provided. Considering different calculation methods for the equivalent parameters of the collector cables, two mechanism models with complementary characteristics are selected.

The remainder of this paper is organized as follows. In Section 2, the ideas and methods based on a data-driven model are introduced. The two sampling methods and training database construction are presented in detail in Section 3. In Section 4, the proposed model is verified with different voltage dips and changing wind scenarios. The discussion and conclusions are discussed in Sections 5 and 6.

2. Ideas and Methods

The following issues need to be taken into account for researching the general equivalent modeling of DFIG-based WFs. One is how to cope with the simulation demand of uncertain operation scenarios and enhance the adaptability of the general model, and the other is how to improve the convenience of using the model while ensuring the accuracy of the model and enhancing the engineering utility value.

To address the two issues, the framework of the model-data-driven general equivalent modeling method for DFIG-based WFs is shown in Figure 1, which includes selecting the mechanism model, setting the weight coefficients, and constructing the training database.

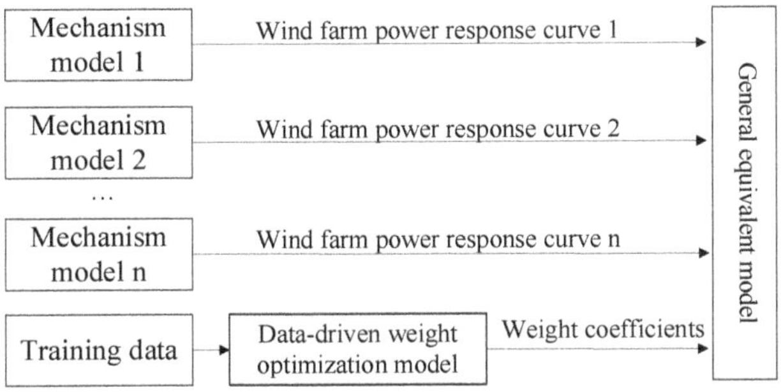

Figure 1. Framework of general equivalent modeling method.

2.1. Mechanism Model Selection

Suitable mechanism models can effectively improve the accuracy of the general equivalent model. The mechanism model of DFIG-based WFs is developed with an aggregation-based method, of which the SEM is the simplest, and the MEM by clustering WTs with similar wind speeds or electrical distances is more common. Therefore, the three equivalent models are tentatively selected as the mechanism models in the paper.

For the three mechanism models, the MATLAB/Simulink platform was used to conduct simulation studies and model accuracy analysis. The simulation results show that the dynamic performances of the SEM and the MEM by clustering WTs with similar electrical distances were almost the same, and the equivalent accuracy of both was lower than that of the MEM by clustering WTs with similar wind speeds. Considering that the selected mechanism model needs to have high accuracy, the MEM by clustering WTs with similar wind speeds was finally chosen as the mechanism model.

In addition, the mechanism model needs to obtain the aggregated parameters of the WTs, transformers, and collector cables. At present, the aggregated parameters of the WTs have been widely agreed upon, i.e., the calculation of aggregated wind speed is based on equal total wind energy, and the calculation of aggregated parameters of WTs is based on the capacity weighting method. In addition, the calculation of aggregated parameters of transformers is also based on the capacity weighting method. However, there are two different calculation methods for the aggregated parameters of the collector cables. One is the equal loss power method [27], and the other is the equal voltage dip method [28]. The corresponding methods are detailed in Equations (1) and (2), respectively:

$$Z_{\text{eq}} = \frac{\sum_{i=1}^{m} (P_{Z\_i}^2 Z_{1\_i})}{\left(\sum_{i=1}^{m} P_i\right)^2} \tag{1}$$

$$Z_{\text{eq}} = \frac{\sum_{i=1}^{m} \left(\sum_{j=1}^{i} (Z_{1\_j} P_{Z\_j}) P_i\right)}{\left(\sum_{i=1}^{m} P_i\right)^2} \tag{2}$$

where Z_{1_i} is the cable impedance of the ith WT branch, P_{Z_j} is the total active power flowing through the impedance Z_{1_j}, P_i is the active power of the ith WT, and m is the number of WTs in the clustering groups.

The aggregated parameters of the collector cables calculated by the two methods had a significant difference, especially for the WTs at the end of one feeder. Therefore, we finally selected the mechanism model by clustering WTs with similar wind speeds, together with the aggregated parameters of the collector cables by the equal power loss method, as mechanism model 1 (MM1). The mechanism model clustering WTs with similar wind speeds, together with the aggregated parameters of the collector cables by the equal voltage dip method, was selected as mechanism model 2 (MM2).

2.2. *The Data-Driven Adaptive Weight Optimization Model*

To set the weight coefficients, the simplest way is to ignore the differences among models and assign the same weight. However, the accuracy of this general WF equivalent model will not significantly improve compared to the traditional model. Therefore, we adopted a time-varying weight coefficient, which means different models had different weight coefficients, and the same model also had different weight coefficients in differen phases.

To find the best weight coefficients at different periods, the particle swarm optimization (PSO) algorithm was adopted [29]. The fitness function of the PSO algorithm was set to the root mean square error of the general WF equivalent model in different phases, that is:

$$\begin{aligned} & optfunction(\omega_1, \omega_2) = \sqrt{\sum_{i=1}^{n} \left(\sum_{k=1}^{m} [\omega_1 y_1(k) + \omega_2 y_2(k) - y_0(k)]^2 \right)} \\ & s.t. \quad \omega_1, \omega_2 \in [0,\ 1] \end{aligned} \tag{3}$$

where ω_1,ω_2 are the weight coefficients of MM1 and MM2 for a specific phase, respectively. $y_1(k)$, $y_2(k)$ and $y_0(k)$ are the output values of MM1, MM2, and detailed WF model at the kth sampling point within a phase, respectively, m is the number of sampling points, and n is the sample size of the training data.

3. Data Source

3.1. *Data Settings*

The wind speed and external fault information are required for WF equivalent modeling. Considering the completeness of data samples, this paper adopts two ways to generate data samples. One is simple random sampling within the selected range, and the other is sampling based on the probability density function. For the first sampling method, six scenarios were considered, i.e., low wind speed range (4.5~8.5 m/s), medium wind speed range (8.5~12.5 m/s), high wind speed range (12.5~22 m/s), medium-low wind speed range (4.5~12.5 m/s), and medium-high wind speed range (8.5~22 m/s) are considered. Next, the input wind speed of each WT was randomly generated with equal probability by ignoring the wake effect. The voltage dip at the point of connection (POC) was drawn with equal probability in the range of 0.1~0.9 p.u. For the second sampling method, considering the statistical characteristics of wind speeds, the wind speed and wind direction of the WF were generated based on the probability density function. Next, the input wind speed of each WT was derived based on the wake effect. In addition, the ground resistance values were also extracted based on the probability density function.

In this paper, for the first sampling method, 50 sets of sample data were generated for each scenario. For the second method, 200 sets of sample data were drawn according to the probability.

3.2. *Training Database*

Taking the large capacity of data samples into account, we adopted the MATLAB 2020b platform to carry out numerical simulations using m-language programming and Simulink to call each other. The simulation results were automatically stored and called for the adaptive weight optimization model based on the PSO algorithm. The specific flow is shown in Figure 2.

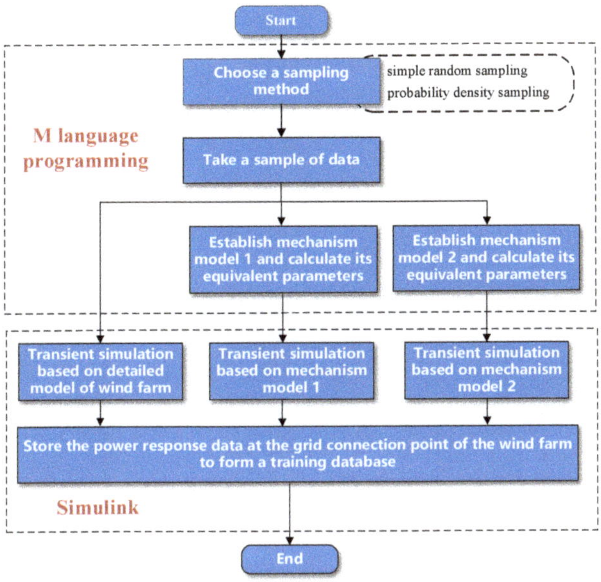

Figure 2. Training database construction flowchart.

4. Example Analysis

4.1. Example of Calculation

A detailed WF consisting of 28 × 1.5 MW DFIGs (DFIG_1—DFIG_28) was set up in the MATLAB/Simulink platform, as shown in Figure 3. The model parameters are given in Table A1.

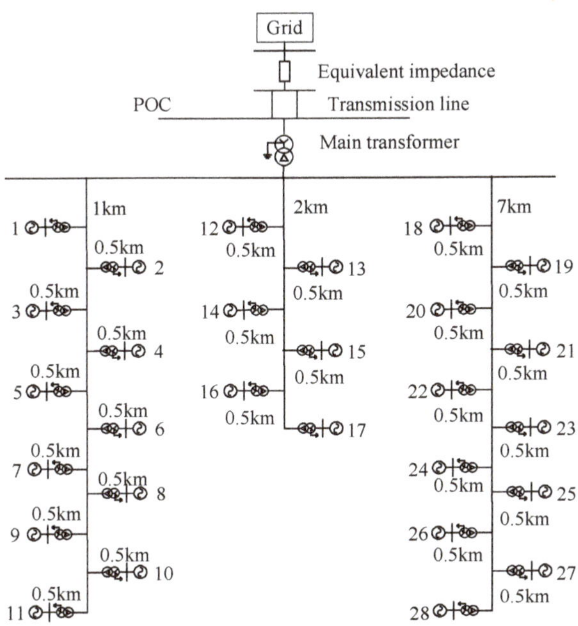

Figure 3. DFIG-based WF topology.

Compared with the simulation results of the detailed WF model (DM), error metrics are defined as:

$$E_r = \frac{1}{m}\sum_{k=1}^{m}\left|\frac{y_{fk} - y_k}{y_k}\right| \times 100\% \tag{4}$$

$$E_a = \frac{1}{m}\sum_{k=1}^{m}\left|y_{fk} - y_k\right| \tag{5}$$

where y_k and y_{fk} are the output values of the detailed WF model and the equivalent model at the kth sampling point, respectively, and m is the number of sampling points.

4.2. Optimization of Adaptive Weight Coefficients

As previously commented in Section 2.2, the PSO algorithm was used to find the optimal weight coefficients of the mechanism models. The numbers of population size and iterations were set to 500 and 400, respectively. Moreover, inertia weight φ is an essential parameter of the PSO algorithm, and we adopted a dynamic adjustment inertia weight strategy, where the inertia weight is dynamically adjusted according to a linear decreasing approach:

$$\varphi(j) = \varphi_{\max} - (\varphi_{\max} - \varphi_{\min})(j/j_{\max}) \tag{6}$$

where $\varphi_{\max} = 0.9$, $\varphi_{\min} = 0.1$, $j_{\max}$ is the maximum number of iterations, and j is the number of current iterations.

According to the time window division method of IEC 61400-27-1 [30], the whole fault process is divided into five phases, namely, pre-fault $(t_0 - 1\text{s},\ t_0)$, early fault $(t_0,\ t_0 + 0.14\text{s})$, the quasi-steady-state phase during the fault $(t_0 + 0.14s,\ t_1)$, early fault recovery $(t_1,\ t_1 + 0.5\text{s})$, and the quasi-steady-state phase during the fault recovery $(t_1 + 0.5\text{s},\ t_1 + 5\text{s})$ (t_0 and t_1 denote the initial time of fault and the time of fault clearance, respectively).

After the iterative optimization of the PSO algorithm, the weight coefficients of the above five time windows corresponding to MM1 and MM2 are shown in Table 1.

Table 1. Adaptive weight coefficients.

Models Category	Active Power Weight Coefficients				
	$(t_0 - 1s, t_0)$	$(t_0, t_0 + 0.14s)$	$(t_0+0.14s, t_1)$	$(t_1, t_1 + 0.5s)$	$(t_1 + 0.5s, t_1 + 5s)$
MM1	0.9713	0.3982	0.7163	0.0065	0.7270
MM2	0.0296	0.6048	0.2855	0.9982	0.2728
Models Category	**Reactive Power Weight Coefficients**				
	$(t_0 - 1s, t_0)$	$(t_0, t_0 + 0.14s)$	$(t_0+0.14s, t_1)$	$(t_1, t_1 + 0.5s)$	$(t_1 + 0.5s, t_1 + 5s)$
MM1	0.0003	0.0871	0.4679	0.0005	0.0043
MM2	0.8898	0.9973	0.6002	0.9555	0.9896

4.3. The Adaptability of the General Equivalent Model

To verify the adaptability of the proposed model (PM), ten groups of input wind speed and fault voltage dip were randomly generated, which are shown in Figure 4. In each scenario, the wind speeds were assigned to the WTs in WF, and the voltage dip was realized by setting a three-phase symmetrical short-circuit fault at the midpoint of one transmission line.

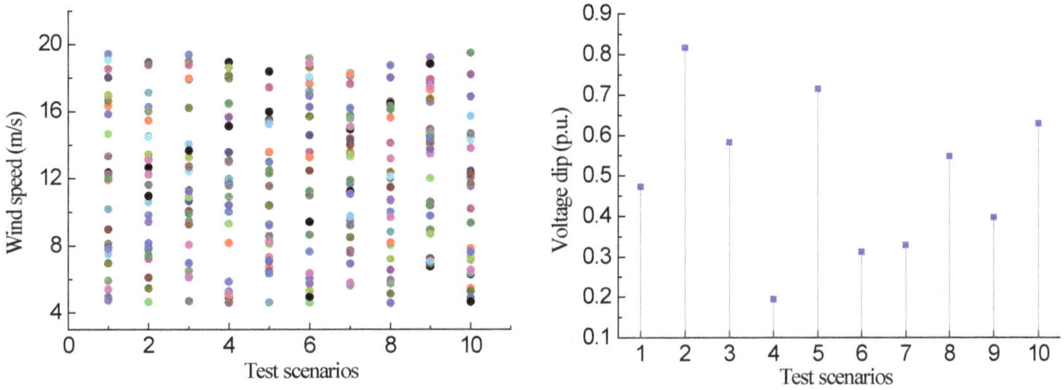

(a) (b)

Figure 4. Ten test scenarios. (**a**) Wind speed scenarios. (**b**) Voltage dip scenarios.

Under the operation conditions of Figure 4, the dynamic responses at the POC of the DM, MM1, MM2 and PM were sampled. The active and reactive power errors of different equivalent WF models were calculated according to Equations (4) and (5), respectively, as presented in Tables 2 and 3. According to the values given in Tables 2 and 3, the differences in accuracy of the three equivalent models are illustrated in Figures 5 and 6, respectively.

Table 2. Active power response error of three equivalent models.

Case	$(t_0 - 1s, t_0)$			$(t_0, t_0 + 0.14s)$			$(t_0 + 0.14s, t_1)$			$(t_1, t_1 + 0.5s)$			$(t_1 + 0.5s, t_1 + 5s)$		
	MM1	**MM2**	**PM**	**MM1**	**MM2**	**PM**	**MM1**	**MM2**	**PM**	**MM1**	**MM2**	**PM**	**MM1**	**MM2**	**PM**
1	0.034%	0.033%	0.022%	1.970%	1.602%	1.721%	1.536%	1.359%	1.412%	0.841%	0.750%	0.787%	0.698%	0.649%	0.665%
2	0.383%	0.384%	0.294%	0.701%	0.387%	0.324%	1.102%	0.822%	0.843%	0.446%	0.523%	0.295%	0.068%	0.093%	0.071%
3	0.161%	0.161%	0.080%	1.317%	1.008%	1.068%	3.373%	1.808%	1.734%	1.218%	0.651%	0.674%	1.028%	0.954%	0.988%
4	0.171%	0.170%	0.082%	4.407%	4.667%	4.282%	1.728%	2.062%	1.748%	4.474%	4.375%	4.202%	0.491%	0.427%	0.479%
5	0.362%	0.368%	0.273%	0.428%	0.477%	0.552%	0.401%	0.141%	0.169%	0.608%	0.717%	0.320%	0.142%	0.155%	0.132%
6	0.360%	0.361%	0.451%	1.921%	1.930%	1.799%	1.856%	1.761%	2.012%	1.231%	1.162%	1.597%	1.802%	1.822%	1.787%
7	0.051%	0.048%	0.045%	3.925%	2.108%	2.431%	0.435%	3.014%	0.883%	1.721%	1.256%	1.022%	1.019%	1.013%	0.998%
8	0.166%	0.179%	0.156%	2.057%	2.606%	2.557%	0.650%	4.800%	1.892%	1.512%	1.279%	0.915%	0.473%	0.436%	0.444%
9	0.406%	0.411%	0.316%	2.357%	1.779%	1.730%	0.103%	0.609%	0.358%	1.245%	0.862%	0.632%	0.095%	0.104%	0.102%
10	0.255%	0.255%	0.166%	0.642%	1.106%	1.062%	0.960%	1.218%	0.302%	1.732%	1.714%	1.280%	0.491%	0.572%	0.533%

Table 3. Reactive power response error of three equivalent models.

Case	$(t_0 - 1s, t_0)$			$(t_0, t_0 + 0.14s)$			$(t_0 + 0.14s, t_1)$			$(t_1, t_1 + 0.5s)$			$(t_1 + 0.5s, t_1 + 5s)$		
	MM1	**MM2**	**PM**	**MM1**	**MM2**	**PM**	**MM1**	**MM2**	**PM**	**MM1**	**MM2**	**PM**	**MM1**	**MM2**	**PM**
1	0.0046	0.0060	0.0002	0.0203	0.0210	0.0129	0.0179	0.0138	0.0064	0.0224	0.0178	0.0194	0.0141	0.0113	0.0127
2	0.0076	0.0087	0.0007	0.0192	0.0148	0.0080	0.0467	0.0112	0.0082	0.0649	0.0290	0.0323	0.0101	0.0053	0.0053
3	0.0057	0.0062	0.0005	0.0407	0.0185	0.0068	0.0212	0.0218	0.0045	0.0990	0.0534	0.0570	0.0176	0.0083	0.0086
4	0.0090	0.0091	0.0003	0.0069	0.0074	0.0081	0.0133	0.0141	0.0161	0.0468	0.0454	0.0397	0.0063	0.0065	0.0060
5	0.0079	0.0078	0.0001	0.0155	0.0139	0.0082	0.0048	0.0032	0.0220	0.0360	0.0319	0.0351	0.0064	0.0054	0.0054
6	0.0076	0.0080	0.0004	0.0084	0.0078	0.0074	0.0215	0.0227	0.0197	0.0396	0.0363	0.0365	0.0048	0.0050	0.0047
7	0.0052	0.0059	0.0001	0.0149	0.0082	0.0069	0.0090	0.0220	0.0102	0.0724	0.0347	0.0365	0.0083	0.0048	0.0049
8	0.0056	0.0057	0.0006	0.0384	0.0178	0.0061	0.0305	0.0307	0.0161	0.0888	0.0459	0.0496	0.0148	0.0070	0.0072
9	0.0089	0.0093	0.0010	0.0141	0.0107	0.0058	0.0245	0.0229	0.0200	0.0334	0.0263	0.0270	0.0050	0.0058	0.0052
10	0.0043	0.0053	0.0022	0.0277	0.0166	0.0059	0.0117	0.0115	0.0057	0.0639	0.0392	0.0424	0.0115	0.0065	0.0068

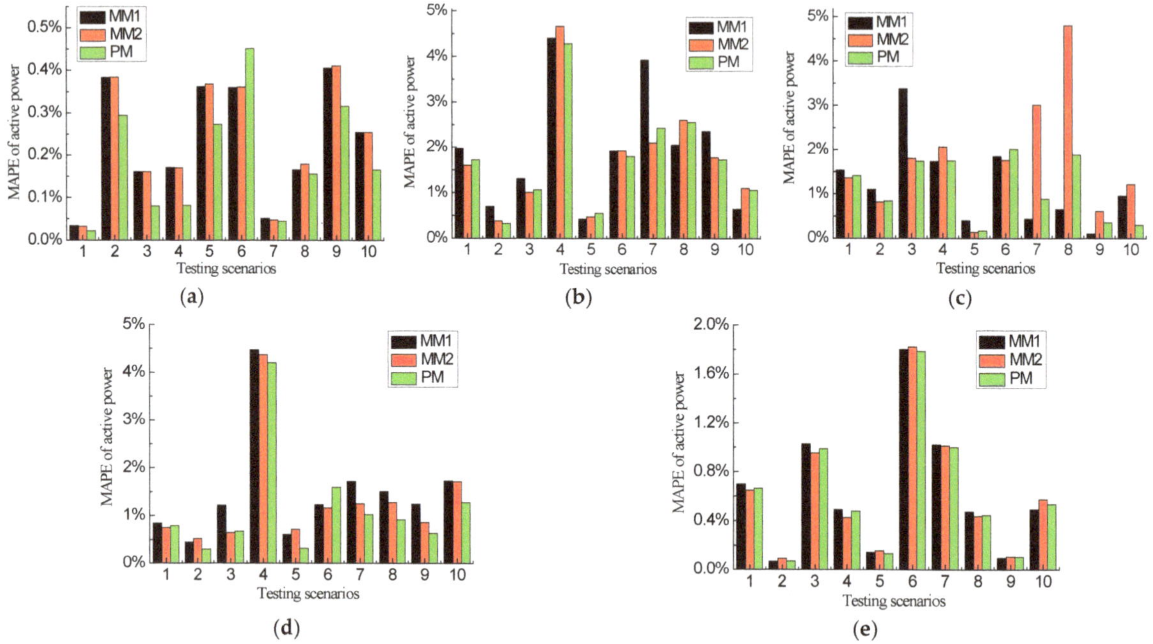

Figure 5. Active power response error of three equivalent models. (**a**) Pre-fault phase. (**b**) Early fault phase. (**c**) Quasi-steady-state phase during the fault. (**d**) Early fault recovery phase. (**e**) Quasi-steady-state phase during the fault recovery.

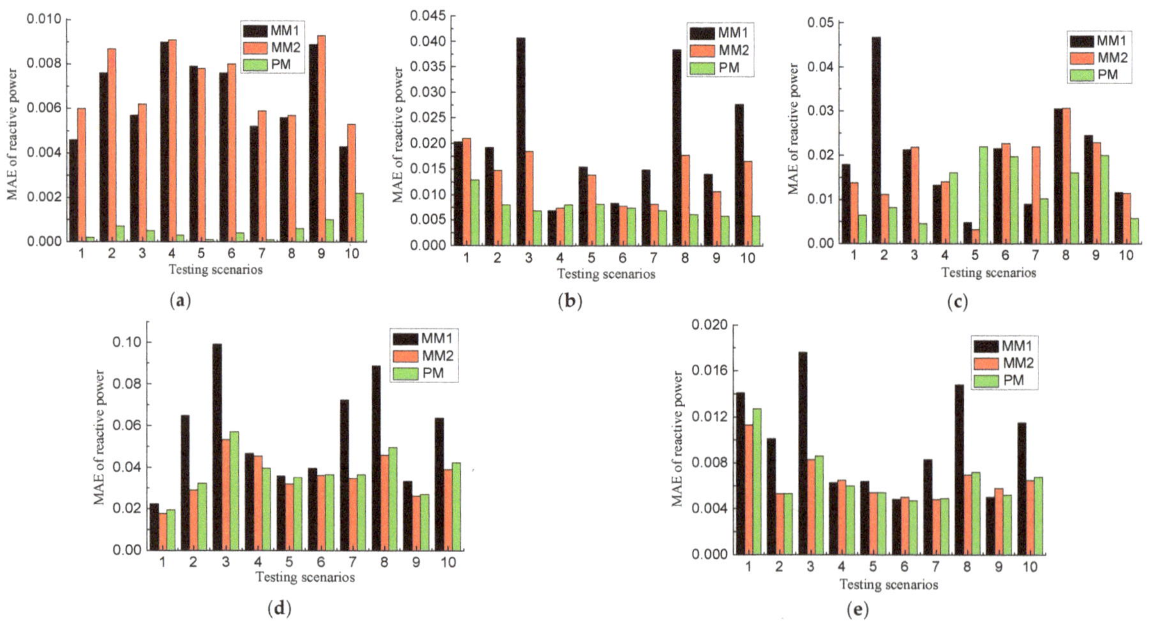

Figure 6. Reactive power response error of three equivalent models. (**a**) Pre-fault phase. (**b**) Early fault phase. (**c**) Quasi-steady-state phase during the fault. (**d**) Early fault recovery phase. (**e**) Quasi-steady-state phase during the fault recovery.

As shown in Table 2 and Figure 5, for the pre-fault phase $(t_0 - 1s, t_0)$, the PM constructed in this paper exhibited higher active power simulation accuracy except in Case 6. In this phase, the active power output of the WF is stable. Moreover, the performance difference of the two mechanism models is relatively fixed for different scenarios. It helps to improve the accuracy of the PM through the optimal configuration of weight coefficients. For the other four phases, the improvement in accuracy decreased slightly, and the accuracy of the PM was between the MM1 and MM2 in some scenarios. The reason lies in that the active power output of WF starts to fluctuate in those phases, and the performance difference of the two mechanism models also begins to change in different scenarios. However, taking the quasi-steady-state phase during the fault $(t_0 + 0.14s,\ t_1)$into account, where the phenomenon that the accuracy of the PM is between the MM1 and MM2 appears more often, the active power errors of each model are presented in Figure 7. It can be seen that the active power performance of the PM is more stable than that of the MM1 and MM2, indicating its better adaptability.

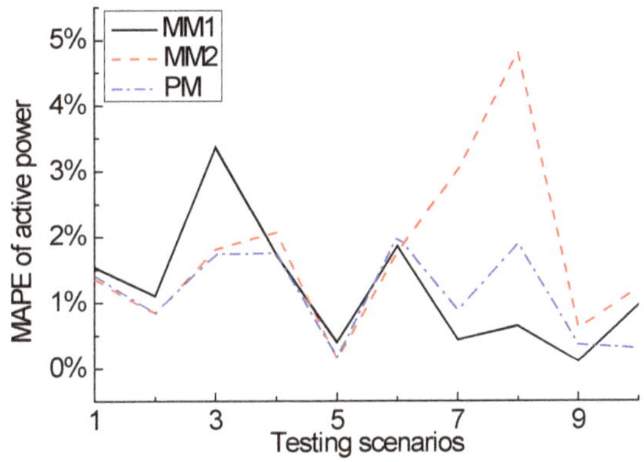

Figure 7. Active power errors of different models in quasi-steady-state process during fault duration.

As shown in Table 3 and Figure 6, for the pre-fault phase $(t_0 - 1s, t_0)$, the PM established in this paper also shows higher reactive power simulation accuracy. For the other four phases, the reactive power performance of the PM is between MM1 and MM2 in some scenarios. However, taking the early fault recovery phase into account, where the above phenomenon appears more often, the reactive power errors of each model are illustrated in Figure 8. It can be seen that the reactive power performance of the PM is also more stable.

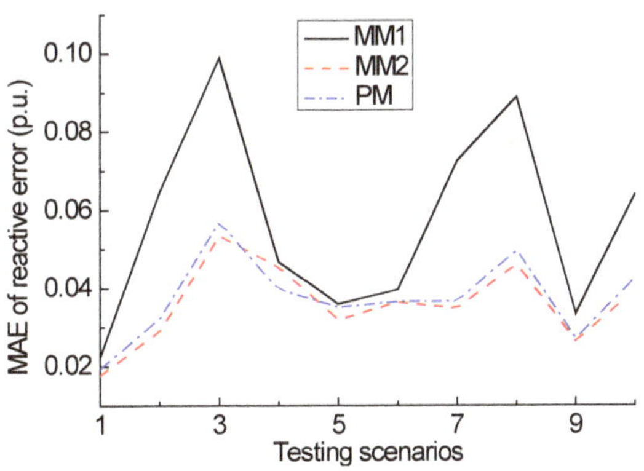

Figure 8. Reactive power errors of different models in quasi-steady-state process during fault duration.

To further demonstrate the effectiveness of the PM, the dynamic responses of the DM, PM, MM1 and MM2 in Cases 6 and 3 are presented in Figures 9 and 10, respectively. As mentioned above, for Case 6, the accuracy of the PM was lower than both the MM1 and MM2. From Figure 9, we can observe that the PM remains accurate and has high correspondence with the electromechanical transient responses of the DM. For the other nine cases, for example, Case 3, it can be seen that the responses of the PM were much closer to that of the DM than the MM1 and MM2, as shown in Figure 10.

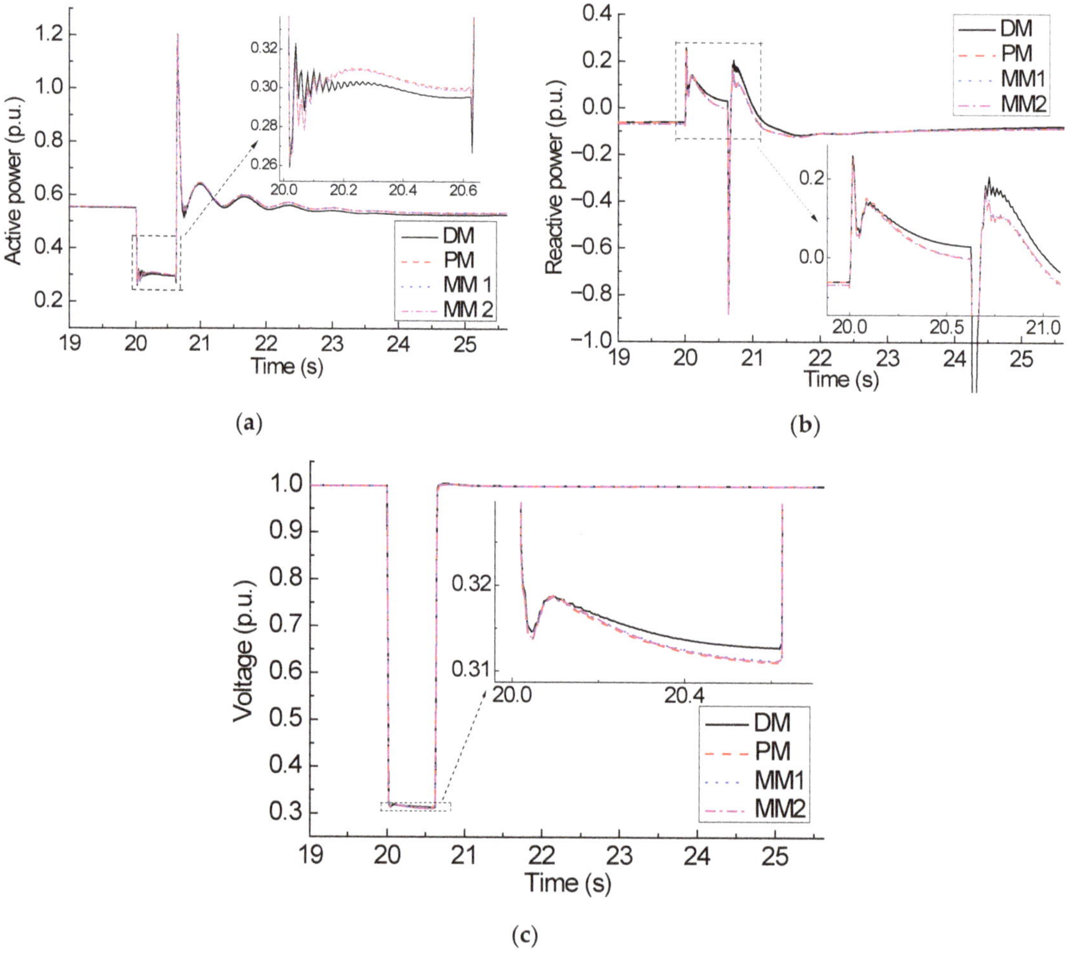

Figure 9. The dynamic responses of WF at POC in Case 6. (**a**) Active power. (**b**) Reactive power. (**c**) Voltage.

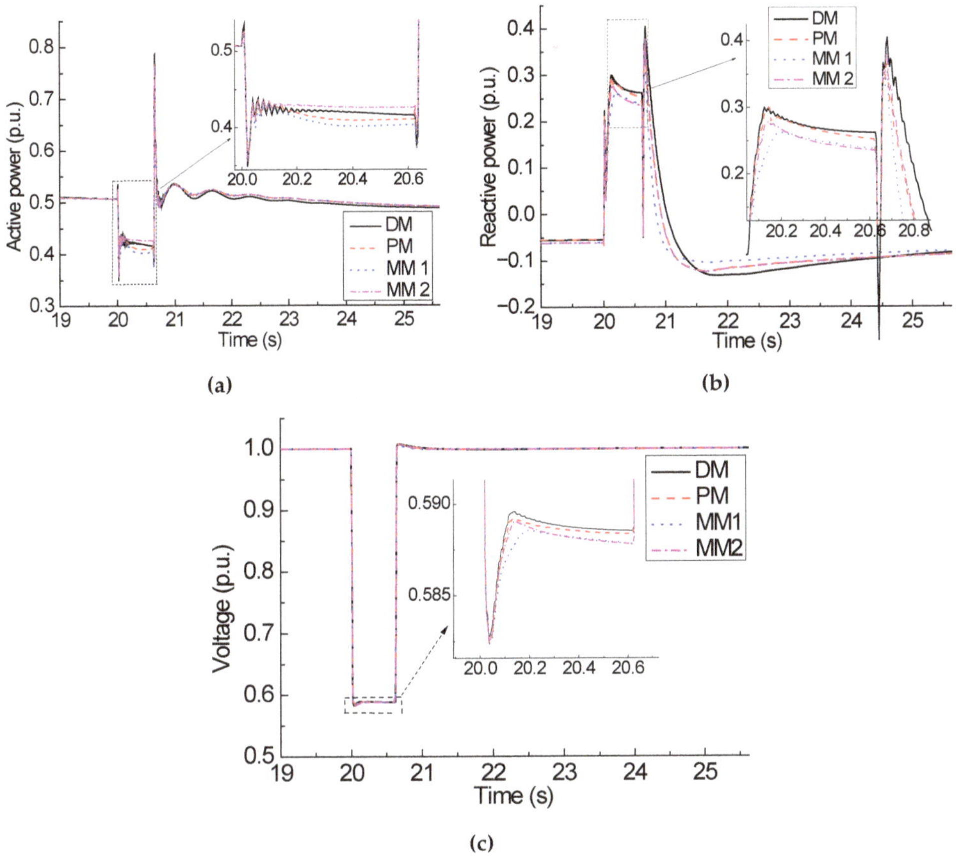

Figure 10. The dynamic responses of WF at POC in Case 3. (**a**) Active power. (**b**) Reactive power. (**c**) Voltage.

5. Discussion

Until now, the general equivalent modeling method for DFIG-based WFs has not been considered from a model-data driven point of view; only a few researchers have analyzed the possibility of identifying the crowbar action based on a data-driven model [31]. This means that the accuracy of WF-equivalent models has been improved, only considering coherent cluster divisions without regard to the adaptability of the equivalent model for different scenarios.

One of the novelties presented in this paper is to enhance the stable performance of WF-equivalent models in uncertain operating scenarios. As the scenario changes, the accuracy of the traditional WF equivalent models will also change, and the performance is often uncertain due to complex influencing factors. As shown in Figures 5 and 6, sometimes the performance of the MM1 is better, and sometimes the performance of the MM2 is better. It is not clear in which scenario the effectiveness of the equivalent model will fail. Therefore, in the view of enhancing the stability of the model performance, we used data-driven mining of complementary characteristics between mechanism models. As shown in Figures 7 and 8, the applicability to uncertain scenarios was significantly improved.

In addition, this paper provides meaningful insights into how to select mechanism models. We analyze the influence of the two calculation methods for the aggregated parameters of the collector cable on the accuracy of the equivalent model, and select two equivalent models with complementary characteristics. This helps to improve the accuracy of the general equivalent model. From Tables 2 and 3, we can observe that except in rare phases, the accuracy of the general equivalent model is smaller than the maximum error between the two mechanism models. Moreover, the average error of the general equivalent model is also smaller than those of the two mechanism models.

The simulations were carried out on a workstation with the following specifications: Intel Xeon(R) Platinum 8375C, 32 CPU @ 2.9 GHz, 128 GB of RAM. The average simulation times of the DM, PM, MM1 and MM2 were 2146 s, 283 s, 135 s and 135 s, respectively. By the PM, the simulation time of the detailed WF can be significantly reduced by about 87 %. Compared to the MM1 and MM2, about two times the simulation time is increased.

6. Conclusions

From the viewpoint of improving the applicability and engineering utility value of the DFIG-based WF equivalent model, this paper proposes a general model-data-driven equivalent modeling method. It verifies the effectiveness of the proposed model for uncertain operation scenarios through simulation.

The two mechanism models are selected based on clustering indicators and calculation methods for the aggregated parameters of the collector cable. The combination of the two models can not only take advantage of the high accuracy of each model, but also take advantage of their complementary nature to reduce simulation errors.

In this paper, the fault process is divided into five stages, and the data-driven adaptive weight optimization model is constructed to optimize the corresponding weight coefficients. It further improves the adaptability of the general WF equivalent model in uncertain scenarios.

However, the main drawback of this general equivalent modeling method is that the accuracy of the PM is sometimes between the MM1 and MM2. In future works, a possible extension could be to apply the time series analysis theory to analyze the fluctuation law of the dynamic responses, and we will further improve the accuracy of the general equivalent model by optimizing the time window division.

Author Contributions: Conceptualization, Q.Z. and J.T.; methodology, Q.Z. and T.D.; software, Q.Z.; validation, J.T. and T.D.; formal analysis, Q.Z., J.T. and M.Z.; investigation, Q.Z. and M.Z.; resources, M.Z..; data curation, T.D.; writing—original draft preparation, Q.Z.; writing—review and editing, Q.Z. and J.T.; visualization, J.T.; supervision, M.Z.; project administration, Q.Z. and J.T. All authors have read and agreed to the published version of the manuscript.

Funding: This work was supported by the Anhui Provincial Natural Science Foundation of China under Grant (2108085QE238), and by the Open Fund of State Key Laboratory of Operation and Control of Renewable Energy & Storage Systems (China Electric Power Research Institute) (No. NYB51202201705), and by the Key Research and Development Program of Anhui Province (202104a05020056).

Data Availability Statement: Data is available upon reasonable request to the corresponding author.

Conflicts of Interest: The authors declare no conflict of interest.

Appendix A

Table A1. Simulation parameters.

Double-fed Asynchronous Wind Turbines	**Wind Turbines**			
	Blade radius /m	31	Shaft system stiffness factor /(pu/rad)	1.11
	Inertia time constant /s	4.32	Rated wind speed /(m/s)	12.5
	Cut-in wind speed /(m/s)	4.5	Cut-out wind speed /(m/s)	22
	Double-fed asynchronous generators			
	Rated power /MW	1.5	Rated frequency /Hz	50
	Rated voltage /kV	0.575	Stator impedance /pu	0.016 + j0.16
	Rotor impedance /pu	0.023 + j0.18	Stator and rotor mutual impedance /pu	j2.9
	Power converters			
	Rated capacity of rotor-side converter /MVA	0.525	Rated capacity of grid-side converter /MVA	0.75
	DC Bus Rated Voltage/kV	1.15	DC side bus capacitance /F	0.01
	Crowbar circuit input threshold /pu	2	Crowbar circuit cut out threshold /pu	0.35
	Crowbar resistance /pu	0.1		
Machine sidetransformer	Rated capacity /MVA	1.75	Rated frequency /Hz	50
	Rated Ratio /kV	25/0.575	Impedance /pu	0.06
Main Transformer	Rated capacity /MVA	150	Rated frequency /Hz	50
	Rated Ratio (kV)	125/25	Impedance /pu	0.135
Cable line	Unit resistance /(Ω/km)	0.1153	Unit inductance (/Ω/km)	j0.3297

References

1. Haces-Fernandez, F.; Cruz-Mendoza, M.; Li, H. Onshore wind farm development: Technologies and layouts. *Energies* **2022**, *15*, 2381. [CrossRef]
2. Council GWE. *GWEC Global Wind Report 2022*; Global Wind Energy Council: Bonn, Germany, 2022; pp. 119–125.
3. Agarala, A.; Bhat, S.; Mitra, A.; Zycham, D.; Sowa, P. Transient stability analysis of a multi-machine power system integrated with renewables. *Energies* **2022**, *15*, 4824. [CrossRef]
4. Skibko, Z.; Hołdynski, G.; Borusiewicz, A. Impact of wind power plant operation on voltage quality parameters—Example from Poland. *Energies* **2022**, *15*, 5573. [CrossRef]
5. Cárdenas, B.; Swinfen-Styles, L.; Rouse, J.; Garvey, S.D. Short-, medium-, and long-duration energy storage in a 100% renewable electricity grid: A UK case study. *Energies* **2021**, *14*, 8524. [CrossRef]
6. Fernández, L.M.; García, C.A.; Saenz, J.R.; Jurado, F. Equivalent models of wind farms by using aggregated wind turbines and equivalent winds. *Energy Convers. Manag.* **2009**, *50*, 691–704. [CrossRef]

7. Zou, J.; Peng, C.; Yan, Y.; Zheng, H.; Li, Y. A survey of dynamic equivalent modeling for wind farm. *Renew. Sustain. Energy Rev.* **2014**, *40*, 956–963. [CrossRef]
8. Han, J.; Li, L.; Song, H.; Liu, M.; Song, Z.; Qu, Y. An equivalent model of wind farm based on multivariate multi-scale entropy and multi-view clustering. *Energies* **2022**, *15*, 6054. [CrossRef]
9. WECC Renewable Energy Modeling Task Force. *WECC Wind Power Plant Dynamic Modeling Guidelines*; ESIG: Reston, VI, USA, 2014.
10. Chao, P.; Li, W.; Liang, X.; Xu, S.; Shuai, Y. An analytical two-machine equivalent method of DFIG-based wind power plants considering complete FRT processes. *IEEE Trans. Power Syst.* **2021**, *36*, 3657–3667. [CrossRef]
11. Wu, Y.K.; Zeng, J.J.; Lu, G.L.; Chau, S.W.; Chiang, Y.C. Development of an equivalent wind farm model for frequency regulation. *IEEE Trans. Ind. Appl.* **2020**, *56*, 2360–2374. [CrossRef]
12. Akhmotov, V.; Knudsen, H. An aggregate model of a grid-connected, large-scale, offshore wind farm for power stability investigations—Importance of windmill mechanical system. *Int. J. Electr. Power Energy Syst.* **2002**, *24*, 709–717. [CrossRef]
13. Brochu, J.; Larose, C.; Gagnon, R. Validation of single- and multiple-machine equivalents for modeling wind power plants. *IEEE Trans. Energy Convers.* **2011**, *26*, 532–541. [CrossRef]
14. Teng, W.; Wang, X.; Meng, Y.; Shi, W. Dynamic clustering equivalent model of wind turbines based on spanning tree. *J. Renew. Sustain. Energy* **2015**, *7*, 063126. [CrossRef]
15. Zhou, Y.; Zhao, L.; Matsuo, I.B.M.; Lee, W.J. A dynamic weighted aggregation equivalent modeling approach for the DFIG wind farm considering the Weibull distribution for fault analysis. *IEEE Trans. Ind. Appl.* **2019**, *55*, 5514–5523. [CrossRef]
16. Jin, Y.; Wu, D.; Ju, P.; Rehtanz, C.; Wu, F.; Pan, X. Modeling of wind speeds inside a wind farm with application to wind farm aggregate modeling considering LVRT characteristic. *IEEE Trans. Energy Convers.* **2020**, *35*, 508–519. [CrossRef]
17. Zhu, Q.; Ding, M.; Han, P. Equivalent modeling of DFIG-based wind power plant considering crowbar protection. *Math. Probl. Eng.* **2016**, *2016*, 8426492. [CrossRef]
18. Wu, Z.; Cao, M.; Li, Y. An equivalent modeling method of DFIG-based wind farm considering improved identification of Crowbar status. *Proc. CSEE* **2022**, *42*, 603–614.
19. Zhu, Q.; Ding, M. Equivalent modeling of PMSG-based wind power plants considering LVRT capabilities: Electromechanical transients in power systems. *SpringerPlus* **2016**, *5*, 2037.
20. Kunjumuhammed, L.P.; Pal, B.C.; Qates, C.; Dyke, K.J. The adequacy of the present practice in dynamic aggregated modeling of wind farm systems. *IEEE Trans. Sustain. Energy* **2017**, *8*, 23–32. [CrossRef]
21. Wang, Z.; Yang, G.; Tang, Y.; Cai, W.; Liu, S.; Wang, X.; Ouyang, J. Modeling method of direct-driven wind generators wind farm based on feature influence factors and improved BP algorithm. *Proc. CSEE* **2019**, *39*, 2604–2614.
22. Ma, Y.; Runolfsson, T.; Jiang, J.N. Cluster analysis of wind turbines of large wind farm with diffusion distance method. *IET Renew. Power Gener.* **2011**, *5*, 109–116. [CrossRef]
23. Ali, M.; Ilie, I.S.; Milanović, J.V.; Chicco, G. Wind farm model aggregation using probabilistic clustering. *IEEE Trans. Power Syst.* **2013**, *28*, 309–316. [CrossRef]
24. Zhu, Q.; Han, P.; Ding, M.; Zhang, X.; Shi, W. Probabilistic equivalent model for wind farms based on clustering-discriminant analysis. *Proc. CSEE* **2014**, *34*, 4770–4780.
25. Zhou, H.; Ju, P.; Xue, Y.; Zhu, J. Probabilistic equivalent model of DFIG-based wind farms and its application in stability analysis. *J. Mod. Power Syst. Clean Energy* **2016**, *4*, 248–255. [CrossRef]
26. Zhu, Q. Research on Equivalent Modeling of Wind Farm for Electromechanical Transient Stability Analysis. Ph.D. Thesis, Hefei University of Technology, Hefei, China, 2017.
27. Muljadi, E.; Butterfield, C.P.; Ellis, A.; Mechenbier, J.; Hochheimer, J.; Young, R.; Miller, N.; Delmerico, R.; Zavadil, R.; Smith, J.C. Equivalencing the collector system of a large wind power plant. In Proceedings of the 2006 IEEE Power Engineering Society General Meeting, Montreal, QC, Canada, 18–22 June 2006; pp. 1–9.
28. Zou, J.; Peng, C.; Xu, H.; Yan, Y. A Fuzzy clustering algorithm-based dynamic equivalent modeling method for wind farm with DFIG. *IEEE Trans. Energy Convers.* **2015**, *30*, 1329–1337. [CrossRef]
29. Lee, D.; Son, S.; Kim, I. Optimal allocation of large-capacity distributed generation with the volt/var control capability using particle swarm optimization. *Energies* **2021**, *14*, 3112. [CrossRef]
30. *IEC 61400-27-1:2015*; Wind Turbines—Part 27-1: Electrical Simulation Models—Wind Turbines. IEC: Geneva, Switzerland, 2015.
31. Wu, L.; Chao, P.; Li, G.; Li, W.; Li, Z. Hybrid data-model-driven aggregation equivalent modeling method for wind farm. *Autom. Electr. Power Syst.* **2022**, *46*, 66–74.

energies

MDPI

Article

Compressed Air Energy Storage System with Burner and Ejector

Dahui Yang [1,2], Xiankui Wen [3,*], Jingliang Zhong [3], Tingyong Feng [1,2], Tongtian Deng [3] and Xiang Li [3]

1 School of Electrical Engineering, Guizhou University, Guiyang 550025, China
2 Postgraduate Workstation of Guizhou Power Grid Co., Ltd., Guiyang 550002, China
3 Electric Power Research Institute of Guizhou Power Grid Co., Ltd., Guiyang 550002, China
* Correspondence: 13985410224@139.com; Tel.: +86-139-8541-0224

Abstract: The timescale of the energy-release process of an energy storage system has put forward higher requirements with the increasing proportion of new energy power generation in the power grid. In this paper, a new type of compressed-air energy storage system with an ejector and combustor is proposed in order to realize short-timescale and long-timescale energy-release processes under the non-supplementary combustion condition and ejector supplementary combustion condition, respectively. A simulation model of the new system is established in APROS software. The results of this study show that the new system can realize continuous power output when energy storage and energy release operate simultaneously, and especially when the ejector coefficient is 0.8 and burner thermal power is 10 MW, the power-generation time is 12.45 h and the total generated power is 140,052 kW·h, which are 15.6 and 17.5 times greater those of the short-timescale condition, respectively. In summary, the compressed-air energy storage system with an ejector and combustor that is proposed in this paper can flexibly meet the demands of multiple timescales' power generation.

Keywords: compressed-air energy storage; multiple timescales; ejector; burner

1. Introduction

In recent years, the clean and low-carbon process of energy utilization is accelerating, and the scale of new energy power generation, such as wind power and photovoltaic power, is growing larger and larger in China [1]. According to the data of the China National Energy Bureau, the installed capacity of wind power and solar power was about 330 million kW and 320 million kW, respectively, by the end of February 2022. However, new energy power generation, such as wind power and photovoltaic power, has the characteristics of strong intermittence, volatility, and randomness [2], which may have adverse effects on the power quality, safety, and stability of the power systems [3].

Large-scale electric energy storage technology is one of the effective ways to solve the above problems [4]. It has been validated that equipping a large amount of energy storage systems can effectively stabilize the gap and volatility of new energy power generation [5]. The energy storage systems will release energy to supplement the power generation when new energy power generation is insufficient, which can ensure the power balance, safe, and stable operation of power grid. Additionally, they will store energy to ensure that the new energy is consumed when the new energy generation is surplus. At present, the main types of large-scale clean power energy storage are pumped storage, compressed-air energy storage (CAES) [6], electrochemical energy storage [7], etc., and their typical rated power and maximum output time [8–10] are shown in Table 1. It can be seen that the longest continuous response time (that is, output power time) is 1–26 h of compressed-air energy storage.

Citation: Yang, D.; Wen, X.; Zhong, J.; Feng, T.; Deng, T.; Li, X. Compressed Air Energy Storage System with Burner and Ejector. *Energies* **2023**, *16*, 537. https://doi.org/10.3390/en16010537

Academic Editors: Wen-Ping Cao, Pinjia Zhang, Xiaoyan Huang and Cungang Hu

Received: 17 September 2022
Revised: 14 November 2022
Accepted: 26 December 2022
Published: 3 January 2023

Table 1. Typical rated power and maximum output time of various energy storage systems [8–10].

Types	Typical Rated Power	Continuous Response Time
Electrochemical energy storage	1 kW~50 MW	1 min~4 h
Pumped storage	100~2000 MW	2~8 h
Compressed-air energy storage	500 kW~300 MW	1–26 h
Flywheel energy storage	5 kW~5 MW	15 s~15 min
Superconducting magnetic energy storage	0.01~1 MW	ms~15 min
Supercapacitor	0.01~1 MW	1 s~15 min

The CAES system has the advantages of a large capacity, low pollution, moment of inertia, long storage cycle, etc. [11], and scholars and researchers have been widely concerned with it because of its broad development prospects. [12]. Since Stal Laval proposed to use underground caves to realize compressed-air energy storage in 1949 [13], domestic and foreign scholars have carried out a lot of research and practice [14] and established a number of CAES commercial power stations and demonstration projects [15]. At present, there are many types of CAES systems, which can be divided into supplementary combustion and non-combustion CAES systems from the perspective of auxiliary combustion [16]. The supplementary combustion CAES system has the advantages of strong reliability, stability, and good flexibility, but the disadvantage is obviously, e.g., consumption of fossil energy and emission of greenhouse gas. Two large-scale CAES power plants have been put into commercial operation in the world, namely the Huntorf power plant in Germany and the McIntosh power plant in the United States [17], both of which are supplementary combustion CAES systems. Compared with the supplementary combustion CAES system, the non-combustion CAES system has no supplementary combustion process, and the heat required in the energy-release stage mainly comes from the compression heat generated by the air-compression process in the energy-storage stage, so it has advantages of being environmentally friendly and pollution free [18,19]. At present, all the CAES demonstration projects in China adopt non-combustion CAES. The advanced adiabatic compressed-air energy storage (AA-CAES) power station of 50 MW in Feicheng, Shandong Province, was put into commercial operation in September 2021, and it is the first CAES commercial power station in China [20]. In the same month, the national demonstration project of 60 MW salt-cave CAES power generation system in Jintan, Jiangsu Province, successfully implemented a grid connection test [21].

As shown in Figure 1, the AA-CAES system is a typical non-combustion CAES system, and it includes two stages, namely energy storage and energy release [22]. In the energy-storage stage, air is compressed by compressors and stored in a storage tank. In the energy-release stage, compressed air enters a turbine from the gas storage tank, expands and releases energy to drive the synchronous generator to generate electricity, and then exhaust air is discharged into atmosphere [23].

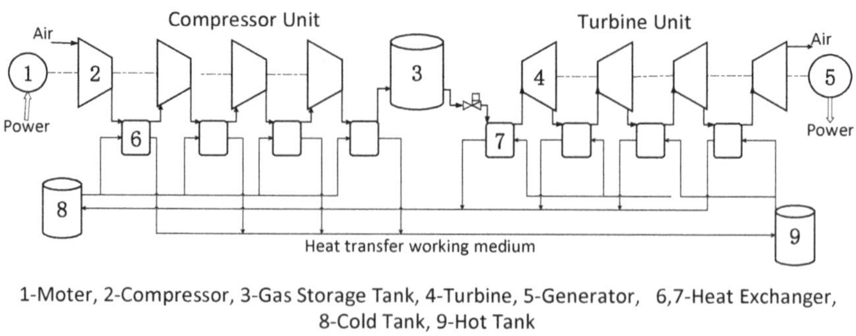

Figure 1. Advanced adiabatic compressed-air energy storage system.

At present, the research on the application of CAES system mainly focuses on two aspects, one is to participate in the source network coupling of grid load frequency regulation and improve the reliability of power supply on the grid side. Kamyar et al. [24] proposed a new design method of CAES system based on the performance requirements of the Ontario power grid by analyzing the actual operation data of a whole year. Based on the characteristics of the CAES system, Wen et al. [25] constructed its primary frequency modulation function and analyzed and set its dead zone, governor droop, clearance, and other parameters, which laid a foundation for CAES to participate in primary frequency modulation of power grid. Yang et al. [26] put forward a load rejection control strategy for an AA-CAES expansion generator by adding a shutoff valve between adjacent expansion units to effectively prevent speed rise. AmirReza. et al. [27] proposed a cogeneration system composed of CAES, organic Rankine cycle, and absorption–compression refrigeration cycle. Considering the multi-generation characteristics of the AA-CAES power plant, Li et al. [28] constructed the joint dispatching constraint model of cooling, heating, and power multi-energy flow for the AA-CAES power station. Hesamoddin et al. [29] proposed a two-stage mathematical optimization model for optimizing the day-ahead operation of generation units, as well as CAESs, in energy and reserve markets in a stochastic way.

The other aspect is to couple with the source storage of new energy power stations, such as wind energy and photovoltaic, on the generation side. Deng et al. [30] proposed a control strategy of a wind-storage combined system on the basis of power stabilization and verified its feasibility by simulation. Amirreza et al. [31] researched an absorption-recompression refrigeration system with CAES and wind turbines, which employed a CAES system to compensate the further energy consumption of the vapor compressor. Alirahmi et al. [32] proposed and thoroughly investigated a novel efficient and environmentally friendly hybrid energy production/storage system comprising a compressed-air energy storage, and the system has an exergy round trip efficiency of 60.4% and a total cost rate of 117.5 $/GJ. Li et al. [33] studied the operation optimization strategy of the wind-storage combined system, considering the dynamic characteristics and operation constraints of CAES. Li et al. [34] studied a grid-connected power-optimization strategy that integrates wind energy and a low-temperature CAES, which can balance the fluctuation of wind energy by reducing energy storage capacity and ensure continuous and stable output power to power grid.

As mentioned above, the current research studies of the AA-CAES system are all based on a single timescale operation, which can only meet the requirements of storing energy when new energy power generation is surplus and releasing energy in a short timescale when new energy power generation is not enough. Nevertheless, the long-timescale fluctuation of new energy and the extreme conditions of zero power generation have a growing impact on the power system with the rapid development of new energy and the increasing proportion of installed capacity. Although increasing the capacity of energy storage system is an important method to effectively solve this problem, it will lead to high system cost and low utilization.

In consideration of the demands of multiple timescales' power generation and economy, a novel AA-CAES system with two main components, namely an ejector and a burner, is proposed in this paper. This system has three operating modes, including a short-timescale mode with adiabatic non-supplementary combustion condition, a long-timescale mode with ejector-supplementary combustion condition, and a continuous-output mode with energy storage and energy release operating simultaneously, which can flexibly adapt to the requirements of multiple timescales' energy release. In addition, a simulation model of the new system was established based on APROS software, and the characteristics of the three operating modes are analyzed.

2. Architecture of Energy Storage System

2.1. Compressed-Air Energy Storage System with Ejector and Combustor

Based on the AA-CAES system, this paper proposes a new CAES system with an ejector and combustor which can release energy in multi-timescales. Its system architecture is shown in Figure 2. The system is composed of a compression energy storage subsystem, a gas storage subsystem, and an expansion energy-release subsystem. Among them, the compression energy storage subsystem is composed of multistage compressors and motors; the gas storage subsystem is a high-pressure gas storage tank; and the expansion energy-release subsystem is composed of an ejector, a combustor, multistage expanders, and a generator.

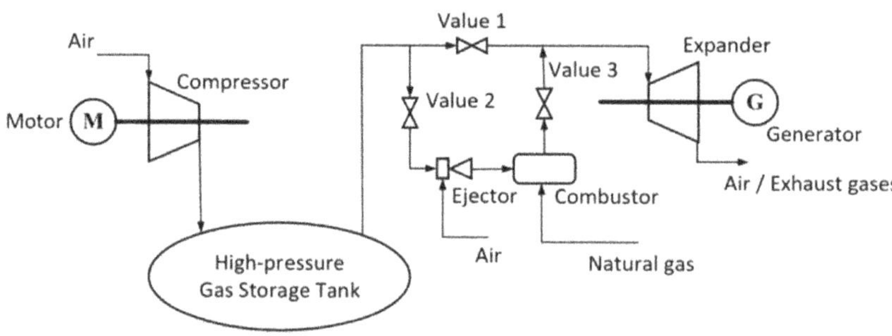

Figure 2. Diagram of compressed-air energy storage system with an ejector and combustor.

In this study, a simulation model of the new CAES system with an ejector and burner is established based on APROS software [35,36]. Considering the characteristics of the CAES system with supplemental combustion and the relevant operating parameters of a 10 MW advanced CAES system demonstration project in Bijie, China [36,37], the detailed design parameters [19,36] are shown in Table 2.

Table 2. Basic design parameters for the new CAES system.

Parameters	Units	Value
Energy-Release Power	MW	10
Energy-Release Pressure	MPa	7
Maximum Storage Pressure	MPa	10
Gas Storage Tank Volume	m^3	6000
Ambient Pressure	MPa	0.1
Ambient Temperature	K	298
10 MW Release Time	s	2880
Hot Tank Temperature	K	403
Hot Tank Pressure	MPa	0.4
Cold Tank Temperature	K	298
Cold Tank Pressure	MPa	0.1
Compressor Motor Power	MW	10
Burner Thermal Power	MW	20
Burner Fuel	/	Natural Gas
Fuel Calorific Value	MJ/Nm^3	36.22
Electric Power Consumed by Other Auxiliary Machines of Energy Storage Subsystem	MW	1.5
Electric Power Consumed by Other Auxiliary Machines of Energy-Release Subsystem	MW	1.5
Generator Power	MW	30

In this paper, a short-timescale simulation method is adopted, and the operation parameters of each expander are shown in Table 3.

Table 3. Operation parameters in the mode of short timescale.

Expander Stages	Unit	1	2	3	4
Inlet Pressure	MPa	7	2.46	0.853	0.283
Outlet Pressure	MPa	2.48	0.873	0.303	0.1
Inlet Temperature	°C	82	82	82	82
Outlet Temperature	°C	14.4	7.6	10.0	18.0
Pressure Loss in Heat Exchanger	MPa	0.02	0.02	0.02	0.02
Air Flow Rate	kg/s	34.7	34.7	34.7	34.7
Expansion Ratio	/	2.82	2.82	2.82	2.82
Isentropic Efficiency	/	0.88	0.88	0.88	0.88
Output power	MW	2.536	2.786	2.691	2.358
speed	r/min	3 000	3 000	3 000	3 000

2.2. Mathematic Model

2.2.1. Expander

The air-expansion process is generally regarded as a polytropic process in the expander, and the output power of the expander [38,39] is calculated as follows:

$$W_i = \frac{k}{k-1} m R_g T_i^{\text{in}} \eta_i \left(1 - \beta_i^{\frac{1-k}{k}}\right), i = 1, 2, \cdots M \tag{1}$$

where W_i is the output power of the ith expanders, kW; m is the air mass flow, kg/s; k is the air adiabatic index; R_g is the gas constant; T_i^{in} is the intake temperature of the ith stage expander, K; η_i is the adiabatic efficiency of the ith stage expander; and β_i is the expansion ratio of the ith stage expander.

2.2.2. Heat Exchanger

In the AA-CAES power generation system, heat exchangers and expanders are arranged in a series. After air flows into the heat exchanger, the heat exchange between the air and tube-wall [36] is as follows:

$$Q_{\text{h}} = A_{\text{h}}(T_{\text{h}} - T_{\text{w}})/(\delta/2K_{\text{w}} + 1/\alpha_{\text{h}}) \tag{2}$$

The heat exchange between heat-transfer medium and pipe-wall is as follows:

$$Q_{\text{c}} = A_{\text{c}}(T_{\text{c}} - T_{\text{w}})/(\delta/2K_{\text{w}} + 1/\alpha_{\text{c}}) \tag{3}$$

where δ is the thickness of the tube-wall, m; T_{w} is the average temperature of the tube-wall, K; T_{h} is the air temperature inside the tube-wall, K; T_{c} is the average temperature of the heat-transfer medium, K; K_{w} is the thermal conductivity of the tube-wall, W/(m·K); α_{c} and α_{h} are the convective heat-transfer coefficients of the inner and outer tube-walls, respectively, $\text{W}/(\text{m}^2\cdot\text{K})$; and A_{c} and A_{h} are the areas of the inner and outer tube-walls, respectively, m^2.

2.2.3. Gas Storage Tank

In the energy-release stage, the discharge process of the gas storage tank is a polytropic process. According to the mass and energy balance equations, the change rules of gas in the gas storage tank during the discharge process [40] can be obtained as follows:

$$\frac{\text{d}p}{\text{d}t} = \frac{R_{\text{g}}}{Vc_{\text{v}}}\left[\frac{\text{d}m}{\text{d}t} c_{\text{p}} T_{\text{ac}} - h_{\text{a}} A (T_{\text{ac}} - T_{\text{a}})\right], \tag{4}$$

$$\frac{\text{d}T_{\text{ac}}}{\text{d}t} = \frac{T_{\text{ac}}}{m}(k-1)\frac{\text{d}m}{\text{d}t} - \frac{1}{mc_{\text{v}}} h_{\text{a}} A (T_{\text{ac}} - T_{\text{a}}), \tag{5}$$

where p is the air pressure of the gas storage tank, Pa; V is the volume of the gas storage tank, m^3; c_p is the air-specific heat at constant pressure, kJ/kg·K; c_v is the air-specific heat at constant volume, kJ/(kg·K); h_a is the convective heat transfer coefficient of the gas storage tank, kW/(m^2·K); A is the internal surface area of the gas storage tank, m^2; m is the air mass in the gas storage tank, kg; T_a is the wall temperature of the gas storage tank, K; and T_{ac} is the air temperature in the gas storage tank, K.

2.2.4. Regulating Valve

The flow equation of the regulating valve [41] is as follows:

$$m_s = \varepsilon C_s f(\mu)\sqrt{\rho \Delta p}, \tag{6}$$

where m_s is the outlet air flow rate of regulating valve, kg/s; ε is the fluid compressibility, C_s is the valve admittance, $f(\mu)$ is the characteristic function of the regulating valve, and ρ is the inlet air density of the regulating valve.

3. Energy Storage Condition

The compression energy-storage subsystem of CAES system has the same system architecture and operation mode under the conditions of energy release at different timescales.

In the stage of energy storage, the air is compressed into high-pressure compressed air by the compressor and stored in the high-pressure gas storage tank to realize the conversion of electric energy to internal energy. The judgment conditions for the completion of compressed-air preparation and the end of the energy-storage phase are as follows:

$$\begin{Bmatrix} p = p_e \\ t = t_s \end{Bmatrix}, \tag{7}$$

where p is the pressure of high-pressure gas storage tank, p_e is the rated pressure of gas storage tank, t is the operation time, and t_s is the end time of energy storage stage.

The end time t_s of energy storage stage is dispatched by the power grid, which is determined according to the peak regulation demand of power grid and dispatched through the day-ahead scheduling plan.

The power consumption of the ith compressor is defined as follows:

$$P_{ci} = G_{ci}(h_{ci,out} - h_{ci,in}) \tag{8}$$

where P_{ci} is the power consumption of the ith compressor, G_{ci} is the air mass flow of the ith compressor during energy storage process, and $h_{ci,in}$ and $h_{ci,out}$ are the specific enthalpies of inlet and outlet of the ith compressor.

The total power consumption of compressor motor can be expressed as follows:

$$P_m = \frac{P_c}{\eta_m \times \eta_c} = \frac{G_c \times \sum_{i=1}^{m}(h_{ci,out} - h_{ci,in})}{\eta_m \times \eta_c} \tag{9}$$

where P_m is the power consumption of compressor motor, P_c is the total power consumption of compressor motor, G_c is the air mass flow of compressor, η_m is the motor efficiency, η_c is the compressor efficiency, and m is the number of compressors.

4. Multi-Timescale Energy-Release Condition

In this paper, the short-timescale power-generation condition is defined as a power-generation time less than 6 h, the long-term scale power-generation condition is defined as a power-generation time between 6 and 30 h, and the continuous-output condition is defined as continuous uninterrupted power generation.

The compressed-air energy storage system with an ejector and combustor has three operation modes in the energy-release stage that can flexibly adapt to three power-generation conditions, namely short timescale, long timescale, and continuous output. The three operating conditions of the energy-release stage of the system are discussed below.

4.1. Short-Timescale Condition

Under the short-timescale condition, the CAES system operates in adiabatic non-supplementary combustion mode.

4.1.1. Energy-Release Stage

The CAES system starts the process of air expansion to generate electricity when electric energy in the power grid is in short supply. As shown in Figure 1, the compressed air enters the expander from the high-pressure gas storage tank to expand and drive the generator to generate electricity by opening Valve 1, closing Valve 2 and Valve 3, sequentially. The compressed air after the work of the upper stage enters the heat exchanger for heating, and then it enters the expander of the next stage for power generation and is discharged into atmosphere after multistage expansion.

The generating power, P_{ti}, of the ith expander is defined as follows:

$$P_{ti} = \frac{G_{ti} \times (h_{ti,in} - h_{ti,out})}{3600} \tag{10}$$

where G_{ti} is the air mass flow of the ith expander, and $h_{ti,in}$ and $h_{ti,out}$ are the specific enthalpies of inlet and outlet of the ith expander.

The total generating power, P_g, of the CAES system can be expressed as Equation (11):

$$P_g = P_t \times \eta_g \times \eta_t = G_g \sum_{i=1}^{m} (h_{ti,in} - h_{ti,out}) \times \eta_g \times \eta_t \tag{11}$$

where P_t is the total generating power of expanders, G_g is the air mass flow of expanders in short scale, η_g is the generator efficiency, η_t is the expander efficiency, and m is the number of expanders.

4.1.2. Timescale Determination

The determination of timescale is mainly related to parameters such as the volume, temperature, and pressure of the high-pressure gas storage tank.

To simplify the calculation, the following assumptions need to be made in this paper:

(1) The compressed air is an ideal gas;
(2) The gas loss at the pipes and valves are ignored;
(3) The temperature change in the process of compression and depressurization of gas storage device is not considered. (The energy storage system has a heat-exchange device, so the internal temperature change of gas storage device is ignored and regarded as an isothermal expansion process.)

According to the ideal gas law, the relationship between the state parameters of gas storage device at the initial time in the energy storage stage is as follows:

$$p_{c,s} V = m_{c,s} R T_{c,s} \tag{12}$$

where $p_{c,s}$ is the air pressure at the initial time, V is the volume of gas storage device, $m_{c,s}$ is the air mass at the initial time, and $T_{c,s}$ is the air temperature at the initial time.

The state parameters of gas storage device at the end time can be expressed as follows:

$$p_{c,f} V = m_{c,f} R T_{c,f} \tag{13}$$

where $p_{c,f}$ is the air pressure at the end time, $m_{c,f}$ is the air mass at the end time, and $T_{c,f}$ is the air temperature at the end time and is equal to $T_{c,s}$.

Equations (12) and (13) can be subtracted to obtain Equation (14):

$$V(p_{c,f} - p_{c,s}) = \Delta m R T_{c,s} \quad (14)$$

where Δm is the total mass of compressed air produced by compressor during energy storage process.

According to the total mass of compressed air and the mass flow of compressor under the rated working conditions, the time, t_c, of the energy storage process can be expressed as follows:

$$t_c = \frac{V(p_{c,f} - p_{c,s})}{R T_{c,s} G_c}, \quad (15)$$

Similarly, we derive the working time, t_g, of the energy-release process as follows:

$$t_g = \frac{V\left(p_{g,s} - p_{g,f}\right)}{R T_{g,s} G_g}, \quad (16)$$

where $p_{g,s}$ is the air pressure at the initial time, $p_{g,f}$ is the air pressure at the end time, and $T_{c,s}$ is the air temperature at the initial time.

4.1.3. Simulation Analysis

As shown in Figure 3a, with the mass flow of compressed air decreased, the generator power decreases, while the working time of the energy-release process, t_g, is increased. Moreover, the relations of different mass flows of compressed air with the generator power, power-generation capacity, and power-generation time are shown in Figure 3b. The compressed air flow is inversely proportional to the power-generation time. Because the total quality of the workable working medium in the gas storage tank is constant, the power-generation capacity tends to decrease with the decrease of the compressed air flow.

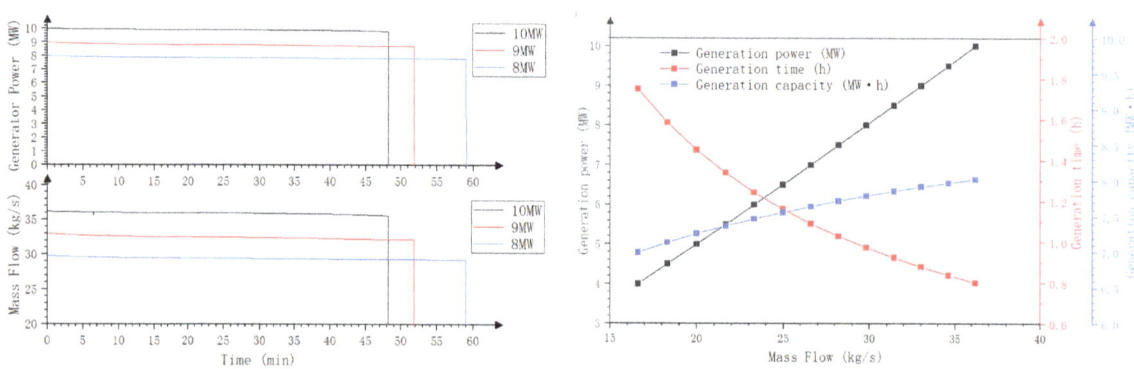

(a) (b)

Figure 3. (**a**) Operating parameters under short-timescale conditions. (**b**) Relation diagram of different mass flow of compressed air with power-generation capacity and power-generation time.

4.2. Long-Timescale Condition

The working condition of the power consumption and energy-storage stage in the long timescale is the same as that in the short timescale.

4.2.1. Energy-Release Stage

As shown in Figure 2, the ambient air is inhaled into the ejector to increase the flow when the high-pressure compressed air passes through the injector by closing Valve 1 and opening Valve 2 and Valve 3, sequentially. By adding natural gas into the combustor for mixed combustion, the temperature, pressure, and expansion work capacity of the compressed air are increased. Finally, the pressurized and heated air enters the turbine and drives the generator to generate electricity, realizing the conversion of internal energy to electrical energy.

This model improves the power-generation capacity through two means. One is to increase the air-mass flow of the expander through the ejector, and the other is to increase the temperature and pressure of the working air through the external combustor. Air enthalpy is positively correlated with air temperature and pressure; that is, raising the enthalpy of working air inlet could increase the enthalpy drop of air inlet and outlet of expander.

4.2.2. Ejector Model

The structure diagram of the ejector is shown in Figure 4.

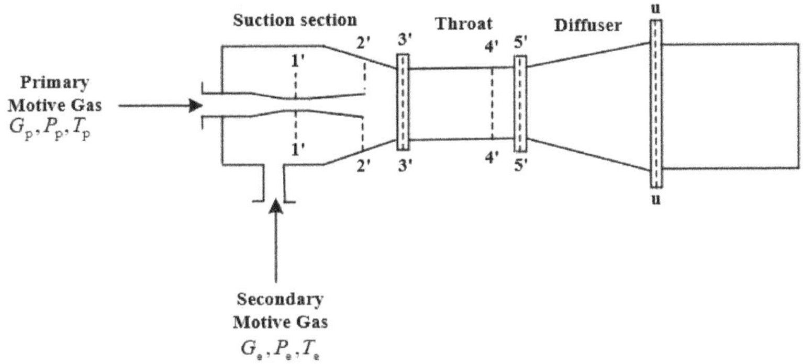

Figure 4. Structure diagram of the ejector.

According to the mass-balance equations, the total mass balance [42] is calculated as follows:

$$G'_g + G_e = G_t \tag{17}$$

The suction coefficient of the ejector is defined [42] as follows:

$$\gamma = \frac{G_e}{G'_g} \tag{18}$$

where G'_g is the mass flow rate of compressed air, and G_e is the mass flow rate of atmosphere air.

Therefore, after high-pressure compressed air passes through the ejector, its mass flow rate changes as follows:

$$G_t = G'_g(1+\gamma) \tag{19}$$

4.2.3. Combustor Model

The structure diagram of the combustor is shown in Figure 5.

In this paper, the new system uses natural gas as the fuel of burner, and its calorific value is 36.22 MJ/Nm^3. The energy changes after burning with natural gas are as follows [43]:

$$Q_b = Q_{Fuel} \times \eta_b \tag{20}$$

where Q_b is the thermal power output of the combustor, Q_{Fuel} is the thermal power of burning natural gas, and η_b is the conversion efficiency of the combustor.

Therefore, under the long-timescale condition, the generating power of the energy-release generation stage is as follows:

$$P'_{\mathrm{g}} = \frac{(G'_{\mathrm{g}}(1+\gamma) \times \sum_{i=1}^{m} (h'_{\mathrm{ti,in}} - h'_{\mathrm{ti,out}}) + Q_b)}{3600} \times \eta_{\mathrm{g}} \times \eta_{\mathrm{t}} \tag{21}$$

where P'_{g} is the generating power of the energy-release generation stage.

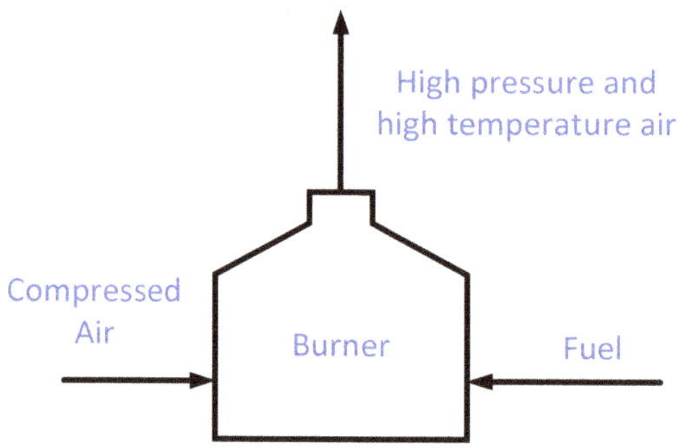

Figure 5. Structure diagram of combustor.

4.2.4. Timescale Determination

There are two aspects to prolonging the energy-release time in this mode: one is to increase the working air flow rate from G'_{g} to $G'_{\mathrm{g}}(1+\gamma)$ through the ejector, and the other is to increase its capacity for work through combustion. When the output power is constant at the rated power of the expansion generator, reducing the compressed-air consumption per unit output power can increase the time of the power generation:

$$P'_{\mathrm{g}} = P_{\mathrm{g}} \tag{22}$$

$$G'_{\mathrm{g}} = \frac{G_{\mathrm{g}} \sum_{i=1}^{m} (h_{\mathrm{ti,in}} - h_{\mathrm{ti,out}}) - Q_b}{(1+\gamma) \times \sum_{i=1}^{m} \left(h'_{\mathrm{ti,in}} - h'_{\mathrm{ti,out}}\right)} \tag{23}$$

Compressed-air energy that can perform expansion work in the high-pressure gas storage tank is constant, so we can obtain the following:

$$G'_{\mathrm{g}} t'_{\mathrm{g}} = G_{\mathrm{g}} t_{\mathrm{g}} \tag{24}$$

$$t'_{\mathrm{g}} = G_{\mathrm{g}} \frac{(1+\gamma) \times \sum_{i=1}^{m} \left(h'_{\mathrm{ti,in}} - h'_{\mathrm{ti,out}}\right)}{G_{\mathrm{g}} \sum_{i=1}^{m} (h_{\mathrm{ti,in}} - h_{\mathrm{ti,out}}) - Q_b} t_g \tag{25}$$

$$W'_{\mathrm{g}} = (1+\gamma) \frac{h_{\mathrm{J,in}}}{h_{\mathrm{J,mix}}} W_{\mathrm{g}} + Q_{\mathrm{b}} t'_{\mathrm{g}} \tag{26}$$

where t'_{g} is the working time of the energy-release process under the long-timescale condition, and W'_{g} is the generation capacity during the energy-release process under the long-timescale condition.

4.2.5. Simulation Analysis

From Equations (25) and (26), it can be seen that the power-generation capacity and power-generation time in a long timescale are affected by the ejection coefficient of the ejector and the thermal power of the burner, and their relationships are shown in Figure 6a,b. It can be seen from Figure 6a,b that, when the ejector coefficient is 0.2 and the burner thermal power is 8 MW, the power-generation time is 3.28 h and the power-generation capacity is 36,114.3 kW·h, which are 4.1 and 4.5 times that of the short-timescale conditions, respectively. When the ejector coefficient and burner thermal power are increased to 0.8 and 10 MW, the power-generation time and power-generation capacity increase by 8.35 h and 103,937.7 kW·h, which are 15.6 and 17.5 times that of those of the short-timescale conditions, respectively.

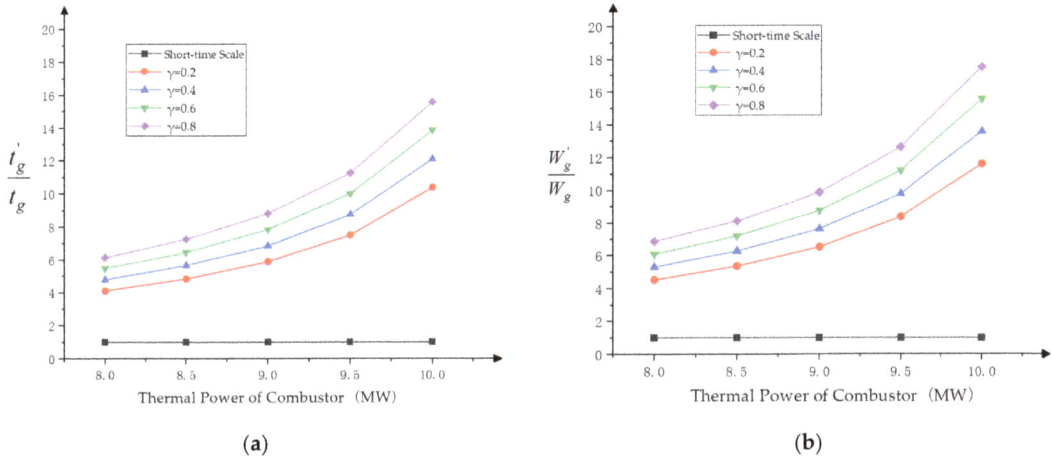

Figure 6. Relation diagram of different injection coefficients and burner thermal power with power-generation capacity and power-generation time: (**a**) power-generation time and (**b**) power-generation capacity.

Under the condition of the long timescale, increasing the ejection coefficient can increase the power-generation time and power-generation capacity of the system. When the compressed-air mass available in the gas storage tank and the flow rate after passing through the jet are constant, increasing the ejection coefficient will make the air flow into the jet decrease and prolong the power-generation time. The power-generation time and power-generation capacity will increase with the increase of the burner thermal power. This is because increasing the burner thermal power will lead to the specific enthalpy of the intake air of the first-stage expander increasing, thus reducing the compressed air flow required to achieve the rated power. Meanwhile, compared with the ejector coefficient, the change of burner thermal power has a more obvious influence on the power-generation time and power-generation capacity.

4.3. Continuous-Output Condition

The continuous-output condition is that the energy-storage stage and energy-release stage are carried out simultaneously. The energy-storage stage can provide high-pressure compressed air for the energy-release stage, while the generator provides electricity energy for the energy-storage subsystem. The power-generation capacity of the whole system is improved by increasing the air mass flow of the expander through the ejector and increasing the air enthalpy at the inlet of the expander by burning natural gas in the external burner.

4.3.1. Constraint Conditions

The construction of continuous-output conditions needs to meet the following constraints:

$$\begin{cases} P_g'' > P_u \\ G_g'' < G_c \\ p_p > p_{g,\lim} \end{cases} \tag{27}$$

where P_g'' is the generating power of the energy-release subsystem, P_u is the system power consumption, G_g'' is the air mass flow of compressed air into the jet, G_c is the air mass flow at the outlet of the last-stage compressor, P_p is the pressure of high-pressure gas storage tank, and $P_{g,lim}$ is the minimum working pressure of the ejector.

Through the above constraints, it is ensured that the mass flow rate of the working air consumed by the ejector is less than or equal to the mass flow rate at the outlet of the last-stage compressor, so that the system has working air that can meet the continuous operating conditions. Meanwhile, it ensures that the ejector can work normally and prevents the ejector from not forming an entrainment effect due to a too-low working pressure.

The generating power of the energy-release subsystem is the same as that under the long-timescale condition:

$$P''_g = P'_g = \frac{\left(G''_g(1+\gamma) \times \sum_{i=1}^{m}\left(h''_{ti,in} - h''_{ti,out}\right) + Q_b\right)}{3600} \times \eta_g \times \eta_t \tag{28}$$

The power consumption of the whole system is as follows:

$$P_u = P_m + \sum P_{o,m} + \sum P_{o,g} \tag{29}$$

where $P_{o,m}$ is the electric power consumed by other auxiliaries in the energy-storage subsystem, and $P_{o,g}$ is the electric power consumed by other auxiliaries in the energy-release subsystem.

The system's generating power is as follows:

$$P_s = P''_g - P_u \tag{30}$$

where P_u is the generating power of the whole system.

4.3.2. Simulation Analysis

Under continuous-output conditions, the compression energy-storage subsystem, gas-storage subsystem, and expansion energy-release subsystem are all put into operation, and their operating parameters are shown in Figure 7. During the duration of 0–12.45 h, the system operates under the long-timescale condition, and the system's output power is 27 MW. When the air pressure at the inlet of the ejector drops to 3 MPa, the compression energy-storage subsystem is put into operation, and gas storage tank's pressure rises. The system operates under continuous-output conditions, and the compressor's motor power is 10 MW, so that the system's output power drops to 17 MW. When the gas storage tank's pressure rises to 10 MPa again, the compression energy-storage subsystem stops running.

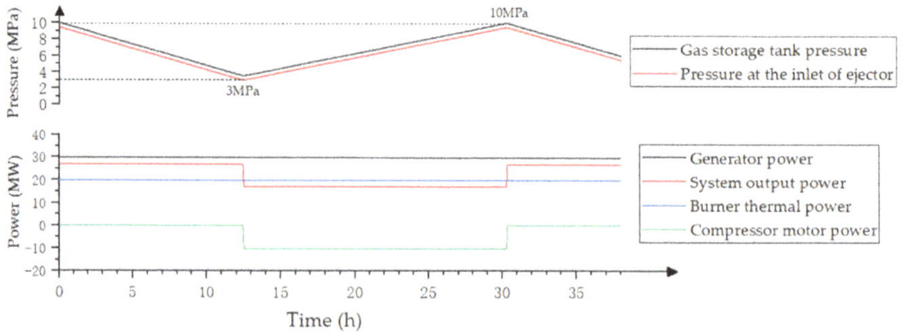

Figure 7. Operating parameters under continuous-output conditions.

5. Conclusions

The current literature has identified the CAES as a potentially important part of coupling renewable energy generation and low-carbon power grids. The extreme conditions of the long timescale's fluctuation and zero-power generation of the new energy power generation require that the CAES system has the ability to generate electricity on multiple timescales. The existing commercial CAES systems cannot meet the needs of multi-timescale power generation well compared to the system studied in this paper. The AA-CAES system can only meet the needs of short-timescale power generation. The supplementary combustion CAES system can meet the needs of short-timescale and long-timescale power generation, but it will cause pollution problems in the short-timescale power generation due to supplementary combustion compared to the system studied here.

In this paper, a new type of compressed-air energy-storage system with an injector and burner was established. The simulation analysis was carried out under short-timescale, long-timescale, and continuous-output conditions, sequentially. Several conclusions were obtained:

(1) The new system can meet the needs of multiple timescales under different operating conditions. The energy-storage stage of the new system is consistent with that of the AA-CAES system, and there are three operation modes in the energy-release stage which can flexibly adapt to three power-generation conditions: short timescale, long timescale, and continuous output.
(2) Under the short-timescale condition, the ejector and combustor are not put into operation, and the CAES system operates as adiabatic compression/expansion processes, which do not need to burn natural gas and are environmentally friendly.
(3) Under medium- and long-timescale conditions, the ejector and combustor are put into operation to prolong the duration of the power generation, so that the new system can meet the needs of long-timescale power generation. Moreover, the power-generation duration and capacity of the system improve with the increase of the ejector coefficient and burner thermal power. When the ejector coefficient is 0.8 and the burner thermal power is 10 MW, the power-generation time is 12.45 h, and the power-generation capacity is 140,052 kW·h, which are 15.6 and 17.5 times that of the short-timescale conditions, respectively.
(4) Under the continuous-output condition, the energy-storage system and energy-release system operate at the same time. The energy-storage subsystem provides high-pressure compressed air for the energy-release stage, while the generator provides electricity for the energy-storage subsystem. By selecting equipment parameters according to the constraint conditions, the new system can realize continuous power output.

Author Contributions: Methodology, D.Y. and X.W.; Software, J.Z.; Validation, D.Y. and J.Z.; Formal analysis, D.Y. and X.W.; Investigation, X.W.; Resources, T.F. and X.L.; Data curation, D.Y. and T.F.; Writing—original draft, D.Y.; Writing—review & editing, X.W.; Visualization, D.Y.; Supervision, X.W. and T.D.; Project administration, D.Y., X.W. and J.Z.; Funding acquisition, X.W. and J.Z. All authors have read and agreed to the published version of the manuscript.

Funding: This work was financially supported by the National Key Research and Development Program of China, grant number 2017YFB0903600; the High Level Innovative Talent Projects of Guizhou Province, grant number GCC [2022]014-1; and the Science & Technology support plan of Guizhou Province, grant number [2020]2Y064.

Data Availability Statement: Not applicable.

Conflicts of Interest: The authors declare no conflict of interest.

Nomenclature

AA-CAES	Advanced adiabatic compressed-air energy storage
CAES	Compressed-air energy storage
A	Internal surface area of gas storage tank $[m^2]$
A_c, A_h	Area of inner and outer pipe-wall $[m^2]$
c_p	Air specific heat at constant pressure $[kJ/(kg{\cdot}K)]$
c_v	Air specific heat at constant volume $[kJ/(kg{\cdot}K)$
C_s	Valve admittance

$f(\mu)$	Characteristic function of regulating valve
G_c	Air mass flow of compressors [kg/s]
G_g	Air mass flow of expander under the short-timescale condition [kg/s]
G'_g	Mass flow of compressed air entering the jet under the long-timescale condition [kg/s]
G''_g	Mass flow of compressed air entering the jet under continuous-output conditions [kg/s]
G_c	Air mass flow at the outlet of last-stage compressor [kg/s]
h_a	Convective heat transfer coefficient of gas storage tank $[\mathrm{kW}/(\mathrm{m}^2{\cdot}\mathrm{K})]$
$h_{ci,\mathrm{in}}, h_{ci,\mathrm{out}}$	Specific enthalpy of inlet and outlet of stage i compressor [kJ/kg]
$h_{ti,\mathrm{in}}, h_{ti,\mathrm{out}}$	Specific enthalpy of inlet and outlet of stage i expander under short-timescale condition [kJ/kg]
$h'_{ti,\mathrm{in}}, h'_{ti,\mathrm{out}}$	Specific enthalpy of inlet and outlet of stage i expander under long-timescale condition [kJ/kg]
$h''_{ti,\mathrm{in}}, h''_{ti,\mathrm{out}}$	Specific enthalpy of inlet and outlet of stage i expander under continuous-output condition [kJ/kg]
$h_{J,\mathrm{in}}$	Specific enthalpy of compressed air entering the jet under long-timescale condition [kJ/kg]
$h_{J,\mathrm{mix}}$	Specific enthalpy of mixed air outflow from the jet in under long-timescale condition [kJ/kg]
k	Air adiabatic index
K_w	Thermal conductivity of pipe-wall [W/(m·K)]
m	Air mass flow of expanders [kg/s]
$m_{c,s}$	Air mass at the initial time [kg]
$m_{c,f}$	Air mass at the end time [kg]
p_e	Rated pressure of gas storage tank [Pa]
P_m	Power consumption of compressor motor
P_g	Generating power of expanders under short-timescale condition
$p_{c,s}$	Air pressure at the initial time [Pa]
$p_{c,f}$	Air pressure at the end time [Pa]
P'_g	Generating power under long-timescale condition [kW]
P''_g	Generating power in continuous-output condition [kW]
P_u	System power consumption [kW]
P_s	System generating power [kW]
P_p	Air pressure of high-pressure gas storage tank [Pa]
$P_{g,\mathrm{lim}}$	Minimum working pressure of ejector [Pa]
Q_h	Heat exchange between air and pipe-wall [kJ]
Q_c	Heat exchange between heat − transfer medium and pipe-wall [kJ]
Q_b	Thermal power output of combustor [kW]
Q_{Fuel}	Thermal power of burning natural gas [kW]
R_g	Gas constant [kJ/(kg·K)]
T_i^{in}	Intake temperature of i stage expander [K]
T_h	Air temperature inside pipe-wall [K]
T_w	Average temperature of pipe-wall [K]
T_c	Average temperature of heat-transfer medium [K]
T_a	Wall temperature of gas storage tank [K]
T_{ac}	Air temperature in gas storage tank [K]
t_s	End time of energy storage stage
$T_{c,s}$	Air temperature at the initial time [K]
$T_{c,f}$	Air temperature at the end time [K]
t_c	Working time of energy storage process
t_g	Working time of energy-release process under short-timescale condition
t'_g	Working time of energy-release process under long-timescale condition
V	Gas storage tank volume $[\mathrm{m}^3]$

W_i	Output power of expanders [kW]
W_g	Generation capacity during under short-timescale condition [kW·h]
W_g'	Generation capacity during under long-timescale condition [kW·h]
α_c, α_h	Convective heat transfer coefficient of inner and outer pipe-wall $[W/(m^2 \cdot K)]$
β_i	Expansion ratio of i stage expander
δ	Thickness of pipe-wall [m]
η_i	Adiabatic efficiency of i stage expander
ε	Fluid compressibility
ρ	Inlet air density of regulating valve $[kg/m^3]$
η_m	Motor efficiency
η_c	Compressor efficiency
η_g	Generator efficiency
η_t	Expander efficiency
γ	Suction coefficient of ejector
η_b	Conversion efficiency of combustor

References

1. Zhang, J.; Daniel, A.; Fang, Y.; Desmond, A.; Antwi, E.O. Review on China's renewable energy and future projections. *Int. J. Smart Grid Clean Energy* **2018**, *7*, 218–224. [CrossRef]
2. Arabkoohsar, A.; Machado, L.; Koury, R.N.N. Operation analysis of a photovoltaic plant integrated with a compressed air energy storage system and a city gate station. *Energy* **2016**, *98*, 78–91. [CrossRef]
3. Denholm, P.; Sioshansi, R. The value of compressed air energy storage with wind in transmission-constrained electric power systems. *Energy Pol.* **2009**, *37*, 3149–3158. [CrossRef]
4. Singh, S.; Singh, M.; Kaushik, S.C. Feasibility study of an islanded microgrid in rural area consisting of PV, wind, biomass and battery energy storage system. *Energy Convers. Manag.* **2016**, *128*, 178–190. [CrossRef]
5. Razmi, A.R.; Soltani, M.; Ardehali, A.; Gharali, K.; Dusseault, M.; Nathwani, J. Design, thermodynamic, and wind assessments of a compressed air energy storage (CAES) integrated with two adjacent wind farms: A case study at Abhar and Kahak sites, Iran. *Energy* **2021**, *221*, 119902. [CrossRef]
6. Zhang, X.J.; Chen, H.S.; Liu, J.C.; Li, W.; Tan, C.Q. Research progress in compressed air energy storage system: A review. *Energy Storage Sci. Technol.* **2012**, *1*, 26–40.
7. Gallo, A.B.; Simões-Moreira, J.R.; Costa, H.K.M.; Santos, M.M.; Moutinho dos Santos, E. Energy storage in the energy transition context: A technology review. *Renew. Sustain. Energy Rev.* **2016**, *65*, 800–822. [CrossRef]
8. Koohi-Fayegh, S.; Rosen, M.A. A review of energy storage types, applications and recent developments. *J. Energy Storage* **2020**, *27*, 101047. [CrossRef]
9. Chen, L.; Zheng, T.; Mei, S.; Xue, X.; Liu, B.; Lu, Q. Review and prospect of compressed air energy storage system. *Clean Energy* **2016**, *4*, 529–541. [CrossRef]
10. Olabi, A.; Wilberforce, T.; Ramadan, M.; Abdelkareem, M.A.; Alami, A.H. Compressed air energy storage systems: Components and operating parameters—A review. *J. Energy Storage* **2021**, *34*, 102000. [CrossRef]
11. Wen, X.K.; Zhang, S.H.; Deng, T.T.; Li, P.; Chen, W. A Summary of Large Capacity Power Energy Storage Peak Regulation and Frequency Adjustment Performance. *Power Gener. Technol.* **2018**, *39*, 487–492.
12. Ibrahim, H.; Ilinca, A.; Perron, J. Energy storage systems—Characteristics and comparisons. *Renew. Sustain. Energy Rev.* **2008**, *12*, 1221–1250. [CrossRef]
13. Venkataramani, G.; Parankusam, P.; Ramalingam, V.; Wang, J. A review on compressed air energy storage—A pathway for smart grid and polygeneration. *Renew. Sustain. Energy Rev* **2016**, *62*, 895–907. [CrossRef]
14. Han, Z.H.; An, P.; Guo, S.C.; Jiang, K.Z. Thermodynamics Performances Analysis of Advanced Adiabatic Compressed Air Energy Storage System Based on Solar Auxiliary Heating. *Acta Energ. Sol. Sin.* **2020**, *41*, 243–250.
15. Tong, Z.; Cheng, Z.; Tong, S. A review on the development of compressed air energy storage in China: Technical and economic challenges to commercialization. *Renew. Sustain. Energy Rev.* **2021**, *135*, 110–178. [CrossRef]
16. Budt, M.; Wolf, D.; Span, R.; Yan, J. A review on compressed air energy storage: Basic principles, past milestones and recent developments. *Appl. Energy* **2016**, *170*, 250–268. [CrossRef]
17. Raju, M.; Khaitan, S.K. Modeling and simulation of compressed air storage in caverns: A case study of the Huntorf plant. *Appl. Energy* **2012**, *89*, 474–481. [CrossRef]
18. Liu, J.L.; Wang, J.H. A comparative research of two adiabatic compressed air energy storage systems. *Energy Convers. Manag.* **2016**, *108*, 566–578. [CrossRef]
19. Guo, H.; Xu, Y.; Zhang, X.; Guo, C.; Chen, H. *Off-design Performance of Compressed Air Energy Storage System with Thermal Storage, Proceedings of the CSEE 2019, Fredericton, Canadian, 18–21 August 2019*; Elsevier: Amsterdam, The Netherlands, 2019; Volume 39, pp. 1366–1377.

20. Du, P.; Cui, H.J. Compressed air energy storage commercial power station connected to the grid in Feicheng, Shandong Province. *State Grid News* **2021**, *08–09*, 3.
21. Li, G.K.; Wang, G.H.; Xue, X.D.; Chen, L.J.; Mei, S.W. Design and Analysis of Condenser Mode for Jintan Salt Cavern Compressed Air Energy Storage Plant of China. *Autom. Electr. Power Syst.* **2021**, *45*, 91–99.
22. Zhang, X.J. Investigation on Compressed Air Energy Storage System. Ph.D. Thesis, Institute of Engineering Thermophysics of Chinese Academy of Sciences, Beijing, China, 2011.
23. Han, Z.H.; Guo, S.C. Analysis of Operation Characteristic on Discharge Process of AA-CAES System. *Acta Energ. Sol. Sin.* **2020**, *41*, 295–301.
24. Kamyar, R.; Ehsan, S.; Roydon, A.F. A comprehensive data-driven study of electrical power grid and its implications for the design, performance, and operational requirements of adiabatic compressed air energy storage systems. *Appl. Energy* **2020**, *257*, 113990.
25. Wen, X.K.; Wu, P.; Zhang, S.H.; Wu, M. Study on Primary Frequency Modulation Parameter Setting of Compressed Air Energy Storage. In Proceedings of the 2nd International Conference on Green Energy and Applications, Singapore, 24–26 March 2018; pp. 143–146.
26. Yang, C.; Li, P.; Sun, L.; Chai, J. Improved Strategy for Load Shedding Anti-overspeed of Compressed Air Energy Storage Expansion Generators. *Acta Energ. Sol. Sin.* **2021**, *42*, 449–454.
27. Razmi, A.; Janbaz, M. Exergoeconomic assessment with reliability consideration of a green cogeneration system based on compressed air energy storage (CAES). *Energy Convers. Manag.* **2020**, *204*, 112320. [CrossRef]
28. Li, Y.W.; Miao, S.H.; Yin, B.X.; Zhang, S.X.; Zhang, S.Y. Optimal Dispatch Model for Micro Integrated Energy System Considering Multi-carrier Energy Generation Characteristic of Advanced Adiabatic Compressed Air Energy Storage. *Power Gener. Technol.* **2020**, *41*, 40–49. [CrossRef]
29. Bafrani, H.A.; Sedighizadeh, M.; Dowlatshahi, M.; Ershadi, M.H.; Rezaei, M.M. Spinning reserve stochastic model of compressed air energy storage in day-ahead joint energy and reserve market using information gap decision theory method. *Int. J. Electr. Power Energy Syst.* **2022**, *141*, 108123. [CrossRef]
30. Deng, M. Research on Operation Control of Wind-Storage Combined System Based on Compressed Air Energy Storage. Master's Thesis, North China Electric Power University, Baoding, China, 2019.
31. Razmi, A.; Soltani, M.; Aghanajafi, C.; Torabi, M. Thermodynamic and economic investigation of a novel integration of the absorption-recompression refrigeration system with compressed air energy storage (CAES). *Energy Convers. Manag.* **2019**, *187*, 262–273. [CrossRef]
32. Alirahmi, S.M.; Razmi, A.R.; Arabkoohsar, A. Comprehensive assessment and multi-objective optimization of a green concept based on a combination of hydrogen and compressed air energy storage (CAES) systems. *Renew. Sustain. Energy Rev.* **2021**, *142*, 110850. [CrossRef]
33. Li, G.K.; Chen, L.J.; Xie, Y.G.; Gao, B.; Zheng, T.W.; Mei, S.W. Coordinated Optimization Strategies of Wind-storage Hybrid System Considering Off-design Characteristics of Compressed Air Energy Storage. *High Volt. Eng.* **2020**, *46*, 511–518.
34. Li, H.Q.; He, J.Q.; Wang, J.; Li, Z.H. Optimization strategy of grid-connected power generation integrating wind energy with low-temperature adiabatic compressed air energy storage. *Renew. Energy Resour.* **2020**, *38*, 1679–1685.
35. APROS Software. Available online: http://www.apros.com.cn/ (accessed on 1 March 2022).
36. Li, P.; Yang, C.; Sun, L.; Xiang, J.; Wen, X.; Zhong, J.; Deng, T. Dynamic characteristics and operation strategy of the discharge process in compressed air energy storage systems for applications in power systems. *Int. J. Energy Res.* **2020**, *44*, 6363–6382. [CrossRef]
37. Li, P. Dynamic Characteristics and Operation Strategy of Discharge Process in Compressed Air Energy Storage System. Master's Thesis, Chongqing University, Chongqing, China, 2020.
38. He, Q.; Fu, H.L.; Kang, H.Q. Thermodynamic analysis of novel advanced adiabatic compressed air energy storage system with variable pressure ratio. *Therm. Power Gener.* **2020**, *49*, 36–42.
39. Razmi, A.R.; Afshar, H.H.; Pourahmadiyan, A.; Torabi, M. Investigation of a combined heat and power (CHP) system based on biomass and compressed air energy storage (CAES). *Sustain. Energy Technol. Assess.* **2021**, *46*, 101253. [CrossRef]
40. Li, X.M. Study on Influence of Components Characteristics for Advanced Adiabatic Compressed Air Energy Storage System. Ph.D. Thesis, University of Chinese Academy of Sciences, Beijing, China, 2015.
41. Xu, H.P. Dynamic Modeling of Compressed Air Energy Storage System in Distributed Microgrid. Master's Thesis, Guizhou University, Guiyang, China, 2018.
42. Wen, X.K.; Zhong, J.L.; Deng, T.T.; Li, P.; Chen, W. The Air Intake Regulation System of Compressed Air Energy System Based on Jet Device. In Proceedings of the 2019 4th International Conference on Electromechanical Control Technology and Transportation (ICECTT), Guilin, China, 26–28 April 2019; pp. 274–277.
43. Jiang, R.H.; Qin, F.G.F.; Chen, B.M.; Yang, X.P.; Yin, H.B.; Xu, Y.J. Thermodynamic performance analysis, assessment and com-parison of an advanced trigenerative compressed air energy storage system under different operation strategies. *Energy* **2019**, *186*, 115862. [CrossRef]

energies

MDPI

Article

Surrogate Model-Based Heat Sink Design for Energy Storage Converters

Gege Qiao [1], Wenping Cao [1,*], Yawei Hu [1], Jiucheng Li [2], Lu Sun [2] and Cungang Hu [1]

[1] School of Electrical Engineering and Automation, Anhui University, Hefei 230031, China
[2] Asunx Semiconductors Co., Ltd., Hefei 230093, China
* Correspondence: wpcao@ahu.edu.cn

Abstract: As forced-air cooling for heat sinks is widely used in the cooling design of electrical and electronic equipment, their thermal performance is of critical importance for maintaining excellent cooling capacity while reducing the size and weight of the heat sink and the equipment as a whole. This paper presents a method based on the combination of computational fluid dynamics (CFD) simulation and surrogate models to optimize heat sinks for high-end energy storage converters. The design takes the thermal resistance and mass of the heat sink as the optimization goals and looks for the best design for the fin height, thickness and spacing, as well as the base thickness. The analytical and numerical results show that the thermal resistance and mass of the heat sink are reduced by the proposed algorithms, as are the temperatures of the heating elements. Test results verify the effectiveness of the optimization method combining CFD simulation with surrogate models.

Keywords: computational fluid dynamics (CFD); energy storage; surrogate model; design optimization; heat sinks; power converters

Citation: Qiao, G.; Cao, W.; Hu, Y.; Li, J.; Sun, L.; Hu, C. Surrogate Model-Based Heat Sink Design for Energy Storage Converters. *Energies* **2023**, *16*, 1075. https://doi.org/10.3390/en16031075

Academic Editor: Alon Kuperman

Received: 31 October 2022
Revised: 30 November 2022
Accepted: 8 December 2022
Published: 18 January 2023

1. Introduction

In the face of global warming and the rapid depletion of fossil-fuel resources, the utilization of renewable energy has become a general trend across the world. With the large-scale and high penetration of renewable energy in power grids, the stable supply of electricity and energy-storage technologies have become technical concerns. Power Conversion Systems (PCS) can improve the utilization quality of clean energy and stabilize the load fluctuations of the power grid. In some cases, PCS can be used as an energy source where power grids are not available (i.e., power islands). Because of these advantages, PCS are gaining in popularity in industrial, as well as domestic, applications. In PCS power electronics converters, switching devices such as metal-oxide-semiconductor field-effect transistors (MOSFETs) and insulated gate bipolar transistors (IGBTs) are major heat sources and the heat is removed typically by heat sinks, which can be of natural cooling or forced-air cooling. The thermal performance of components such as IGBTs is directly related to the reliability of the whole converter. Recognizing that the main cause of electronic equipment failure is overheating [1], heat sink design for converters is a key focus of this work.

In the literature, quantitative analysis of the key parameters is considered useful. Typically, the analysis is based on calculating the thermal resistance of the heat sink, which is related to the thickness of the substrate, and the number and thickness of the heat sink fins. However, this does not give an optimization method for the structural design of the heat sink [2]. In some cases, the geometric parameters that affect the thermal resistance of the heat sink are optimized by using a genetic algorithm method [3]. Ref. [4] studies the influence of a single parameter of the heat sink on the performance and then develops the orthogonal method to assess the influence of different parameters on the performance of the heat sink. However, the design cycle is tedious, and the experimental cost is relatively high. Generally, when studying the heat dissipation performance of a heat sink, it is necessary to consider not only the thermal resistance but also the volume and the cost of the heat sink. There are several parameters that affect the heat dissipation performance of the heat sink, such as the thickness of the base, and the length, thickness and number of fins. The optimization in this paper considers the influence of multiple parameters for multiple optimization targets. In this paper, the heat sink and forced air cooling are used in an IGBT-based energy storage converter where IGBTs are considered the main heat sources for cooling design. CFD simulation and surrogate model algorithms are combined to realize the double objective optimization (the thermal resistance and mass of the heat sink). The optimal parameter design can be obtained by bringing the optimized results into the CFD model to verify the correctness of the designs.

2. Mathematical Model

This work carries out a thermal analysis of a high-end PCS converter which involves an extensive understanding of heat transfer within a heat sink.

2.1. Design Variables

In this work, there are four independent variables: the thickness of the base, and fin height, thickness and spacing. The structural diagram of the optimized forced-air-cooled heat sink is shown in Figure 1.

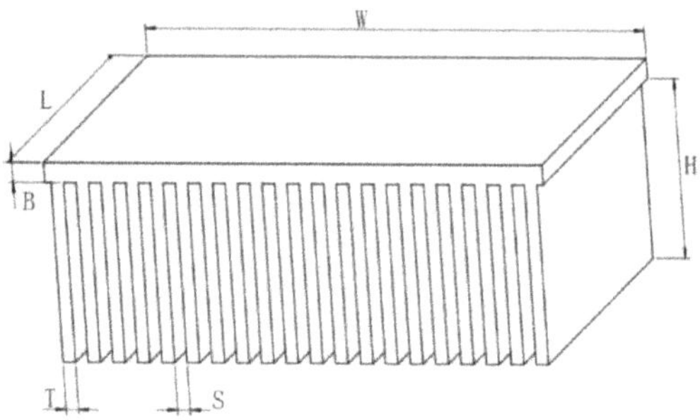

Figure 1. Structural diagram of the forced-air-cooled heat sink.

Where B is the thickness of the base, and H, T and X are the height, thickness and spacing of the fins, respectively, all of which taken together are defined as x_1, x_2, x_3, x_4, W is the width of the heat sink, L is the length of the heat sink and H is the height of the heat sink. Based on these variables, the thermal resistance and mass of the heat sink can be obtained and used as the objective functions of the optimization design.

2.1.1. Boundary Conditions

The optimization variables of the heat sink are associated with IGBT size, fan size, model and airflow rate. For effective heat transfer, the height of the fins shall not be greater than that of the fan, otherwise, airflow misses part of the fins and reduces the heat dissipation performance. Similarly, the spacing between fins shall be greater than 1 mm. The thickness of the heat sink fin shall be greater than 1 mm for easy manufacture. In summary, the constraints of design optimization are given by:

$$\begin{cases} 1\text{mm} \leq x_1 \leq 20\text{mm} \\ 50\text{mm} \leq x_2 \leq 100\text{mm} \\ 1\text{mm} \leq x_3 \leq 4\text{mm} \\ 1\text{mm} \leq x_4 \leq 8\text{mm} \end{cases} \quad (1)$$

2.1.2. Objective Functions

Considering the heat dissipation in the heat sink, the optimization objectives of the heat sink are low thermal resistance, low mass, high heat transfer efficiency and overall cost. In this paper, the thermal resistance and mass of the heat sink are selected as the design targets, and the entropy weight method [5] and linear weighting method [6] are combined to process the numerical data. This processing method can not only solve the problem of different variable units but also eliminate the influence of different variable values on the optimization results, making the data dimensionless for comparison [7]. The specific process of data processing has five steps:

- Standardize the data;

$$T_{ij} = \frac{\max(x_{1j}, \ldots, x_{nj}) - x_{ij}}{\max(x_{1j}, \ldots, x_{nj}) - \min(x_{1j}, \ldots, x_{nj})} + 0.0001 \tag{2}$$

- Calculate the entropy value of the jth index, which ranges between 0 and 1;

$$e_j = -\left(\frac{1}{\ln n}\right)\sum_{i=1}^{n} \frac{T_{ij}}{\sum\limits_{i=1}^{n} T_{ij}} \ln \frac{T_{ij}}{\sum\limits_{i=1}^{n} T_{ij}} \tag{3}$$

- Calculate the weighting of each data set (I = 1,2 . . . ,n; j = 1,2 . . . ,m);

$$w_j = \frac{1 - e_j}{\sum\limits_{j=1}^{m} (1 - e_j)} \tag{4}$$

- Establish the priority function of the targets, based on the linear weighting method;

$$P_i = \sum_{j=1}^{m} w_j x_{ij} \tag{5}$$

- Obtain the multi-objective optimization function of the heat sink;

$$\min F(x) = \min[w_1 R(x) + w_2 M(x)] \tag{6}$$

From the above calculations, the weightings of the thermal resistance and the mass are:

$$w_1 = 0.588,$$
$$w_2 = 0.412.$$

2.2. Thermal Resistance Model

The thermal resistance of the heat sink consists of four parts [8], the thermal resistance between the junction and the device case, the thermal resistance between the shell and the environment, the thermal resistance between the shell and the heat sink, and the thermal resistance between the heat sink and the environment [9]. Figure 2 shows the equivalent thermal resistance network diagram of the heat sink.

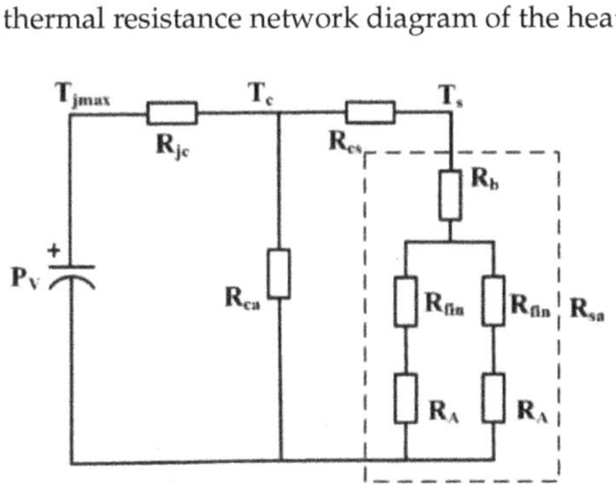

Figure 2. Thermal resistance model.

In Figure 2, P_V is the total heat loss of the IGBT, R_{jc} is the thermal resistance between the junction and shell, R_{ca} is the thermal resistance between the shell and the environment, R_{cs} is the thermal resistance between the shell and heat sink and R_{sa} and is the thermal resistance between the heat sink and the environment. R_b is the heat conduction resistance of the heat sink base, R_{fin} is the heat conduction resistance of the fins and R_A is the heat convection resistance of the fins. Similar to electrical resistances in electrical circuits, the thermal resistance can be calculated by adding up the component resistances according to their series or parallel connections. The heat transfer in the forced-air-cooled heat sink follows Ohm's law for thermal circuits. Compared with R_{cs} and R_{sa}, R_{ca} is very large and can be ignored in parallel connection; then:

$$\Delta T = P\left(R_{jc} + R_{cs} + R_{sa}\right) \tag{7}$$

According to Equation (7), the temperature rise of the device is determined by the loss and thermal resistance of the power device. The losses of power devices and the thermal resistance of the junction–shell are mainly determined by the production process, packaging materials and performance index. These two parameters are constant values [10]. The thermal resistance between the shell and heat sink is very small and can be ignored. The thermal resistance between the heat sink and the environment can be reduced by optimizing the geometric structure, materials and external environment of the heat sink. Compared to other parameters, the thermal resistance between the heat sink and the environment is also relatively simple and convenient to optimize. For this reason, the thermal resistance between the heat sink and the environment is the objective that majorly corresponds to the total thermal resistance.

2.3. Optimization Model

The thermal resistance of the heat sink in this design is mainly composed of the heat convection thermal resistance and the heat conduction resistance of the heat sink base and fins, which can be calculated by Equations (8)–(10). The total thermal resistance of the heat sink can be calculated by Equation (11).

$$R_b = \frac{B}{k_s WL} \tag{8}$$

$$R_{fin} = \frac{H}{k_s TL} \tag{9}$$

$$R_A = \frac{1 - 0.152(vL)^{-\frac{1}{10}}}{5.12(vL)^{\frac{4}{5}}(H+S)\cdot 2N} \tag{10}$$

$$R_{sa} = R_b + \frac{R_{fin}}{2N} + R_A \tag{11}$$

where N is the number of air ducts, ks is the thermal conductivity of the heat sink material and v is the airflow velocity.

In this design, the size of the heat sink is 230 × 420 mm and the height is variable. According to Equations (8)–(11), the optimization model of the heat sink resistance is:

$$\min F_1 = \frac{x_1}{k_s * 420 * 230} + \frac{x_2}{k_s * x_3 * 230}8 + \frac{1 - 0.152(230v)^{-\frac{1}{10}}}{5.12(230v)^{\frac{4}{5}}(x_2 + x_4) * 2 * \{\left[\frac{420 - x_3}{x_3 + x_4}\right] + 1\}} \tag{12}$$

From Figure 2, the heat sink mass optimization model is:

$$\min F_2 = 420 * 230 * x_1\rho_m + 230 * x_2x_3x_5\rho_m \tag{13}$$

where ρ_m is the density of the heat sink material, which is aluminum in this case.

3. The Proposed Optimization Methods

In the literature, surrogate model algorithms are adopted, which do not rely on actual models, thus reducing computational time and cost. The surrogate model algorithms use a Kriging algorithm to establish a simplified model by using basic parameters such as independent variables, objective functions and limiting conditions [11]. The success of the surrogate model depends on its sampling strategy and the number and location of sampling points [12]. The generated sample points should accurately reflect the distribution characteristics of sampling points [13,14]. A Kriging algorithm is then applied to search for the optimization point.

3.1. Kriging Models

At present, the methods of establishing models are divided into two categories: one is a parametric model, and the other is a non-parametric model. The former is based on the assumption that the known parameters obey a given population distribution. The latter model does not assume any particular distribution of the population. In the case of a given sample, it is calculated according to non-parametric statistics [15].

In the optimization, a Kriging model based on a parametric model is selected, which is also an alternative model widely used at present. The Kriging model is given by [16–18]:

$$y(x) = \mu + z(x) \tag{14}$$

where $y(x)$ is the response function, μ is the constant, $z(x)$ is the random process, respectively. Its expectation, variance and covariance are:

$$\begin{cases} E[z(x)] = 0 \\ Var[z(x)] = \sigma^2 \\ cov[z(x^j), z(x^k)] = \sigma^2 R[R(x^j, x^k)] \end{cases} \tag{15}$$

where R is the symmetric correlation matrix on the diagonal, $R(x^i, x^j)$ and is the correlation parameter between sample points x^j, x^k. In this optimization, a Gaussian correlation function is selected, then

$$R(x^j, x^k) = \exp[-\sum_{i=1}^{n} \theta_i \left| x_i^j - x_i^k \right|^2] \tag{16}$$

where n is the number of groups of design variables. In this optimization, $n = 50$. $\left| x_i^j - x_i^k \right|$ is the distance between the kth components of sample points x^j and x^k, and θ_i is the unknown related parameter used to fit the model, which can be solved by Equation (17).

$$\max F(\theta_i) = -\frac{[n_s \ln(\widehat{\sigma}^2) + \ln|R|]}{2} \tag{17}$$

Then, the estimated value at sample point x is:

$$\widehat{y}(x) = \widehat{\mu} + r^T(x) R^{-1} (y - \widehat{\mu} m) \tag{18}$$

$$r^T(x) = [R(x, x^1), R(x, x^2), \ldots R(x, x^{n_s})]^T \tag{19}$$

where y is the column vector with length n_s and m is the unit vector.

Figure 3 shows six Kriging models constructed with different variables and target parameters. It can be seen that the Kriging models differ from their 2D parameters and objectives. Considering different two-dimensional parameters and the minimum value of the objective function, a set of optimization values can be generated. This design needs to consider that the target value is the lowest in the case of four-dimensional parameters, which requires a search algorithm to find the target minimum value [19].

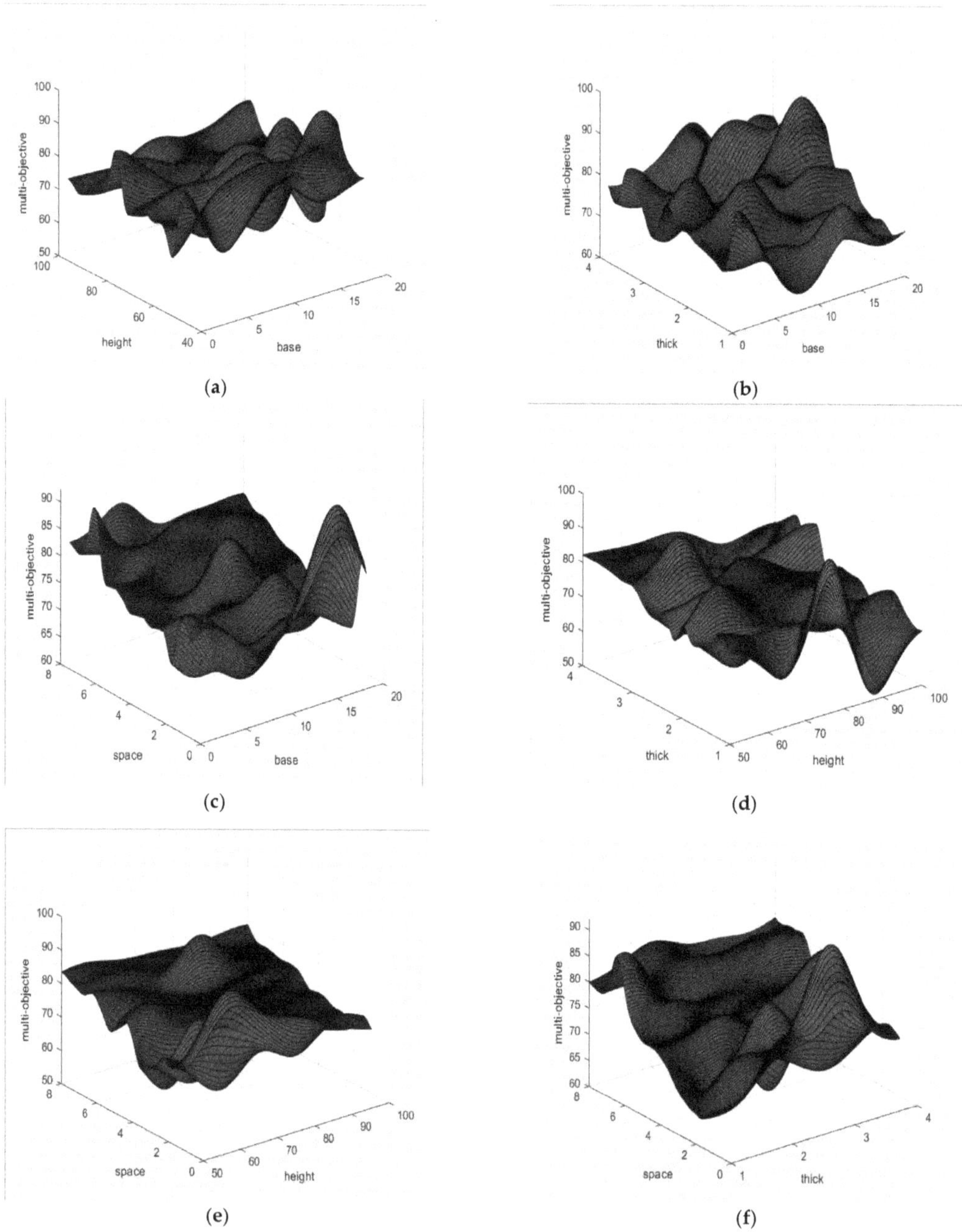

Figure 3. Kriging model constructed with different variables and target parameters. (**a**) Kriging model constructed with B, H and multi-objective. (**b**) Kriging model constructed with B, T and multi-objective. (**c**) Kriging model constructed with B, S and multi-objective. (**d**) Kriging model constructed with H, T and multi-objective. (**e**) Kriging model constructed with H, S and multi-objective. (**f**) Kriging model constructed with T, S and multi-objective.

3.2. Particle Swarm Optimization (PSO) Algorithm

In the literature, particle swarm optimization (PSO) algorithms are favored due to their advantages of easy programming, high efficiency and fast convergence [20], the optimization in this paper selects PSO as the search algorithm. Each particle in the particle swarm itself can find the optimal solution of its current position and record it as an individual extreme value. Particles can share an individual extremum with other particles, and the optimal individual extremum is regarded as the global optimal solution of the whole particle swarm [21]. Then, the particle adjusts its speed and position according to its individual extreme value and the current global optimal solution, and constantly approaches the optimal solution [22]. The PSO algorithm is shown in Equation (20).

$$\begin{cases} V_i^{k+1} = wV_i^k + C_1 r_1 (P_i^k - X_i^k) + C_2 r_2 (P_g^k - X_i^k) \\ X_i^{k+1} = X_i^k + V_i^{k+1} \end{cases} \tag{20}$$

where V_i^k and X_i^k are the velocity and position of the ith particle at the kth iteration, P_i^k represents the individual extreme value of the ith particle at the kth iteration, P_g^k represents the global optimal solution of the ith particle at the kth iteration, C_1 and C_2 are individual learning factors and global learning factors of particles, r_1 and r_2 are random constants between $[0, 1]$ and w is the inertia weight, respectively. In this design, the value of w decreases linearly according to the number of iterations k. The equation is given by [23]:

$$w(k) = w_{max} - \frac{w_{max} - w_{min}}{k_{max}} \times k \tag{21}$$

where w_{max}, w_{min} are the upper and lower limits of the inertia weight and k_{max} is the maximum number of iterations, respectively.

3.3. Process of Surrogate Algorithms

The whole optimization process is shown in Figure 4. First, design the variables and determine the range of independent variables, and then take Latin hypercube sampling [24]. After obtaining a limited number of sample points, choose whether to supplement points in some areas or redesign the design according to the sampling conditions of the sample points [25]. The sampling condition of this design is good, so the supplement of sample points is not considered. The second step is the selection of target parameters. This can choose single objective optimization or multi-objective optimization according to the design requirements. In general, the objectives of multi-objective optimization are in conflict with each other. The improvement of one sub-objective may cause the reduction of another sub-objective, and the solution result cannot make all sub-objectives optimal. This requires that, when carrying out multi-objective optimization, the weight proportion between different objectives should be considered [26]. In the multi-objective optimization problem, the solution is not unique, which requires specific analysis of specific problems in the optimization design to find solutions that meet more sub-objectives as much as possible. In this process, the optimal design should meet the requirements of independent variables, constraints and target parameters at the same time. The third step is to establish a finite element model and use the finite element model to obtain relevant parameters. Then, the Kriging model is established using 80% of the sampled data for an approximate model, which is used to replace the established finite element model. The remaining 20% is used for error detection. If the error between the data obtained from the approximate model and the simulation is within the 20% threshold, the surrogate model is considered successful. After error detection, the PSO algorithm is used to obtain the optimal solution, which is finally brought into the finite element model to verify the results, to complete the optimization design [27].

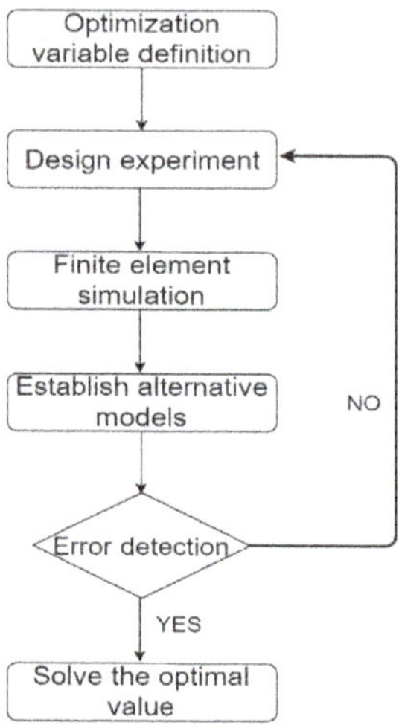

Figure 4. Flowchart of the process of surrogate algorithms.

4. Simulation Results

The structure of the PCS converter is complex for 3D thermal modeling and analysis. In this work, a simplified model is developed for design optimization, and the 3D finite element model only includes key components, as shown in Figure 5.

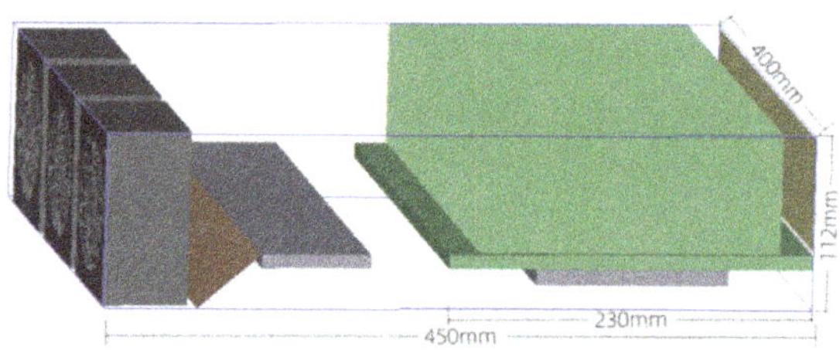

Figure 5. 3D finite element model.

In this converter, four heating element IGBTs are installed on the base of the heat sink, and the three fans blow air to the fins of the heat sink through the air deflector. In the model, the size of the cabinet is 450 mm × 400 mm × 114 mm, the radius of the fan is 44 mm, the size of the IGBT is 122 mm × 62 mm × 14 mm, and the spacing between IGBTs is 38 mm. The length and width of the heat sink base are 420 mm and 230 mm, the height of the base is 12 mm, the height and thickness of the fins are 64 mm and 3.4 mm, and the spacing between heat sinks is 7 mm, respectively. The power loss of each IGBT is 800 W, and the total power loss is 3200 W. The ambient temperature is 50 °C. The simulation results are plotted in Figures 6–9. It can be seen that the maximum temperature of the heating element can reach 133.43 °C. The calculated results are brought into the finite element model for verification. The finite element simulation and optimization results are shown in Figures 6–9, respectively.

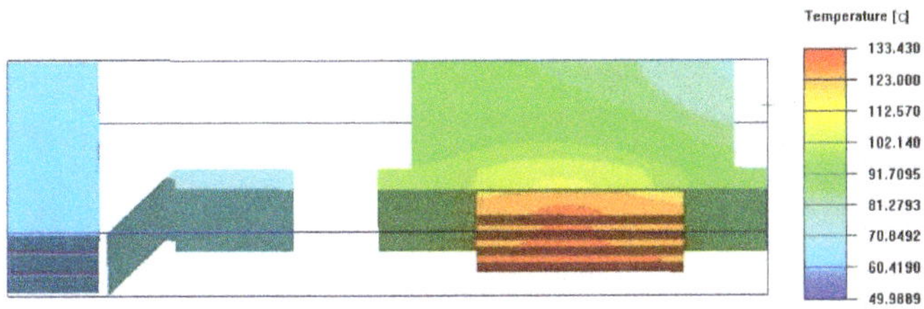

Figure 6. 3D Finite element simulation results before optimization.

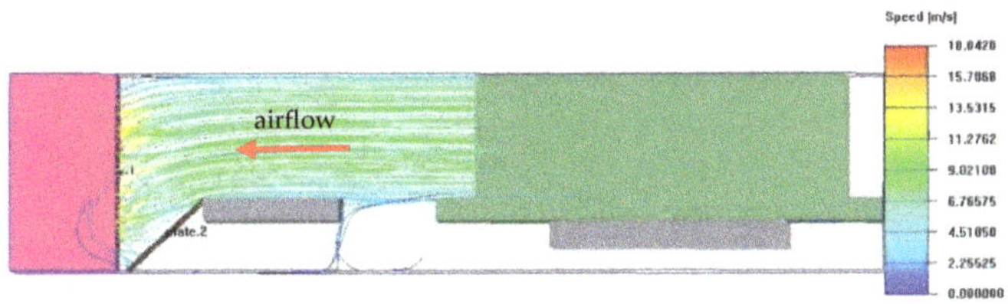

Figure 7. Airflow speed vector results before optimization.

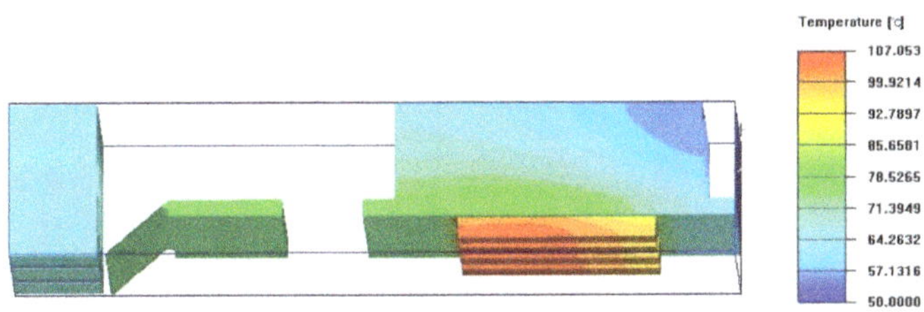

Figure 8. 3D Finite element simulation results after optimization.

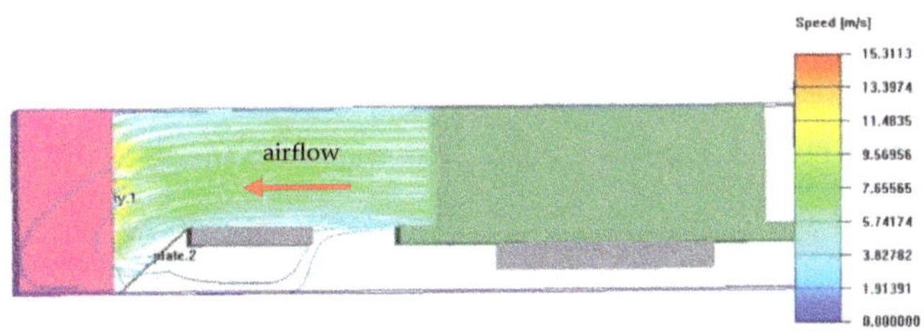

Figure 9. Airflow speed vector results after optimization.

It can be seen that the maximum airflow velocity after optimization is significantly reduced. The changes in the heat sink ducts and the relative positions of the heat sink and fans can vary the airflow path and contact surface of the heatsink, which is confirmed by Figures 6 and 8.

In contrast, a traditional method with response surface optimization in ANSYS adopts an optimization algorithm, as illustrated in Table 1. The proposed surrogate optimization results with the surrogate model are better. After the optimization design, the thermal resistance and mass are reduced from 0.025 °C/W and 7.562 kg to 0.017 °C/W and 7.312 kg, respectively. From the temperature cloud diagrams in the 3D finite element simulation, it can also be seen that the hot-spot temperature is reduced from the 133.430 °C value to 107.053 °C, proving the effectiveness of the proposed surrogate optimization algorithms.

Table 1. Optimum design of the heat sink.

	B	H	T	S	Thermal	Mass	Temperature
Base design	12.0 mm	64.0 mm	3.4 mm	7.0 mm	0.0253 °C/W	7.562 kg	133.430 °C
Traditional design	8.0 mm	80.0 mm	2.0 mm	4.4 mm	0.0194 °C/W	7.423 kg	114.733 °C
Optimization design	10.8 mm	65.6 mm	1.1 mm	2.3 mm	0.0171 °C/W	7.312 kg	107.053 °C
Reduction (%)	10.0	−2.5	67.6	67.1	32.4	3.3	19.8

5. Gray Correlation Analysis

Gray correlation analysis can evaluate the importance of input parameters, calculate the impact of different parameters on multiple objectives, and verify the selection of parameters in the design [28]. After all data are made dimensionless [29], the correlation coefficient is calculated by Equation (22). Then the average value is calculated to obtain the gray correlation degree of each parameter with the target [30].

$$\begin{cases} r_i^k = \frac{\Delta_{min}+\lambda\Delta_{max}}{\Delta_i^k+\lambda\Delta_{max}} \\ \Delta_i^k = \left|y_0^k - y_i^k\right| \end{cases} \quad (22)$$

where r_i^k represents the correlation coefficient of the kth component of the ith parameter, y_0^k and y_i^k represents the results of dimensionless processing of the reference sequence and the comparison sequence, Δ_i^k represents the deviation sequence, Δ_{min} and Δ_{max} represents the minimum and maximum deviation, and λ is the resolution coefficient, respectively. In this design, the resolution coefficient is taken as 0.5. The gray correlation degrees of the four variables relative to thermal resistance, mass and weighted multi-objective are shown in Table 2.

Table 2. Summary of Gray correlation analysis.

Dimensionless	Base	Height	Thick	Space
Temperature	0.637699	0.590268	0.668855	0.697728
Mass	0.622044	0.673455	0.698375	0.605095
Multi-objective	0.639337	0.586499	0.672616	0.726296

Within the given range of variables, the four design variables have a high degree of correlation with the target parameters, and the correlation degree can basically reach 0.6, which also proves the rationality of the design variables selected. Of these, for the temperature single target, the spacing of the heat sink fins has the greatest influence, and the correlation degree is as high as 0.7. The correlation degree of the thickness of the fins is the second, which is 0.67. Compared with other parameters, the correlation between the thickness of the fins and the temperature is the smallest, but it can also reach 0.59. It can also be seen that the airflow path is very crucial for the heat dissipation optimization design, while the influence of the surface area of the fins is less obvious. For the mass in the target parameters, the parameter with the greatest correlation is the thickness of the fins, and the correlation reaches 0.7. For the dual objectives of thermal resistance and mass, the most relevant parameter in this optimization is fin spacing, which is as high as 0.73, and the second important parameter is fin thickness, with a correlation degree of 0.67. By comparison, fin height has the smallest correlation degree with the double targets, with a correlation coefficient of 0.59. In this double objective optimization, fin spacing and thickness are the most significant factors affecting the performance of the heat sink.

6. Conclusions

This paper has presented an optimization method for heat sinks based on the CFD method and surrogate models. Firstly, the heat dissipation process is simulated by using the 3D finite element model. Next, using the sampling points selected by Latin hypercube sampling and the simulation results of CFD, a surrogate model is established to assess the correlation of four design variables (B, H, T, S) and the two design targets (thermal resistance and mass), according to Kriging models. Then a PSO algorithm is used to obtain the optimal solution. The test results show that the optimized thickness of the base and the height, thickness and spacing of the fins can effectively reduce the thermal resistance of the heat sink and the mass of the heat sink. In turn, the proposed method can reduce hotspots (such as an IGBT junction) and improve the thermal efficiency of the converters. The developed technology can provide guidance for heat sink designs and optimize the energy efficiency of industrial products.

Author Contributions: Conceptualization, J.L. and Y.H.; methodology, W.C.; software, G.Q.; validation, G.Q., Y.H. and J.L.; investigation, C.H.; resources, L.S.; writing—original draft preparation, G.Q.; writing—review and editing, W.C.; supervision, C.H.; project administration, L.S.; funding acquisition, W.C. All authors have read and agreed to the published version of the manuscript.

Funding: This research received no external funding.

Institutional Review Board Statement: Not applicable.

Informed Consent Statement: Not applicable.

Data Availability Statement: Not applicable.

Conflicts of Interest: The authors declare no conflict of interest.

References

1. Shen, L.; Jiang, J.; Fang, Y.; Yang, Z.; Hou, Y. Design and optimization of IGBT heat dissipation scheme based on air cooling. *Refrigeration* **2018**, *46*, 95–98. Available online: https://kns.cnki.net/kcms/detail/detail.aspx?dbcode=CJFD&dbname=CJFDLAST2018&filename=DWYC201808019&uniplatform=NZKPT&v=nUx0m-QPZhytRoXds8TOiuHzmCdP5uH1I_Uzb7RhTEH_qRCRtFADpE4g2Y3YYuAG (accessed on 2 October 2022).
2. Zhao, H.; Zhu, Y.; Yin, Z. Thermal design of high-power inverter. *Electr. Eng.* **2018**, *19*, 149–156. Available online: https://kns.cnki.net/kcms/detail/detail.aspx?dbcode=CJFD&dbname=CJFDLAST2018&filename=DQJS201808038&uniplatform=NZKPT&v=n_mEH02HHLKWCbqbWeO9nd-3v3886CNvSUdQOT5OKvevOzmFThtUhOHsl5EupxBP (accessed on 2 October 2022).
3. Zhou, Z.; Bing, L.; Jian, L.; Wan, Y.; Du, L.; Chao, W. Optimization design of heat sink fins based on genetic algorithm. *Electro-Mech. Eng.* **2019**, *35*, 29–33. Available online: https://kns.cnki.net/kcms/detail/detail.aspx?dbcode=CJFD&dbname=CJFDLAST2019&filename=DZJX201905008&uniplatform=NZKPT&v=nt8dGV7hmt5S8ncV0n3---hVbYTvtRD2CAAPp6_QE1vOV-oE9SnpMWDTccGxpbl8 (accessed on 2 October 2022).
4. Ding, C.; Zhang, T.; Luo, J.; Cui, Y.; Yu, L. Optimization design research on high power LED lamps of radiator. *Chin. J. Electron Device* **2016**, *39*, 750–754. Available online: https://kns.cnki.net/kcms/detail/detail.aspx?dbcode=CJFD&dbname=CJFDLAST2016&filename=DZQJ201603049&uniplatform=NZKPT&v=sT7RuFw0Jcb0cMM5ZwdbJG9TvAVwhHAs5PcbCqqzrv7fmyTxGBmNqIA3EtFPazCn (accessed on 2 October 2022).
5. Zhang, M.; Du, J.; Luo, J.; Nie, B.; Xiong, W.; Liu, M.; Zhao, S. Research on feature selection of multi-objective optimization. Available online: https://kns.cnki.net/kcms/detail/detail.aspx?dbcode=CAPJ&dbname=CAPJLAST&filename=JSGG20221020002&uniplatform=NZKPT&v=ZhiUyUsnOHCdm5VBxZlxys-l7vytq9hLD_NRIC7eFcufvCP13qxIm0EIE2mu6c1- (accessed on 2 October 2022).
6. Li, J.; Yang, Q.; Niu, P.; Xian, Z.; Zhang, J.; Liang, J.; Li, Y. Optimization design and analysis of high-power LED heat radiator based on genetic algorithm and matlab simulation. *Trans. China Electrotech. Soc.* **2013**, *28*, 213–220.
7. Zhao, J.; Zhang, M.; Zhu, Y.; Cheng, R.; Li, X.; Wang, L.; Hu, C. Thermofluid modeling for concurrent size-topology optimization of heat sinks for planar motors. *J. Tsinghua Univ. (Sci. Technol.)* **2022**, *62*, 400–407.
8. Wang, L.; Li, L.; Zhou, X.; Zhu, S.; Sun, C. Multi-objective optimization of forced air-cooled heatsink based on particle swarm optimization algorithm. *Electr. Eng.* **2022**, *23*, 20–25. Available online: https://kns.cnki.net/kcms/detail/detail.aspx?dbcode=CJFD&dbname=CJFDLAST2022&filename=DQJS202202004&uniplatform=NZKPT&v=8V4xD_7-9NRAGCJQNf7AZZNJ0KrFU2GHCrBqHXQ71FF7V1e4gGBdMwh10_NeK_w4 (accessed on 2 October 2022).
9. Zhou, Z.; Li, X.; Wang, S.; Yin, F.; Lu, Y.; Xin, L.; Wang, L.C. Optimal design of heat sink for CLLLC resonant converter. *J. Power Sources* **2021**, *45*, 382–385. Available online: https://kns.cnki.net/kcms/detail/detail.aspx?dbcode=CJFD&dbname=CJFDLAST2021&filename=DYJS202103030&uniplatform=NZKPT&v=AktyuqzNqLsb8YPYZs74uvICbCqugPQ35skddfKV4NqOAVoQp1KffrFdyq9QoCoM (accessed on 2 October 2022).
10. Lu, X.; Xie, L.; Hou, W.; Ling, Xie; Wen, Hou. Two-Stage Operational Amplifier Based on Multipath Zero Point Elimination. *Ind. Technol. Innov.* **2020**, *7*, 45–49. Available online: https://kns.cnki.net/kcms2/article/abstract?v=3uoqIhG8C44YLTlOAiTRKibYlV5Vjs7iy_Rpms2pqwbFRRUtoUImHW0PuUQKlilXd9Fxpb85H68pH_Sw3R_1vgsyRoGwC-cm&uniplatform=NZKPT (accessed on 2 October 2022).

11. Hu, G.; Chen, Z.; Zhang, D. Optimization design of diaphragm profile based on Kriging model. *J. Mech. Transm.* **2022**, *46*, 75–79. Available online: https://kns.cnki.net/kcms2/article/abstract?v=3uoqIhG8C44YLTlOAiTRKibYlV5Vjs7iJTKGjg9uTdeTsOI_ra5_Xceslx1ry8096OvZ5q1IbQBEzsn1ubniRA27XgbxfeFA&uniplatform=NZKPT (accessed on 2 October 2022).
12. Cai, D.; Shi, D.; Chen, J. Probabilistic load flow calculation method based on polynomial normal transformation and latin hypercube sampling. *Proc. CSEF* **2013**, *13*, 92–100. Available online: https://kns.cnki.net/kcms2/article/abstract?v=3uoqIhG8C44YLTlOAiTRKgchrJ08w1e7xAZywCwkEEK4n59o2U1FxvRsc3oI8uuQBUK6uJ_wVNHLPqnb-QWhlU-_a6KQ-7cA&uniplatform=NZKPT (accessed on 2 October 2022).
13. Shin, P.S.; Woo, S.H.; Zhang, Y.; Koh, C.S. An application of latin hypercube sampling strategy for cogging torque reduction of large-scale permanent magnet motor. *IEEE Trans. Magn.* **2008**, *44*, 4421–4424. Available online: https://ieeexplore.ieee.org/document/4717645 (accessed on 2 October 2022).
14. Zhang, Z.; Jiao, C.; Sang, T.; Zhang, Q. Latin Hypercube Sampling Simulation of Response Surface Model Parameters of vehicle body. *Comput. Simul.* **2021**, *38*, 123–127. Available online: https://kns.cnki.net/kcms2/article/abstract?v=3uoqIhG8C44YLTlOAiTRKibYlV5Vjs7iy_Rpms2pqwbFRRUtoUImHVM9wC0ZpO3S8Q-VDu0JUxw3LVKKDFLOhDjEy3_7mHLF&uniplatform=NZKPT (accessed on 2 October 2022).
15. Feng, X.; Hua, J.; Zhou, S.; Yin, Y.; Shuai, Z.; Wang, X. Optimal design for geological sedimentation simulation platform based on Kriging model. *China Sci.* **2022**, *17*, 1162–1166. Available online: https://kns.cnki.net/kcms2/article/abstract?v=3uoqIhG8C44YLTlOAiTRKibYlV5Vjs7iJTKGjg9uTdeTsOI_ra5_XQ2lBi0_44Jk9CGGM_nlUP6OELW1Fykmky4OWHFEjXG-&uniplatform=NZKPT (accessed on 2 October 2022).
16. Wang, H. Kriging parameter estimation algorithm based on combination optimization. *Comput. Mod.* **2021**, *9*, 51–56.
17. Wang, W.; Li, J.; Liu, G.; Wei, J.; Zhang, Z.; Cheng, M. Optimization design of drum brake stability based on kriging surrogate model. *J. Vib. Shock* **2021**, *40*, 134–138. Available online: https://kns.cnki.net/kcms2/article/abstract?v=3uoqIhG8C44YLTlOAiTRKibYlV5Vjs7iy_Rpms2pqwbFRRUtoUImHZQQ2VJjpVrs5jV9FUVCoDNKnd-ew0sODo7ZKVX-tdM1&uniplatform=NZKPT (accessed on 2 October 2022).
18. Le, C.; Ma, Y. Parallel surrogate-based Optimization algorithm based on kriging model using adaptive multi-phases strategy. *Comput. Integr. Manuf. Syst.* **2021**, *27*, 3227–3235. Available online: https://kns.cnki.net/kcms2/article/abstract?v=3uoqIhG8C44YLTlOAiTRKibYlV5Vjs7iy_Rpms2pqwbFRRUtoUImHSIxD9BZFa_VvmZUniUwZU95L3EmKuzmE-ddjJXTZJxe&uniplatform=NZKPT (accessed on 2 October 2022).
19. Wen, X.; Tian, F.; Xie, F. Calculation method of fuzzy reliability of structures. *Mod. Ind. Econ. Inf.* **2022**, *12*, 263–265. Available online: https://kns.cnki.net/kcms2/article/abstract?v=3uoqIhG8C44YLTlOAiTRKibYlV5Vjs7iJTKGjg9uTdeTsOI_ra5_Xa2GWI2tEmmEDoI4GOZ2hXLdMXakEw2EcU6soX7qBnfb&uniplatform=NZKPT (accessed on 2 October 2022).
20. Chen, Z.; Wang, Y.; Chen, L.; Li, X. Structural damage identification based on improved particle swarm optimization algorithm with learning factors. *J. North China Univ. Water Resour. Electr. Power (Nat. Sci. Ed.)* **2022**, *43*, 43–47. Available online: https://kns.cnki.net/kcms2/article/abstract?v=3uoqIhG8C44YLTlOAiTRKibYlV5Vjs7iJTKGjg9uTdeTsOI_ra5_XdQLk (accessed on 2 October 2022).
21. Wang, Y.; Guo, R.; Liu, R.; Cui, H. Optimum design of double layer shaped charge structure based on PSO and SVM. *J. Ordnance Equip. Eng.* **2022**, *43*, 244–249. Available online: https://kns.cnki.net/kcms2/article/abstract?v=3uoqIhG8C44YLTlOAiTRKibYlV5Vjs7iJTKGjg9uTdeTsOI_ra5_XVwcEBwpKqCwKy-c-_FsnnAtveDuQf9WiSOClozTILiF&uniplatform=NZKPT (accessed on 2 October 2022).
22. Yan, Q.; Dong, X.; Mu, J.; Ma, Y. Optimal configuration of energy storage in an active distribution network based on improved multi-objective particle swarm optimization. *Power Syst. Prot. Control* **2022**, *50*, 11–19. Available online: https://kns.cnki.net/kcms2/article/abstract?v=3uoqIhG8C44YLTlOAiTRKibYlV5Vjs7iJTKGjg9uTdeTsOI_ra5_XReFrJuhIC5K2PEyefb-Yt9GnpyagDFG6PBhIVNvujbV&uniplatform=NZKPT (accessed on 2 October 2022).
23. Guo, P.; Luo, X.; Wang, Y.; Bai, L.; Li, H. Identification of shaft centerline orbit for hydropower units based on particle swarm optimization and improved BP neural network. *Proc. CSEE* **2011**, *31*, 93–97. Available online: http://www.alljournals.cn/view_abstract.aspx?pcid=5B3AB970F71A803DEACDC0559115BFCF0A068CD97DD29835&cid=6CDB4E49EF88F71A&jid=828E9BBC945455BDDF9DA6953F5D0883&aid=CADBAE05CAAADCCAE2148AD3563EDAC8&yid=9377ED8094509821 (accessed on 2 October 2022).
24. Morrow, J.; Littler, T.; Song, X.; Tan, Z.; Cao, W.; Yang, N. Asymmetrical rotor design for a synchronous machine based on surrogate optimization algorithm. In Proceedings of the 8th IET International Conference on Power Electronics, Machines and Drives (PEMD 2016), Glasgow, UK, 19–21 April 2016.
25. Song, X.; Zhang, J.; Kang, S.; Ma, M.; Ji, B.; Cao, W.; Pickert, V. Surrogate-based analysis and optimization for the design of heat sinks with jet impingement. *J. North China Univ. Water Resour. Electr. Power (Nat. Sci. Ed.)* **2014**, *4*, 429–437. [CrossRef]
26. He, T.; Shao, M.; Bai, X.; Cao, Z.; Jie, Y.L.F.; Wen, X. Research on multi-objective optimization of production line based on genetic algorithm. *Manag. Informatiz.* **2022**, *11*, 177–182. Available online: https://kns.cnki.net/kcms2/article/abstract?v=3uoqIhG8C44YLTlOAiTRKibYlV5Vjs7iJTKGjg9uTdeTsOI_ra5_XZIsdaxgLxFIeQqdxmAxav3r4qvVJzHxqfNawSEaMrCa&uniplatform=NZKPT (accessed on 2 October 2022).
27. Tan, Z.; Song, X.; Cao, W.; Liu, Z.; Tong, Y. DFIG Machine Design for Maximizing Power Output Based on Surrogate Optimization Algorithm. *IEEE Trans. Energy Convers.* **2015**, *30*, 1154–1162. Available online: https://ieeexplore.ieee.org/document/7078905 (accessed on 2 October 2022). [CrossRef]

28. Marco, L.; Cristiano, N.; Antonio, B. Survey and sensitivity analysis of critical parameters in lithium-ion battery thermo-electrochemical modeling. *Electrochim. Acta* **2021**, *394*, 139098. Available online: https://www.sciencedirect.com/science/article/abs/pii/S0013468621013888 (accessed on 2 October 2022).
29. Li, H.; Huang, D.; Xiong, Z.; Li, X.; Zhang, Z. Structural optimization of extrusion die for TA1 internal thread tube based on grey correlation degree. *J. Plast. Eng.* **2022**, *29*, 56–63. Available online: https://kns.cnki.net/kcms2/article/abstract?v=3uoqIhG8C44YLTlOAiTRKibYlV5Vjs7iJTKGjg9uTdeTsOI_ra5_XXbrszYM6X6oN9lDKVGDfsL_i4pEEA3G06dDVyHjB95e&uniplatform=NZKPT (accessed on 2 October 2022).
30. Wang, P.; Wang, X.; Liu, Z.; Wang, H.; Fu, H. Optimization of process parameters of ultrasonic rolling extrusion based on grey correlation analysis method. *J. Plast. Eng.* **2022**, *29*, 36–43. Available online: https://kns.cnki.net/kcms2/article/abstract?v=3uoqIhG8C44YLTlOAiTRKibYlV5Vjs7iJTKGjg9uTdeTsOI_ra5_XXbrszYM6X6o1DqxaPSM2JBbjiTpJzWFNvFlbX5kneu3&uniplatform=NZKPT (accessed on 2 October 2022).

energies

MDPI

Article

Comparative Study of the Transmission Capacity of Grid-Forming Converters and Grid-Following Converters

Bojun Kong [1], Jian Zhu [1], Shengbo Wang [1], Xingmin Xu [1], Xiaokuan Jin [2,*], Junjie Yin [2] and Jianhua Wang [2]

[1] State Grid Yangzhou Power Supply Company, Yangzhou 210019, China; ygdkongbj@js.sgcc.com.cn (B.K.); ygzjian@js.sgcc.com.cn (J.Z.); gywsb@js.sgcc.com.cn (S.W.); ygxxm@js.sgcc.com.cn (X.X.)
[2] School of Electrical Engineering, Southeast University, Nanjing 210096, China; yinjunjie@seu.edu.cn (J.Y.); wangjianhua@seu.edu.cn (J.W.)
* Correspondence: jinxiaokuan@seu.edu.cn

Abstract: The development trend of high shares of renewables and power electronics has increased the demand for new energy converters in the power system, but there is a lack of systematic research on the stability of different types of converters when transmitting power, which is worth exploring in depth. In this study, the power transfer capabilities of grid-forming and grid-following converters are investigated separately through an equivalent circuit diagram and phasor diagram when connected to the grid, and a quantitative relationship between converters' power transmission limit and short circuit ratio under static stability conditions is obtained, leading to the conclusion that, in terms of power transmission, grid-forming converters are more suitable for weak grids with high damping and low inertia, whereas grid-following converters are more suitable for strong grids with high inertia. The conclusions are further verified by constructing the converter grid-connected models for different grid strengths through the PLECS simulation platform and the real-time simulation RTBOX1 and F28379D launchpad platform.

Keywords: grid-forming converter; grid-following converter; static power transmission limit; power coupling; short circuit ratio

Citation: Kong, B.; Zhu, J.; Wang, S.; Xu, X.; Jin, X.; Yin, J.; Wang, J. Comparative Study of the Transmission Capacity of Grid-Forming Converters and Grid-Following Converters. *Energies* **2023**, *16*, 2594. https://doi.org/10.3390/en16062594

Academic Editor: Lorand Szabo

Received: 3 February 2023
Revised: 2 March 2023
Accepted: 7 March 2023
Published: 9 March 2023

1. Introduction

The penetration rate of renewable energy and power electronic equipment has increased, forming a "double high" development trend [1]. The high transmission capacity requirement of converters also leads to problems in grid stability [2].

When a new energy converter is connected to the power grid as a power source, there may be an imbalance between its actual output power and its input power on the DC side when disturbance occurs, which will affect the capacitance voltage on the DC side of the converter, then change its output current through the control system; finally, its actual output power changes to adopt a new balance. If one part fails during this circulation, different kinds of power grid instability will occur. Among them, the static stability can initially be judged according to whether the converter has a stable static operating point [3,4]. However, power electronic converters have different response characteristics compared with traditional equipment, and there is still scant research on the stability of different types of converters when transmitting power and quantitative analyses of their power transmission limit, which are highly related to the power grid stability of power-angle, voltage, and frequency [5,6].

Among them, power-angle stability refers to the fact that the output power of the generator to the grid can still be maintained within a constant range after the grid is subjected to minimal interference (interference close to zero). Considering the synchronization consistency between converters and synchronous machines, the stability problem can be divided into static stability, small disturbance stability, and transient stability [7–9]. However, static stability is the basis for the study of dynamic small disturbance, transient stability, and other problems; the transient instability problem presents the characteristics of time-varying nonlinearity and needs to take into account all elements constituting the grid, which is relatively complex. Therefore, this paper mainly studies the influence of converter power transmission on the static stability of the grid.

Power electronic converters, as non-synchronous sources, can be divided into grid-forming (GFM) and grid-following (GFL) converters [10]. Among them, GFM converters can be seen as voltage sources, while GFL converters can be seen as current sources [11]. Power electronic converters should not only meet the power-angle stability like traditional converters when transmitting power, but also need to prevent PLL loss [12], delay [13], or other effects causing static or small interference instability, which need further research.

In this study, the equivalent circuit models of GFM and GFL converters are established, the power transmission limits of the two converters under static stability are obtained, and the conclusions and conditions that they are equal are determined. Then, the short circuit ratio (SCR) is used to measure the power grid strength of the system under static stability [14], utilizing the control strategy block diagram of both converters; we conclude that the GFM converter is more suitable for weak grids, and the GFL converter is more suitable for strong grids in terms of power transmission and smooth regulation. Finally, the PLECS platform is used to construct the grid-connected system suitable for different power grid strengths, and the validity of this theoretical analysis is verified using the simulation platform PLECS, RTBOX1, and the F28379 launchpad.

2. Static Stability Limit of the Converter's Power Transmission

The main focus of this section is the power transmission limit of the converter under static stability. The converter is connected to the grid as a power source; therefore, in addition to providing active power, a large amount of reactive power and harmonics may be output to the grid, complicating analyses of the stability. Thus, it seems reasonable to simplify this by considering static power transmission limits when converters operate in unity power factor first [15]. Equivalent circuits and phasor diagrams of converters are constructed to facilitate further analyses of power transmission limits and the static stability.

2.1. Equivalent Circuit of the GFM Converter

In Figure 1a, U_c is the RMS voltage value of the converter, and U_g is the RMS value of the grid side voltage. Assuming that the voltage at the point of common coupling (PCC) is the reference phase, the angle between the grid side voltage and the voltage at the PCC is the power-angle δ, and the line impedance $R_g + jX_g = Z_g\angle\varphi$. To facilitate control and improve stability, all the reactive power is provided by reactive power compensators. The power factor of the converter is one, expressed as $I_t\angle 0°$.

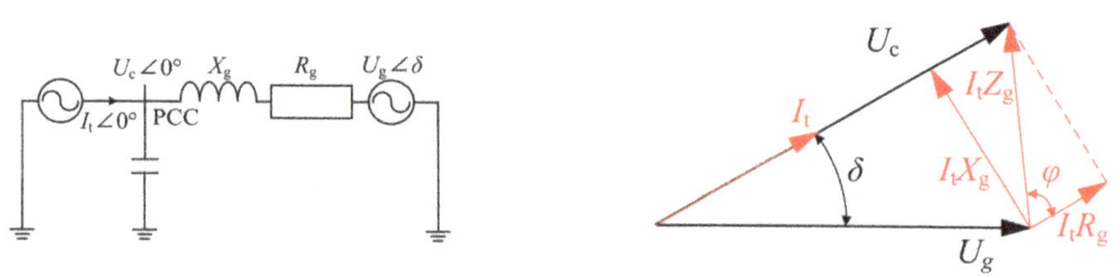

Figure 1. (**a**) Equivalent circuit diagram of the GFM converter; (**b**) phasor diagram of the GFM converter with unity power factor.

The phase diagram shown in Figure 1b was drawn by combining the relationships of the variables in Figure 1a, which, in turn, gives U_c and the active power, P, transmitted by the system in three symmetrical phases:

$$U_c = \sqrt{U_g^2 - I_t^2 X_g^2} + I_t R_g \tag{1}$$

$$P = 3U_c I_t = 3\left(\sqrt{U_g^2 - I_t^2 X_g^2} + I_t R_g\right) \cdot I_t \tag{2}$$

As shown in Figure 2a, the P–I_t curve can be plotted for different line impedance ratios after the transmission of active power P and current I_t has been normalized. When I_t increases, P exhibits an overall trend of increasing and then decreasing; thus, there is an extreme point which is the maximum value, P_{max}, referring to the maximum power that the converter can transmit in the unity power factor state. As the line impedance angle, φ, increases, the resistive component of the line impedance decreases, and the maximum power, P_{max}, that can be transmitted by a GFM converter will also decrease. As shown in Figure 2a, when the converter is commanded to transmit power P_{set}, for R_g/X_g = 1 and R_g/X_g = 0.5 grids, there is a stable static operating point, i.e., the left intersection of curves; for the R_g/X_g = 0 grid, there is no intersection, so static instability and other severe phenomena will occur.

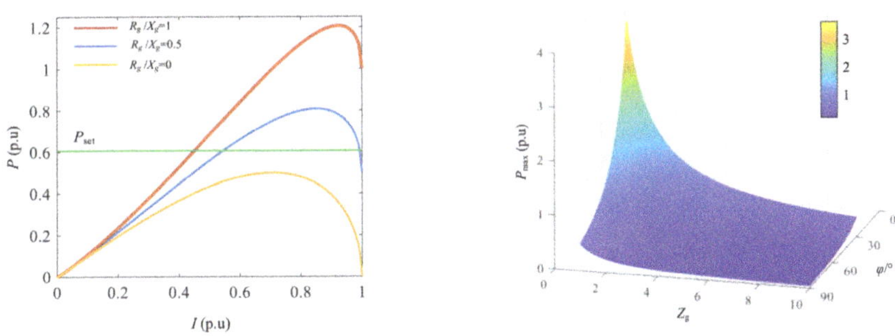

(a) **(b)**

Figure 2. (**a**) P–I_t curves for different line impedance ratios; (**b**) plot of P_{max} and Z_g, φ.

Replacing X_g and R_g in Equation (2) with the impedance modulus Z_g and the impedance angle φ by equivalents $X_g = \tan\varphi/(1 + \tan^2\varphi)^{(1/2)}$ and $R_g = (1 + \tan^2\varphi)^{(-1/2)}$, Equation (3) can be derived, and with the aid of the mathematical application Wolfram Alpha, the explicit analytical solution for P_{max} can be displayed as in Equation (4):

$$P = 3\left(I_t\sqrt{U_g^2 - I_t^2 Z_g^2 \frac{\tan^2\varphi}{1+\tan^2\varphi}} + \frac{Z_g}{\sqrt{1+\tan^2\varphi}} I_t^2\right) \tag{3}$$

$$P_{max} = P(I_t)\Big|_{I_t = I_{pmax}} = \frac{3U_g^2}{2Z_g} \cdot \frac{\sqrt{1+\tan^2\varphi}}{\sqrt{1+\tan^2\varphi} - 1},\ I_{pmax} \approx \frac{U_g}{\sqrt{2}Z_g} \tag{4}$$

From Equation (4), we can derive Figure 2b; when Z_g and φ increase, P_{max} will be reduced, and the specific analysis will be combined with the SCR in Section 3.1.

However, in real operation conditions, especially in weak grids, converters may be called upon to provide the necessary reactive power to support the power system and compensate the voltage drop of the original transmission line using locally installed reactive power compensators [14]; notably, the required reactive power from compensators can be considerably high which makes the installation of them costly.

As shown in Figure 3a, the phasor diagram illustrates how GFM converters provide approximately $0.7P$ reactive power, which is required by grids with θ which means a power factor angle of I_t. Similarly, we can derive P in the variable power factor in Equation (5).

$$\begin{aligned} P &= U_c I_{tp} = (\sqrt{U_g^2 - (I_{tp}X_g - I_{tq}R_g)^2} + I_{tp}R_g + I_{tq}X_g)I_{tp} \\ &= I_t \cos\theta \sqrt{U_g^2 - I_t^2(X_g\cos\theta - R_g\sin\theta)^2} + I_t^2(R_g\cos^2\theta + \frac{X_g}{2}\sin 2\theta) \end{aligned} \quad (5)$$

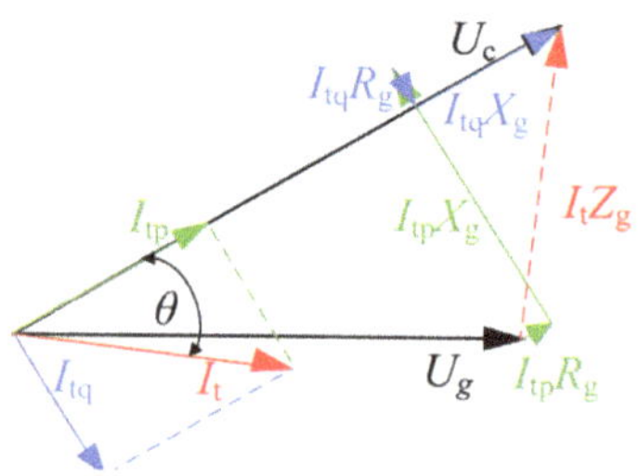

(a)

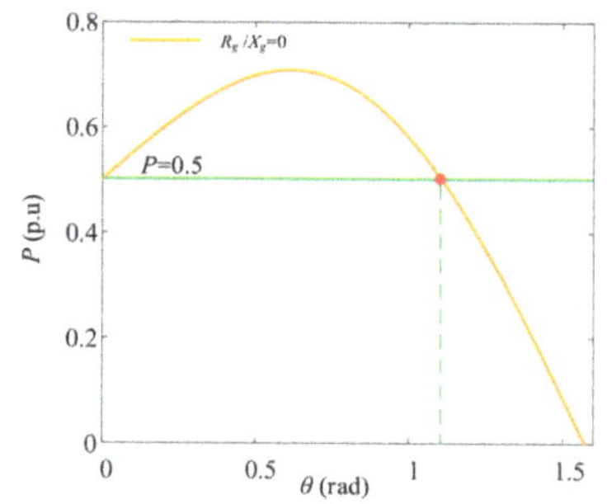

(b)

Figure 3. (**a**) Phasor diagram of the GFM converter provides the required reactive power; (**b**) plot of P and θ.

As shown in Figure 3b, for the $R_g/X_g = 0$ grid, P reached the highest at 0.5 (p.u.) when I was near 0.7 (p.u.). Keep the I and change the converter's power factor, and it turns out that P reached the highest value when the θ neared 0.6 rad and was higher than 0.5 (p.u.) when the θ value was smaller than 1.0 rad. By decreasing the power factor in a large scale, GFM converters can provide an even higher active power. Therefore, although power transmission with variable power factors is very complicated, some aspects of which are beyond the scope of this study, it can be concluded that converters can meet the requirements of rated active power with or without providing sufficient reactive power; this paper focuses more on the state of unity power factor to simplify the question.

2.2. *Equivalent Circuit of the GFL Converter*

As shown in Figure 4a, compared with Figure 1a, due to the presence of the reactive power compensation capacitor, X_c [16], the output current is not equal to the output current, Ig, from the PCC to the grid side when the GFL converter is considered as a current source; the output voltage is still equal to the voltage at the PCC when the GFL converter is considered as a voltage source, so the model of the GFL converter is slightly more complex. The GFL converter is regarded as a current source whose output current is decoupled by dq through Park transformation and the PLL control strategy in the GFL converter [10]. The dq decoupling analysis method of currents is more suitable for GFL than GFM converters, not only because the GFL converter is a current source which should be focused more in its output current, but also due to its current inner loop control, which will be further discussed in Section 3.2. The quantitative relationship is $I_{gd} = I_{td}$, $I_{gq} = U_{td}/X_c$.

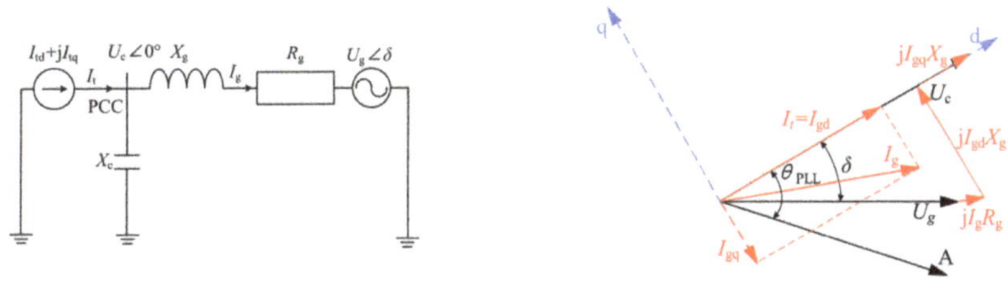

(a) (b)

Figure 4. (**a**) Equivalent circuit diagram of a GFL converter; (**b**) phasor diagram of GFL converter.

Combining the relationship and the diagram in Figure 4b, the voltage at the PCC can be derived as shown in Equation (6). Under static stability conditions, PLL accurately tracks the current; thus, $U_{td} = U_t$ can be substituted into Equation (6) to obtain Equation (7). In general, the reactance to ground at the PCC X_c is much greater than the inductive reactance in the X_g grid, i.e., $X_c/(X_c - X_g) = 1$; thus, Equation (7) is formally identical to Equation (2). Similarly, the maximum transmission power of a GFL converter can be presented as Equation (8).

$$P_{\max} = \sqrt{U_g^2 - I_{gd}^2 X_g^2} + I_{gd}R_g + I_{gq}X_g = \sqrt{U_g^2 - I_{td}^2 X_g^2} + I_{td}R_g + \frac{U_{td}}{X_c}X_g \tag{6}$$

$$U_c = \frac{X_c}{X_c - X_g}\left(\sqrt{U_g^2 - I_{td}^2 X_g^2} + I_{td}R_g\right) \tag{7}$$

$$P_{\max} = \frac{X_c}{X_c - X_g} \cdot \frac{3U_g^2}{2Z_g} \cdot \frac{\sqrt{1 + \tan^2\varphi}}{\sqrt{1 + \tan^2\varphi} - 1} \tag{8}$$

The form of Equation (8) is the same as Equation (4); judging from the equivalent circuit diagram, the ultimate transmission capacity in the unity power factor state is the same for both GFM and GFL converters under static stability conditions. It is a useful conclusion and means that the type of converter can be ignored when considering the ultimate transmission capacity of the converter alone.

3. Relationship between Static Power Transmission of the Converter and Grid Strength

3.1. Influence of Power Grid Strength Variation on the Steady Working Area of the System

According to the relative definition of SCR used to measure power grid strength in traditional power systems, the power grid strength of a grid-connected system with GFM and GFL converters can also be expressed by SCR [15]: the larger its value, the higher the strength of the system.

Equation (9) defines the SCR in renewable energy power grids: P_{sc} is the short circuit capacity of the grid, $P_{sc} = 3U_g^2/Z_g$; U_g is the RMS value of the grid phase voltage; Z_g is the impedance modulus value of the grid; and P_N is the rated transmission power of the converters, which is equal to the P_{set} value in the GFM and GFL converters mentioned above. Thus, the strength of the power grid is not only related to the power grid itself, but also to the converter, which is jointly determined by both.

It seems that the definition of the SCR requires analyses of both converters and grids; however, because the transmission power limit of GFM converters is similar to that of GFL converters, it can reasonably be assumed that the rated transmission power of the two converters is equal, to determine the steady working area by only discussing the characteristics of the power grid itself using the SCR theory.

According to Equation (8) and Figure 2b, as Z_g increases continuously, SCR decreases, and P_{max} also decreases rapidly. However, when φ increases, the power grid strength increases, and P_{max} increases. Moreover, power grids connected to the converter are mostly of high voltage, whose φ is almost 90°; in this case, the line impedance angle has little effect on the improvement in P_{max}. In summary, as Z_g increases, the grid's stiffness reduces, the SCR decreases, and P_{max} decreases.

Furthermore, the steady working area requires $P_{set} \leq P_N$; according to Equations (8) and (9), the constraint relationship between the transferable power (steady working area) of converters and power grid strengths under static stability can be calculated as Equation (10).

Notably, although the minimum SCR increases with the increase in impedance angle, it always exists at $SCR_{min} \leq 2$, which is consistent with the conclusion that the SCR of a very weak power grid is less than 2, and further verifies the accuracy of the conclusion.

$$\mathrm{SCR} = \frac{P_{sc}}{P_N} = \frac{3U_g^2}{Z_g \cdot P_{set}} \tag{9}$$

$$\begin{cases} P_{set} \leq \frac{3U_g^2}{2Z_g} \cdot \frac{\sqrt{1+\tan^2\varphi}+\tan^2\varphi+1}{\tan^2\varphi} \\ \mathrm{SCR} \geq 2 - \frac{2(\sqrt{1+\tan^2\varphi}+1)}{\sqrt{1+\tan^2\varphi}+\tan^2\varphi+1} \end{cases} \tag{10}$$

3.2. *Influence of Power Grid Strength on the Power Transmission of Converters*

As shown in Figure 5a, the block diagram of a GFM converter was used to further analyze the influencing factors of its stability during transmission power, which is not only conducive to exploring whether the converter is stable when it reaches P_{max}, but also when exploring the stability characteristics of the converter when it transmits arbitrary power under a certain strength grid. GFM converters regulate the active and reactive power output, P and Q, to the system by controlling the output voltage, U_c, and power-angle, δ, having $P = 3U_cU_g\sin\delta/X_g$ and $Q = 3U_c(U_c - U_g\cos\delta)/X_g$.

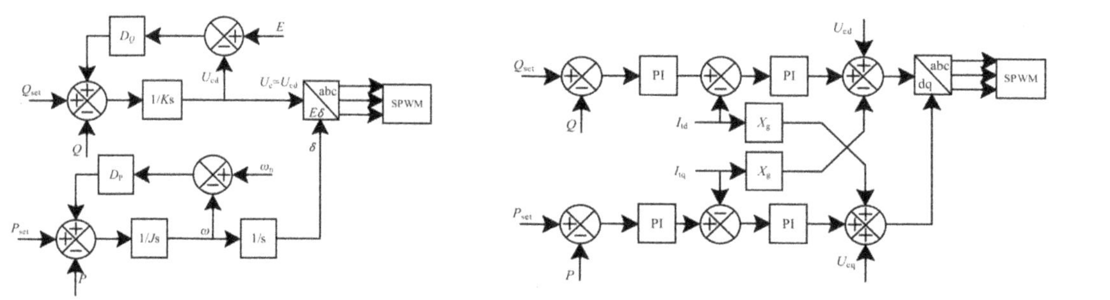

(a) (b)

Figure 5. (**a**) Block diagram of a GFM converter; (**b**) block diagram of a GFL converter.

Thus, by taking the partial derivative with respect to P and Q, with $\partial P/\partial U_c = 3U_g\sin\delta/X_g$, $\partial Q/\partial\delta = 3U_cU_g\sin\delta/X_g$, U_c and δ are expected to be taken as control quantities, P and Q are taken as controlled quantities, and the control GFM converter is a two independent single-input single-output system; thus, power coupling is not conducive to GFM converter control, which demonstrably does not meet the objective of the independent control of P and Q.

When the power grid strength decreases, SCR decreases, X_g increases, and $\partial P/\partial U_c$ decreases; therefore, the coupling relationship between P and Q decreases, and the stability of the GFM converter when transmitting P_{max} is further guaranteed. Notably, when P changes, the coupling relationship between P and Q does not change. Therefore, the characteristics of a GFM converter are applicable to weak grids in the steady working area of the system.

At the same time, because GFM converters continue the traditional control strategy of virtual synchronous generators (VSGs) and introduce gains, such as D_P and D_Q, when they are set to a certain value, the damping and standby inertia of the system is sufficient, and the power transmission is smoothly regulated within the limit range because the impedance of the weak grid is large. However, when the grid strength increases, the impedance decreases. Due to the limitation of VSG parameters, high-frequency oscillations may occur during power transmission due to parameter mismatch, which is not conducive to system stability. This further verifies that GFM converters are not suitable for strong grids.

As shown in Figure 5b, GFL converters regulate P and Q by controlling the d- and q-axis components of the output current, I_{td} and I_{tq}, respectively. As shown in Figure 4b, the converter is in a state of unity power factor; thus, $I_t = I_{td}$, and its P–I_{td} curve is similar to the curve in Figure 2a. When I_{td} increases, the operation is stable if P increases monotonically, and may be statically unstable if the monotonicity of P changes. The P–I_{td} curve is a convex function; therefore, it is concluded that the ultimate transmission power of GFL converters is the same as P_{max} in Equation (4), and as such, GFL converters are more suitable for higher-strength grids. Additionally, when the grid strength is high, due to the small impedance of the system, the time constant requirement is not high, and the power can be smoothly adjusted within the specified time; when the grid strength decreases, due to the rising impedance of the system, the power may not be adjusted within the specified time. If PI parameters are not set reasonably and the converter itself is adjusted too fast, it will produce an overshoot and a shock on the grid, which further verifies that GFL converters are more suitable for strong grids.

4. Simulation Waveform Analysis

Simulated waveforms are used to further verify the characteristics of GFM and GFL converters in transmitting power at different grid strengths. Assuming that the systems connected to the converter are all high-voltage grids, the grid strength is changed by varying the inductive reactance parameters in the line. More details about the simulation in PLECS are shown in Tables 1–3.

Table 1. Parameters of systems of three grid strengths.

Phase Voltage U_g (V)	R_g (Ω)	X_g (Ω)	SCR	P_{max} (W)
220	0.1	0.50	29.04	145,200
		3.00	4.84	25,000
		10.00	1.45	7300

Table 2. Parameters of systems of the GFM converter.

J/(kg·m^2)	D_p	D_q	K
0.057	5	321	7.1

Table 3. Parameters of systems of the GFL converter.

K_{p_PLL}	K_{i_PLL}	K_{p_i}	K_{i_i}
0.7978	99.0138	0.0043	0.7143

As shown in Table 1, it is assumed that the rated transmission power of both GFM and GFL converters is constant, and since the SCR is generally considered to be stronger than 20 for a strong grid and less than 2 for a very weak grid, it was reasonable to set the SCR to 29.04, 4.84, and 1.45 for this simulation.

Tables 2 and 3 show the parameters of GFM and GFL converter systems; the simulation was performed to verify the transmission capacity rather than modify converters, so these parameters were kept the same during the simulation.

4.1. Simulation of GFM Converter Power Transfer When SCR = 4.84

The main purpose of the grid-connected simulation model with an SCR of 4.84 is to reduce the influence of the grid strength on both converters, and thus, to observe the influence of the systems own power transmission limit on the converters' transmission power. As shown in Figure 6a, the actual value of P_{max} is slightly lower than the theoretical value (25,000 W) due to the limitations of GFM converter parameter adjustments. When the commanded power transmission is 22,000 W, the converter can deliver a rapid response within 1 s and reach the static stable equilibrium, as expected.

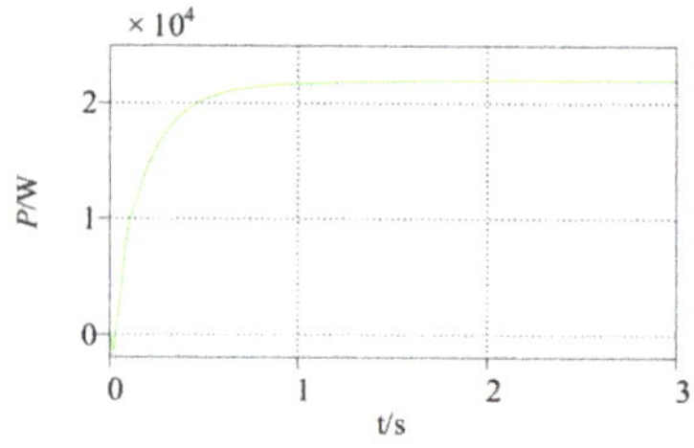

(**a**)

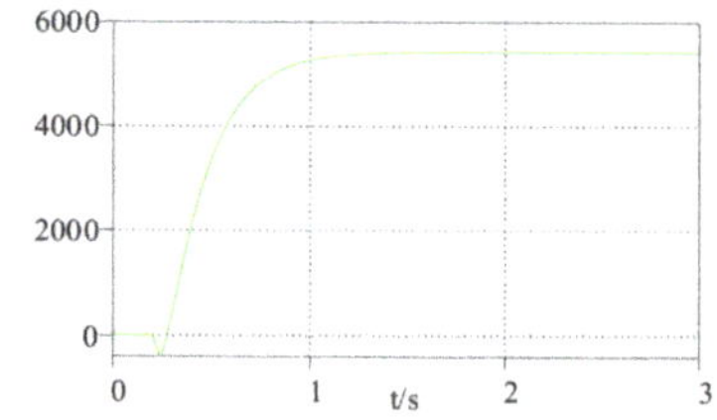

(**b**)

Figure 6. (**a**) Active power of a GFM converter (P_{set} = 22,000 W, SCR = 4.84); (**b**) reactive power of a GFM converter (P_{set} = 25,000 W, SCR = 4.84).

As shown in Figure 7, the GFM converter has high output current sinusoidality, low harmonics, and high immunity to interference at a statically stable transmission of 22,000 W, further verifying its good static stability performance. When the converter command rose from 22,000 W to 25,000 W, the actual power transmitted by the converter will rise steadily to 25,000 W; however, at this time, the converter cannot work in the unity power factor state, so it will output a certain amount of reactive power steadily to the outside. At this time, the converter output current is stable, as shown in Figure 7.

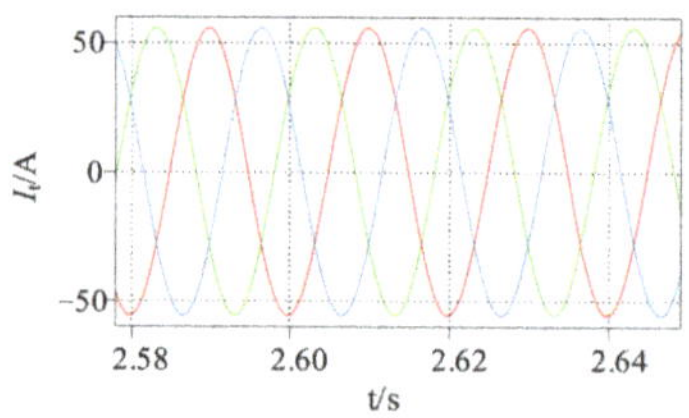

(**a**)

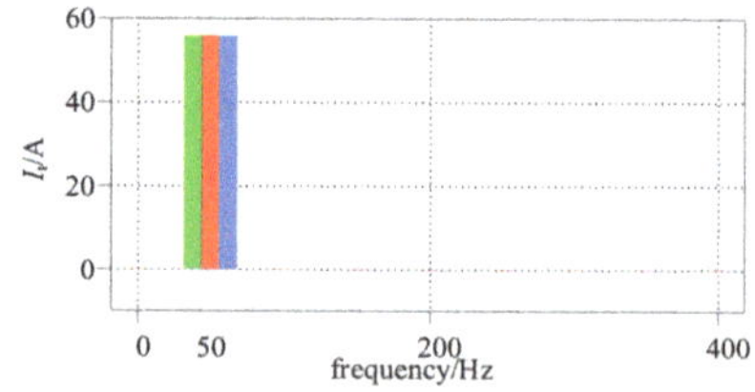

(**b**)

Figure 7. (**a**) Current curve of a GFM converter (P_{set} = 25,000 W, SCR = 4.84); (**b**) current THD of a GFM converter (P_{set} = 25,000 W, SCR = 4.84). (Green for phase A, red for phase B, blue for phase C).

When the active power command continues to increase, the converter cannot transmit the active power evenly when compensated by infinite reactive power. However, as shown in Figure 8a, when the active power command increases from 25,000 W to 50,000 W, the converter can only transmit about 40,000 W with high reactive power, so the system will be in a state of static instability and the output active power will be in a state of low-frequency oscillation (at a frequency of approximately 1 Hz). The reason is that the change in power-angle, δ, at this time cannot meet the requirements of active power instruction, so the power angle oscillates at low frequency, and the reactive power oscillates synchronously with the active power. In this case, although the converter's output current oscillates, it has a good sinusoidal degree and no obvious harmonics, as shown in Figure 8b.

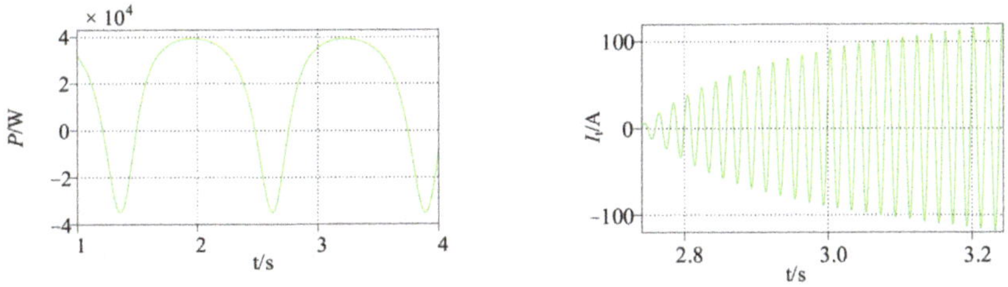

(a) (b)

Figure 8. (**a**) Active power of a GFM converter (P_{set} = 50,000 W, SCR = 4.84); (**b**) current of a GFM converter (P_{set} = 50,000 W, SCR = 4.84).

4.2. Simulation of GFL Converter Power Transfer When SCR = 4.84

Similarly, when a GFL converter is commanded to transmit at 22,000 W, it still responds quickly and has little harmonics. When P_{set} is raised to 25,000 W, P slowly rises to slightly above the command, then becomes out of control and quickly rises to another stable operating point; the value of P is determined by the converter itself as long as P_{set} exceeds the limit power. The reason is shown in Figure 9b, as Q first operates out of control and cannot be stabilized in a smaller range, resulting in a sharp increase in Q, which further leads to a runaway P value, but at this point, Q can be as high as 40,000 var and a certain degree of harmonics is injected into the grid, as shown in Figure 10. The fifth and seventh harmonics indicate that Q is so high that it exceeds the GFL converter's transmission limit and causes instability. Due to the power coupling of a GFM converter, the reactive power does not become out of control; only a stable offset occurs, as shown in Figure 6.

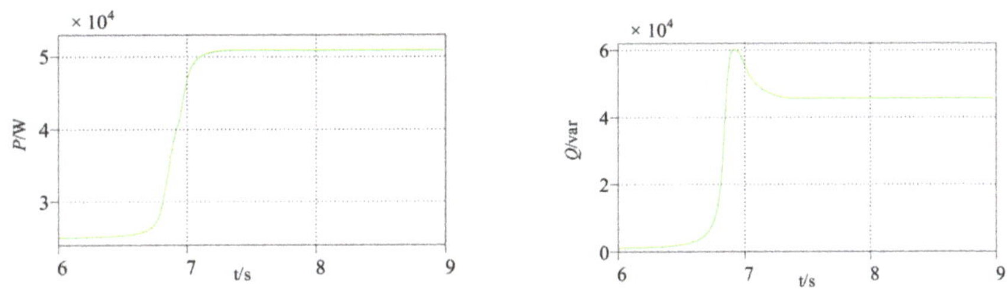

(a) (b)

Figure 9. (**a**) Active power of a GFL converter (P_{set} = 25,000 W, SCR = 4.84); (**b**) reactive power of a GFL converter (P_{set} = 25,000 W, SCR = 4.84).

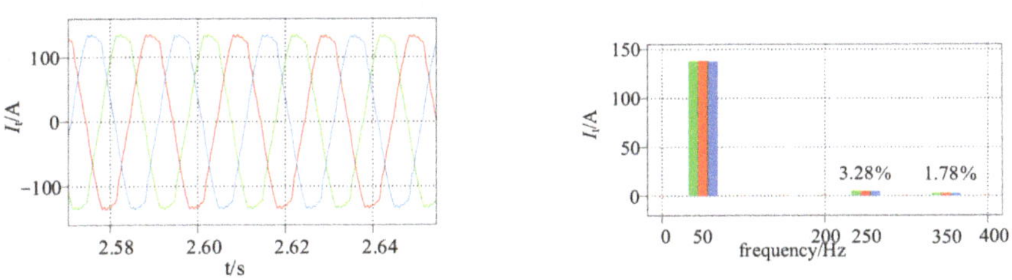

(a) (b)

Figure 10. (**a**) Current of a GFL converter (P_{set} = 25,000 W, SCR = 4.84); (**b**) current THD of a GFL converter (P_{set} = 25,000 W, SCR = 4.84). (Green for phase A, red for phase B, blue for phase C).

4.3. Comparison of the Power Transfer Capability of Converters at Strong and Weak Grid Strengths

A simulation model for a strong grid with a SCR greater than 20 and a weak grid with SCR less than 2 was established to compare the influence of the grid strength on the transmission of power within the limit range of GFM and GFL converters.

As shown in Figure 11, when the P_{set} of the very weak grid was 7300 W, it could reach the target requirement within 1–2 s, whereas I_t and U_c have little harmonic similarity to that depicted in Figure 7: when the P_{set} of the strong grid reached 140,000 W (slightly below the ideal limit), it could respond rapidly in about 0.1 s, whereas U_c had high fifth and seventh harmonics, as shown in Figure 12, which is due to the large influence of power coupling to make the control quantity U_{cd} exceed normal levels and the PWM of the GFM converter to become far from sinusoidal. High harmonics from both GFM and GFL converters are caused by the runaway of control quantities, although specific reasons are different. GFM converters set a fixed virtual impedance in the face of a strong grid with a small impedance, which does not match the grid, resulting in a high frequency oscillation of U_c with a frequency of approximately 100 Hz; in contrast, the weaker grid impedance is larger, which offsets the virtual impedance setting problem and smooths the power output of the converter.

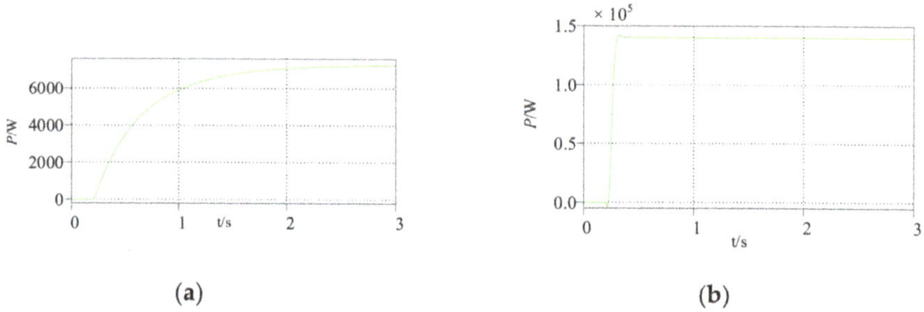

Figure 11. (**a**) Active power of a GFM converter (P_{set} = 7000 W, SCR = 1.45); (**b**) active power of a GFM converter (P_{set} = 140,000 W, SCR = 29.04).

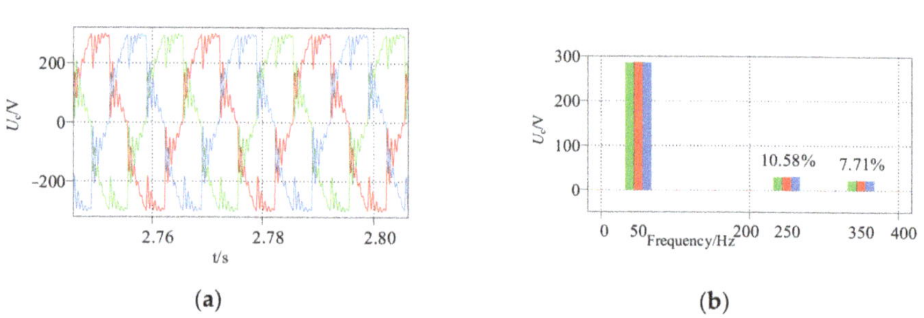

Figure 12. (**a**) Voltage of a GFM converter (P_{set} = 140,000 W, SCR = 29.04); (**b**) THD of a GFM converter (P_{set} = 140,000 W, SCR = 29.04). (Green for phase A, red for phase B, blue for phase C).

As shown in Figure 13, GFL converters can output P_{max} relatively more rapidly under both weak and strong grid conditions, with I_t and U_c having small harmonics. The PI controller time constant here is fixed at approximately 0.5 s; when the GFL converter runs on the weak grid, its regulation speed is higher than the grid, so P has obvious oscillation overshoot. At the same time, there is no virtual impedance setting in the GFL converter, so the regulation time is not too fast but relatively smooth in strong grids.

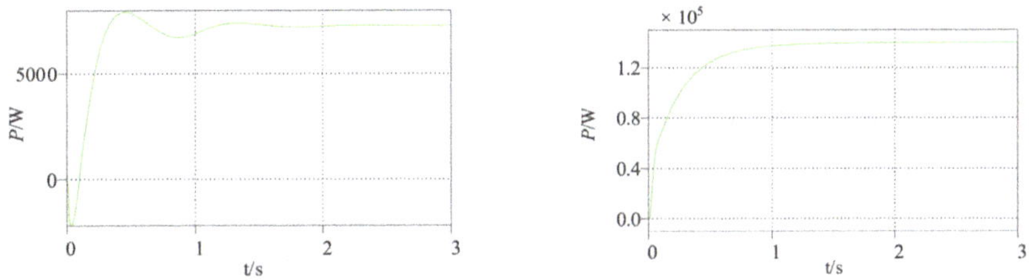

(**a**) (**b**)

Figure 13. (**a**) Active power of the GFL converter (P_{set} = 7000 W, SCR = 4.84); (**b**) active power of the GFL converter (P_{set} = 140,000 W, SCR = 4.84).

4.4. Hardware in-Loop Simulation Using RTBOX1

Figure 14 shows the real-time simulation devices RTBOX1 and F28379D launchpad. Using this platform facilitates the verification of the above conclusions. Figure 15 shows two states of transmitting power by GFM or GFL converters; the simulation waveform characteristics, such as harmonics, are similar to those in PLECS, which further verifies that their transmission capacity could be realized in actual power grids.

Figure 14. RTBOX1 and F28379D launchpad.

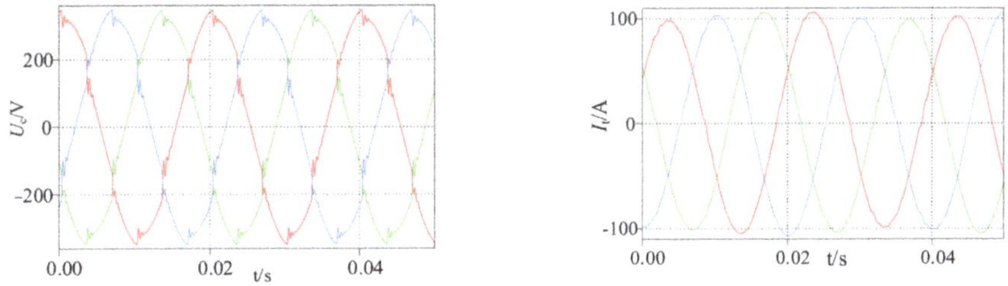

(**a**) (**b**)

Figure 15. (**a**) Voltage of the GFM converter (P_{set} = 140,000 W, SCR = 29.04); (**b**) current of the GFL converter (P_{set} = 22,000 W, SCR = 4.84). (Green for phase A, red for phase B, blue for phase C).

5. Conclusions

This study investigated the transfer capabilities of the two main types of non-synchronous machine power sources connected to GFM and GFL grids, focusing on the transfer capability of converters at different grid strengths based on their static and stable operation.

1. There is a quantitative relationship between the statically stable power transmission limit and the line impedance of GFM and GFL converters at a unity power factor. At variable power factors, the converters can even transmit active rated power sufficiently. For high-voltage grids, the reactance, X_g, increases; its impedance modulus, Z_g, increases. SCR decreases, P_{max} decreases, and the operating range is reduced. However, when X_g increases, its impedance angle, φ, increases, and P_{max} increases, but the impact of φ on the transmission capacity is much smaller than that of Z_g. Therefore, the larger the impedance, the smaller the P_{max}; thus, the two are approximately inversely related.
2. The grid strength of a system is determined by SCR, which is jointly influenced by Z_g and the rated transmission power of the converter. When the reactance of PCC to ground is much greater than X_g, the P_{max} values of GFM and GFL converters are equal, and it can be considered that the grid strength is only related to X_g, i.e., the greater the grid strength, the greater the P_{max}.
3. Within P_{max} (in a steady working area), GFM converters are more suitable for weak grids because of the power coupling problem, which is more obvious in strong grids, whereas GFL converters are more suitable for strong grids because of the monotonic nature of the current in the P–I_t curve. Moreover, in smoothing power transfer, the parameters of GFM converters, which require virtual impedance settings, make them more suitable for weak grid regulation because they are prone to high-frequency oscillations under strong grids. GFL converters, however, with their parameters of time constants and PI controllers, are more suitable for the regulation of strong grids because they are prone to overshooting oscillations in weak grids.

Author Contributions: Conceptualization, J.W.; Methodology, J.Y.; Software, X.J.; Validation, S.W.; Investigation, B.K.; Resources, J.Z.; Data curation, X.X. All the authors contributed equally to this manuscript. All authors have read and agreed to the published version of the manuscript.

Funding: This research was funded by Science and Technology Program of State Grid Jiangsu Electric Power Co., Ltd. (J2021012).

Data Availability Statement: No new data were created or analyzed in this study. Data sharing is not applicable to this article.

Conflicts of Interest: The authors declare no conflict of interest. The funders had no role in the design of the study; in the collection, analyses, or interpretation of data; in the writing of the manuscript; or in the decision to publish the results.

References

1. Huang, C.; Zheng, Z.; Ma, X.; Yan, H.; Wang, M.; Fan, Z. Research on the influence of resilient grid learning ability on power system. *Energy Rep.* **2022**, *8*, 43–50. [CrossRef]
2. Granata, S.; Di Benedetto, M.; Terlizzi, C.; Leuzzi, R.; Bifaretti, S.; Zanchetta, P. Power Electronics Converters for the Internet of Energy: A Review. *Energies* **2022**, *15*, 2604. [CrossRef]
3. Amin, M.; Molinas, M. Small-signal stability assessment of power electronics based power systems: A discussion of impedance- and eigenvalue-based methods. *IEEE Trans. Ind. Appl.* **2017**, *53*, 5014–5030. [CrossRef]
4. Iioka, D.; Kusano, K.; Matsuura, T.; Hamada, H.; Miyazaki, T. Appropriate Volt–Var Curve Settings for PV Inverters Based on Distribution Network Characteristics Using Match Rate of Operating Point. *Energies* **2022**, *15*, 1375. [CrossRef]
5. Khan, S.A.; Wang, M.; Su, W.; Liu, G.; Chaturvedi, S. Grid-Forming Converters for Stability Issues in Future Power Grids. *Energies* **2022**, *15*, 4937. [CrossRef]
6. Yap, K.Y.; Sarimuthu, C.R.; Lim, J.M.-Y. Virtual Inertia-Based Inverters for Mitigating Frequency Instability in Grid-Connected Renewable Energy System: A Review. *Appl. Sci.* **2019**, *9*, 5300. [CrossRef]
7. Vadi, S.; Padmanaban, S.; Bayindir, R.; Blaabjerg, F.; Mihet-Popa, L. A Review on Optimization and Control Methods Used to Provide Transient Stability in Microgrids. *Energies* **2019**, *12*, 3582. [CrossRef]

8. Zhang, S.; Zhu, Z.; Li, Y. A Critical Review of Data-Driven Transient Stability Assessment of Power Systems: Principles, Prospects and Challenges. *Energies* **2021**, *14*, 7238. [CrossRef]
9. Hosseinzadeh, N.; Aziz, A.; Mahmud, A.; Gargoom, A.; Rabbani, M. Voltage Stability of Power Systems with Renewable-Energy Inverter-Based Generators: A Review. *Electronics* **2021**, *10*, 115. [CrossRef]
10. Pattabiraman, D.; Lasseter, R.H.; Jahns, T.M. Comparison of grid following and grid forming control for a high inverter penetration power system. In Proceedings of the 2018 IEEE Power & Energy Society General Meeting (PESGM), Portland, OR, USA, 5–10 August 2018; pp. 1–5.
11. Poolla, B.K.; Groß, D.; Dörfler, F. Placement and implementation of grid-forming and grid-following virtual inertia and fast frequency response. *IEEE Trans. Power Syst.* **2019**, *34*, 3035–3046. [CrossRef]
12. Awal, M.A.; Husain, I. Unified virtual oscillator control for grid-forming and grid-following converters. *IEEE J. Emerg. Sel. Top. Power Electron.* **2020**, *9*, 4573–4586. [CrossRef]
13. Chen, X.; Zhang, Y.; Wang, S.; Chen, J.; Gong, C. Impedance-phased dynamic control method for grid-connected inverters in a weak grid. *IEEE Trans. Power Electron.* **2016**, *32*, 274–283. [CrossRef]
14. Yang, D.; Wang, X.; Liu, F.; Xin, K.; Liu, Y.; Blaabjerg, F. Adaptive reactive power control of PV power plants for improved power transfer capability under ultra-weak grid conditions. *IEEE Trans. Smart Grid* **2017**, *10*, 1269–1279. [CrossRef]
15. Kang, Y.; Lin, X.; Zheng, Y.; Quan, X.L.; Hu, J.B.; Yuan, S. The static stable-limit and static stable-working zone for single-machine infinite-bus system of renewable-energy grid-connected converter. *Proc. CSEE* **2020**, *40*, 4506–4515.
16. Chen, M.; Zhou, D.; Tayyebi, A.; Prieto-Araujo, E.; Dörfler, F.; Blaabjerg, F. Generalized multivariable grid-forming control design for power converters. *IEEE Trans. Smart Grid* **2022**, *13*, 2873–2885. [CrossRef]

energies

Article

A Model Independent Predictive Control of PMSG Wind Turbine Systems with a New Mechanism to Update Variables

Yuzhe Zhang [1], Xiaodong Liu [2], Haitao Li [1] and Zhenbin Zhang [1,*]

[1] School of Electrical Engineering, Shandong University, Jinan 250061, China
[2] State Grid Wuxi Power Supply Company, Wuxi 214000, China
* Correspondence: zbz@sdu.edu.cn; Tel.: +86-0531-8839-2002

Abstract: Permanent magnet synchronous generator (PMSG) wind power system with full power rating converter configuration is especially suitable for wind energy applications. Direct model predictive control (DMPC) has led to more possibilities in terms of choice because of its straightforward concept for PMSG wind turbine systems in *high-power* off-shore wind farms. However, due to complete dependence on the model knowledge, parameter mismatches will seriously deteriorate the system control performances. This work presents a model/parameter-independent predictive control method with a novel mechanism to update current/power variations online. The proposed method makes use of only *two* measurements from the former intervals and the selected control vectors to estimate all variations of the candidate vectors in the present interval. Benefiting from this updating mechanism, the proposed method is completely independent of the model parameters in the state prediction. However, it still has a very low calculating requirement and smooth current/power variation waveforms. The proposed method is compared with classical DMPC. The results validate that the proposed solution outperforms the classical DMPC with model deviations, with considerably improved robustness.

Keywords: PMSG wind turbines; back-to-back power converters; model-independent predictive control; robust control

Citation: Zhang, Y.; Liu, X.; Li, H.; Zhang, Z. A Model Independent Predictive Control of PMSG Wind Turbine Systems with a New Mechanism to Update Variables. *Energies* **2023**, *16*, 3764. https://doi.org/10.3390/en16093764

Academic Editors: Davide Astolfi and Lorand Szabo

Received: 11 March 2023
Revised: 15 April 2023
Accepted: 26 April 2023
Published: 27 April 2023

1. Introduction

Electricity generated by renewable energy has grown significantly since the end of the 20th century. Among the variations, high-power offshore wind energy has become increasingly competitive with other energy sources in terms of rich reserves, long generation time, and "cost-per-generated kilowatt hour" [1,2]. Large-capacity wind turbines have become essential in offshore wind energy installation in the last few years [3]. Permanent magnet synchronous generator (PMSG) with direct-drive configuration has considerable advantages in such a system in terms of higher energy density, gearbox elimination/reduction, and less maintenance [4,5]. As one such configuration, a very potent system based on full power rating back-to-back converter is spreading rapidly for high-power offshore wind energy generation [5]. A simplified circuit diagram of such a PMSG system is introduced in Figure 1. Considering both the nonlinear characters and switching nature of the power converters, the direct model predictive control (DMPC, also called finite-control-set model predictive control, FCS-MPC) is a promising method to implement with multiple-target optimization in a single step [6]. Therefore, DMPC makes the time-averaged modulation stage unnecessary in the control of power converters [7]. It offers fast dynamic performance and a straightforward design. The literature shows that the DMPC has already emerged as a widely-used control option for the system shown in Figure 1 [8–10].

Due to the reliance on model-based prediction of system states, DMPC suffers from system model deviations (caused by, e.g., inaccurate modeling and incorrect component parameters) [6,11]. In practice, the considerable *mismatches* between the values in the actual equipment and the values used in the controller often occur on *unavailable* (e.g., stator flux) and *time-varying* parameters (e.g., coil inductance in generator or filter) of the wind energy system. Inaccurate prediction of the system's future behaviors would result in this situation and induce performance deterioration (e.g., large steady-state tracking bias and increased control variable ripples [12]).

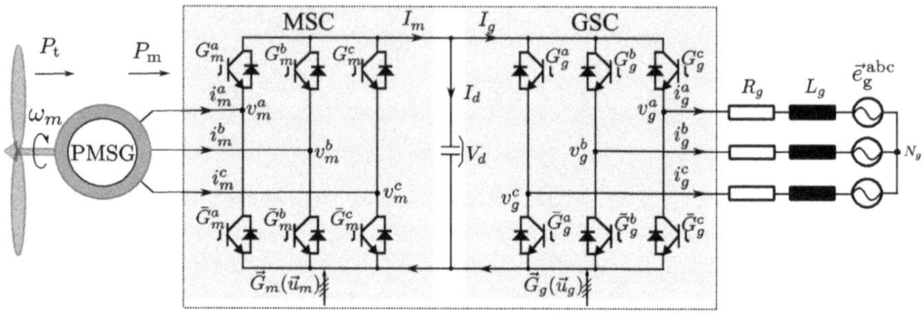

Figure 1. Simplified electrical circuit of a two-level back-to-back power converter-based PMSG wind turbine system with equivalent RL-filter.

Researchers have proposed several methods to improve the model predictive control's robustness. Three methods can be grouped as three different concepts. The first concept is *observer-based methods* [13,14]. In [15,16], an observer for perturbations caused by mismatch was employed in a deadbeat control solution. Reference [17] presents a full parameter disturbance and load observer to simultaneously estimate both electrical and mechanical parameters of PMSG for model predictive speed control. However, observer-based solutions are usually complicated, requiring higher tuning efforts and compromising system stability by adding redundant control loops. The second concept is *error compensation* [18]. In [19,20], the prediction model compensates for the state tracking bias caused by parameter mismatches, assuming that the prediction error for a given switching state remains unchanged within a few control intervals. In [21], a new cost function in proportional integral (PI) form is designed to eliminate steady-state errors. The improved parameter stability region is theoretically derived. However, since the prediction stage still uses the mismatched parameters, the control performance improvement is not satisfactory under large parameter deviation conditions. The third group is *model-free or model-independent predictive control (MIPC)* [22]. Reference [23] first presented such a solution for a PMSG drive. Different from the previous two groups of solutions, it designed look-up tables (LUTs) to maintain status variation for a previous control interval as information for the prediction of the system behaviors in the future interval. This eliminates the model information and, therefore, the controller gains greater robustness to significant parameter deviation. Nevertheless, if certain specific voltage vectors have not been applied for multiple control intervals, most of the previously stored current variation information will be outdated and unreliable for future prediction [23,24], resulting in large *updating lag*. This will significantly affect the control performance, causing unacceptable ripples in the output waveforms and even posing a significant risk to the stability of the system during transient states. Authors of the recent contributions [25] proposed a solution in which previously applied voltage vectors and current measurements are used to estimate the current variation under other voltage vectors to form a smoother updating waveform with less lag. However, such an update mechanism can only be activated when *three* previously applied voltage vectors differ. Moreover, a total number of 210 switching sequences have to be categorized to seek the optimal one, which significantly increases the computational burden of the controller.

To conquer the problems analyzed above, this paper proposes a new MIPC method based on the estimation of all possible variations to quickly update the LUTs. We validate this method on the control of both the generator and grid sides of a two-level back-to-back converter-based PMSG wind power system. It requires only *two* measurements from the former intervals and the relevant information regarding control vectors. Compared to the method that only updates the relevant variation of the selected vector, it estimates the necessary state variations for all possible voltage vectors in the present interval. The main contributions of this work include:

1. A new effective model-independent predictive control for both the generator and the grid side of high-power PMSG wind turbine systems is presented (see Section 4). The proposed method is immune to both generator- and grid-side parameter mismatches and model deviations. Robustness improvement of the proposed method outperforms the classical model-based MPC technique (see Section 5.2);
2. A new state variable variation updating mechanism is proposed, which assures smooth current/power variation waveforms and non-lag updating. The proposed solution is analytically developed and requires many fewer online calculations in comparison with the recently reported approaches (see Section 4);
3. The proposed solution is tested in various scenarios (see Section 5), which show promising results in enhanced robustness.

The contents of this article are organized as follows. In Section 2, the basic modeling of the grid-tied two-level back-to-back converter system is presented. Section 3 reviews the classical model-based and model-free methods for such a system, and in Section 4, we introduce the detailed proposed mechanism and the design of the controller. Section 5 reports on the verification and analysis of the proposed method. Finally, Section 6 concludes this paper.

2. System Description and Modeling

A two-level voltage source back-to-back power converter PMSG wind turbine system with direct-drive configuration is presented in Figure 1. A machine-side converter (MSC) and grid-side converter (GSC) are connected by the DC-link capacitor. During normal operation phases, MSC is used as the power interface to control the generator, while GSC aims to regulate the DC-link and grid-side power. The aerodynamics and turbine are modeled in, e.g., ref. [6]. These are not repeated here. Meanwhile, we assume a speed reference ω_{m}^* has been obtained as the maximum power point tracking (MPPT) requirement by another external controller which is not the focus of this study. In the following, the generator side, grid side, MSC, and GSC are modeled. Note that variables $\boldsymbol{x}^{\alpha\beta}$ in the stationary frame and $\boldsymbol{x}^{\mathrm{dq}}$ in the rotary reference frame are derived, invoking the corresponding (power invariant) Clarke and Park transformations, respectively.

The current dynamics of the surface-mounted permanent magnet generator (SPMSG) model in rotary reference coordinates can be expressed as

$$\begin{pmatrix} g_{i_{\mathrm{m}}^{\mathrm{d}}} \\ g_{i_{\mathrm{m}}^{\mathrm{q}}} \end{pmatrix} = \frac{1}{L_{\mathrm{s}}}\begin{pmatrix} v_{\mathrm{m}}^{\mathrm{d}} \\ v_{\mathrm{m}}^{\mathrm{q}} \end{pmatrix} - \frac{R_{\mathrm{s}}}{L_{\mathrm{s}}}\begin{pmatrix} i_{\mathrm{m}}^{\mathrm{d}} \\ i_{\mathrm{m}}^{\mathrm{q}} \end{pmatrix} - \omega_{\mathrm{e}}\begin{pmatrix} -i_{\mathrm{m}}^{\mathrm{q}} \\ i_{\mathrm{m}}^{\mathrm{d}} \end{pmatrix} - \frac{1}{L_{\mathrm{s}}}\begin{pmatrix} 0 \\ \omega_{\mathrm{e}} \cdot \psi_{\mathrm{pm}} \end{pmatrix}, \tag{1}$$

where $\boldsymbol{G}_{\mathrm{m}} = (g_{i_{\mathrm{m}}^{\mathrm{d}}}, g_{i_{\mathrm{m}}^{\mathrm{q}}})^\top$ is the gradient of the PMSG stator currents and $\boldsymbol{v}_{\mathrm{m}}^{\mathrm{dq}}$ and $\boldsymbol{i}_{\mathrm{m}}^{\mathrm{dq}}$ denote MSC output voltage and PMSG stator currents in the dq frame, respectively.

Using measured values of the voltage and current from the *point of coupling*, the instantaneous power is calculated as [26]

$$\begin{pmatrix} P \\ Q \end{pmatrix} = \underbrace{\begin{pmatrix} e_{\mathrm{g}}^{\alpha} & e_{\mathrm{g}}^{\beta} \\ e_{\mathrm{g}}^{\beta} & -e_{\mathrm{g}}^{\alpha} \end{pmatrix}}_{\boldsymbol{E}_g}\begin{pmatrix} i_{\mathrm{g}}^{\alpha} \\ i_{\mathrm{g}}^{\beta} \end{pmatrix}. \tag{2}$$

Power dynamics in stationary coordinates for a balanced grid are

$$\begin{pmatrix} g_P \\ g_Q \end{pmatrix} = \frac{1}{L_g} \boldsymbol{E}_g \begin{pmatrix} v_g^\alpha - e_g^\alpha \\ v_g^\beta - e_g^\beta \end{pmatrix} - \begin{pmatrix} \frac{R_g}{L_g} P + \omega_g Q \\ \frac{R_g}{L_g} Q - \omega_g P \end{pmatrix}, \tag{3}$$

where $\boldsymbol{G}_g = (g_P, g_Q)^\top$ represents the grid-side power gradient and v_g^α, v_g^β, and e_g^α, e_g^β represent GSC output voltage vectors and grid voltages in the $\alpha\beta$ frame, respectively.

Introducing G_y^x as the switch signal for the IGBTs in Figure 1, where $y \in \{m, g\}$ and $x \in \{a, b, c\}$, the complementary signal for the opposite IGBTs in the same converter leg can be written as $\bar{G}_y^x$. The switching state u_y^x can be defined accordingly as

$$u_y^x := \mathcal{G}(G_y^x) = \begin{cases} P & \text{if}: G_y^{x1} = 1 \\ N & \text{if}: G_y^{x1} = 0 \end{cases} \tag{4}$$

for phase x. The 3-phase has 2^3 vector options for each side converter to meet the control requirements, presented as

$$\boldsymbol{u}_y^{abc} = (u_y^a, u_y^b, u_y^c)^\top \in \mathcal{S}_8 := \{NNN, NNP, \cdots, PPN, PPP\}. \tag{5}$$

Taking switching states and DC-link voltage V_d into consideration, the phase voltages of the converter can be obtained as [27]

$$\boldsymbol{v}_y^{abc} = \begin{pmatrix} v_y^a \\ v_y^b \\ v_y^c \end{pmatrix} = \frac{V_d}{3} \begin{pmatrix} 2 & -1 & -1 \\ -1 & 2 & -1 \\ -1 & -1 & 2 \end{pmatrix} \boldsymbol{u}_y^{abc}. \tag{6}$$

Known from Figure 1, DC-link voltage depends on the current flow of the MSC and GSC and can be modeled as

$$\frac{dV_d(t)}{dt} = \frac{1}{C} I_d(t) = \frac{1}{C}\big(I_g(t) - I_m(t)\big), \tag{7}$$

where $I_g = \boldsymbol{i}_g^{abc\top} \cdot \boldsymbol{u}_g^{abc}$ and $I_m = \boldsymbol{i}_m^{abc\top} \cdot \boldsymbol{u}_m^{abc}$ present the DC-link current components of the grid and generator side, respectively.

3. Classical Direct Model Predictive Control Methods

3.1. System Requirements and Cost Function Design

The control objectives for the GSC and MSC in a normal back-to-back converter-based wind turbine system, shown in Figure 1, are listed as (see, e.g., [28]):

- *Torque/current control:* The primary torque/current control must have a promising dynamic performance to generate proper reference considering: (a) MPPT of the wind turbine system or (b) suitable torque generation for supercritical wind speed. Stress on the mechanical components needs to be reduced by minimizing torque ripples and current THDs;
- *Complex power control:* The GSC must regulate the grid-side complex power quickly and dynamically to reduce fluctuations in the DC link caused by the intermittent feed-in of wind power. To meet the requirements of the Grid Code, a low-current distortion factor should be ensured;
- *DC-link voltage control:* A stable DC-link voltage is required for the proper operation of the system;
- *Switching frequency regulation:* As the wind turbine power level rises, needed improvement of efficiency at any point requires reducing switching losses by low switching frequencies.

Closely related to these requirements, state variables, i.e., the generator stator current $\boldsymbol{x}_m = (i_m^d, i_m^q)^\top$ and grid power $\boldsymbol{x}_g = (P, Q)^\top$ are taken into consideration in two cost functions (y = m for generator side, y = g for grid side), as

$$J_y = \left\| x_y(k+2) - x_y^* \right\|_2^2 + \lambda_y \Delta u_y(k+1), \tag{8}$$

where $\|x\|_2$ is the status of the system x and $\Delta u_y(k+1) = u_y(k+1) - u_y(k)$ represents the switching change in this control period. The weighting factor, λ_y, is designed for the optimization of multiple control objectives, which consist of the desired status and switching frequency here. For a surface-mounted PMSG wind turbine control, i_m^{d*} shall be zero to ensure the maximum torque per ampere (MTPA) control of PMSG. A unity power factor can be achieved by setting Q^* as zero in the grid-side control. The outer PI loops calculate the reference i_m^{q*} and P^* to meet the above requirements of the generator's speed and DC-link voltage. The delay of the sampling process in the real digital controller can be compensated by introducing the state variables at $k+2$, instead of variables at $k+1$, during the traversal of all possible control vectors [29].

3.2. Classical Direct Model Predictive Control

To get the state variables at the $k+1$ instant, the classical DMPC calculates the future stator currents and power for a finite set of voltage vectors by utilizing the system model, as

$$x_y(k+1) = x_y(k) + T_s \cdot G_y(k), \tag{9}$$

where $y \in \{m, g\}$, x_m is the stator current vector i and x_g is the grid power vector S. $T_s \cdot G_y(k)$ denotes the state variation caused by this control interval. The same idea gives the prediction at the $k+2$ instant of the state variables as

$$x_y(k+2) = x_y(k+1) + T_s \cdot G_y(k+1). \tag{10}$$

The controller selects and records the optimal vector from all control vectors, which minimizes the cost function (8). In the next control interval, this will be applied, and relevant delay compensation will be achieved as mentioned above. This procedure is repeated while new measurements arrive continuously [30]. However, it is observed that prediction of state variables in Equation (10) requires accurate information on the system parameters. The controller will derive the wrong prediction of the system utilizing mismatched parameters, resulting in the selection of the non-optimal control vectors. This has serious effects on the control of the stator current or the grid power until the system no longer functions properly and stably. To this end, we will present model-independent solutions in the following part.

3.3. A Latest Model-Independent Predictive Control Solution

Reference [24] presented an efficient MIPC algorithm for synchronous reluctance motor control. In this work, its principles are extended to the underlying PMSG wind turbine system as a benchmark. The dq-axis stator currents (of PMSG) and complex power (of GSC) variations under different voltage vectors are calculated and stored in four LUTs. The LUTs are updated online with new measurements. The future behavior of the system is calculated using the LUTs' data as

$$x_y(k+1) = x_y(k) + \Delta x_y(k), \tag{11}$$

where $\Delta x_y(k)$ denote the currents (for $y = m$) and power variations (for $y = g$) caused by the application of the voltage vector at time step k, which is determined in the $k-1$ control interval. $\Delta x_y(k)$ is considered the same as the last ones stored in the LUTs. Then the possible status at time step $k+2$ is derived as

$$x_y(k+2) = x_y(k+1) + \Delta x_y(k+1). \tag{12}$$

The current and power prediction of MIPC is hence accomplished using Δx_y stored in the LUTs, avoiding the use of system parameters.

This is easy to understand because, for a voltage vector that has not been selected during a long period, the stored information will be obsolete, which will influence the prediction accuracy and the control performance. The MIPC in [24] forces the output of this vector that has not been applied for a given time period, regardless of the cost function minimization optimization principle, which results in increased ripple and degraded steady-state performance. An improved MIPC with a new state variable update principle is proposed in the next section.

4. Proposed Model-Independent Predictive Control Solution

This section introduces an improved model-independent predictive control method with fast updating look-up tables (LUTs). The overall diagram is given in Figure 2. A grid-side power analysis based on instantaneous power clearly describes the idea of the method. The basis for this begins with a small signal model analysis.

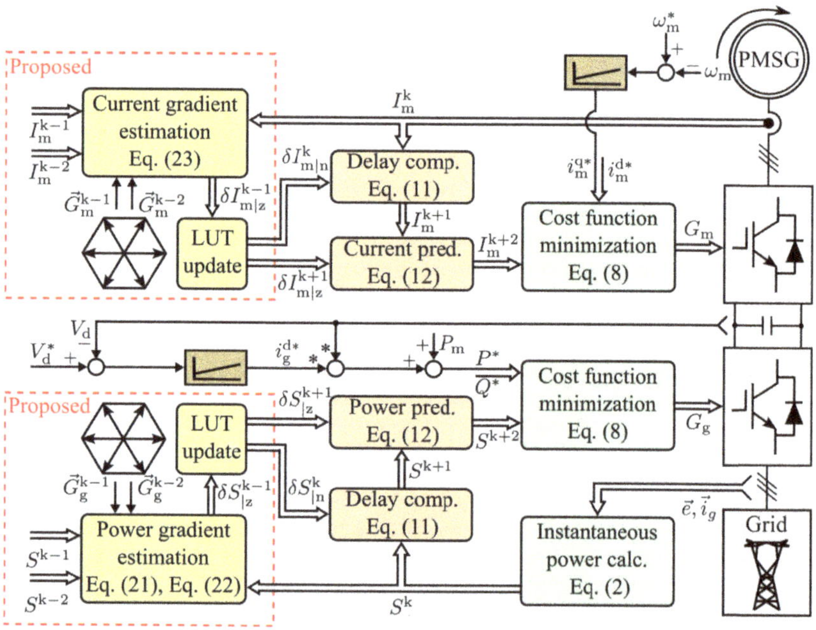

Figure 2. Control scheme of the proposed model-independent solution for back-to-back power converter based PMSG wind turbine system.

4.1. Small-Signal Modeling of the System

Grid active and reactive power in steady-state is described as

$$\boldsymbol{G}_{\mathrm{m}} = \begin{pmatrix} 0 \\ 0 \end{pmatrix} = \frac{1}{L_{\mathrm{g}}} \boldsymbol{E}_{g} \begin{pmatrix} v_{\mathrm{g}}^{\alpha} - e_{\mathrm{g}}^{\alpha} \\ v_{\mathrm{g}}^{\beta} - e_{\mathrm{g}}^{\beta} \end{pmatrix} - \begin{pmatrix} \frac{R_{\mathrm{g}}}{L_{\mathrm{g}}} P + \omega_{\mathrm{g}} Q \\ \frac{R_{\mathrm{g}}}{L_{\mathrm{g}}} Q - \omega_{\mathrm{g}} P \end{pmatrix}. \tag{13}$$

Equation (3) subtracts Equation (13), deriving the small-signal model of the grid side as

$$\begin{pmatrix} g_{\delta P} \\ g_{\delta Q} \end{pmatrix} = \frac{1}{L_{\mathrm{g}}} \boldsymbol{E}_{g} \begin{pmatrix} \delta v_{\mathrm{g}}^{\alpha} \\ \delta v_{\mathrm{g}}^{\beta} \end{pmatrix} - \begin{pmatrix} \frac{R_{\mathrm{g}}}{L_{\mathrm{g}}} \delta P + \omega_{\mathrm{g}} \delta Q \\ \frac{R_{\mathrm{g}}}{L_{\mathrm{g}}} \delta Q - \omega_{\mathrm{g}} \delta P \end{pmatrix}, \tag{14}$$

where $(\delta P, \delta Q)^{\top} = \delta \boldsymbol{S}$ represents the complex power variation and $\delta v_{\mathrm{g}} = v_{\mathrm{g}} - v_{\mathrm{g}}^{\mathrm{s}}$ indicates the relationship between control voltage vectors and the steady-state voltage vector. For instance, $\delta v_{\mathrm{g|i}} = v_{\mathrm{g|i}} - v_{\mathrm{g}}^{\mathrm{s}}$ represents the vector between the steady-state voltage and the ith converter voltage vector.

Converting to discrete form, Equation (14) yields

$$\begin{pmatrix} \delta P \\ \delta Q \end{pmatrix} = \frac{T_s}{L_g} \boldsymbol{E}_g \begin{pmatrix} \delta v_g^{\alpha} \\ \delta v_g^{\beta} \end{pmatrix}, \tag{15}$$

where T_s denotes the sampling interval. Now the small-signal model can be introduced into the controller with measurements of the grid-side voltages. Note that $T_s \cdot \delta \boldsymbol{S}$ are small enough to be neglected. Similarly, the discrete small-signal model of PMSG can be derived as

$$\begin{pmatrix} \delta i_m^d \\ \delta i_m^q \end{pmatrix} = \frac{T_s}{L_s} \begin{pmatrix} \delta v_m^d \\ \delta v_m^q \end{pmatrix}. \tag{16}$$

The small-signal equations illustrate the relationship between the state variables' variation and the relevant control vectors. With these equations, the LUTs have the opportunity to update variation caused by all control vectors at the same interval.

4.2. Look-Up Table Update Principle

The steady-state vector v_g^s can not easily be obtained by the sampling process. The elimination of this vector from the small-signal model (see Equation (15)) relies on the iterative calculation among the converter voltage vectors. Based on Equation (15), two former voltage vectors ($\delta v_{g|k-1}$ and $\delta v_{g|k-2}$) with their relevant power variation ($\delta P_{|k-1}, \delta Q_{|k-1}$ and $\delta P_{|k-2}, \delta Q_{|k-2}$) can be derived as

$$\begin{pmatrix} \delta P_{|k-1} - \delta P_{|k-2} \\ \delta Q_{|k-1} - \delta Q_{|k-2} \end{pmatrix} = \frac{T_s}{L_g} \boldsymbol{E}_g \begin{pmatrix} \delta v_{g|i}^{\alpha} - \delta v_{g|j}^{\alpha} \\ \delta v_{g|i}^{\beta} - \delta v_{g|j}^{\beta} \end{pmatrix}. \tag{17}$$

Note that, given $\delta v_{g|k-1} = v_{g|k-1} - v_g^s$, $\delta v_{g|k-2} = v_{g|k-2} - v_g^s$, we obtain $\delta v_{g|k-1} - \delta v_{g|k-2} = v_{g|k-1} - v_{g|k-2}$.

Hence, the steady-state vector v_g^s can be eliminated from Equation (17), and Equation (17) can be rewritten as

$$\begin{pmatrix} \delta P_{|k-1} - \delta P_{|k-2} \\ \delta Q_{|k-1} - \delta Q_{|k-2} \end{pmatrix} = \frac{T_s}{L_g} \boldsymbol{E}_g \begin{pmatrix} v_{g|i}^{\alpha} - v_{g|j}^{\alpha} \\ v_{g|i}^{\beta} - v_{g|j}^{\beta} \end{pmatrix}. \tag{18}$$

Based on Equation (18), the state variations can be derived using information from interval $k-1$ and $k-2$ as

$$\begin{pmatrix} \Delta P_{|i}(k-1) - \Delta P_{|j}(k-2) \\ \Delta Q_{|i}(k-1) - \Delta Q_{|j}(k-2) \end{pmatrix} = \frac{T_s}{L_g} \boldsymbol{E}_g \times \begin{pmatrix} v_{g|i}^{\alpha}(k-1) - v_{g|j}^{\alpha}(k-2) \\ v_{g|i}^{\beta}(k-1) - v_{g|j}^{\beta}(k-2) \end{pmatrix}, \tag{19}$$

where $\boldsymbol{v}_{g|i}(k-1)$ and $\boldsymbol{v}_{g|j}(k-2)$ are the ith and jth converter vector, which are applied at the $k-1$ and $k-2$ instants, respectively; $\Delta \boldsymbol{S}_{|i}(k-1) = \boldsymbol{S}(k) - \boldsymbol{S}(k-1), \Delta \boldsymbol{S}_{|j}(k-2) = \boldsymbol{S}(k-1) - \boldsymbol{S}(k-2)$ represent state variation calculated with sampling values.

We then change the special ith vector at $k-1$ instant to all vectors (zth, $z \in \{1, 2, \cdots, 8\}$) that may be selected, and their relevant power variation can be estimated as

$$\begin{pmatrix} \Delta P_{|z}(k-1) - \Delta P_{|j}(k-2) \\ \Delta Q_{|z}(k-1) - \Delta Q_{|j}(k-2) \end{pmatrix} = \frac{T_s}{L_g} \boldsymbol{E}_g \times \begin{pmatrix} v_{g|z}^{\alpha}(k-1) - v_{g|j}^{\alpha}(k-2) \\ v_{g|z}^{\beta}(k-1) - v_{g|j}^{\beta}(k-2) \end{pmatrix}. \tag{20}$$

So far, a relationship has been found between the selected voltage vector at the instant $k-2$, with its relevant change in complex power, and the possible complex power variation with all candidate control vectors at instant $k-1$.

Note that the calculation of Equation (20) still relies on the inductance parameter. To remove this parameter from status estimation, Equations (19) and (20) can be combined and Equations (21) and (22) can be derived, only introducing the control vector applied at instant $k-1$ and the corresponding status variation, as

$$\Delta P_{|z}(k-1) = \frac{\left(v^{\alpha}_{g|z}(k-1) - v^{\alpha}_{g|j}(k-2)\right)e^{\alpha}_{g} + \left(v^{\beta}_{g|z}(k-1) - v^{\beta}_{g|j}(k-2)\right)e^{\beta}_{g}}{\left(v^{\alpha}_{g|i}(k-1) - v^{\alpha}_{g|j}(k-2)\right)e^{\alpha}_{g} + \left(v^{\beta}_{g|i}(k-1) - v^{\beta}_{g|j}(k-2)\right)e^{\beta}_{g}} \times \left(\Delta P_{|i}(k-1) - \Delta P_{|j}(k-2)\right) + \Delta P_{|j}(k-2), \quad (21)$$

$$\Delta Q_{|z}(k-1) = \frac{\left(v^{\alpha}_{g|z}(k-1) - v^{\alpha}_{g|j}(k-2)\right)e^{\beta}_{g} - \left(v^{\beta}_{g|z}(k-1) - v^{\beta}_{g|j}(k-2)\right)e^{\alpha}_{g}}{\left(v^{\alpha}_{g|i}(k-1) - v^{\alpha}_{g|j}(k-2)\right)e^{\beta}_{g} - \left(v^{\beta}_{g|i}(k-1) - v^{\beta}_{g|j}(k-2)\right)e^{\alpha}_{g}} \times \left(\Delta Q_{|i}(k-1) - \Delta Q_{|j}(k-2)\right) + \Delta Q_{|j}(k-2). \quad (22)$$

For the generator side, following a similar theoretical calculation process, stator current variation update equations are estimated using the last two current measurements as

$$\Delta \boldsymbol{I}_{m|z}(k-1) = \frac{\left(\boldsymbol{v}^{dq}_{m|z}(k-1) - \boldsymbol{v}^{dq}_{m|j}(k-2)\right)}{\left(\boldsymbol{v}^{dq}_{m|i}(k-1) - \boldsymbol{v}^{dq}_{m|j}(k-2)\right)} \times \left(\Delta \boldsymbol{I}_{m|i}(k-1) - \Delta \boldsymbol{I}_{m|j}(k-2)\right) + \Delta \boldsymbol{I}_{m|j}(k-2), \quad (23)$$

where $\Delta \boldsymbol{I}_{m|i}(k-1), \Delta \boldsymbol{I}_{m|j}(k-2)$ are the calculated current variation corresponding to applied vectors $\boldsymbol{v}^{dq}_{m|i}(k-1), \boldsymbol{v}^{dq}_{m|j}(k-2)$; and where $\Delta \boldsymbol{I}_{m|i}(k-1) = \boldsymbol{I}_{m}(k) - \boldsymbol{I}_{m}(k-1), \Delta \boldsymbol{I}_{m|j}(k-2) = \boldsymbol{I}_{m}(k-1) - \boldsymbol{I}_{m}(k-2)$. $\Delta \boldsymbol{I}_{m|z}(k-1)(z \in \{1,2,\cdots,8\})$ denote estimated current variation under all candidate MSC output vectors at instant $k-1$.

Furthermore, the denominators in Equations (21)–(23) need to remain non-zero while estimating the status variation update (Equations (21)–(23)). In this work, the LUTs are updated when the denominators in these equations outweigh a certain value. Otherwise, the LUTs maintain as their values the previous interval. By obtaining the variation updating mechanism under all control vectors of back-to-back power converters, the controller guarantees the LUTs' update frequency and skips *updating-lag* completely.

The process of this method is summarized as follows. Firstly, the controller obtains the current state value of the system by sampling and calculates the variation between instant $k-2$ and $k-1$ by reading the value of the previous moment. Secondly, it uses Equations (21)–(23) to update all candidate control vectors with their possible variation in one control interval. Finally, the controller uses the generator side and grid side LUTs instead of the system model to calculate the future stator currents and to power-predict the system variables for the finite set of voltage vectors, completing the compensation and prediction and selecting the optimal vector that minimizes the cost function.

5. Verification

This section investigates the control performances of the proposed MIPC and the classical DMPC, which validate the effectiveness of the proposed method. The system parameters are listed in Table 1.

Table 1. System configuration.

Parameter	Value	Parameter	Value
DC-link Voltage V_d(V)	600	DC Capacitance C (F)	1100×10^{-6}
PMSG Inductance $L^d_s = L^q_s$ (H)	19.43×10^{-3}	PMSG Resistance R_s (Ω)	0.14
Nominal Torque T^n_e ($N \cdot m$)	29	Nominal Power P^n (kVA)	3.475
PM Flux ψ_{pm} (Wb)	0.43	PMSG Pole Pairs N_p (-)	3
PMSG Inertia J ($kg \cdot m^2$)	0.01	Grid Voltage $\boldsymbol{e}^{abc}_g$ (V)	$210/\sqrt{2}$
Grid Frequency ω_g (rad/s)	100π	Filter Resistance R_g (Ω)	1.56×10^{-3}
Filter Inductance L_g (H)	16×10^{-3}	Sampling Time T_s (μs)	50

5.1. Overall Validation of the Proposed Method

The test at various operating points is conducted in order to confirm the effectiveness of the system's overall control. Depending on the wind speed, an optimal torque reference T_e^* or speed reference ω_{m}^* is determined by a proper "maximum power point tracking" (MPPT) control. While the reference generation techniques are not the central focus of this work, for simplicity, the test scenario is designed as follows: the "MPPT speed reference" ω_{m}^* has multiple changes, with a particularly steep slope to test the roughest conditions; the DC-link voltage reference V_{d}^* remains 600 V during the whole process; reactive power reference is set at 0 var to achieve unity power factor control.

Figure 3a shows the overall performances of the proposed method, and the zoomed performances are given in Figure 3b. The waveforms show that the proposed MIPC achieved good steady and transient state performances globally. The smooth and good tracking of speed, current, and power for both sides of the back-to-back converter is obtained, and the DC-link voltage remains stable in both the steady state and the transient state.

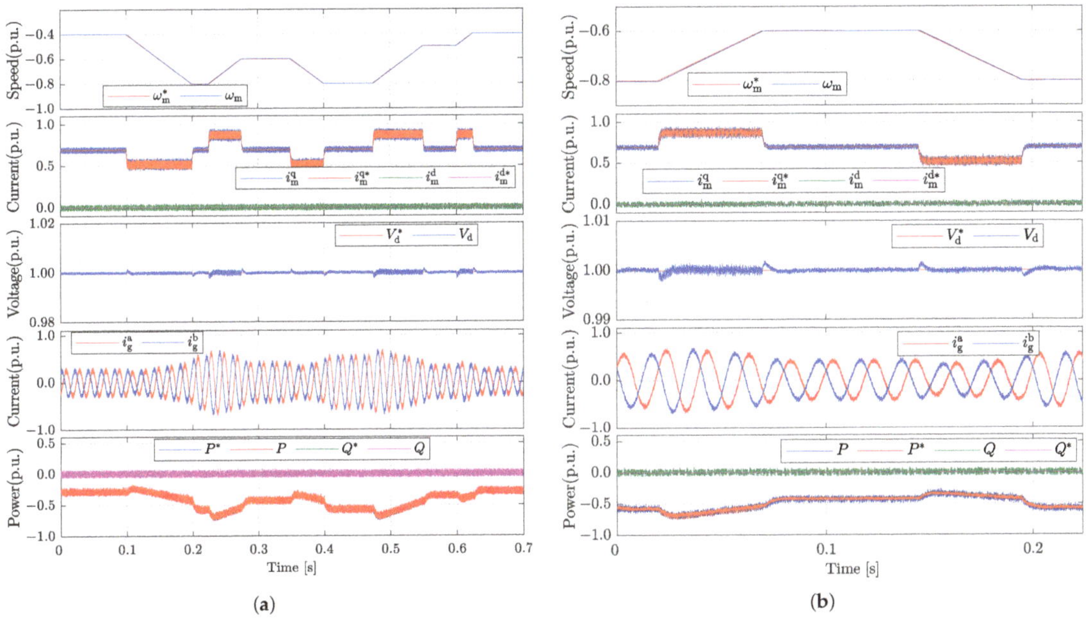

Figure 3. Performance of the proposed MIPC method. (**a**) Overall control performance. (**b**) Zoomed control performance. From top to bottom are the PMSG mechanical speed, stator dq-axis currents, DC-link voltage, grid-side currents, active and reactive power, and their references, respectively. The speed base is 125 r/min. The current base is 15 A.

5.2. Robustness Comparison

Under various parameter mismatch conditions, the control performances of the proposed MIPC and the classical DPMC method are compared in this section. For a fair evaluation, the same test scenarios were created for both control methods in each condition.

In the first test, the permanent-magnet flux linkage in the controller is varied to 50% and 200% of the actual value (ψ_{pm}) to investigate the influence of flux variation. As can be seen from Figure 4, flux mismatch will mainly lead to torque tracking bias. For 50% flux error, the torque is 4.5% larger than the reference. For 200% flux error, the torque is 5.8% smaller than the reference. This phenomenon is in accordance with the analytical analysis presented in [16].The control performance of the proposed MIPC is unaffected for the permanent-magnet flux linkage variations, which is in line with the principle that ψ_{pm} is not introduced throughout the control process of this method.

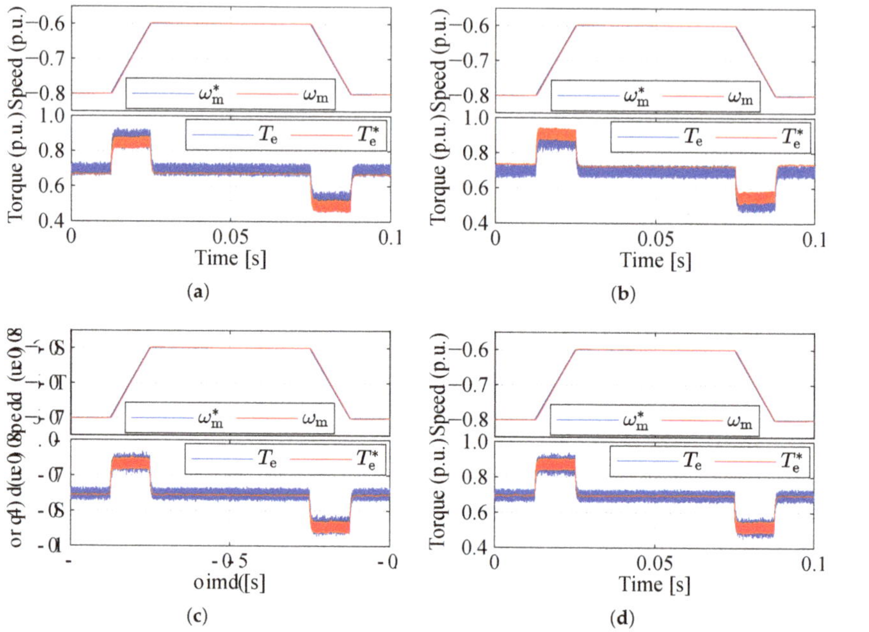

Figure 4. Performances under permanent-magnet flux linkage variations: (**a**) Classical DMPC (50% ψ_{pm}); (**b**) Classical DMPC (200% ψ_{pm}); (**c**) Classical DMPC (100% ψ_{pm}); (**d**) Proposed MIPC. For all sub-figures, from top to bottom are the generator speed (base value 125 [rad/s]) and the generator torque (base value 29 [N · m]), respectively.

In the controller of the classical DMPC solution, we set the inductance of the filter and PMSG stator to vary from 50% to 200% compared to the actual value in the wind turbine system plant. The results are shown in Figure 5 (generator side) and Figure 6 (grid side). Obviously, inaccurate inductance parameters will cause increased ripples (both dq-axis currents and active and reactive power) and enlarge current THDs. The stator current THD increased from 2.14% with the nominal parameter to 3.321% with 0.5 L_s and to 2.968% with 2 L_s. The grid-side performances in terms of current THD both exceed 4% with 0.5 L_g and with 2 L_g, which represents deterioration in power quality compared to the performance with the nominal parameter (3.712%). The control performances under various parameters' mismatches are all collected in Table 2. The results are in accordance with the theoretical analysis, i.e., the controller will derive the wrong prediction of the system utilizing mismatched parameters relating to the selection of the non-optimal control vectors. This will seriously affect the control performance. The control performance of the proposed MIPC is unaffected for all parameter variations, which is in line with the principle that no parameters are introduced throughout the control process of this method. In this section, we verify the good robustness of the proposed model-independent approach compared to the classical DMPC.

Table 2. Comparative test data of classical DMPC and proposed MIPC.

Control Method	Maximum Torque Error	Generator Current THD	Grid Current THD
Classical DMPC	4.50% (50% ψ_{pm})	3.32% (50% L_s)	4.05% (50% L_g)
	5.80% (200% ψ_{pm})	2.97% (200% L_s)	6.28% (200% L_g)
	0.73% (100% ψ_{pm})	2.15% (100% L_s)	3.71% (100% L_g)
Proposed MIPC	0.75%	2.09%	3.66%

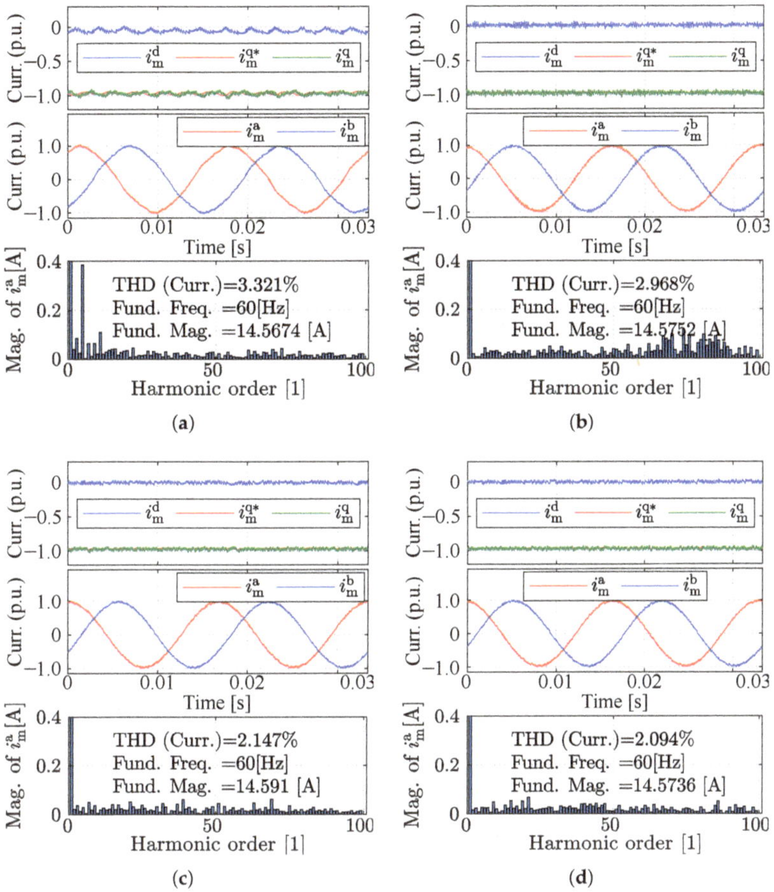

Figure 5. Generator side: (**a**) Classical DMPC (50% L_s); (**b**) Classical DMPC (200% L_s); (**c**) Classical DMPC (100% L_s); (**d**) Proposed MIPC. For all sub-figures, from top to bottom are the dq-axis stator current and their references (base value 15 [A]), phase currents, and the current spectrum, respectively.

In addition, the Classical DMPC and proposed MIPC under several simultaneous parameter mismatches are also tested. In the DMPC controller, the permanent-magnet flux linkage and the inductance of the PMSG stator are set to 50% as the actual value, while the filter inductance is set to 200% as the actual value. The performances are given in Figure 7. The tracking of torque and power as well as the current distortion show that the control performance of the proposed MIPC does not deteriorate even in the face of multiple mismatch at the same time and still shows higher robustness than the classical DMPC.

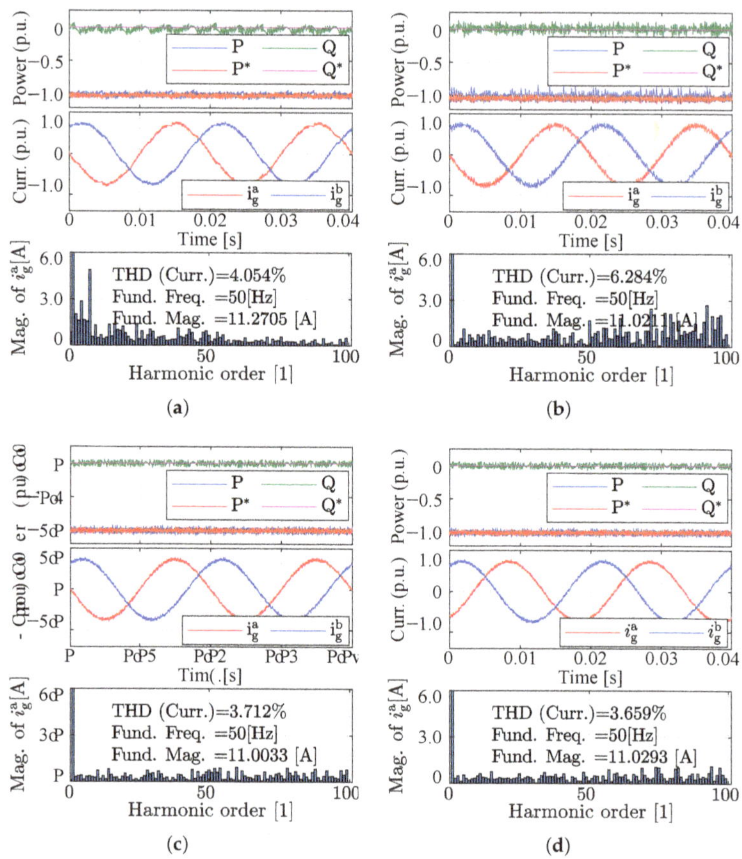

Figure 6. Grid side: (**a**) Classical DMPC (50% L_g); (**b**) Classical DMPC (200% L_g); (**c**) Classical DMPC (100% L_g); (**d**) Proposed MIPC. For all sub-figures, from top to bottom are the active and reactive power (base value 3475 [W]) and their references, phase currents, and the current spectrum (base value 15 [A]), respectively.

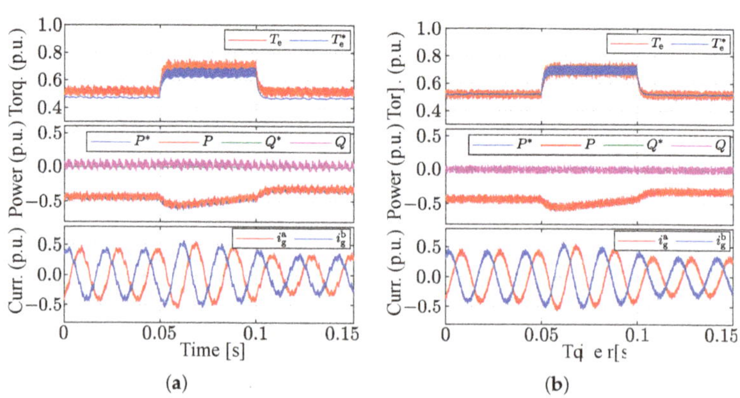

Figure 7. The control performance in the face of multiple mismatches: (**a**) Classical DMPC; (**b**) Proposed MIPC. For all sub-figures, from top to bottom are the generator torque (base value 30 [Nm]), the active and reactive power (base value 3475 [W]) and their references, and the phase currents on the grid side (base value 15 [A]), respectively.

5.3. Current/Power Update Mechanism Comparison between the Proposed and the Classical MIPC

System prediction is the key to the control performance of the predictive controller. In the proposed MIPC, the fast and accurate state variable variation estimation contributes to the accurate prediction of the system (see Figure 8). The proposed MIPC updates current and power variation for all possible voltage vectors during one control interval by means of the measured $k-1$ and $k-2$ instant values. Fast variable variation update frequency can be assured; see Figure 9a. The classical MIPC scheme in [24] updates the current and power variation only once for one voltage vector during the whole control interval. *Stagnant* current and power variation appear when one voltage vector is not applied for long consecutive control intervals; see Figure 9b. Comparing the sampled values of the system with the predicted values calculated using the LUTs in Figure 8, the two overlap, indicating that the proposed model-independent predictive control accurately predicts the trajectory of the system at future moments, which guarantees the control performance.

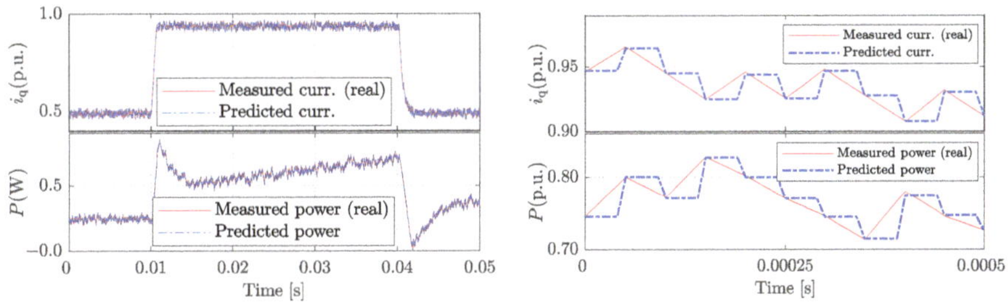

Figure 8. Prediction accuracy of the proposed MIPC. (**a**) Q-axis stator current and grid power prediction validation. (**b**) Zoomed comparison between measurement and prediction.

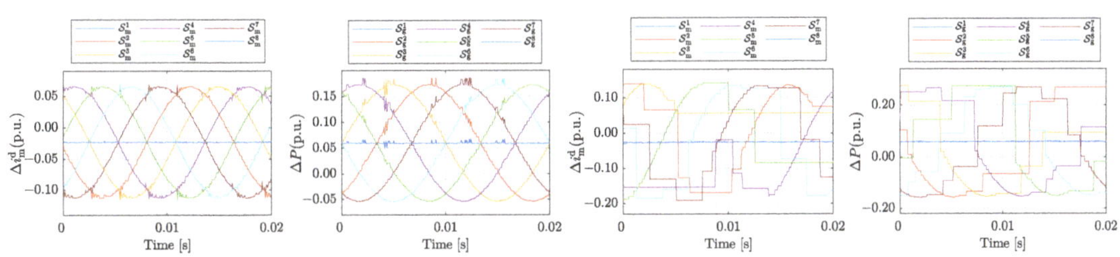

Figure 9. Estimated d-axis current variation and active power variation caused by different voltage vectors using (**a**) the proposed MIPC and (**b**) classical MIPC [24]. S_x^1–S_x^8 denote the available voltage vectors of the machine side ($x=m$) and the grid side ($x=g$), respectively.

6. Conclusions

Constrained by its complete dependence on the model, conventional direct model predictive control easily exhibits deterioration in control performance when the model parameters are mismatched. State-of-the-art model-independent predictive control (MIPC) introduces historical operation data in the prediction of future statuses. Nevertheless, it suffers from low look-up table (LUT) update frequency, unsteady state variable changing rate, and extensive computational burden. This work proposed an improved MIPC with a new look-up table update method that only introduces the information from the former two instants to estimate all needed variations in the same period. Compared with the traditional finite-set model predictive control and the existing MIPC, the proposed solution achieves robustness to unmeasurable and time-varying parameters without sacrificing control performance. The proposed method can be applied to other power converter topologies with minor modifications. Future work will focus on addressing measurement robustness and extending the proposed methods to multilevel power conversion systems.

Author Contributions: Conceptualization, Z.Z. and X.L.; methodology, Y.Z. and X.L.; software, X.L. and H.L.; validation, H.L. and Y.Z.; formal analysis, Y.Z.; investigation, Y.Z. and X.L.; resources, Y.Z.; data curation, Y.Z.; writing—original draft preparation, Y.Z. and X.L.; writing—review and editing, Y.Z. and Z.Z.; visualization, Y.Z. and X.L.; supervision, Z.Z.; project administration, Z.Z.; funding acquisition, Z.Z. All authors have read and agreed to the published version of the manuscript.

Funding: This research was funded by National Key R & D Program of China under 2022YFB4201700, and in part by the General Program of National Natural Science Foundation of China under Grants 51977124, 52007107, and 52277191.

Data Availability Statement: Data is available upon reasonable request to the corresponding author.

Conflicts of Interest: The authors declare no conflict of interest.

References

1. Nadeem, M.; Zheng, X.; Tai, N.; Gul, M.; Tahir, S. Analysis of Propagation Delay for Multi-Terminal High Voltage Direct Current Networks Interconnecting the Large-Scale Off-Shore Renewable Energy. *Energies* **2018**, *11*, 2115. [CrossRef]
2. Sun, Y.; Zhang, Z.; Zhang, Y.; Li, Y.; Li, Z. A Time-Domain Virtual-Flux Based Predictive Control of Modular Multilevel Converters for Offshore Wind Energy Integration. *IEEE Trans. Energy Convers.* **2021**, *37*, 1803–1814. [CrossRef]
3. Steffen, J.; Lengsfeld, S.; Jung, M.; Ponick, B.; Herranz Gracia, M.; Spagnolo, A.; Klöpzig, M.; Schleicher, K.; Schäfer, K. Design of a Medium Voltage Generator with DC-Cascade for High Power Wind Energy Conversion Systems. *Energies* **2021**, *14*, 3106. [CrossRef]
4. Blaabjerg, F.; Ma, K. Future on Power Electronics for Wind Turbine Systems. *IEEE J. Emerg. Sel. Top. Power Electron.* **2013**, *1*, 139–152. [CrossRef]
5. Liserre, M.; Cardenas, R.; Molinas, M.; Rodriguez, J. Overview of Multi-MW Wind Turbines and Wind Parks. *IEEE Trans. Ind. Electron.* **2011**, *58*, 1081–1095. [CrossRef]
6. Zhang, Z.; Li, Z.; Kazmierkowski, M.P.; Rodriguez, J.; Kennel, R. Robust Predictive Control of Three-Level NPC Back-to-Back Power Converter PMSG Wind Turbine Systems with Revised Predictions. *IEEE Trans. Power Electron.* **2018**, *33*, 9588–9598. [CrossRef]
7. Singh, V.; Tripathi, R.; Hanamoto, T. HIL Co-Simulation of Finite Set-Model Predictive Control Using FPGA for a Three-Phase VSI System. *Energies* **2018**, *11*, 909. [CrossRef]
8. Zhang, J.; Sun, T.; Wang, F.; Rodriguez, J., Kennel, R. A Computationally Efficient Quasi-Centralized DMPC for Back-to-Back Converter PMSG Wind Turbine Systems without DC-Link Tracking Errors. *IEEE Trans. Ind. Electron* **2016**, *63*, 6160–6171. [CrossRef]
9. Zhang, Z.; Wang, F.; Wang, J.; Rodriguez, J.; Kennel, R. Nonlinear Direct Control for Three-Level NPC Back-to-Back Converter PMSG Wind Turbine Systems: Experimental Assessment With FPGA. *IEEE Trans. Ind. Inform.* **2017**, *13*, 1172–1183. [CrossRef]
10. Zhang, Z.; Fang, H.; Gao, F.; Rodriguez, J.; Kennel, R. Multiple-Vector Model Predictive Power Control for Grid-Tied Wind Turbine System with Enhanced Steady-State Control Performance. *IEEE Trans. Ind. Electron.* **2017**, *64*, 6287–6298. [CrossRef]
11. Lyu, Z.; Wu, X.; Gao, J.; Tan, G. An Improved Finite-Control-Set Model Predictive Current Control for IPMSM under Model Parameter Mismatches. *Energies* **2021**, *14*, 6342. [CrossRef]
12. Xu, Y.; Ding, X.; Wang, J.; Wang, C. Robust Three-vector-based Low-complexity Model Predictive Current Control with Supertwisting-algorithm-based Second-order Sliding-mode Observer for Permanent Magnet Synchronous Motor. *IET Power Electron.* **2019**, *12*, 2895–2903. [CrossRef]
13. Zhang, Y.; Wu, Z.; Yan, Q.; Huang, N.; Du, G. An Improved Model-Free Current Predictive Control of Permanent Magnet Synchronous Motor Based on High-Gain Disturbance Observer. *Energies* **2022**, *16*, 141. [CrossRef]
14. Gao, P.; Zhang, G.; Lv, X. Model-Free Control Using Improved Smoothing Extended State Observer and Super-Twisting Nonlinear Sliding Mode Control for PMSM Drives. *Energies* **2021**, *14*, 922. [CrossRef]
15. Abdelrahem, M.; Hackl, C.M.; Zhang, Z.; Kennel, R. Robust Predictive Control for Direct-Driven Surface-Mounted Permanent-Magnet Synchronous Generators without Mechanical Sensors. *IEEE Trans. Energy Convers.* **2018**, *33*, 179–189. [CrossRef]
16. Zhang, X.; Zhang, L.; Zhang, Y. Model Predictive Current Control for PMSM Drives with Parameter Robustness Improvement. *IEEE Trans. Power Electron.* **2019**, *34*, 1645–1657. [CrossRef]
17. Zhang, X.; Cheng, Y.; Zhao, Z.; He, Y. Robust Model Predictive Direct Speed Control for SPMSM Drives Based on Full Parameter Disturbances and Load Observer. *IEEE Trans. Power Electron.* **2020**, *35*, 8361–8373. [CrossRef]
18. Hu, M.; Yang, F.; Liu, Y.; Wu, L. Finite Control Set Model-Free Predictive Current Control of a Permanent Magnet Synchronous Motor. *Energies* **2022**, *15*, 1045. [CrossRef]

19. Siami, M.; Khaburi, D.A.; Abbaszadeh, A.; Rodriguez, J. Robustness Improvement of Predictive Current Control Using Prediction Error Correction for Permanent-Magnet Synchronous Machines. *IEEE Trans. Ind. Electron.* **2016**, *63*, 3458–3466. [CrossRef]
20. Siami, M.; Khaburi, D.A.; Rodriguez, J. Torque Ripple Reduction of Predictive Torque Control for PMSM Drives With Parameter Mismatch. *IEEE Trans. Power Electron.* **2017**, *32*, 7160–7168. [CrossRef]
21. Liu, X.; Zhou, L.; Wang, J.; Gao, X.; Li, Z.; Zhang, Z. Robust Predictive Current Control of Permanent-Magnet Synchronous Motors with Newly Designed Cost Function. *IEEE Trans. Power Electron.* **2020**, *35*, 10778–10788. [CrossRef]
22. Sabzevari, S.; Heydari, R.; Mohiti, M.; Savaghebi, M.; Rodriguez, J. Model-Free Neural Network-Based Predictive Control for Robust Operation of Power Converters. *Energies* **2021**, *14*, 2325. [CrossRef]
23. Lin, C.-K.; Liu, T.-H.; Yu, J.; Fu, L.-C.; Hsiao, C.-F. Model-Free Predictive Current Control for Interior Permanent-Magnet Synchronous Motor Drives Based on Current Difference Detection Technique. *IEEE Trans. Ind. Electron.* **2014**, *61*, 667–681. [CrossRef]
24. Lin, C.-K.; Yu, J.; Lai, Y.-S.; Yu, H.-C. Improved Model-Free Predictive Current Control for Synchronous Reluctance Motor Drives. *IEEE Trans. Ind. Electron.* **2016**, *63*, 3942–3953. [CrossRef]
25. Carlet, P.G.; Tinazzi, F.; Bolognani, S.; Zigliotto, M. An Effective Model-Free Predictive Current Control for Synchronous Reluctance Motor Drives. *IEEE Trans. Ind. Appl.* **2019**, *55*, 3781–3790. [CrossRef]
26. Zhou, D.; Jiang, C.; Quan, Z.; Li, Y. Vector Shifted Model Predictive Power Control of Three-Level Neutral-Point-Clamped Rectifiers. *IEEE Trans. Ind. Electron.* **2020**, *67*, 7157–7166. [CrossRef]
27. Zhang, Z.B. On Control of Grid-Tied Back-to-Back Power Converters and Pmsg Wind Turbine Systems. Ph.D. Dissertation, Technical University of Munich, Munich, Germany, 2016.
28. Zhang, Z.; Wang, F.; Sun, T.; Rodriguez, J.; Kennel, R. FPGA-Based Experimental Investigation of a Quasi-Centralized Model Predictive Control for Back-to-Back Converters. *IEEE Trans. Power Electron.* **2016**, *31*, 662–674. [CrossRef]
29. Cortes, P.; Rodriguez, J.; Silva, C.; Flores, A. Delay Compensation in Model Predictive Current Control of a Three-Phase Inverter. *IEEE Trans. Ind. Electron.* **2012**, *59*, 1323–1325. [CrossRef]
30. Wang, F.; Zhang, Z.; Mei, X.; Rodríguez, J.; Kennel, R. Advanced Control Strategies of Induction Machine: Field Oriented Control, Direct Torque Control and Model Predictive Control. *Energies* **2018**, *11*, 120. [CrossRef]

energies

MDPI

Article

Annual Energy Production Design Optimization for PM Generators Considering Maximum Power Point Trajectory of Wind Turbines

Huaping Yang [1], Wenjuan Zhang [2,*], Litao Dai [1], Wan Feng [2] and Haixia Zhang [2]

[1] College of Electrical and Information Engineering, Hunan University, Changsha 410022, China; k4231622@163.com (H.Y.); lito@hnu.edu.cn (L.D.)
[2] School of Electronic Information and Electrical Engineering, Changsha University, Changsha 410022, China; fengwan@hnu.edu.cn (W.F.); z20190423@ccsu.edu.cn (H.Z.)
* Correspondence: z20141074@ccsu.edu.cn; Tel.: +86-18692208506

Abstract: Efficiency optimization is an important goal in the design of permanent magnet generators. However, traditional design optimization methods only focus on improving the rated efficiency without considering the annual cycle for overall efficiency improvement. To overcome this drawback, this paper presents a design optimization method for improving annual energy production (AEP) of wind direct-drive permanent magnet generators. Unlike the conventional efficiency optimization method that only improves the rated point efficiency, the proposed method improves the overall efficiency of the generator during the operating cycle by matching the maximum power point trajectory of the wind turbine. The periodic loss model of the permanent magnet generator is established and further constituted as the objective function to perform the optimization search using a genetic algorithm. Through simulation and experimental verification, the proposed method can obtain a higher AEP compared with the conventional design optimization method, and the proposed method can be extended to other variable speed power generation fields.

Keywords: annual energy production; permanent magnet synchronous generator; design optimization; maximum power point tracking; wind turbine

Citation: Yang, H.; Zhang, W.; Dai, L.; Feng, W.; Zhang, H. Annual Energy Production Design Optimization for PM Generators Considering Maximum Power Point Trajectory of Wind Turbines. *Energies* **2023**, *16*, 4120. https://doi.org/10.3390/en16104120

Academic Editor: Frede Blaabjerg

Received: 25 April 2023
Revised: 12 May 2023
Accepted: 13 May 2023
Published: 16 May 2023

1. Introduction

As a clean energy resource, wind power has received great research attention and achieved continuous development [1]. Because of their low failure rate and high-power-density, permanent magnet synchronous generators (PMSGs) are increasingly used in wind energy conversion systems (WECSs) [2]. In order to obtain high efficiency, reliable operation, and low cost-effectiveness of WECSs, the generator design optimization is an important aspect [3]. Power electronic converters allow synchronous generators to operate with variable speed while guaranteeing a constant electricity generation frequency [4]. In the period of variable speed operation, the PMSG needs to match the maximum energy trajectory of the turbine by regulating the mechanical speed under different operating conditions. Therefore, for PMSGs of WECSs, it is a new challenge to realize high efficiency operation under variable speed and load during the working cycle.

Currently, intelligent stochastic algorithms such as the genetic algorithm (GA) [5], evolutionary algorithm [6], and other algorithms [7] are commonly utilized to design and optimize the performances of PMSGs. They are mainly based on establishing an analytical or numerical link between structural parameters and generator performance, and then evolving the optimal combination of structural parameters to maximize or minimize the target value by random search and successive iterations.

The optimal design of a PMSG considering the uncertainty of raw material cost is proposed in [8] using a GA, and a hybrid process of GA and pattern search was used to optimize a 6 MW direct-drive PMSG to reduce the cost of energy [9]. In [10], a sensitivity analysis of structural parameters on the performance of a hybrid-excited dual-PM wind generator was carried out, and some leading parameters were identified and further optimized using a combination of finite elements and GA. High efficiency is a valuable optimization goal for generators to increase system energy production and reduce the temperature rise due to operating losses, which has been investigated in many studies [11,12].

However, most of the existing papers are only optimized for rated efficiency, i.e., the objective function only includes the generator efficiency at the rated point, without considering the improvement of the cycle operation efficiency. This may not be applicable to WECSs, where wind turbines are less likely to operate near the rated conditions due to climatic and seasonal influences. Therefore, the classical approach of design optimization with rated efficiency may instead lead to a lower AEP.

In [13], a generator's operation range is divided into four parts, these four parts are equated into four feature points, and the efficiencies of all four points are optimized collaboratively, allowing the overall efficiency to be enhanced. However, its operating conditions are relatively simple, with a fixed rotational speed of 4200 rpm. For variable speed WECS, the PMG speed needs to rise with increasing input power to improve the energy conversion efficiency, which increases the difficulty of design optimization.

In this paper, a new design optimization method is proposed to maximize the AEP of the PMSG considering the maximum power point trajectory (MPPT) of the wind turbine. First, the aerodynamic model of the wind turbine is introduced to develop the relationship between its MPPT and optimal shaft speed, and the wind speed distribution is investigated to obtain the feature operating points with multiple wind speeds. Secondly, the loss models of the PMSG with multiple feature points are formulated as the objective function, and the GA is utilized to improve the overall efficiency. Finally, the optimization scheme to improve the AEP is obtained and verified by the simulation and experiment.

2. Direct-Drive WECS

2.1. Wind Speed Distribution

With the elimination of gearboxes in direct-drive WECS, the efficiency and reliability of the system can be improved greatly. The system features a PMSG driven directly by the wind turbine and synchronizes the power generation with the grid frequency through a full-size AC/DC/AC converter, and is capable of variable speed operation in the full speed range, the structure of which is shown in Figure 1.

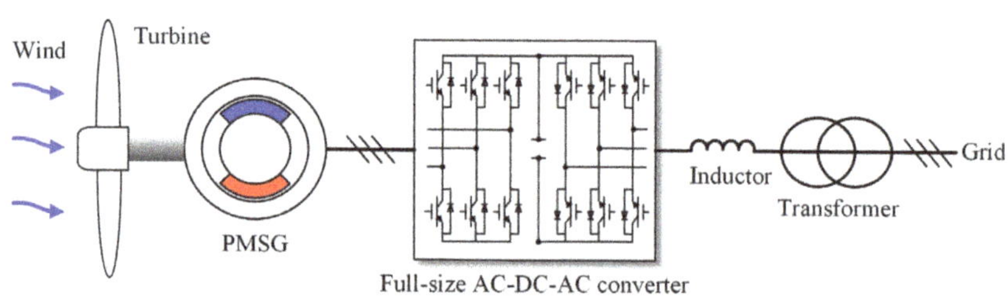

Figure 1. Structure of direct-drive WECS.

The wind speed of the wind farm conforms to the Rayleigh distribution [14], and the probability distribution is as follows:

$$p(V_\omega) = \frac{\pi}{2}\frac{V_\omega}{V_{\text{avg}}}e^{-\frac{\pi}{4}\left(\frac{V_\omega}{V_{\text{avg}}}\right)^2} \quad (1)$$

where V_ω is the wind speed, V_{avg} is the average value of the wind speed, and the annual average wind speed of the case study is about 8.5 m/s. The cut-in speed is 3 m/s, and the cut-out speed is 12 m/s. The cut-in and cut-out wind speeds are made according to the comprehensive situation of site conditions, the mechanical strength of the generation unit, operating condition characteristics, etc. On the basis of lightweight and large blade design, the cut-out wind speed is therefore relatively lower in this study.

Five feature points are selected to represent the overall distribution of wind speed for design optimization trade-offs, since too many feature points would lead to convergence failure and too few would not be representative. According to the calculation of the definite integral, the probability at each typical wind speed can be obtained as

$$\alpha_i = \frac{\int_{i_{\min}}^{i_{\max}} p(V_\omega)dV_\omega}{\int_3^{12} p(V_\omega)dV_\omega} \qquad i = 1,\ 2,\ \ldots,\ 5 \tag{2}$$

It can be calculated that α_1 is equal to 0.1017 (corresponding to the wind speed range of 3–4 m/s); α_2 is 0.2468, (4–6 m/s); α_3 is 0.2580 (6–8 m/s); α_4 is 0.2247 (8–10 m/s); and α_5 is 0.1688 (10–12 m/s). The plotted wind speed curve and the probability fitting results are shown in Figure 2.

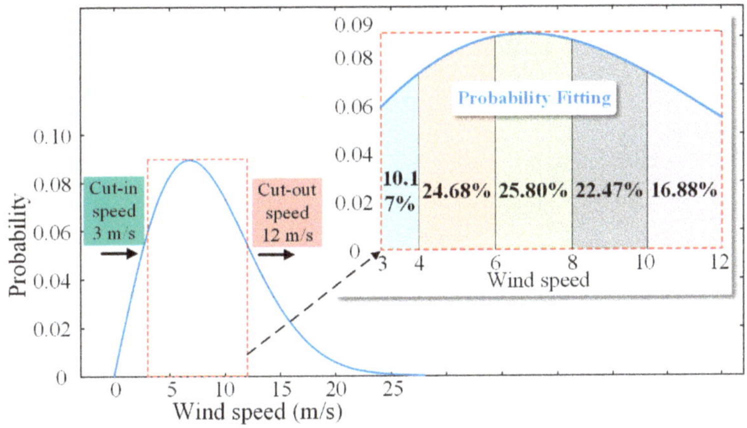

Figure 2. Wind speed Rayleigh distribution and the probability fitting of representative wind speeds.

2.2. Aerodynamic Model of Wind Turbines

The output power (mechanical power P_{mech}) of the wind turbine is given by [14]

$$P_{\text{mech}} = \frac{1}{2}\rho\pi R^2 V_\omega^3 C_p \tag{3}$$

where ρ is the air density, and R is the turbine radius. C_p is the wind turbine power coefficient, which is a function of the tip speed ratio λ and the pitch angle β, and it can be expressed as

$$C_p(\lambda,\ \beta) = 0.35\left(\frac{151}{\lambda_i} - 0.58\beta - 0.002\beta^{2.14} - 13.2\right)^{-\left(\frac{18.4}{\lambda_i}\right)} \\ \text{with } \frac{1}{\lambda_i} = \frac{1}{\lambda + 0.08\beta} - \frac{0.035}{\beta^3 + 1} \text{ and } \lambda = R\frac{\omega_r}{V_\omega} \tag{4}$$

where ω_r is the turbine rotational speed. The energy available to the generator is determined by both the wind speed as well as the turbine speed, and their relationship is shown in Figure 3.

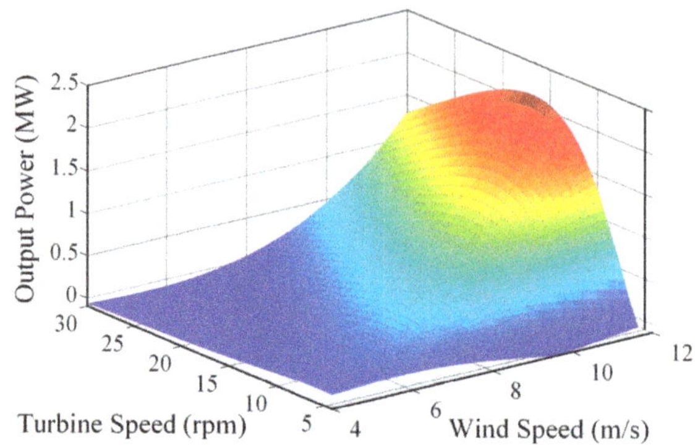

Figure 3. The three-dimensional characteristics of wind turbine energy conversion.

The variable-speed operation of the wind turbine is beneficial to improving the energy conversion efficiency of the system, from which the MPPT curve of the turbine can be obtained, as shown in Figure 4a, and the optimal turbine shaft speed corresponding to the five feature points can also be derived, as shown in Figure 4b.

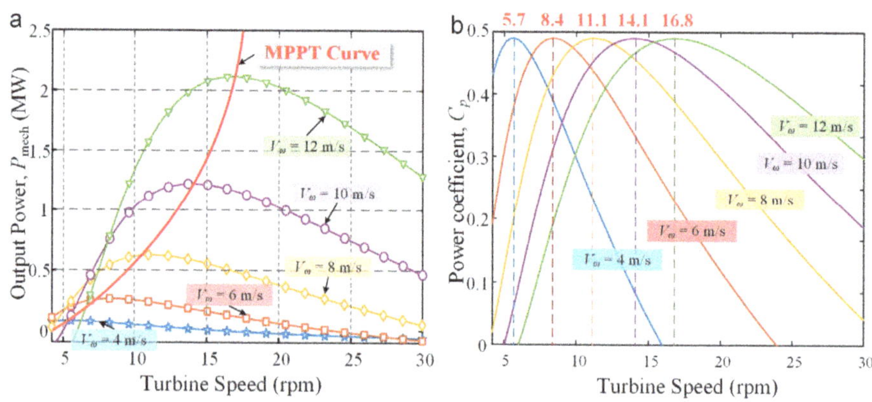

Figure 4. Operating characteristics of the wind turbine at representative wind speeds. (**a**) Output power vs. turbine speed. (**b**) Power coefficient vs. turbine speed.

Therefore, the actual operating conditions of the generator during the annual cycle can be predicted (including shaft speed, mechanical power, and operating hours) and are concisely represented as five feature operating points, as shown in Table 1. It should be noted that these characteristic points are the basis of the generator's cycle loss modeling and are closely related to the objective function in the optimization procedure.

Table 1. Feature operating points of the designed PMSG.

Point-i	Wind Speed-$V_{\omega.i}$ (m/s)	Optimal Shaft Speed-$\omega_{r.i}$ (rpm)	Input Power-$P_{int.i}$ (MW)	Weights-α_i (%)
1	4	5.7	0.1	10.17
2	6	8.4	0.3	24.68
3	8	11.1	0.6	25.80
4	10	14.1	1.2	22.47
5	12	16.8	2.1	16.88

3. Design Optimization

3.1. Optimization Process

Design optimization is performed by constructing a mathematical correlation between generator structural parameters and losses, and using GA to find the optimal weighted minimum value of losses at multiple feature points so as to obtain the structural parameters of PMSG that achieve the maximum cycle efficiency. The optimization process is shown in Figure 5.

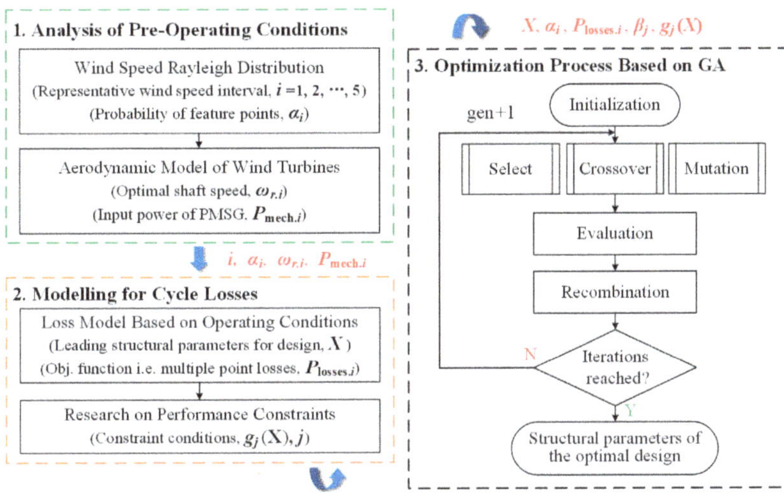

Figure 5. Flowchart of cycle efficiency design optimization.

As seen from the workflow, the optimization is divided into: (1) operating conditions pre-analysis, (2) cycle loss calculation, and (3) the stochastic algorithm optimization process, of which the operating conditions pre-analysis has been introduced in Section 2. The rest will be introduced in this section, and the optimization process is completed when the number of iterations reaches a set value.

3.2. Modelling for Cycle Losses

The accuracy of the loss model of the PM generators is crucial, and directly determines the validity of the optimization results. The model can provide a straightforward link between structural parameters, operating conditions, and losses.

The losses of permanent magnet synchronous generators come from many sources, and the total losses are expressed as follows [15]

$$P_{\text{losses}} = P_{\text{cu}} + P_{\text{iron}} + P_{\text{edPM}} + P_{\text{wind}} + P_{\text{fr}} \tag{5}$$

where P_{cu} is the copper Joule loss of stator windings, P_{iron} is the iron core loss including the stator and rotor core, P_{edPM} is the eddy current loss of PMs, P_{wind} is the windage loss, and P_{fr} is the friction loss.

First, the Joule loss of copper windings is the main component of PM machine loss, which can be expressed as [16]:

$$P_{\text{cu}} = \int_{V_{\text{w}}}^{\text{Tot vol}} \rho J^2 dV \tag{6}$$

where J is the current density, indicating the winding current density per unit area, and is affected by the skin effect and proximity effect; the current density in the winding is not uniformly distributed, especially under high frequency current; V_w is the copper area; and ρ is the electrical resistivity, which is related to the material properties and proportional to the temperature.

$$\rho = \rho_{20}(1 + \alpha_T(T_w - 20)) \tag{7}$$

where ρ_{20} is the resistivity at 20 °C, about 1.724×10^{-8} [Ohm·m] for pure copper, α_T is the temperature coefficient, which equals about 0.003862 per degree C, T_w is the winding temperature, and can be solved analytically by the thermal equivalent circuit method or numerically, where the analytical method has a significant computational speed advantage and the numerical method has a more reliable computational accuracy. In order to speed up the optimization process, the thermal network method of ANSYS Motor-CAD is used in this paper. During the iteration of the algorithm, electromagnetic and thermal calculations are required in each design case, and an example of the thermal network calculation diagram for the maximum operating condition of the generator under the natural cooling method is shown in Figure 6.

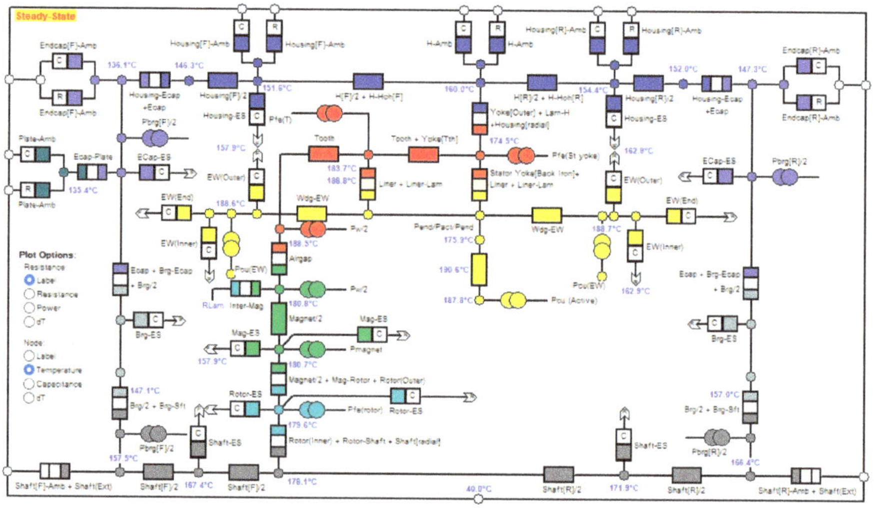

Figure 6. Diagram of thermal network circuit solving.

Next, the core loss is also the main component of the PM motor loss, especially in the case of high electrical speed and high magnetic saturation. This accounts for a significant proportion, and the core loss is expressed as follows:

$$P_{\text{iron}} = P_{\text{ironH}} + P_{\text{ironE}} \tag{8}$$

where P_{ironH}, P_{ironE} are the iron hysteresis and eddy-current loss, respectively, and they are generated in the stator and rotor iron core, and dominated in the stator. They can be expressed as [16].

$$P_{\text{ironH}} = \int_{V_{\text{ic}}}^{\text{Tot vol}} \left(\sum_{n=1}^{N} k_h n \omega_r B_n(n)^2 \right) dV \tag{9}$$

$$P_{\text{ironE}} = \int_{V_{\text{ic}}}^{\text{Tot vol}} \left(\sum_{n=1}^{N} k_c (n \omega_r)^2 B_n(n)^2 \right) dV \tag{10}$$

where V_{ic} is the volume of the iron core, k_h and k_c are the coefficients of hysteresis and eddy-current losses, n is the harmonic order of iron flux density, and $B_n(n)$ is the amplitude of the nth harmonic order of flux density. It can be seen that the iron core loss is mainly related to the parameters such as core flux density, operational speed, and iron core volume, etc.

In addition, as a surface-mounted PM machine structure, the eddy current loss of the PMs is also not negligible, and it can be calculated as [17].

$$P_{\text{edPM}} = \frac{n\omega_r}{\pi}\int_0^{\frac{2\pi}{\omega_r}}\int_{R_r}^{R_m}\int_{-\frac{\alpha p}{2}}^{\frac{\alpha p}{2}} \rho_{PM} J_m^2 r dr d\theta dt \tag{11}$$

where ρ_{PM} is the electrical resistivity of PMs, R_r, R_m are the radius of the inner and outer PM, respectively, and J_m is the induced eddy current in PMs due to the time-varying armature reaction field, which can be calculated by the 2D finite-element method.

Furthermore, P_{wind} and P_{fr} are the windage and friction losses of the PM generator, respectively, and they are related to the generator structure and the operating speed, etc. Their calculation formula are as follows [18]:

$$P_{\text{wind}} = 2k_{\text{wd}} D_{ro}^2 l_{\text{ef}} \omega_r^3 10^{-6} \tag{12}$$

$$P_{\text{fr}} = k_{\text{fb}} m_r \omega_r 10^{-3} \tag{13}$$

where k_{wd}, k_{fb} are the factors of the windage and friction loss, respectively. D_{ro} is the rotor diameter; l_{ef} is the effective length of the core; m_r is the rotor mass. It can be seen that almost all types of losses are closely related to the generator speed ω_r and the input power of the generator.

According to the investigation of the wind turbine operating conditions in Section 2.1, the loss model with multiple feature points can be developed, and the cycle loss function $P_{\text{cyc}}(X)$ is as follows:

$$P_{\text{cyc}}(X) = \sum_{i=1}^{5} \alpha_i P_{\text{losses}.i}(X) \tag{14}$$

where X is the set of design variables, and $P_{\text{losses}.i}$ is the total loss for each feature point.

3.3. Relevant Parameters for Optimization

The fixed parameters of the PMSG to be designed through the preliminary research are shown in Table 2. The material used for the stator and rotor is a silicon steel sheet with 50 mm thickness, and the material used for the PMs is NdFeB of the brand N38UH and has a residual magnetic density of 1.26 T at 20 °C.

Table 2. Fixed parameters of the designed PMSG.

Symbol	Parameter	Value
P	Rated power (kW)	2100
p/Q_S	Poles/slots	60/288
ω_M	Rated speed (rpm)	16.8
V_R	Rated voltage (V)	660
V_{dc}	DC bus voltage (V)	1100

In order to improve the operating efficiency of the PMSG at multiple wind speeds, a stochastic algorithm is used to find the optimal solution between multiple efficiency objective functions. Furthermore, multiple constraints are considered to rationalize the optimization process. The processing of the objectives and constraints is as follows:

$$F(X) = \begin{cases} P_{\text{cyc}}(X) & \text{Meet the constraints} \\ P_{\text{cyc}}(X) + \sum_{j=1}^{6} \beta_j g_j(X) & \text{Does not meet the constraints} \end{cases} \tag{15}$$

where $g_j(X)$ is the constraint function derived by the design constraints, including the flux density, current density, power density, and slot filling factor; and β_j is the penalty coefficient of the constraint conditions. Table 3 shows the parameters of the constraint conditions.

Table 3. Parameters of constraint conditions.

Parameter	Range	Parameter	Range
Power density (kW/kg)	$\geq$0.1	Current density (A/mm^2)	$\leq$3.5
Slot filling factor (%)	$\leq$80	Airgap flux density (T)	$\leq$0.8
Tooth flux density (T)	$\leq$1.85	Yoke flux density (T)	$\leq$1.65

Eight design variables with specific ranges are selected in this paper, as shown in Table 4. Figure 7 shows the PMSG structure and the geometric description of the design variables.

Table 4. Parameters of the design variables.

Symbol	Parameter	Range
D_{i1}	Stator inner diameter (mm)	3500–4500
L_1	Core length (mm)	1200–1800
h_S	Slot depth (mm)	80–120
w_T	Tooth width (mm)	15–30
h_Y	Stator yoke thickness (mm)	30–70
α_P	Pole-arc coefficient	0.5–1
h_M	PM thickness (mm)	15–30
h_δ	Air-gap thickness (mm)	4–10

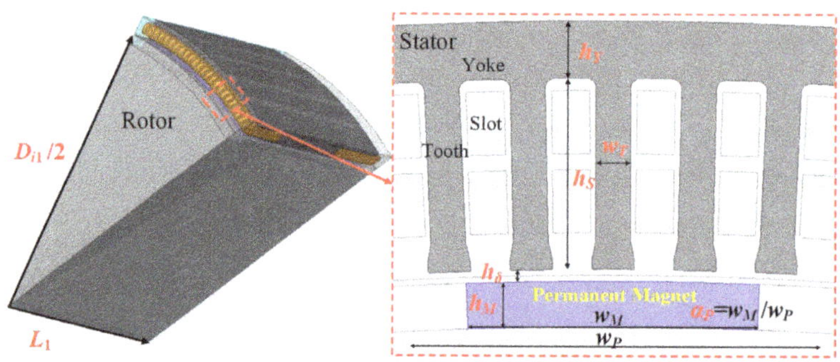

Figure 7. Generator structure and its design parameters.

4. Results and Verification

4.1. Results from Algorithm Calculation

In order to demonstrate the effectiveness of the proposed method for cycle efficiency improvement, a comparison experiment based on rated efficiency optimization (REO) is provided in this paper. The parameters of the two optimization schemes are the same except for the difference of the objective function (the objective function of the comparison experiment is the rated loss of the generator). The genetic algorithm was used to optimize the two optimization schemes in turn. For the trade-off between convergence and computational time, the number of iterations is set to 200 in this paper, and it can be seen in Figure 8 that multiple efficiencies have converged. The two optimization schemes are optimized by GA in turn. Each population of GA contains 500 individuals, and the number of generations is 200. After the iterative calculations, the variation of the fitness values for each feature operating condition is shown in Figure 8. The optimized design parameters of the proposed method are shown in Table 5.

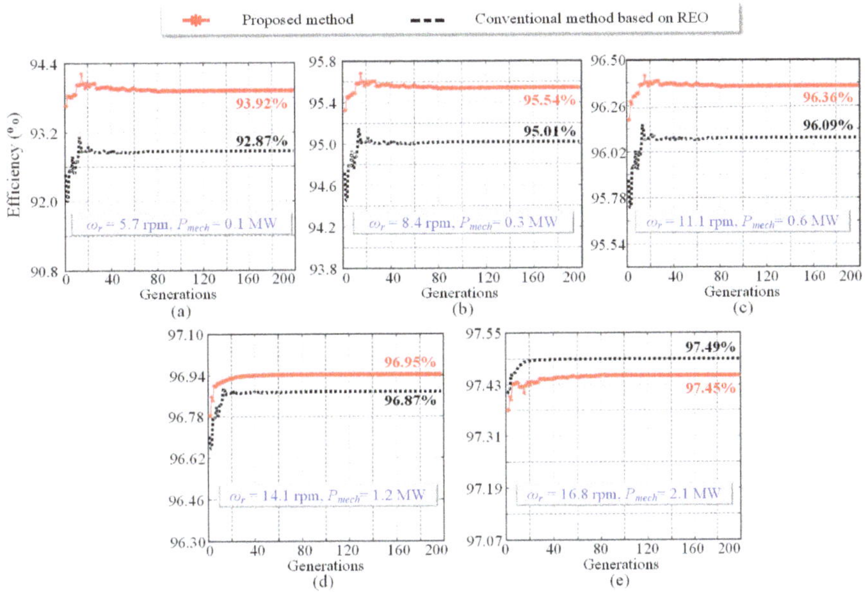

Figure 8. Iterative optimization of the proposed method and the conventional method based on REO. (**a**) Generator efficiency at the wind speed of 4 m/s. (**b**) Generator efficiency at the wind speed of 6 m/s. (**c**) Generator efficiency at the wind speed of 8 m/s. (**d**) Generator efficiency at the wind speed of 10 m/s. (**e**) Generator efficiency at the wind speed of 12 m/s.

Table 5. Optimized design parameters.

Symbol	Value	Symbol	Value
D_{i1}	3778.2 mm	h_Y	40.6 mm
L_1	1261.8 mm	α_P	0.779
h_S	96.7 mm	h_M	19.8 mm
w_T	20.5 mm	h_δ	6.2 mm

From the comparison in Figure 8, it can be seen that the conventional optimization method is effective in optimizing the rated point efficiency, which is 97.49% for a wind speed of 12 m/s, 96.87% for 10 m/s, 96.09% for 8 m/s, 95.01% for 6 m/s, and 92.87% for a wind speed of 4 m/s, while its cycle-weighted efficiency is 95.91%. On the other hand, the proposed optimization method has a higher cycle efficiency, which is 97.45% for a wind speed of 12 m/s, 96.95% for 10 m/s, 96.36% for 8 m/s, 95.54% for 6 m/s, 93.92% for a cycle-efficiency of 4 m/s, and a cycle-weighted efficiency of 96.23%. In summary, the proposed method has a higher annual cycle efficiency than the conventional REO-based method, with a weighted average efficiency of 0.32% higher.

4.2. Simulation Verification

The optimized design parameters were brought into the finite-element electromagnetic Ansys Maxwell software, and the simulated performances of the designed PMSG under various operating conditions is shown in Table 6. The results show that the performance indicators are well within the constraint conditions, and the generator has high operating efficiency under different operating conditions.

Table 6. Simulation performance of the optimized PMSG.

Performance	Value	Efficiency	Value
Power density	0.1 kW/kg	At wind speed of 4 m/s	94.32%
Current density	2.3 A/mm^2	At wind speed of 6 m/s	95.62%
Slot fill. factor	79.8%	At wind speed of 8 m/s	96.29%
Flux density in airgap/tooth/yoke	0.79 T/1.82 T/1.58 T	At wind speed of 10 m/s	96.84%
Power factor	0.97	At wind speed of 12 m/s	97.41%

4.3. Experimental Verification

The designed PMSG was fabricated and assembled, and its performance was tested on a dragging experimental platform. The photos of the test site and the performance demonstration are shown in Figure 9. The efficiencies obtained from the tests and model calculations at different speeds and powers are shown in Table 7. The results show that the experimental data are closer to the model calculations, and the effectiveness of the proposed method can be verified.

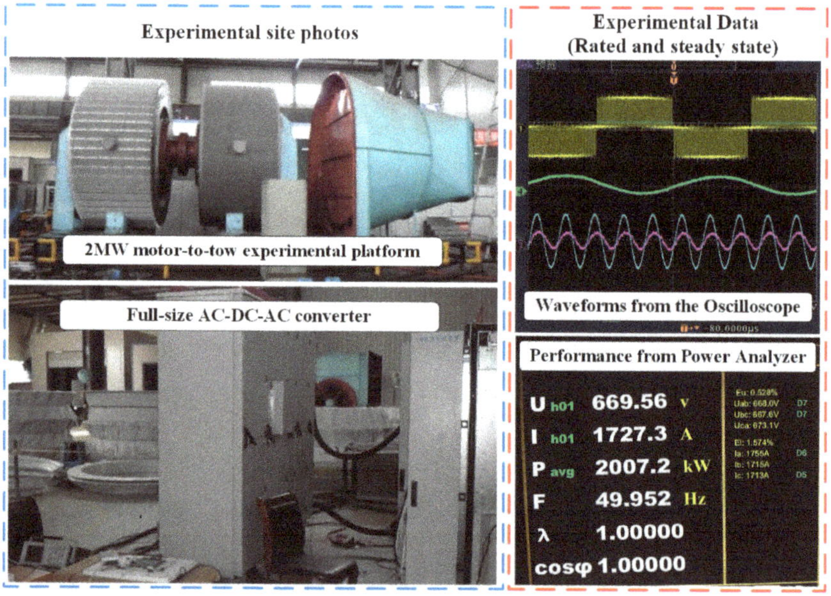

Figure 9. Experimental site photos and performance demonstration.

Table 7. Comparison of the efficiency data between model calculations and experiments.

	ω_r=5.7, P_{mech}=0.1	ω_r=8.4, P_{mech}=0.3	ω_r=11.1, P_{mech}=0.6	ω_r=14.1, P_{mech}=1.2	ω_r=16.8, P_{mech}=2.1
Model	93.92%	95.54%	96.36%	96.95%	97.45%
Test	94.01%	95.81%	96.64%	97.13%	97.67%

5. Conclusions

This paper presented a design optimization method for direct-drive wind PMSGs to improve the AEP of WECSs. The proposed method considered the optimal trajectory of the wind turbine and optimized the efficiency of several feature operating points in an integrated manner, and the method was compared with the conventional design optimization method based on the rated point efficiency. The computational results of the algorithm showed that the proposed method effectively improved the annual weighted efficiency by 0.32%, and the designed generator was validated by simulation and an experimental platform.

The proposed method is not only applicable to the design of wind turbines, but also has implications for the design and optimization of other electric machines with variable speeds and variable operating conditions.

Author Contributions: Methodology, W.Z.; Software, W.F. and H.Z.; Validation, H.Y., W.Z. and H.Z.; Formal analysis, W.Z. and L.D. All authors have read and agreed to the published version of the manuscript.

Funding: This work was supported in part by the Natural Science Foundation of China (52277034) and the Changsha Major Science and Technology Special Project of China (kq2205003).

Conflicts of Interest: The authors declare no conflict of interest.

References

1. Yaramasu, V.; Wu, B.; Sen, P.C.; Kouro, S.; Narimani, M. High-power wind energy conversion systems: State-of-the-art and emerging technologies. *Proc. IEEE* **2015**, *103*, 740–788. [CrossRef]
2. Gao, J.; Dai, L.T.; Zhang, W.J. Improved genetic optimization algorithm with subdomain model for multi-objective optimal design of SPMSM. *CES Trans. Elect. Mach. Syst. Mar.* **2018**, *2*, 160–165. [CrossRef]
3. Xu, Y.; Maki, N.; Izumi, M. Overview Study on Electrical Design of Large-Scale Wind Turbine HTS Generators. *IEEE Trans. Appl. Supercond.* **2018**, *28*, 1–5. [CrossRef]
4. Borkowski, D.; Wegiel, T. Small Hydropower Plant with Integrated Turbine-Generators Working at Variable Speed. *IEEE Trans Energy Conv. Jun.* **2013**, *28*, 452–459. [CrossRef]
5. Zhang, P.; Sizov, G.Y.; Li, M.; Ionel, D.M.; Demerdash, N.A.O.; Stretz, S.J.; Yeadon, A.W. Multi-Objective Tradeoffs in the Design Optimization of a Brushless Permanent-Magnet Machine with Fractional-Slot Concentrated Windings. *IEEE Trans. Ind. Appl.* **2014**, *50*, 3285–3294. [CrossRef]
6. Sindhya, K.; Manninen, A.; Miettinen, K.; Pippuri, J. Design of a Permanent Magnet Synchronous Generator Using Interactive Multiobjective Optimization. *IEEE Trans. Ind. Electron. Dec.* **2017**, *64*, 9776–9783. [CrossRef]
7. de Paula Machado Bazzo, T.; Kölzer, J.F.; Carlson, R.; Wurtz, F.; Gerbaud, L. Multiphysics Design Optimization of a Permanent Magnet Synchronous Generator. *IEEE Trans. Ind. Electron. Dec.* **2017**, *64*, 9815–9823. [CrossRef]
8. Pinilla, M.; Martinez, S. Optimal design of permanent-magnet direct-drive generator for wind energy considering the cost uncertainty in raw materials. *Renew. Energy* **2012**, *41*, 267–276. [CrossRef]
9. McDonald, A.; Bhuiyan, N.A. On the Optimization of Generators for Offshore Direct Drive Wind Turbines. *IEEE Trans. Energy Convers.* **2017**, *32*, 348–358. [CrossRef]
10. Zhao, X.; Niu, S.; Fu, W. Sensitivity Analysis and Design Optimization of a New Hybrid-Excited Dual-PM Generator With Relieving-DC-Saturation Structure for Stand-Alone Wind Power Generation. *IEEE Trans. Magn.* **2020**, *56*, 1–5. [CrossRef]
11. Wang, Q.; Niu, S.; Yang, L. Design Optimization of a Novel Scale-Down Hybrid-Excited Dual Permanent Magnet Generator for Direct-Drive Wind Power Application. *IEEE Trans. Magn.* **2018**, *54*, 1–4. [CrossRef]
12. Sabioni, C.L.; Ribeiro, M.F.O.; Vasconcelos, J.A. Robust Design of an Axial-Flux Permanent Magnet Synchronous Generator Based on Many-Objective Optimization Approach. *IEEE Trans. Magn.* **2018**, *54*, 1–4. [CrossRef]
13. Gao, J.; Dai, L.; Zhang, W.; Huang, S.; Wu, X. Multi-interval Efficiency Design Optimization for Permanent Magnet Synchronous Generators Used in Hybrid Electric Special Vehicles. *IEEE Trans. Ind. Electron.* **2021**, *68*, 4646–4656. [CrossRef]
14. Eriksson, S.; Bernhoff, H. Loss evaluation and design optimisation for direct driven permanent magnet synchronous generators for wind power. *Appl. Energy* **2011**, *88*, 265–271. [CrossRef]
15. Liu, C.; Chau, K.T.; Lee, C.H.T.; Song, Z. A Critical Review of Advanced Electric Machines and Control Strategies for Electric Vehicles. *Proc. IEEE* **2020**, *109*, 1004–1028. [CrossRef]
16. Belahcen, A.; Arkkio, A. Permanent magnets models and losses in 2D FEM simulation of electrical machines. In Proceedings of the XIX International Conference on Electrical Machines—ICEM 2010, Rome, Italy, 6–8 September 2010; pp. 1–6. [CrossRef]
17. Ishak, D.; Zhu, Z.Q.; Howe, D. Eddy-current loss in the rotor magnets of permanent-magnet brushless machines having a fractional number of slots per pole. *IEEE Trans. Magn.* **2005**, *41*, 2462–2469. [CrossRef]
18. Zarko, D. A Systematic Approach to Optimized Design of Permanent Magnet Motors with Reduced Torque Pulsations. Ph.D. Dissertation, Univ. Wisconsin—Madison, Madison, WI, USA, 2004.

energies

MDPI

Article

Permanent Magnet Flux Linkage Analysis and Maximum Torque per Ampere (MTPA) Control of High Saturation IPMSM

Chengxu Li [1], Wenjuan Zhang [2,*], Jian Gao [1] and Shoudao Huang [1]

[1] College of Electrical and Information Engineering, Hunan University, Changsha 410082, China; lichengxu@hnu.edu.cn (C.L.); gaojian0895@hnu.edu.cn (J.G.); hsd1962@hnu.edu.cn (S.H.)

[2] Department of Electronic and Electrical Engineering, Changsha University, Changsha 410022, China

* Correspondence: zwjs0909@sina.com

Abstract: The maximum torque per ampere (MTPA) control is significant for improving the efficiency of the interior permanent magnet synchronous motor (IPMSM). However, for the high saturation IPMSM, the change of the permanent magnet (PM) flux linkage is more complicated, which can cause the MTPA control to deviate from the optimal solution. Therefore, an improved MTPA control method for the high saturation IPMSM is proposed in this paper. Compared with other methods, the proposed method improves the conventional models of flux linkage and torque by analyzing the nonlinear variation of the PM flux linkage with the *dq*-axis currents. Subsequently, an expression suitable for the MTPA control of high saturated IPMSM is derived based on the improved models. The proposed parameter fitting models are then fitted using data from 11 operating points and incorporated into the MTPA optimization algorithm to obtain the MTPA curve. Finally, the effectiveness of the proposed method in enhancing the control accuracy of the MTPA angle is verified through simulations and experiments.

Keywords: maximum torque per ampere; high saturation; permanent magnet flux linkage; interior permanent magnet synchronous motor; nonlinear fitting model

Citation: Li, C.; Zhang, W.; Gao, J.; Huang, S. Permanent Magnet Flux Linkage Analysis and Maximum Torque per Ampere (MTPA) Control of High Saturation IPMSM. *Energies* **2023**, *16*, 4717. https://doi.org/10.3390/en16124717

Academic Editor: Ryszard Palka

Received: 17 April 2023
Revised: 4 June 2023
Accepted: 12 June 2023
Published: 15 June 2023

1. Introduction

Interior permanent magnet synchronous motor (IPMSM) has been an attractive choice in practical applications due to its excellent features of high efficiency and high power density, such as robotics, elevators, air conditioners, and compressors [1,2]. Meanwhile, due to the asymmetry of the *dq*-axis magnetic circuits, the reluctance torque and the magnet torque exist simultaneously in the electromagnetic torque of the IPMSM. For a given output torque, there are different combinations of the *dq*-axis currents [3]. In order to fully utilize the reluctance torque through the optimal combination of *dq*-axis currents, the maximum torque per ampere (MTPA) control has become a preferred control strategy of the IPMSM. The purpose of the MTPA control is to trace an optimal current angle under a given output torque to minimize the stator current amplitude, and the optimal current angle is known as the MTPA angle [4–24].

The conventional method regards the parameters of the motor as constant values in the *dq*-axis model. An optimized equation is then used to calculate the MTPA angle for the different torque [4]. However, the inductances will change nonlinearly with the current, which affects the reluctance torque. The permanent magnet (PM) flux linkage will change with the temperature, which affects the magnet torque [5–7]. Hence, when the parameters of the motor are regarded as constants to implement the MTPA control, the real-time optimal distribution of the *dq*-axis currents cannot be achieved. For the problem of parameter uncertainty in MTPA control, many scholars have proposed various methods to identify the changing parameters under different operating loads. For instance, in Ref. [8], the finite element method is used to obtain the variation of parameters with current. In

Ref. [9], the high frequency injection method is employed to identify the parameters during motor operation. Based on different parameter identification methods, MTPA control can be categorized into offline and online methods.

The offline methods are used to obtain the MTPA curve before the motor is operated in an actual environment. In Ref. [10], the finite element method is used to obtain the MTPA curve after the motor design is completed. Although the finite element method is accurate in theory, and the MTPA angle can be found, it requires knowing the detailed mechanical dimensions of the motor. Furthermore, some technical and assembly errors usually exist in the prototype manufacturing process, resulting in a mismatch between the actual and simulated values of the motor parameters [11]. The look-up table (LUT) method can solve the shortcomings of the finite element method [12–14]. In Ref. [14], the optimal *dq*-axis currents are measured corresponding to different working conditions through experiments; the data are saved to the controller storage space, and the optimal given values are accurately indexed according to different working conditions during the IPMSM operation. However, the LUT method requires a time-consuming experimental process and a large number of experimental data.

To compensate for the shortcomings of offline methods, online methods can usually be used. Refs. [15–17] present the search-based methods to online search the optimal current angle corresponding to the minimum stator current by injecting a signal. Although the search-based methods do not depend on the *dq*-axis model, the response performance of the system is unsatisfactory, and the algorithm may fail in the process of torque variation, thus affecting the stability of the system. In Refs. [18–22], the parameters of the motor are estimated by the high frequency signal injection or advanced algorithm, and the estimated parameters are brought into the MTPA algorithm to obtain the MTPA angle. However, Ref. [23] has proved that although the parameter estimation method can compensate for the problem of parameter inaccuracy, each optimization iteration needs to update the parameters rather than directly consider the changes of parameters in the optimization algorithm. This problem will still lead to deviation in the calculated MTPA angle. In Refs. [23–26], the MTPA methods based on surface fitting are proposed. The proposed fitting models of the flux linkage and the inductance are integrated into the MTPA algorithm, which avoids updating parameters outside the MTPA algorithm.

However, the above methods all ignore the effect of the high saturation on the PM flux linkage. For the high saturation IPMSM, due to the restriction of weight and volume, its power density and saturation degree are much higher than those of commercial motors. Hence, in the actual operation process, the change of the motor parameters is more complicated than that of the low saturation motor [27–29]. Ref. [28] presents that the vector direction of the PM flux linkage will change with the saturation degree of the motor. Ref. [29] presents that the amplitude of the PM flux linkage will change nonlinearly with the *dq*-axis currents. These variations of the PM flux linkage will also cause the MTPA control to deviate from the optimal solution, while the above methods ignore these variations. Hence, the above methods are not applicable to the high saturation IPMSM.

To avoid these problems and develop an accurate MTPA control for the high saturation IPMSM, this paper proposes an improved MTPA control method based on a modified *dq*-axis model, and the organization is as follows. The conventional *dq*-axis model and the MTPA control method are described in Section 2. The nonlinear change of PM flux linkage with the current and the influence of PM flux linkage on torque are analyzed, and the proposed MTPA is explicated in Section 3. In Sections 4 and 5, the simulation and experimental results are presented to verify the proposed method. Finally, the conclusions are given in Section 6.

2. Conventional IPMSM Model and MTPA Control

In the conventional *dq*-axis model of the IPMSM, the PM flux linkage and the inductance are usually regarded as constants, and the flux linkage equations are denoted as

$$\begin{bmatrix} \psi_d \\ \psi_q \end{bmatrix} = \begin{bmatrix} L_d & 0 \\ 0 & L_q \end{bmatrix} \begin{bmatrix} i_d \\ i_q \end{bmatrix} + \begin{bmatrix} \psi_f \\ 0 \end{bmatrix} \tag{1}$$

where ψ_d, ψ_q, i_d, i_q, L_d, and L_q are the *d*-axis and *q*-axis flux linkages, currents, and inductances, respectively; ψ_f is the PM flux linkage. The total torque of the motor is expressed by

$$Te_{total} = \frac{3}{2}p(\psi_d i_q - \psi_q i_d) \tag{2}$$

where p is the pole pair; Te_{total} is the total torque. Substituting (1) into (2), the total torque can then be divided into the magnet torque caused by the PM flux linkage and the reluctance torque caused by the inductance difference.

$$Te_{total} = Te_{pm} + Te_{rel} = \frac{3}{2}p i_q \psi_f + \frac{3}{2}p(L_d - L_q)i_d i_q \tag{3}$$

where Te_{pm} is the magnet torque; Te_{rel} is the reluctance torque.

In addition, the relationships between the phase current and the *dq*-axis currents are shown in Figure 1a, and the equations are expressed as

$$\begin{bmatrix} i_d \\ i_q \end{bmatrix} = \begin{bmatrix} -i_s & 0 \\ 0 & i_s \end{bmatrix} \begin{bmatrix} \sin\beta \\ \cos\beta \end{bmatrix} \tag{4}$$

where i_s is the maximum phase current; β is the current angle between the phase current vector and the *q*-axis. The relationships between each torque and the current angle are shown in Figure 1b.

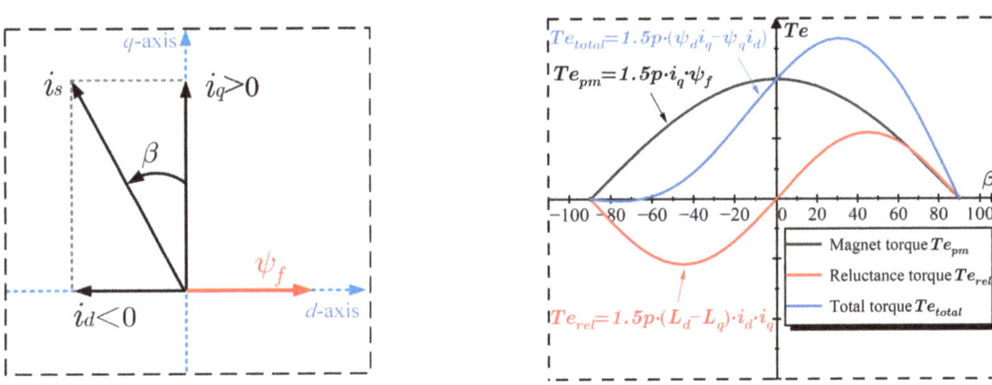

Figure 1. (**a**) Relationships between phase current and *dq*-axis currents. (**b**) Relationships between each torque and current angle.

Substituting (4) into (3), the torque Equation (2) can be expressed as

$$Te_{total} = \frac{3}{2}p(\psi_f i_s \cos\beta - \frac{1}{2}L_\Delta i_s^2 sin2\beta),\ L_\Delta = L_d - L_q \tag{5}$$

The purpose of the MTPA control is to trace a current angle to make the ratio of total torque and current maximize, which is expressed as

$$\max_{\beta} \frac{Te_{total}}{i_s} \tag{6}$$

To find an appropriate current angle β from (6), make the partial derivative of the total torque expression (5) with respect to the current angle be zero $\partial Te_{total}/\partial\beta = 0$.

$$\frac{\partial\psi_f}{\partial\beta}cos\beta - \psi_f \sin\beta - \frac{1}{2}\frac{\partial L_\Delta}{\partial\beta}i_s \sin 2\beta - L_\Delta i_s \cos 2\beta = 0 \tag{7}$$

The conventional MTPA control method regards the inductance and the PM flux linkage as constants. Hence, the first and third terms on the left side of (7) are zero, and the current angle can be obtained by

$$\beta = \sin^{-1}\frac{\psi_f\sqrt{\psi_f^2 + 8L_\Delta^2 i_s^2}}{4L\Delta i_s} \tag{8}$$

However, the inductance and the PM flux linkage will change with different operating conditions of the motor, resulting in a significant error between the MTPA angle calculated by Equation (8) and the actual MTPA angle. Moreover, in the conventional *dq*-axis model, the change of the PM flux linkage with saturation is not considered; this will further increase the error.

3. PM Flux Linkage Analysis and Proposed MTPA Control

3.1. Proposed IPMSM Model and Fitting Models

For the high saturation IPMSM, the magnetic density of the motor is high, which will lead to the PM flux linkage change nonlinearly with the *dq*-axis currents. When i_d remains unchanged, the core is gradually magnetized with the increase of i_q, which will cause the vector direction of the PM flux linkage ψ'_f to gradually shift to the negative direction of the *q*-axis, and the *q*-axis component of the PM flux linkage ψ'_{fq} will be generated, as shown in Figure 2 (superscript$'$ indicates that magnetic saturation is considered).

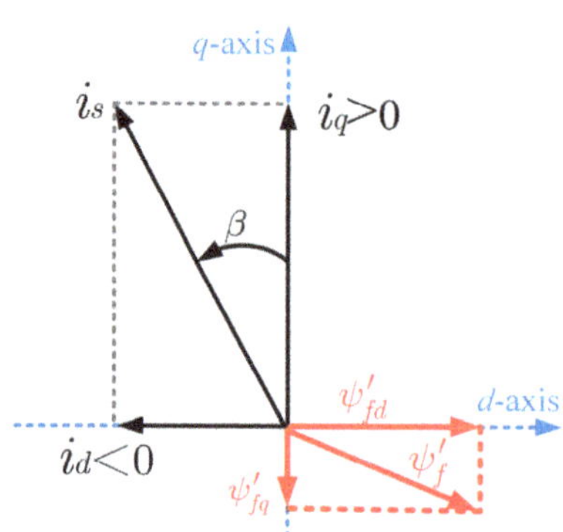

Figure 2. Vector direction of PM flux linkage.

With the increase of i_q, the saturation of the core increases, and the inclination degree of PM flux linkage also increases, which causes the *d*-axis component of the PM flux linkage ψ'_{fd} to decrease and ψ'_{fq} to increase along the negative direction of the *q*-axis, as shown in Figure 3a. Meanwhile, when i_q remains unchanged, with the decrease of i_d, due to the influence of the demagnetization current, desaturation will occur in the *d*-axis direction, which causes ψ'_{fd} to increase and ψ'_{fq} to decrease along the negative direction of the q-axis, as shown in Figure 3b.

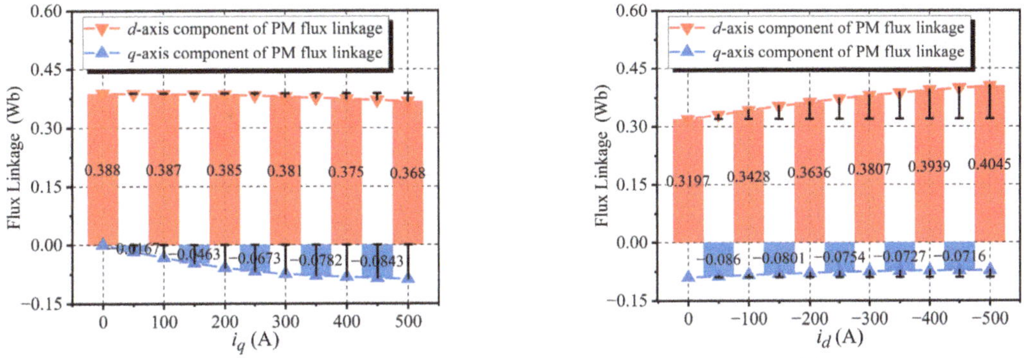

Figure 3. Variation of PM flux linkage: (**a**) Increase i_q when i_d remains unchanged. (**b**) Decrease i_d when i_q remains unchanged.

Therefore, not only the inductance changes but also the change of PM flux linkage should be considered in the flux linkage equations, and (1) should be rewritten as

$$\begin{bmatrix} \psi'_d \\ \psi'_q \end{bmatrix} = \begin{bmatrix} L'_d & 0 \\ 0 & L'_q \end{bmatrix} \begin{bmatrix} i_d \\ i_q \end{bmatrix} + \begin{bmatrix} \psi'_{fd} \\ \psi'_{fq} \end{bmatrix} \tag{9}$$

where ψ'_{fd} and ψ'_{fq} are the dq-axis components of the PM flux linkage, respectively. Then, the magnet torque can also be divided into the dq-axis components, and (3) is rewritten as

$$\begin{aligned} Te'_{total} &= Te'_{pm} + Te'_{rel} = Te'_{pmd} + Te'_{pmq} + Te'_{rel} \\ &= \tfrac{3}{2}p\psi'_{fd}i_q - \tfrac{3}{2}p\psi'_{fq}i_d + \tfrac{3}{2}pL'_{\Delta}i_d i_q \end{aligned} \tag{10}$$

where Te'_{pmd} and Te'_{pmq} are the dq-axis components of the magnet torque, respectively. The relationships between each torque and current angle are shown in Figure 4.

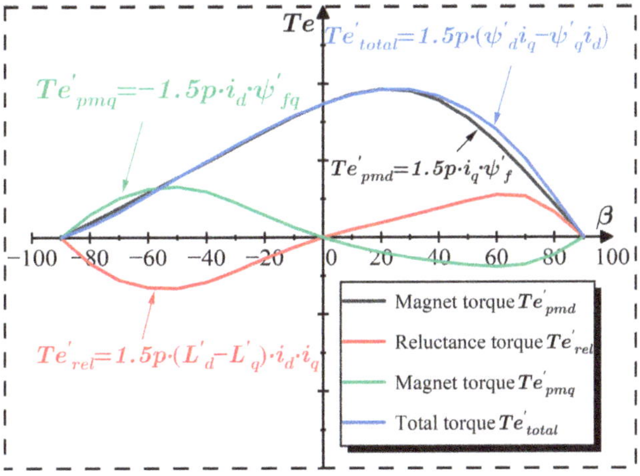

Figure 4. Relationships between each torque and current angle.

Figure 4 presents that when the current angle is 0 to 90 degrees ($i_d \leq 0$, $i_q \geq 0$), Te'_{pmq} will cause Te'_{total} to decrease, while the MTPA control needs to trace the appropriate current angle according to the change of Te'_{total} (6). Therefore, in the actual MTPA control, the influence of Te'_{pmq} on Te'_{total} cannot be ignored. It means that for the MTPA control of high saturation IPMSM, it is necessary to consider the changes of the PM flux linkage amplitude and vector direction.

Substituting (4) into (10), make the partial derivative of Te'_{total} with respect to β be zero.

$$\begin{aligned} \frac{\partial Te'_{total}}{\partial \beta} &= \frac{\partial \psi'_{fd}}{\partial \beta} cos\beta - \psi'_{fd} \sin\beta - \frac{1}{2}\frac{\partial L'_{\Delta}}{\partial \beta} i_s \sin 2\beta \\ &- L'_{\Delta} i_s \cos 2\beta + \frac{\partial \psi'_{fq}}{\partial \beta} \sin\beta + \psi'_{fq} \cos\beta = 0 \end{aligned} \tag{11}$$

According to ref [23], $\partial\psi'_{fd}/\partial\beta$, $\partial\psi'_{fq}/\partial\beta$, and $\partial L'_{\Delta}/\partial\beta$ cannot be ignored in the MTPA algorithm (11). Thus, this paper proposes the fitting models of ψ'_{fd}, ψ'_{fq}, and L'_{Δ} in regard to the dq-axis currents, respectively.

$$\begin{aligned} &L'_{\Delta} = a_0 + a_1 i_d + a_2 i_q + a_3 i_d i_q \\ &\psi'_{fd} = b_0 + b_1 i_d + b_2 i_q + b_3 i_d i_q + b_4 i_q^2,\ \psi'_{fq} = c_0 i_q + c_1 i_d i_q + c_2 i_q^2 \end{aligned} \tag{12}$$

where a_i, b_i, and c_i are constant coefficients (i = 0, 1, 2, 3, 4).

3.2. Determination of the Coefficients and Implementation of the MTPA Control

This paper only considers the motoring mode of the IPMSM. Assuming that the dq-axis currents ranges are $-i \leq i_d \leq 0$ and $i \geq i_q \geq 0$, plot four quarter circles with amplitudes of 1/4, 1/2, 3/4, and 1 of the current i, respectively. Divide the quarter circle with radius i into four sectors equally, and select points from 1 to 11. Point 1 is on a circle with a radius of 0.25i; points 2 and 3 are on a circle with a radius of 0.5i; points 4, 5, and 6 are on a circle with a radius of 0.75i; points 7, 8, 9, 10 and 11 are on a circle with a radius of i; and each point is on the boundary of the sector. The specific coordinates of each point are shown in Figure 5.

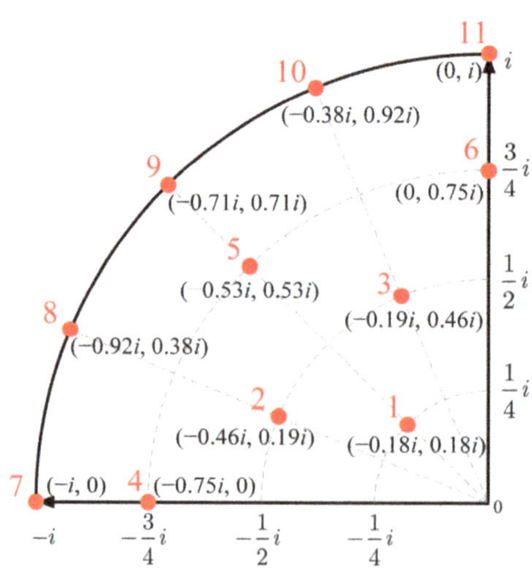

Figure 5. The specific coordinates of each point.

Points 4, 7, and 8 are selected to occupy more weight in the desaturation area of the motor, and points 6, 10, and 11 are selected to occupy more weight in the saturation area of the motor. The inductance parameters of 11 points can be firstly obtained by parameter estimation. Then, ψ'_{fd} and ψ'_{fq} at each point can be calculated by using the steady-state voltage Equation (13) according to the monitored dq-axis voltage.

$$\psi'_{fd} = \frac{v_q - R_s i_q}{w_e} - L'_d i_d \, , \psi'_{fq} = \frac{R_s i_d - v_d}{w_e} - L'_q i_q \tag{13}$$

where R_s is the stator resistance; v_d and v_q are the dq-axis voltage, respectively; w_e is the electrical speed.

Finally, the coefficients in (12) can be determined based on the data of the 11 operating points in Figure 5 and the proposed optimization algorithm (14).

$$\text{Minimize}: \sum_{i=1}^{11} (L^*_\Delta - L^\wedge_\Delta)^2 + \sum_{i=1}^{11} (\psi^*_{fd} - \psi^\wedge_{fd})^2 + \sum_{i=1}^{11} (\psi^*_{fq} - \psi^\wedge_{fq})^2 \tag{14}$$

where L^*_Δ, ψ^*_{fd}, and ψ^*_{fq} are the actual value; $L^\wedge_\Delta$, $\psi^\wedge_{fd}$ and $\psi^\wedge_{fq}$ are the fitted value.

After determining the coefficients of the fitting models, the total torque expression without the PM flux linkage and the inductance can be expressed as

$$Te'_{total} = \frac{3}{2} p (b_0 i_q + k i_d i_q + b_2 i_q^2 + m i_d i_q^2 + b_4 i_q^3 + n i_d^2 i_q + a_3 i_d^2 i_q^2) \tag{15}$$

where

$$k = b_1 + a_0 + c_0 \, , m = b_3 + a_2 + c_2 \, , n = a_1 + c_1 \tag{16}$$

Substituting (4) into (15), make the partial derivative of Te'_{total} with respect to β be zero.

$$\begin{aligned} \frac{\partial Te'_{total}}{\partial \beta} &= k i_s \sin^2\beta - b_0 \sin\beta - k i_s \cos^2\beta - 2 b_2 i_s \cos\beta \sin\beta \\ &- m i_s^2 \cos^3\beta + 2 m i_s^2 \sin^2\beta \cos\beta - 3 b_4 i_s^2 \cos^2\beta \sin\beta \\ &+ 2 n i_s^2 \cos^2\beta \sin\beta - n i_s^2 \sin^3\beta + 2 a_3 i_s^3 \sin\beta \cos^3\beta \\ &- 2 a_3 i_s^3 \sin^3\beta \cos\beta = 0 \end{aligned} \tag{17}$$

It can be seen from (17) that the equation has only two variables (i_s, β), which can be solved by numerical solution or nonlinear equation solver. For real-time applications, the proposed algorithm can also be utilized to generate LUT in advance to improve the dynamics of the control system. Compared with the conventional LUT method, the proposed method can generate the LUT offline only by using a PC without sampling the entire working range of the motor, which significantly reduces the workload. Eventually, the dq-axis currents trajectories can be obtained by substituting the current angle β into (4). This method not only considers the changes of the inductance in the optimization process but also considers the changes of the PM flux linkage in the high saturation state. The control block diagram is shown in Figure 6.

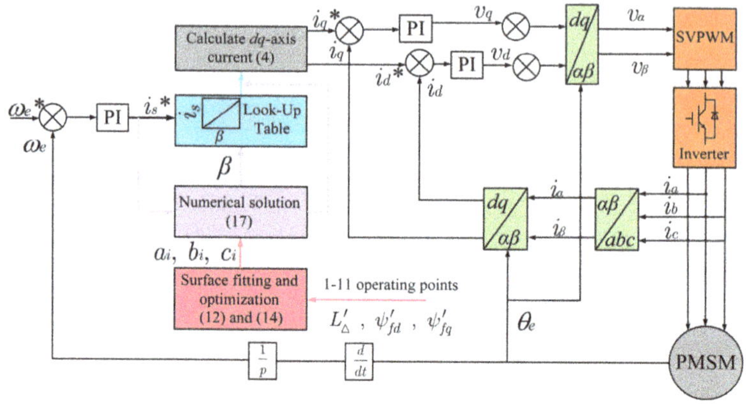

Figure 6. The control block diagram(* represents the corresponding set value).

4. Simulation Analysis and Verification

To verify the accuracy of the proposed method, this paper establishes a finite element model (FEM) of the high saturation IPMSM, as shown in Figure 7. The main parameters of the model are listed in Table 1, including the *dq*-axis inductance and the PM flux linkage when the *dq*-axis currents are zero.

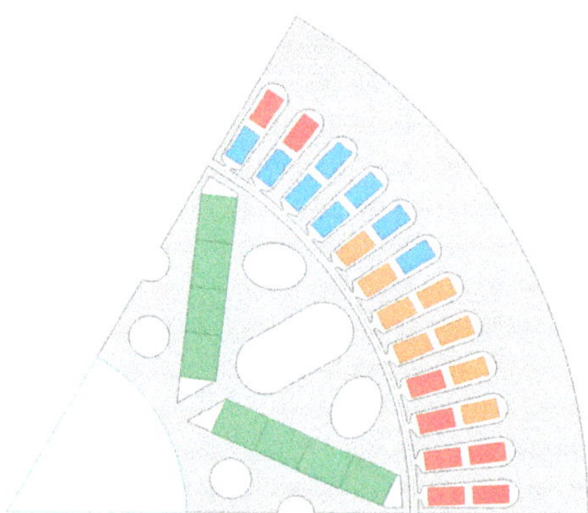

Figure 7. The FEM of IPMSM.

Table 1. Key parameters of IPMSM.

Parameters	Quantity	Parameters	Quantity
Rated power	300 kW	Pole pairs	3
Rated current	470 A	PM flux linkage	0.332 Wb
Rated speed	3000 r/min	Stator resistance	0.005 Ω
Rated torque	955 Nm	*d*-axis inductance	0.4723 mH
Rated frequency	150 Hz	*q*-axis inductance	1.0228 mH

For the high saturation IPMSM at the different load conditions, the stator core is usually in the high saturation state. Especially, the maximum magnetic density (MMD) of stator teeth is generally greater than 1.8 T. To verify the saturation degree of stator teeth, the motor speed is set to the rated speed. Then, the MMD of the stator teeth is simulated at the different load conditions, as shown in Figure 8a.

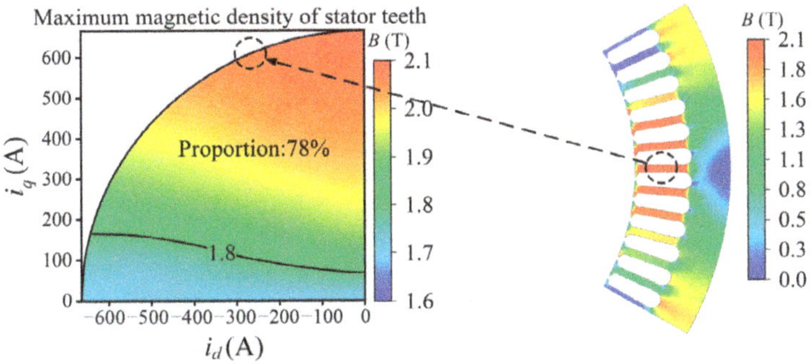

(a) (b)

Figure 8. Saturation degree of IPMSM. (**a**) MMD of stator teeth at different load conditions. (**b**) Distribution of stator magnetic density.

Figure 8a presents that the MMD of the stator teeth exceeds 1.8 T for 78% of the rated current range. When $i_d = -275$ A and $i_q = 605$ A, the IPMSM can operate at the rated torque with the minimum current. Meanwhile, the MMD of the stator teeth is 2.05 T, as shown in Figure 8b. Thus, Figure 8 indicates that the saturation of the IPMSM model is high. In the high saturation state, the inductance and the PM flux linkage will change nonlinearly with the *dq*-axis currents, as shown in Figure 9.

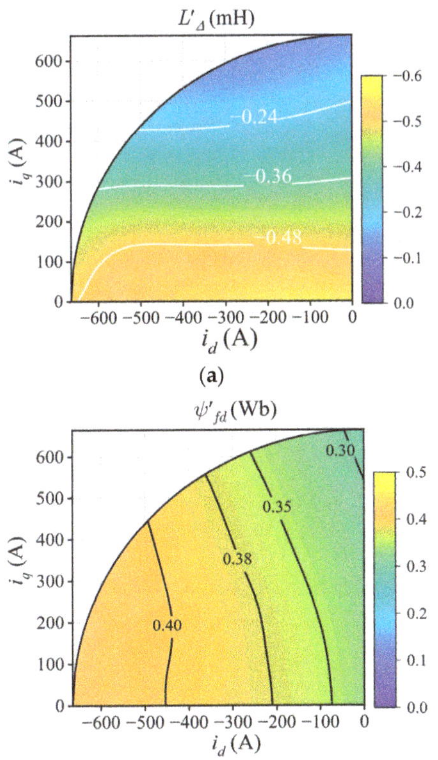

(a)

(b)

Figure 9. *Cont.*

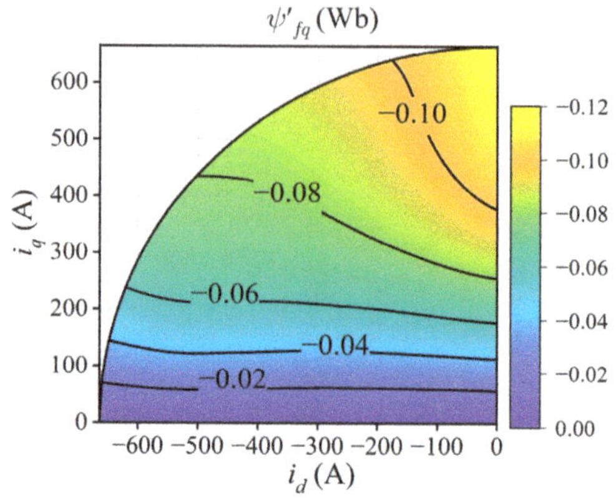

(c)

Figure 9. Variation of parameters. (**a**) Inductance difference. (**b**) *d*-axis component of PM flux linkage. (**c**) *q*-axis component of PM flux linkage.

According to the analysis in Section 3, the variation of the inductance and PM flux linkage will directly affect the amplitude of the reluctance and magnet torque, respectively, as shown in Figure 10.

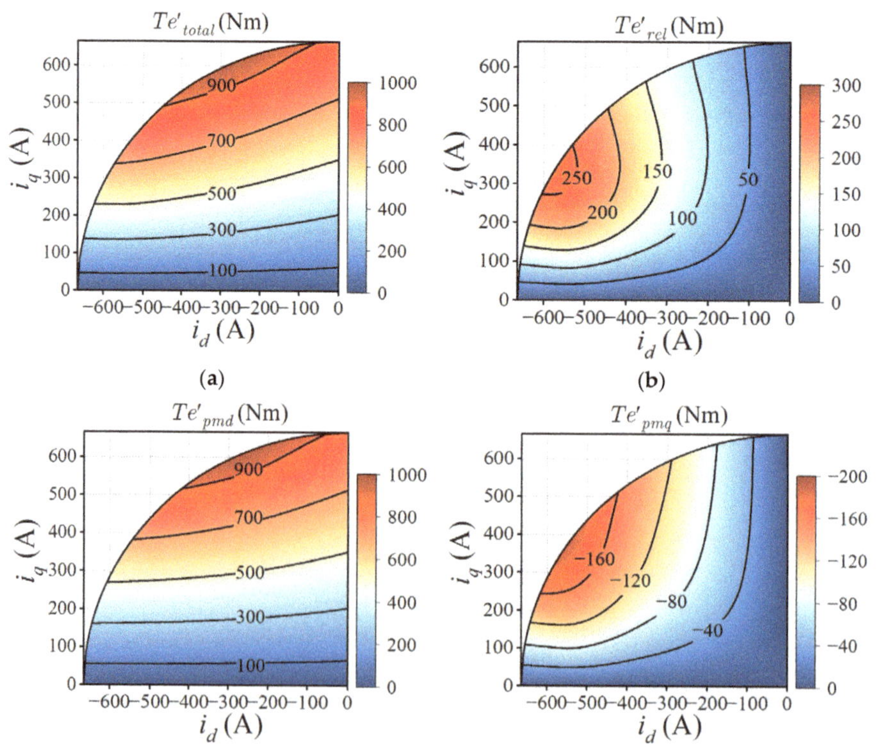

Figure 10. Variation of each torque within rated current range. (**a**) Total torque. (**b**) Reluctance torque. (**c**) *d*-axis component of magnet torque. (**d**) *q*-axis component of magnet torque.

Figure 10 presents that for the high saturation IPMSM, the total torque is composed of three components. Each torque component significantly impacts the total torque, and all of them change nonlinearly with the change of the *dq*-axis current. For this reason, the variation of these three torque components should be considered simultaneously in the MTPA optimization algorithm. According to the coordinates of 11 working points in Figure 5, the simulated inductance and the PM flux linkage are substituted into (14), and the coefficients of fitting models are obtained, as shown in Table 2.

Table 2. Results of coefficient fitting based on simulation.

Coefficient	a_i	b_i	c_i
$i = 0$	-5.55×10^{-4}	0.342	-3.91×10^{-4}
$i = 1$	-1.12×10^{-8}	-1.32×10^{-4}	-1.23×10^{-7}
$i = 2$	6.23×10^{-7}	-4.37×10^{-5}	3.43×10^{-7}
$i = 3$	-1.46×10^{-10}	-1.34×10^{-7}	0
$i = 4$	0	-5.41×10^{-8}	0

Then, the coefficients are substituted into (17), and the equation results are directly substituted into (4) by numerical solution, as shown in Figure 6. Finally, compared with other methods, (1) the conventional method substitutes the constant parameters at no-load condition into (8); (2) the parameter estimation method [22] substitutes the variable parameters at different conditions into (8); (3) the conventional fitting method [23]. The MTPA curves obtained by different methods are shown in Figure 11. It is obvious from Figure 11 that the proposed method is closer to the actual MTPA curve than other methods, which verifies the validity of the analysis and the method proposed in this paper.

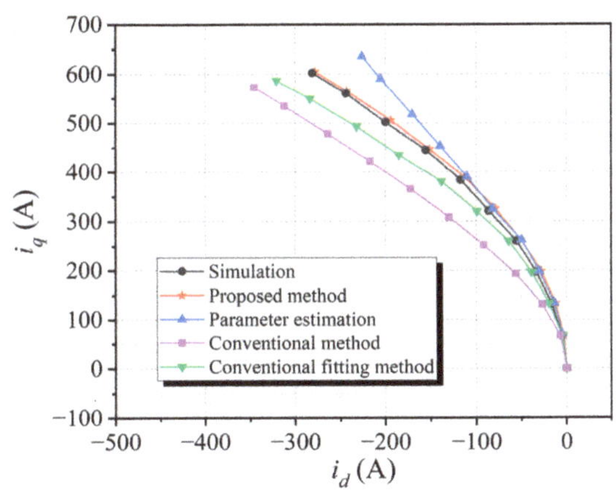

Figure 11. Comparison of proposed method and other methods.

5. Experimental Results

A 300 kW high saturation IPMSM has been prototyped to verify the proposed approach, and the design parameters of the prototype are given in Table 1. The experiment platform is shown in Figure 12a. The prototype is driven by a traction motor at 3000 rpm and operates in the torque control mode. Meanwhile, a torque sensor (ATESTEO: DF3) is also installed to measure the torque at different load conditions. In addition, in order to reduce the influence of temperature on parameters, so as to better verify the influence of high saturation on parameter, the cooling mode of the prototype is water-cooling, the temperature of cooling water is 30 °C, and the water flow is 60 L/min. The Hi-Techniques' high-speed logger is utilized to observe and reserve the experimental information in real time, and the PC software is used to export and plot the experimental data, as shown in Figure 12b.

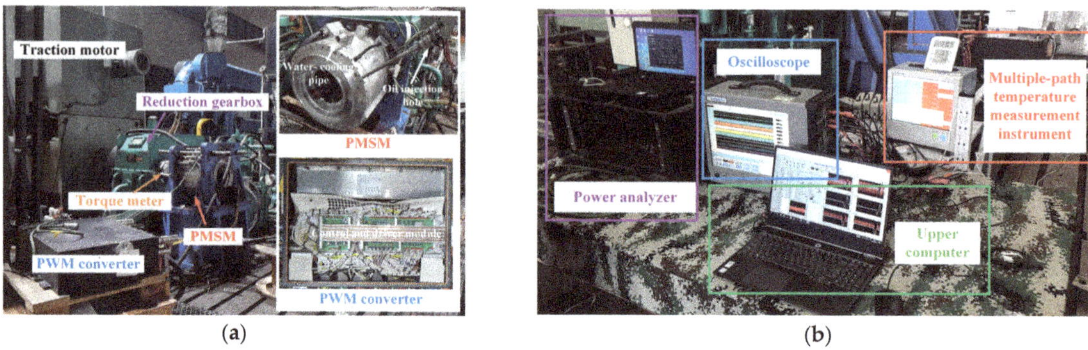

Figure 12. Experimental platform. (**a**) Prototype and driving system. (**b**) Instrument for recording data.

Firstly, the prototype is tested to find the actual MTPA curve. The detailed test steps are as follows. Set i_d to make the prototype operate at no-load condition. Then, as the load gradually increases, the voltage, current, and torque are recorded at different load conditions. Finally, gradually decrease i_d and repeat the previous steps. The measured torque within the rated current range is shown in Figure 13, and the MTPA curve is *OB*.

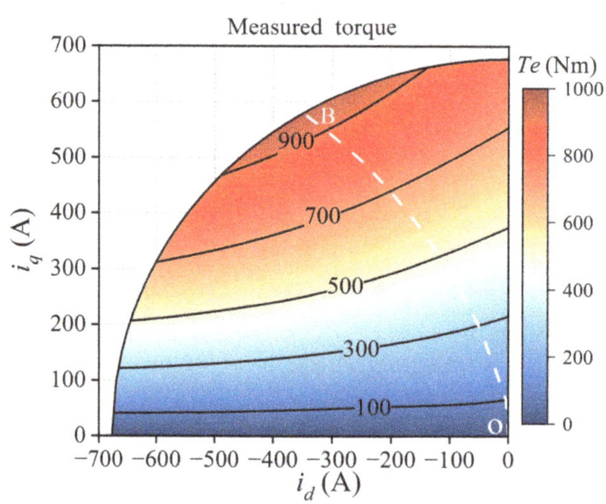

Figure 13. Measured torque and MTPA curve.

Due to certain technical and assembly errors in the prototype manufacturing process, the actual value and the simulated value are different. For instance, the minimum current at the rated torque is different from the simulation. When the IPMSM is operated at rated torque, the *dq*-axis currents are $i_d = -350$ A and $i_q = 580$ A, respectively. The recorded waveforms of the DC bus voltage, three-phase voltage, current, and torque are shown in Figure 14.

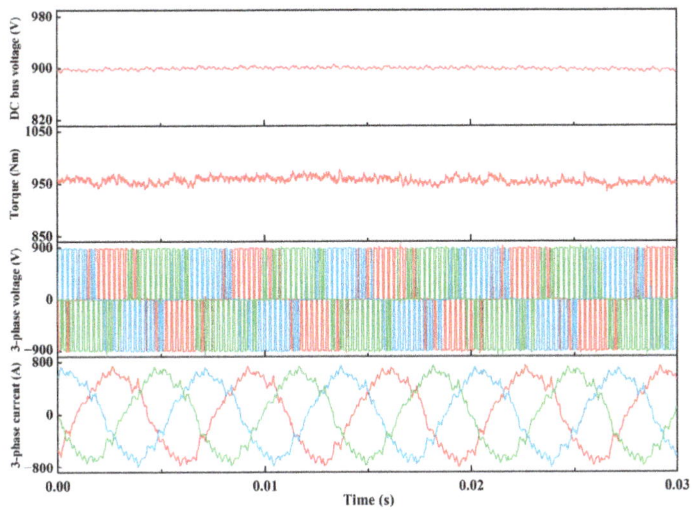

Figure 14. Test waveforms at rated torque.

Then, the parameter estimation method is used to estimate the inductance parameters of the 11 operating points, and the *dq*-axis components of PM flux linkage can be calculated according to the monitored *dq*-axis voltage, as shown in Figure 15. According to the algorithm (14), the fitting coefficients of the prototype are listed in Table 3.

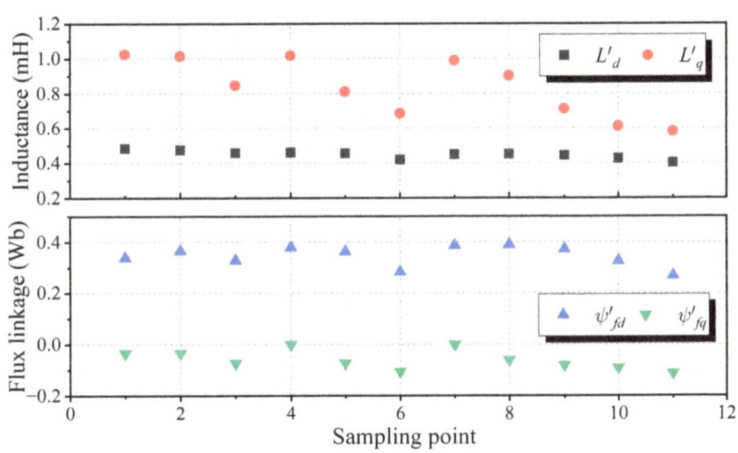

Figure 15. Parameters of 11 operating points.

Table 3. Results of coefficient fitting based on experiments.

Coefficient	a_i	b_i	c_i
$i = 0$	-6.28×10^{-4}	0.317	-3.34×10^{-4}
$i = 1$	-1.13×10^{-7}	-6.19×10^{-5}	-9.24×10^{-8}
$i = 2$	6.97×10^{-7}	-1.1×10^{-4}	2.54×10^{-7}
$i = 3$	1.92×10^{-10}	-2.3×10^{-7}	0
$i = 4$	0	-8.23×10^{-9}	0

Based on the fitting coefficients of the prototype, the offline numerical calculation is used to obtain the relationship between i_s and β, and generates a LUT for real-time applications, as shown in Figure 6. Eventually, the MTPA curves and MTPA angle errors of the different methods are shown in Figure 16.

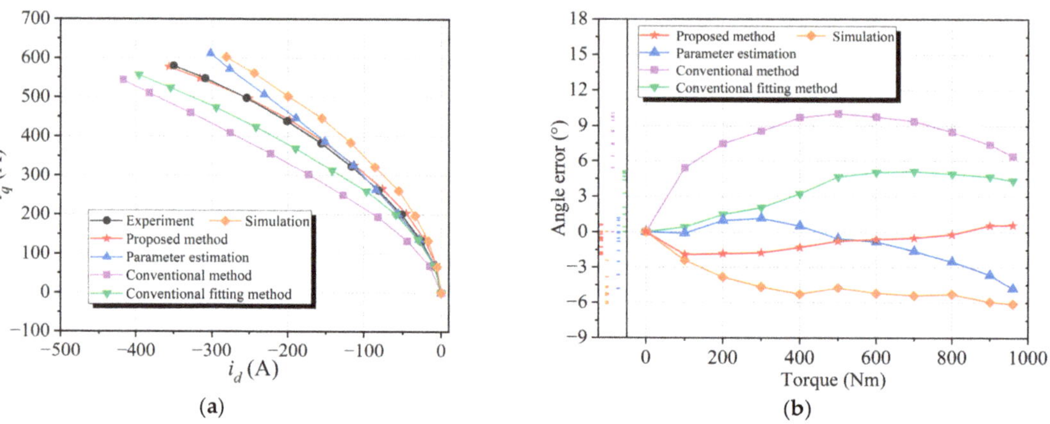

Figure 16. Experimental results. (**a**) MTPA curves; (**b**) MTPA angle errors.

Figure 16b presents that the error of the proposed method is relatively stable, and the error does not increase with the increase of the saturation degree. For example, when the motor is operated at the rated load, although the saturation degree of the motor is large, the error of the proposed method is still small. The errors of different methods at the rated load are listed in Table 4.

Table 4. Comparison of different methods.

Methods	Errors at Rated Load
Proposed	0.6°
Simulation	−6.1°
Conventional	6.4°
Conventional fitting	4.3°
Parameter estimation	−4.8°

In the rated load range, the maximum errors caused by the parameter estimation and conventional fitting method are −4.8° and 5.1°, respectively. The main reason for these errors is that the influence of the high saturation on the PM flux linkage is not considered. Furthermore, the maximum error of the conventional method and the simulation are 10° and −6.1°, respectively. The maximum error of the proposed method is −1.8°. Therefore, the proposed method can achieve better accuracy compared with other methods. Meanwhile, these experimental results indicate that for the MTPA control of high saturation IPMSM, the nonlinear change of the PM flux linkage with the *dq*-axis current cannot be ignored.

6. Conclusions

The MTPA control is a regular method for the IPMSM to find an optimal current angle to minimize the stator current amplitude for given output torque. However, for the high saturation IPMSM, the change of the motor parameters is more complicated than that of the low saturation motor, and the nonlinear relationship between the PM flux linkage and the *dq*-axis currents becomes more distinct. When the conventional MTPA control method is used, the accuracy of the MTPA angle often fails to meet the requirements.

Thus, an improved MTPA control method is proposed in this paper. By analyzing the change of the PM flux linkage and the influence of the PM flux linkage on the total torque, the conventional flux linkage and torque models are improved. Based on the improved dq-axis models, an expression more suitable for MTPA control of high saturation IPMSM is derived. Compared to other methods, the proposed method takes into account the nonlinear variation of the PM flux linkage with magnetic saturation. Additionally, this method also considers the q-axis component generated by the PM flux linkage and the partial derivatives of the parameters in the MTPA optimization algorithm. Simulation and experimental results demonstrate that the proposed method can better follow the MTPA angle of the high saturation IPMSM at different load conditions.

Author Contributions: Methodology, C.L.; software, W.Z.; validation, J.G.; formal analysis, S.H. All authors have read and agreed to the published version of the manuscript.

Funding: This work was supported in part by the National Natural Science Foundation of China (52277034) and the Changsha Major Science and Technology Special Project of China (kq2205003).

Data Availability Statement: Not applicable.

Conflicts of Interest: The authors declare no conflict of interest.

References

1. Martinez, M.; Fernandez, D.; Reigosa, D.; Guerrero, J.M.; Briz, F. Wireless torque pulsations measurement system for PMSMs. *IEEE Trans. Ind. Appl.* **2020**, *56*, 6467–6476. [CrossRef]
2. Taherzadeh, M.; Hamida, M.A.; Ghanes, M.; Koteich, M. A new torque observation technique for a PMSM considering unknown magnetic conditions. *IEEE Trans. Ind. Electron.* **2021**, *68*, 1961–1971. [CrossRef]
3. Uddin, M.N.; Rahman, M.M. Online torque-flux estimation-based nonlinear torque and flux control scheme of IPMSM drive for reduced torque ripples. *IEEE Trans. Power Electron.* **2019**, *34*, 636–645. [CrossRef]
4. Pan, C.-T.; Sue, S.-M. A linear maximum torque per ampere control for IPMSM drives over full-speed range. *IEEE Trans. Energy Convers.* **2005**, *20*, 359–366. [CrossRef]
5. Liu, K.; Feng, J.; Guo, S.; Xiao, L.; Zhu, Z.-Q. Identification of flux linkage map of permanent magnet synchronous machines under uncertain circuit resistance and inverter nonlinearity. *IEEE Trans. Ind. Inform.* **2018**, *14*, 556–568. [CrossRef]
6. Raja, R.; Sebastian, T.; Wang, M. Online stator inductance estimation for permanent magnet motors using PWM excitation. *IEEE Trans. Transp. Electrif.* **2019**, *5*, 107–117. [CrossRef]
7. Ge, H.; Jiang, J.W.; Ye, J.; Emadi, A. Behavior study of permanent magnet synchronous machines based on a new normalized model. *IEEE Trans. Ind. Electron.* **2019**, *66*, 7539–7550. [CrossRef]
8. Sokolov, E.; Mihov, M. Parameter Estimation of an Interior Permanent Magnet Synchronous Motor. In Proceedings of the 16th Conference on Electrical Machines, Drives and Power Systems (ELMA), Varna, Bulgaria, 6–8 June 2019; pp. 1–5.
9. Feng, G.; Lai, C.; Mukherjee, K.; Kar, N.C. Online PMSM Magnet Flux-Linkage Estimation for Rotor Magnet Condition Monitoring Using Measured Speed Harmonics. *IEEE Trans. Ind. Appl.* **2017**, *53*, 2786–2794. [CrossRef]
10. Ge, H.; Bilgin, B.; Emadi, A. Global loss minimization control of PMSM considering cross-coupling and saturation. In Proceedings of the 2015 IEEE Energy Conversion Congress and Exposition (ECCE), Montreal, QC, Canada, 20–24 September 2015; pp. 6139–6144.
11. Ahmad, H.; Ro, J.-S. Analysis and design optimization of V-shaped permanent magnet vernier motor for torque density improvement. *IEEE Access* **2021**, *9*, 13542–13552. [CrossRef]
12. Sun, T.; Wang, J.; Chen, X. Maximum torque per ampere (MTPA) control for interior permanent magnet synchronous machine drives based on virtual signal injection. *IEEE Trans. Power Electron.* **2015**, *30*, 5036–5045. [CrossRef]
13. Babel, A.S.; Cintron-Rivera, J.G.; Strangas, E.G. A multiple look-up table torque controller for improved performance of IPM machines. In Proceedings of the 2013 International Electric Machines & Drives Conference, Chicago, IL, USA, 12–15 May 2013; pp. 521–525.
14. Kim, S.M.; Kwon, T. A simple method to minimize effects of temperature variation on IPMSM control in real-time manner. In Proceedings of the 2014 IEEE Energy Conversion Congress and Exposition (ECCE), Pittsburgh, PA, USA, 14–18 September 2014; pp. 4212–4217.
15. Xia, J.; Guo, Y.; Li, Z.; Jatskevich, J.; Zhang, X. Step-signal-injection-based robust MTPA operation strategy for interior permanent magnet synchronous machines. *IEEE Trans. Energy Convers.* **2019**, *34*, 2052–2061. [CrossRef]
16. Dianov, A.; Anuchin, A. Adaptive maximum torque per ampere control for IPMSM drives with load varying over mechanical revolution. *IEEE J. Emerg. Sel. Top. Power Electron.* **2022**, *10*, 3409–3417. [CrossRef]
17. Lai, C.; Feng, G.; Mukherjee, K.; Tjong, J.; Kar, N.C. Maximum torque per ampere control for IPMSM using gradient descent algorithm based on measured speed harmonics. *IEEE Trans. Ind. Inform.* **2018**, *14*, 1424–1435. [CrossRef]

18. de Castro, A.G.; Guazzelli, P.R.U.; de Oliveira, C.M.R.; Pereira, W.C.D.A.; de Paula, G.T.; Monteiro, J.R.B.D.A. Optimized current waveform for torque ripple mitigation and MTPA operation of PMSM with back EMF harmonics based on genetic algorithm and artificial neural network. *IEEE Lat. Am. Trans.* **2020**, *18*, 1646–1655. [CrossRef]
19. Tang, Q.; Shen, A.; Luo, P.; Shen, H.; He, X. IPMSMs sensorless MTPA control based on virtual q-axis inductance by using virtual high-frequency signal injection. *IEEE Trans. Ind. Electron.* **2020**, *67*, 136–146. [CrossRef]
20. Sun, T.; Koç, M.; Wang, J. MTPA control of IPMSM drives based on virtual signal injection considering machine parameter variations. *IEEE Trans. Ind. Electron.* **2018**, *65*, 6089–6098. [CrossRef]
21. Kim, H.; Lee, Y.; Sul, S.; Yu, J.; Oh, J. Online MTPA control of IPMSM based on robust numerical optimization technique. *IEEE Trans. Ind. Appl.* **2019**, *55*, 3736–3746. [CrossRef]
22. Yu, Y.; Huang, X.; Li, Z.; Wu, M.; Shi, T.; Cao, Y.; Yang, G.; Niu, F. Full parameter estimation for permanent magnet synchronous motors. *IEEE Trans. Ind. Electron.* **2022**, *69*, 4376–4386. [CrossRef]
23. Rabiei, A.; Thiringer, T.; Alatalo, M.; Grunditz, E.A. Improved maximum-torque-per-ampere algorithm accounting for core saturation, cross-coupling effect, and temperature for a PMSM intended for vehicular applications. *IEEE Trans. Transp. Electrif.* **2016**, *2*, 150–159. [CrossRef]
24. Tinazzi, F.; Zigliotto, M. Torque estimation in high-efficency IPM synchronous motor drives. *IEEE Trans. Energy Convers.* **2015**, *30*, 983–990. [CrossRef]
25. Miao, Y.; Ge, H.; Preindl, M.; Ye, J.; Cheng, B.; Emadi, A. MTPA fitting and torque estimation technique based on a new flux-linkage model for interior-permanent-magnet synchronous machines. *IEEE Trans. Ind. Appl.* **2017**, *53*, 5451–5460. [CrossRef]
26. Feng, G.; Lai, C.; Han, Y.; Kar, C. Fast maximum torque per amper (MTPA) angle detection for interior PMSMs using online polynomial curve fitting. *IEEE Trans. Power Electron.* **2022**, *37*, 2045–2056.
27. Lee, K.; Ha, J.; Simili, D.V. Analysis and suppression of slotting and cross-coupling effects on current control in PM synchronous motor drives. *IEEE Trans. Power Electron.* **2019**, *34*, 9942–9956. [CrossRef]
28. Shi, Y.; Chai, J.; Sun, X.; Mu, S. Detailed description and analysis of the cross-coupling magnetic saturation on permanent magnet synchronous motor. *J. Eng.* **2018**, *2018*, 1855–1859. [CrossRef]
29. Thul, A.; Groschup, B.; Hameyer, K. Influences on the Accuracy of Torque Calculation for Permanent Magnet Synchronous Machines. *IEEE Trans. Energy Convers.* **2020**, *35*, 2261–2268. [CrossRef]

energies

MDPI

Article

Adaptive Control of a Virtual Synchronous Generator with Multiparameter Coordination

Bixing Ren [1], Qiang Li [1], Zhiyuan Fan [2] and Yichao Sun [2,*]

[1] State Grid Jiangsu Electric Power Company Ltd., Research Institute, Nanjing 211103, China; renbixing@126.com (B.R.); liq4@js.sgcc.com.cn (Q.L.)
[2] NARI School of Electrical and Automation Engineering, Nanjing Normal University, Nanjing 210023, China; 211846042@njnu.edu.cn
* Correspondence: yichao.sun1987@gmail.com

Abstract: This paper proposes an adaptive strategy of co-regulating the three parameters—P/ω droop coefficient, virtual inertia, and damping coefficient—for the virtual synchronous generator (VSG). This approach is able to solve the uncoordinated performance between the virtual inertia and the damping using the conventional adaptive control in which the system may experience serious frequency fluctuations. Through the mathematical modeling of the VSG grid-connected system, the segmental analysis of the VSG transient process is carried out, and the parameter adjustment law of each stage is obtained. The VSG angular velocity change and the angular velocity instantaneous change rate are associated with the inertia to realize the adaptive adjustment of the inertia, and the adaptive adjustment of the P/ω droop coefficient is carried out in real time according to the VSG angular velocity change. A functional relationship is established between the P/ω droop coefficient, virtual inertia, and damping coefficient so that the P/ω droop coefficient, virtual inertia, and damping coefficient are coordinated to keep the system in the best damping ratio state all the time. Finally, the superiority of the proposed strategy is proved by simulation comparison.

Keywords: virtual synchronous generator grid-connected system; adaptive co-regulation strategy; optimal damping ratio

Citation: Ren, B.; Li, Q.; Fan, Z.; Sun, Y. Adaptive Control of a Virtual Synchronous Generator with Multiparameter Coordination. *Energies* **2023**, *16*, 4789. https://doi.org/10.3390/en16124789

Academic Editor: Antonio Cano-Ortega

Received: 24 April 2023
Revised: 1 June 2023
Accepted: 5 June 2023
Published: 19 June 2023

1. Introduction

With the increasingly serious environmental pollution and the exhaustion of traditional fossil energy, clean energy, such as wind and light, has become an indispensable alternative [1]. New energy power generation is generally connected to the AC microgrid through the inverter device, but since the inverter device does not have the inertia and damping of the synchronous generator when it is connected in a large proportion, the inertia and damping of the system will be insufficient. When disturbed, its ability to suppress interference becomes weak and even causes the system frequency to collapse in severe cases [2,3].

The virtual synchronous generator control [4] simulates the inertia and damping characteristics of the synchronous generator, while the inverter equipment can also provide inertia and damping support for the system. Virtual inertia and damping are the core control parameters of VSG, which are flexible and adjustable, and proper adjustment of these parameters can effectively improve the control performance of VSG. With the increase in the proportion of new energy connected to the power system, VSG technology has received more and more attention from researchers. Ref. [5] proposes a virtual inertial control strategy for the microgrid system with energy storage, which effectively improves the frequency characteristics of the microgrid system. However, when the proportion of new energy inside the system changes, its inertia takes a fixed value, and the frequency response characteristics of the system deteriorate. To solve this problem, adjustment techniques based on adaptive strategies have been proposed by researchers [6–13]. Ref. [8]

proposed a VSG virtual inertia adaptive control algorithm based on stick control. When the rate of change of the angular frequency is less than a certain threshold, the inertia takes a smaller value; otherwise, it takes a larger value. However, stage adjustment and the optimal tracking of inertia to frequency changes cannot be achieved. Ref. [9] proposes a virtual inertia adaptive control strategy that is jointly determined by the VSG rotor angular frequency change rate and deviation, which solves the adaptive effective tracking to a certain extent, but the selection range of virtual inertia and the basis for the selection of key parameters are not given in the paper. Refs. [11,12] proposed a coordinated adaptive control strategy of inertia and damping. When the VSG angular frequency change rate increases, the inertia value is higher, and when the angular frequency deviation increases, the damping value is higher, which effectively improves the dynamic and static state of the system frequency performance; however, only the adaptive expression is given, and the selection basis of the correlation coefficient in the expression cannot be given. Ref. [13] proposed a cooperative adaptive control strategy for VSG parameters. The strategy uses an exponential adaptive algorithm to determine the virtual inertia, which reduces the sensitivity of the relevant control parameters in the adaptive algorithm and combines the performance index constraints to achieve the coordination of the damping coefficient, which realizes the optimal tracking of inertia and damping for frequency changes. However, the paper ignores the effect of the P/ω droop coefficient. The literature [14] proposes an adaptive inertial damping integrated control (SA-RIDC) method, which decides whether to adjust the virtual inertia or the damping coefficient according to the derivative of the angular frequency differential during frequency oscillations in order to achieve alternating control of the virtual inertia and the damping coefficient. The literature [15] proposes a control strategy based on a joint adaptive virtual rotational inertia and damping coefficient with an optimal damping ratio. The literature [16] proposes an adaptive VSG control strategy for battery energy storage systems to ensure the stability of the power system. The literature [17] draws on the work angle and angular frequency curves of synchronous generators to design a refined virtual inertia fuzzy regulation law, while considering the four performance indicators of active overshoot, the frequency rate of change, regulation time, and rise time to select a suitable damping ratio, and the virtual damping is coordinated and adaptively adjusted according to the selected damping ratio with the change of virtual inertia. In the paper [18], virtual synchronous generator technology was introduced in the IIDG control system to optimize the active-frequency control, the reactive-voltage control, and the voltage-current control, respectively. The impact of the damping coefficient on the output of the microgrid system is analyzed by means of a small signal model, and a self-adaptive damping control strategy is proposed.

In this paper, an improved VSG multiparameter optimal cooperative control strategy is proposed. Based on the optimal damping ratio, the initial value of the virtual inertia and the adaptive control of the P/ω droop coefficient are set, and combined with the performance index constraints, the coordination between the P/ω droop coefficient, inertia coefficient, and damping coefficient is realized. This not only realizes the optimal tracking of the P/ω droop coefficient, inertia coefficient, and damping coefficient for frequency changes but also avoids the influence of parameter incoordination on the quality and stability of the system.

2. Topology and Mathematical Model of VSG

2.1. Topology of VSG

A traditional VSG system is shown in Figure 1. The main circuit of the system consists of a DC source, voltage-type converter, filter circuit, load, and power grid. The control loop collects the output voltage, current, active power, reactive power, and other signals of the main circuit; generates modulation signals through the virtual governor, excitation controller, voltage and current double-loop controller; and finally generates the pulse signal to control the converter [19,20]. Topology and control block diagram of Figure 1: L_{abc} is the filter inductor; C_{abc} is the filter capacitor; P_{e} and Q_{e} are the active and reactive power output by the converter, respectively; $i_{\mathrm{abc\_inv}}$ and $u_{\mathrm{abc\_inv}}$ are the output current and voltage, K is the integral loop coefficient, U_n is the amplitude of the given voltage, U_{inv} is the RMS value of the three-phase voltage output from the converter, R_r is the stator resistance; L_r is the inductance, R_g is the line resistance, L_g is the line inductance; u_g is the grid voltage, i_{Labc} is the inductor current, u_{abc_ref} is the modulation voltage reference, g_{PWM} is the modulation signal.

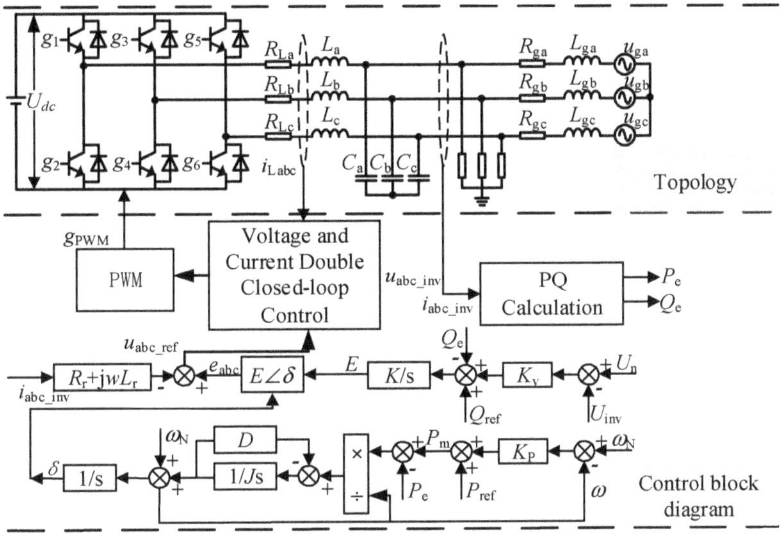

Figure 1. Topology and control block diagram of VSG.

2.2. Mathematical Model of VSG

The stator electrical equation and the typical second-order rotor motion equation of the synchronous generator are [6]:

$$\begin{cases} e = u + i(R + j\omega L) \\ T_m - T_e = \frac{P_m - P_e}{\omega} = J\frac{d\omega}{dt} + D\Delta\omega \\ \Delta\omega = \omega - \omega_N \\ \frac{d\delta}{dt} = \omega \end{cases} \tag{1}$$

Among them, R is the stator resistance; L is the inductance; u is the armature terminal voltage; T_m is the mechanical torque; T_e is the electromagnetic torque; J is the moment of inertia; D is the damping coefficient; ω is the mechanical angular velocity; ω_N is the rated angular velocity of the system; δ is the output power angle.

As shown in Figure 1, drawing on the principles of the synchronous generator governor and excitation regulator, the active power regulation and reactive power regulation equations are designed so that the entire converter system can truly simulate the characteristics of the synchronous generator.

$$\begin{cases} \omega = 2\pi f \\ E = \frac{K}{s}[Q_{ref} + K_V(U_n - U_{inv}) - Q_e] \\ \Delta T = [P_{ref} + K_P(\omega_N - \omega) - P_e]/\omega \end{cases} \tag{2}$$

In (2), P_{ref} and Q_{ref} are the system active and reactive commands respectively; K_V is the voltage regulation coefficient; K_P is the active power droop coefficient; K is the integral loop coefficient; E is the virtual potential command; U_n is the amplitude of the given voltage; U_{inv} is the effective value of the three-phase voltage output by the converter; f is the frequency of the terminal voltage of the virtual synchronous generator.

Finally, after combining the above-obtained voltages in the voltage synthesis link reference value and phase angle, the output voltage of VSG can be obtained as

$$e = \begin{cases} \sqrt{2}E\sin\delta \\ \sqrt{2}E\sin\left(\delta - \frac{2\pi}{3}\right) \\ \sqrt{2}E\sin\left(\delta + \frac{2\pi}{3}\right) \end{cases} \tag{3}$$

3. Multiparameter Cooperative Adaptive Control of VSG

3.1. The Influence of VSG Parameters on the System

According to the equivalent diagram in Figure 2, the output active power of the grid-connected inverter of VSG can be expressed as:

$$P_e = \frac{EU}{X}\sin\delta \tag{4}$$

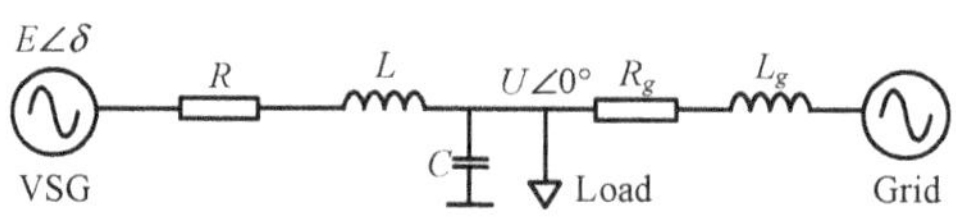

Figure 2. Equivalent diagram of VSG-connected to the grid.

Combining Formulas (1) and (4), we can obtain:

$$\frac{\Delta\omega}{\Delta P} = \frac{\omega_N - \omega}{P_{ref} - P_e} = -\frac{1}{J\omega_N s + D\omega_N + K_P} \tag{5}$$

Therefore, the closed-loop transfer function of the VSG active loop can be deduced as:

$$\varphi(s) = \frac{P_e}{P_{ref}} = \frac{\frac{EU}{J\omega_N X}}{s^2 + \frac{D\omega_N + K_P}{J\omega_N}s + \frac{EU}{J\omega_N X}} \tag{6}$$

The corresponding natural oscillation angular frequency and damping ratio are:

$$\begin{cases} \xi = \frac{0.5(D\omega_N + K_P)}{\sqrt{\frac{J\omega_N EU}{X}}} \\ \omega_n = \sqrt{\frac{EU}{J\omega_N X}} \end{cases} \tag{7}$$

When $0 < \xi < 1$, the power-frequency system is an underdamped system; when $\xi = 1$, the power-frequency system is a critically damped system; When $\xi > 1$, the power-frequency system is an overdamped system. Considering the two dynamic indicators of response speed and overshoot, the "Siemens second-order optimal system" control strategy is adopted, that is, the damping ratio is set to 0.707. Among them, in the underdamped state, within a certain allowable error, the adjustment time t_s and $\sigma\%$ are:

$$\begin{cases} t_s = \frac{4}{\xi\omega_n} = \frac{8J\omega_N}{D\omega_N + K_P} \\ \sigma\% = e^{-\frac{\pi\xi}{\sqrt{1-\xi^2}}} \times 100\% \end{cases} \tag{8}$$

When the droop coefficient is constant, it can be seen from Figure 3a that the damping ratio ξ of the VSG system is proportional to the damping coefficient D and inversely proportional to the virtual inertia J.

While the system adjustment time in Figure 3b is the opposite, it is proportional to the virtual inertia J and inversely proportional to the damping coefficient D.

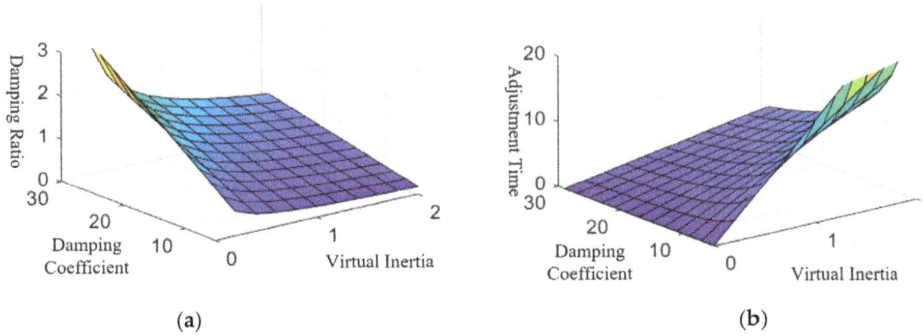

Figure 3. (**a**) Plot of damping ratio versus virtual inertia and damping coefficient. (**b**) Plot of adjustment time versus virtual inertia and damping coefficient.

The above analysis does not consider the influence of the droop coefficient on the dynamic response of the system. Combining Formulas (7) and (8), it can be known that when the virtual inertia and damping coefficient are constant, the damping ratio increases with the increase of the droop coefficient, and the adjustment time decreases with the increase of the droop coefficient. Relative to the damping ratio, the system adjustment time is more deeply affected by the droop factor.

It can be seen from the above analysis that the virtual inertia J, damping coefficient D, and droop coefficient K_P in the traditional virtual synchronous machine control strategy remain unchanged. If any one of the values is changed alone, although the transient characteristics of the VSG grid connection can be significantly improved, it cannot take into account the stability and rapidity of the transient process after the system is disturbed.

3.2. Cooperative Adaptive Selection Strategy of Control Parameters

In this paper, the frequency oscillation process of VSG is divided into four different stages, as shown in Figure 4. The characteristics of the different phases are analyzed, the virtual inertia J and the droop coefficient K_P are adjusted in real time, and the damping coefficient D is coordinated according to the relationship between the three to achieve stable control of the grid-connected VSG transient process.

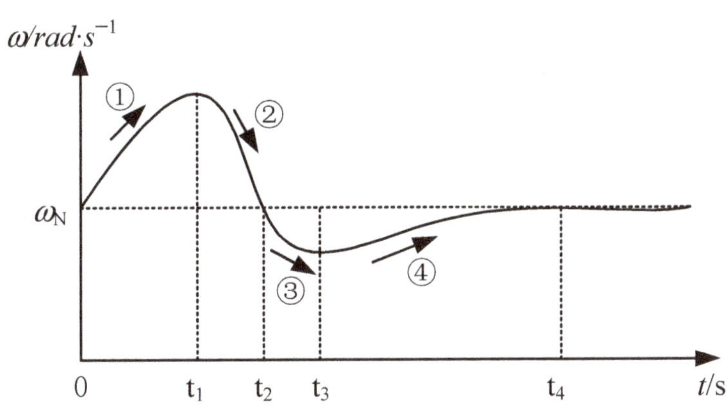

Figure 4. Angular velocity fluctuation graph.

$\Delta\omega \cdot d\omega/dt > 0$ exists in both stage 1 and stage 3 and $|\Delta\omega|$ gradually increases in both stages. These two stages are defined as the acceleration stage of rotor angular velocity. This stage requires larger virtual inertia J and larger droop coefficient K_P to reduce the amplitude of rotor angular velocity offset.

$\Delta\omega \cdot d\omega/dt < 0$ exists in both stage 2 and stage 4. Since $|\Delta\omega|$ gradually decreases, the two stages are defined as the deceleration stage of rotor angular velocity. In this stage, the virtual inertia J and the droop coefficient K_P need to be reduced to make the rotor angular velocity return to a stable value as soon as possible.

However, in the acceleration stage of the rotor, although the increase in inertia can improve the anti-interference, it will reduce the response speed. In this stage, the damping method can be used to improve the response speed.

In the deceleration stage, the virtual inertia J and the droop coefficient K_P are reduced, and the system's suppression of the fluctuation of the rotor angular velocity is weakened to speed up the decay rate of the rotor angular velocity; nevertheless, it will cause an increase in the amplitude of the fluctuation of the rotor angular velocity. Therefore, at this stage, the damping coefficient can be increased to reduce the overshoot of the system and make the frequency return to stability as soon as possible.

The selection of J is determined by $\Delta\omega$ and $d\omega/dt$ at the same time. In order to avoid the complicated control strategy, the change rule is set as:

- When $\Delta\omega$ and $d\omega/dt$ change in the same direction, J needs to be increased;
- When $\Delta\omega$ and $d\omega/dt$ change in opposite directions, J should be kept unchanged.

The virtual inertia J and the droop coefficient K_P are associated with the VSG angular velocity and the instantaneous value of the angular velocity to obtain the control parameter adaptive strategy, as shown in Formulas (9) and (10).

$$J = \begin{cases} J_0 + k_J \left| \Delta\omega \frac{d\omega}{dt} \right|, \Delta\omega \cdot \frac{d\omega}{dt} > 0 \\ J_0, \Delta\omega \cdot \frac{d\omega}{dt} \leq 0 \end{cases} \tag{9}$$

$$K_P = K_{P0} + k_\omega \cdot |\Delta\omega| \tag{10}$$

In the formulas, J_0 and K_{P0} are the virtual inertia and droop coefficient of VSG fixed parameters, respectively; k_J is the inertia adjustment coefficient and k_ω is the adjustment coefficient of the droop coefficient.

The coordinated control design of the droop coefficient, virtual inertia, and damping coefficient is carried out. Combining Formulas (7), (9) and (10), the damping coefficient D design under the correlation can be obtained:

$$D = 2\xi \sqrt{J \frac{EU}{\omega_N X}} - \frac{1}{\omega_N} K_P \tag{11}$$

Based on automatic control theory, in order to keep the system in the optimal control operation state, ξ can be set to 0.707. It can be known from Formula (11) that when other parameters in the system are constant, the droop coefficient, virtual inertia, and damping coefficient can be jointly designed according to the requirements of system characteristics.

3.3. The Setting of Parameter Value Range

The adaptive adjustment coefficient of VSG inertia can be selected according to the value range of inertia. According to the setting principle of the virtual inertia value of the VSG scheme of the University of Leuven [21], the maximum value of the virtual inertia must satisfy:

$$J_{max} < \frac{P_{max}}{\max\left\{ \omega \left(\frac{d\omega}{dt} \right) \right\}} \tag{12}$$

In Formula (12), P_{max} is the maximum power that the system can withstand. In order to ensure the stable operation of the system, the angular frequency of the system is limited with reference to the current national standard power system frequency deviation (50 ± 0.2 Hz).

The maximum and minimum frequencies are ω_{max} and ω_{min}, respectively; then, the damping coefficient selection in Formula (10) should satisfy Formula (13):

$$0 \leq \frac{1}{D\omega_N + K_P} \leq \frac{\omega_{max} - \omega_{min}}{P_{max} - P_{min}} \tag{13}$$

In Formula (13), P_{min} is the minimum output power of VSG. Therefore, the minimum value of the damping coefficient D is:

$$\left(\frac{P_{max} - P_{min}}{\omega_{max} - \omega_{min}} - K_P\right)/\omega N \leq D \tag{14}$$

And the droop factor shall also be taken to satisfy Formula (15):

$$K_P \leq P_{max}/\Delta\omega_{max} \tag{15}$$

In addition, in order to make the system have good response rapidity, each transient component of its response should have a large decay factor—that is, the closed-loop pole of the system should be far away from the imaginary axis, and the closed-loop pole should satisfy [22–25]:

$$\text{Re}(s_i) = -\omega_n\xi = -\frac{D + K_P/\omega_N}{2J} \leq -10 \tag{16}$$

In VSG control, moderate damping, a fast response, and a small overshoot are usually desired in the control system. The damping ratio ξ is therefore chosen to be in the range (0.7, 1). In summary, the range of virtual inertia J and droop factor K_P can be obtained as shown in Figure 5.

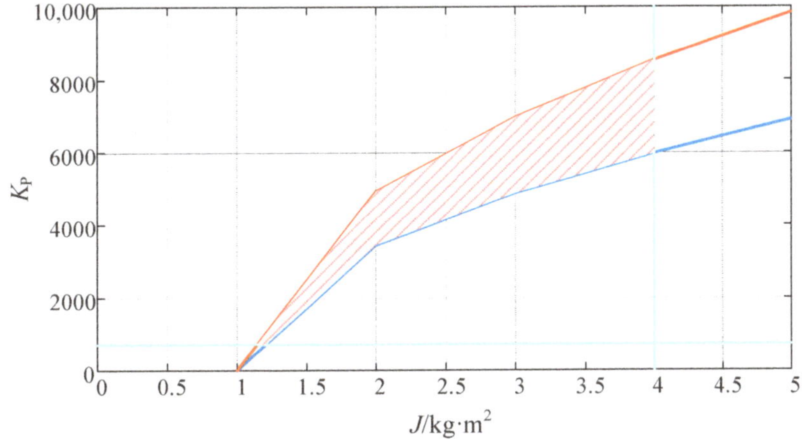

Figure 5. The range of virtual inertia J and droop factor K_P.

To sum up, the flow chart of the core algorithm in this paper is shown in Figure 6, where N_ω is the threshold for setting a triggering adaptive function.

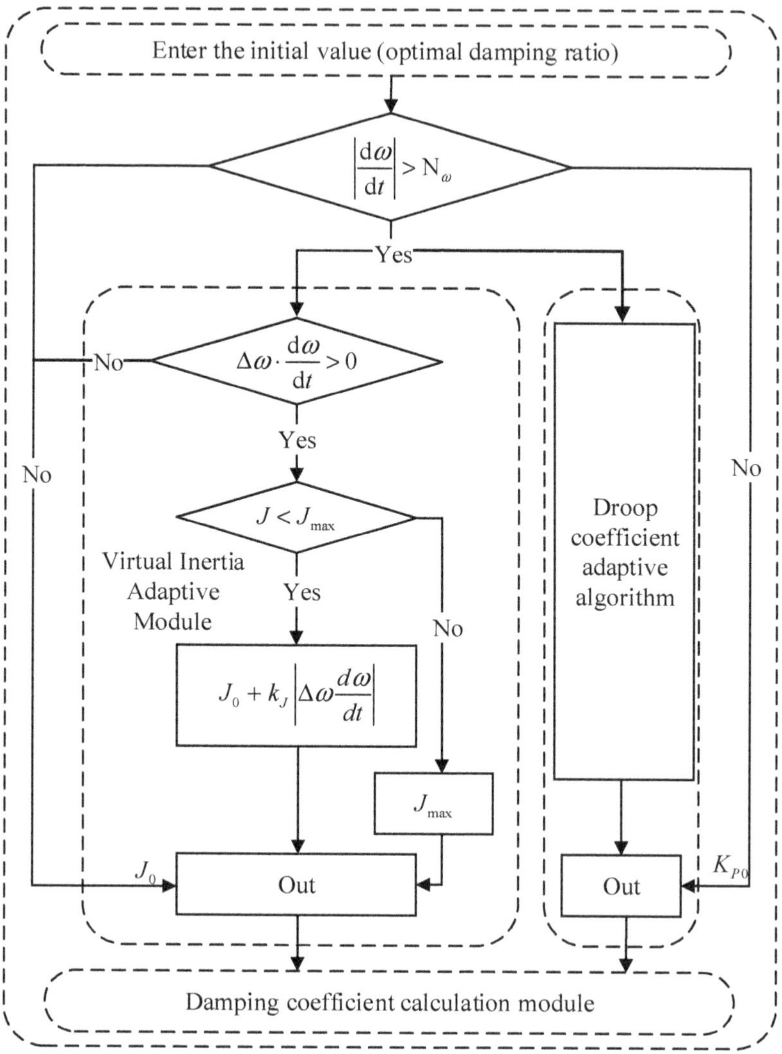

Figure 6. Core algorithm flow chart.

Based on the above design, the bode diagram of the whole system is shown in Figure 7. In the figure, the amplitude margin as well as the phase margin of the control strategy proposed in this paper is the largest, so that the error value is the smallest and not prone to damped oscillations, resulting in the highest stability and the best system performance. Fuzzy adaptive optimal control is the next most effective control strategy. Traditional adaptive control is the least effective.

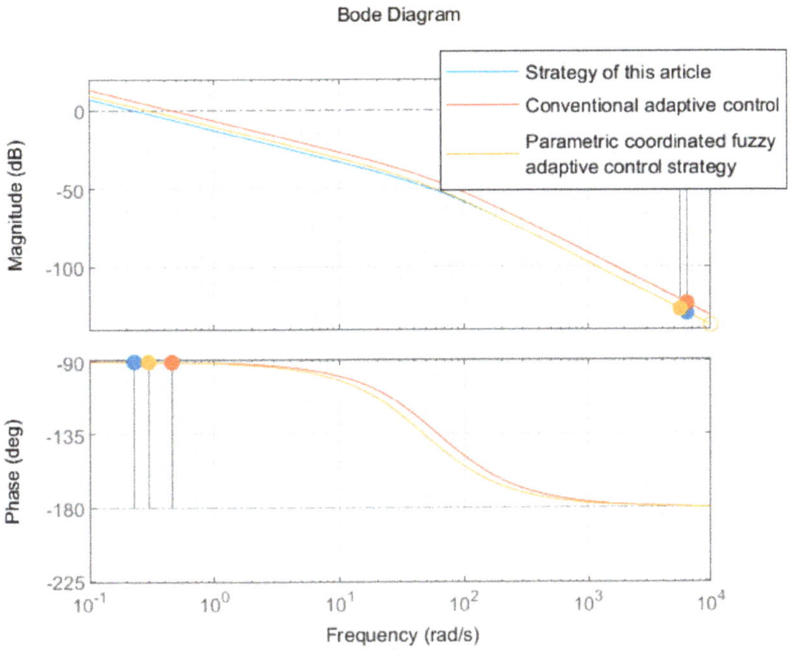

Figure 7. Bode diagram of the system.

4. Simulation Analysis

In order to verify the control strategy proposed in this paper, an improved VSG system simulation model on the MATLAB/Simulink platform is built into this work, and the relevant control parameters used are shown in Table 1.

The system is in a grid-connected operation state, the simulation duration is set to 2 s, the initial steady state is assumed, and the grid frequency is equal to the rated frequency. In order to simulate the change in load power, the load power is set to increase by 10 kW at 1 s, and the reactive power is constant at 0 kVar during this period.

Table 1. Simulation parameters of VSG.

Parameter	Numerical Value	Parameter	Numerical Value
Rated voltage on AC grid u_g (V)	380	Filter inductor L_{abc}(mH)	0.8
Rated voltage on DC side Udc (V)	800	Filter capacitor C_{abc} (uF)	10
Rated active power (W)	50,000	Initial value of virtual inertia J_0 (kg·m^2)	1.127
Rated reactive power (Var)	0	Initial value of droop coefficient K_{P0}	5000

4.1. Effect of Changes in System Parameters on System Frequency

Figures 8–10 show the frequency variation curve of the system with different inertia, damping, and droop coefficient. When the virtual damping D and droop factor K_P are fixed, the greater the virtual inertia J is, the longer the dynamic adjustment time is, but the frequency overshoot is reduced. The system transitions from an overdamped state to an underdamped state, as shown in Figure 8.

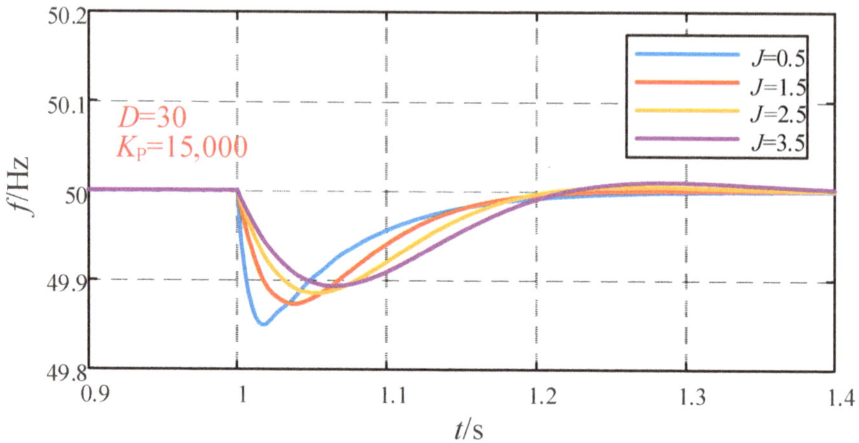

Figure 8. Frequency waveform under *J* change.

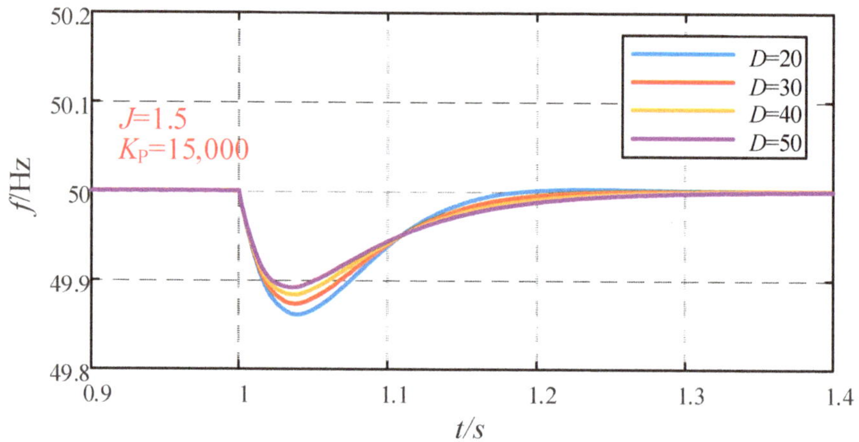

Figure 9. Frequency waveform under *D* change.

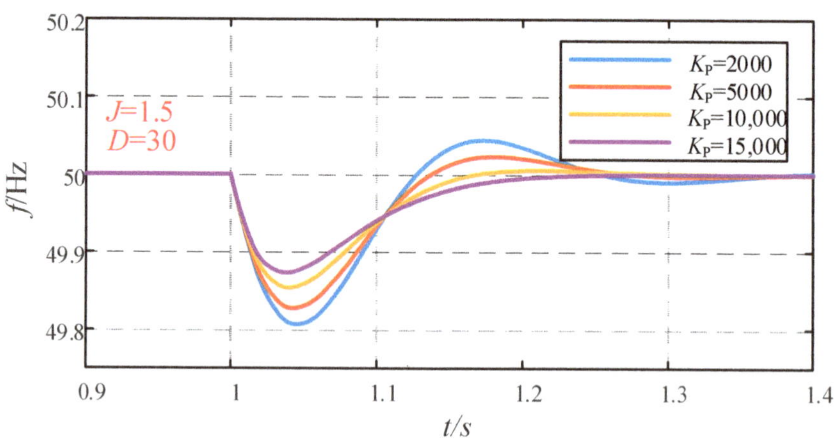

Figure 10. Frequency waveform under K_P change.

When the virtual inertia J and sag factor K_P are fixed, the smaller the virtual damping D is, the longer the dynamic adjustment time is, and the greater the frequency overshoot is, the system transitions from an overdamped state to an underdamped state, as shown in Figure 9.

When the virtual inertia J and virtual damping D are fixed, the larger the sag coefficient K_P is, the smaller the frequency overshoot is, as shown in Figure 10. The correctness of the theoretical analysis in this paper is verified.

4.2. Simulation under Varying System Loads

4.2.1. Simulation Comparison of Different Control Strategies

In order to verify the superiority of this control strategy in grid-connected mode, it is compared with coordinated inertia damping (ξ = 0.707), J and D coordinated adaptive control and parametric coordinated fuzzy adaptive control strategy, respectively. The frequency change is shown in Figure 11.

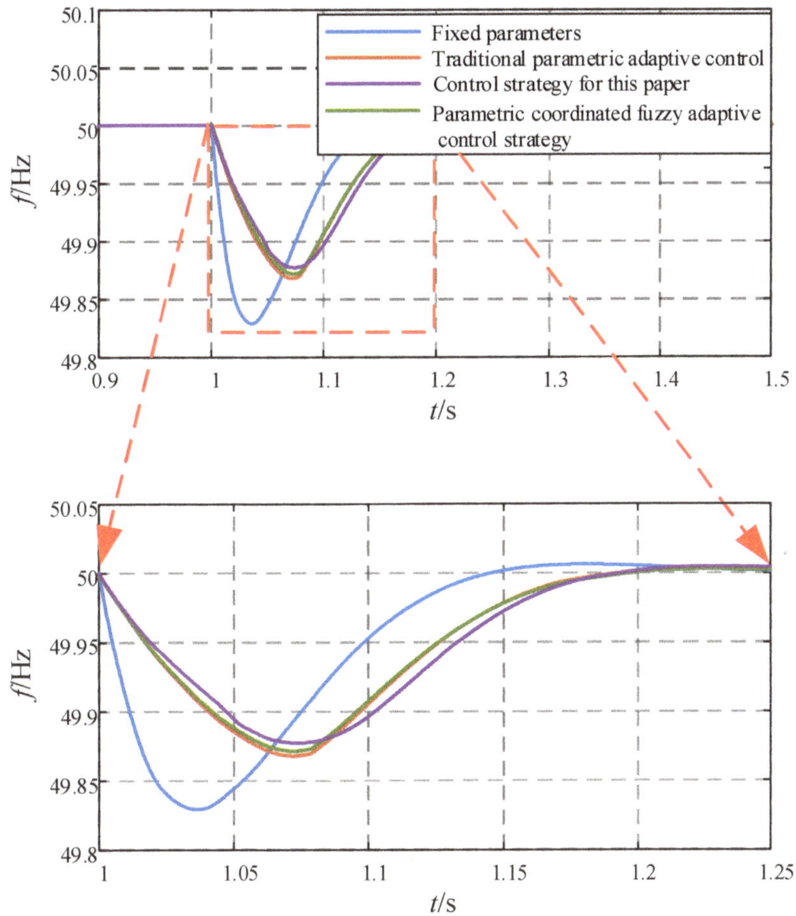

Figure 11. Frequency variation under different control strategies.

The frequency change curve and a partially enlarged view can be seen in Figure 11. It can be seen from Figure 11 that the coordinated inertia damping (ξ = 0.707) control strategy is adopted—when the parameters are fixed, the maximum offset of the frequency fluctuation reaches 0.17 Hz, and it takes about 0.3 s to reach the steady state. When using

J and *D* coordinated adaptive control, the maximum frequency offset is reduced to 0.133 Hz, and the adjustment time is about 0.22 s. When using a parametric coordinated fuzzy adaptive control strategy, the maximum frequency offset is reduced to 0.132 Hz, and the adjustment time remains the same. When the multiparameter cooperative adaptive control strategy is adopted, the maximum frequency offset is 0.123 Hz, and the adjustment time is 0.22 s. The specific performance indicators are shown in Table 2.

The maximum deviation of the multiparameter cooperative adaptive control is optimized by 7.52% compared with the *J* and *D* coordinated adaptive control and is optimized by 27.65% compared with the coordinated inertia damping ($\xi = 0.707$) control strategy. The adjustment time of the multiparameter cooperative adaptive control is equivalent to that of the *J* and *D* coordinated adaptive control and is 26.67% higher than that of the coordinated inertia damping control.

Table 2. Control strategy performance indicators.

Control Strategy	Nadir (Hz)	Deviation (Hz)	Adjustment Time (s)	ROCOF
Fixed parameters	49.83	0.17	0.3	maximum
J, *D* adaptive control	49.867	0.133	0.22	medium
Parametric coordinated fuzzy adaptive control strategy	49.868	0.132	0.22	medium
Multiparameter cooperative adaptive control	49.877	0.123	0.22	minimum

4.2.2. Comparison of Parameter Changes

Figures 12 and 13 show the variation of the *J* and *D* coordinated adaptive control parameters and the variation of the multiparameter collaborative adaptive control parameters in this paper, respectively.

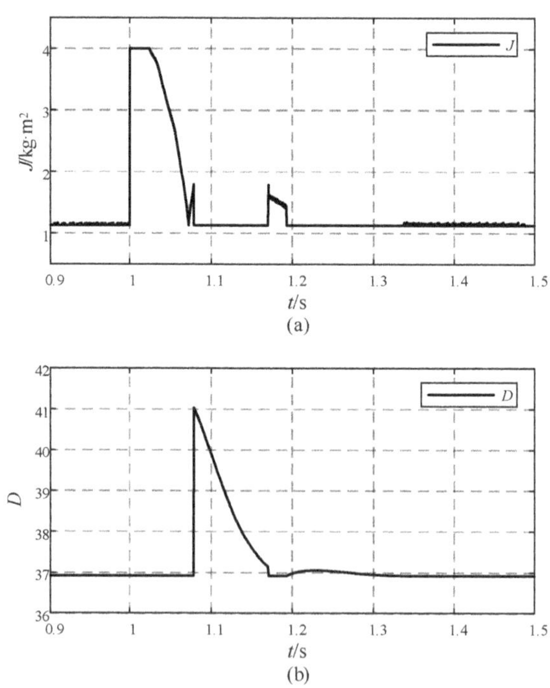

Figure 12. *J* and *D* coordinated adaptive control parameter variation: (**a**) virtual inertia change curve; (**b**) damping coefficient change curve.

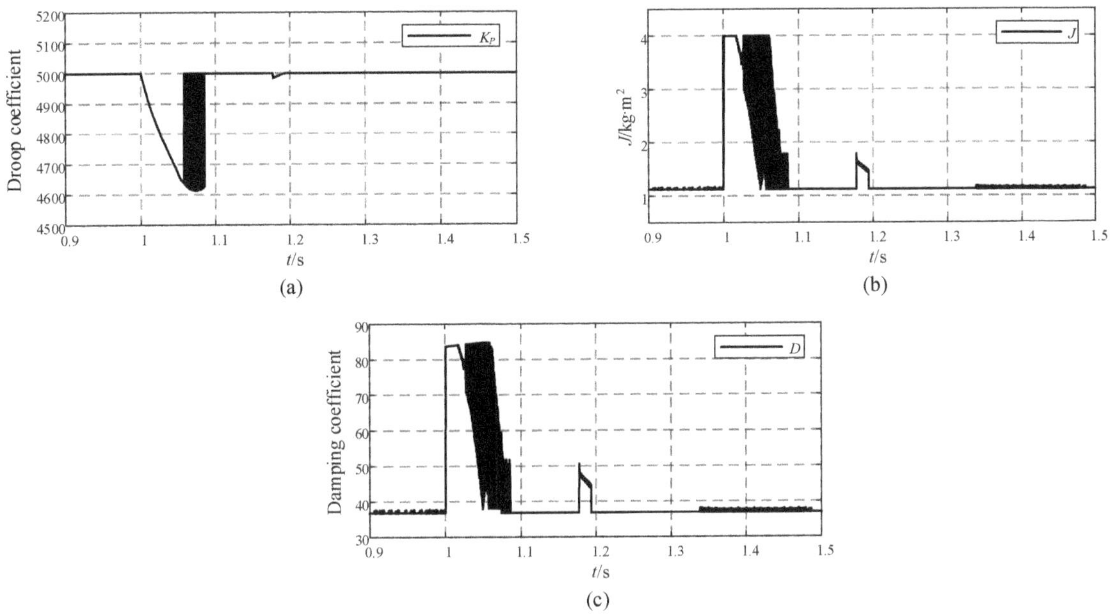

Figure 13. (**a**) Droop coefficient change curve. (**b**) Virtual inertia change curve. (**c**) Damping coefficient change curve.

Figure 12, as well as Figure 13b,c, shows the frequency change caused by the load change at 1 s; the angular frequency rate of change and frequency offset are both negative, and in the angular frequency acceleration phase, the virtual inertia and damping coefficient increase rapidly. During the frequency fluctuation, when the angular frequency rate of change and frequency offset have different signs, the virtual inertia and damping coefficient decrease rapidly in the angular frequency deceleration stage.

However, when J and D adaptive control is in a perturbation cycle of the change process, during the same moment only one quantity changes (that is, the control for alternate), other quantities are not affected. This control system is simple, ignoring the influence of the droop coefficient.

It also can be seen from Figure 13a that the droop coefficient will have a corresponding adaptive change when the frequency changes caused by the load change, to further synergistically optimize the damping coefficient. During dynamic changes, the droop coefficient varies continuously in the range of 4600 to 5000, and the virtual inertia varies in the range of 1 to 4, while the damping coefficient varies in the range of 38 to 85 under the combined effect of both.

4.3. Simulation under Grid Frequency Variation

To further verify the performance of the proposed control strategy, the grid frequency was simulated: at 0.5 s, the grid frequency increased by 0.2 Hz, and after 1 s, the frequency returned to 50 Hz.

4.3.1. System Frequency Performance

The frequency change curve and a partially enlarged view can be seen in Figure 14. As can be seen from the graph, frequency overshoot exists using the fixed parameter control strategy; when using conventional adaptive control, the overshoot is reduced. The rate of change in frequency is greater for both of these approaches. When using fuzzy adaptive control and the strategy proposed in this paper, there is no overshoot in the frequency, and the frequency variation rate of the system is minimal under the control strategy proposed in this paper.

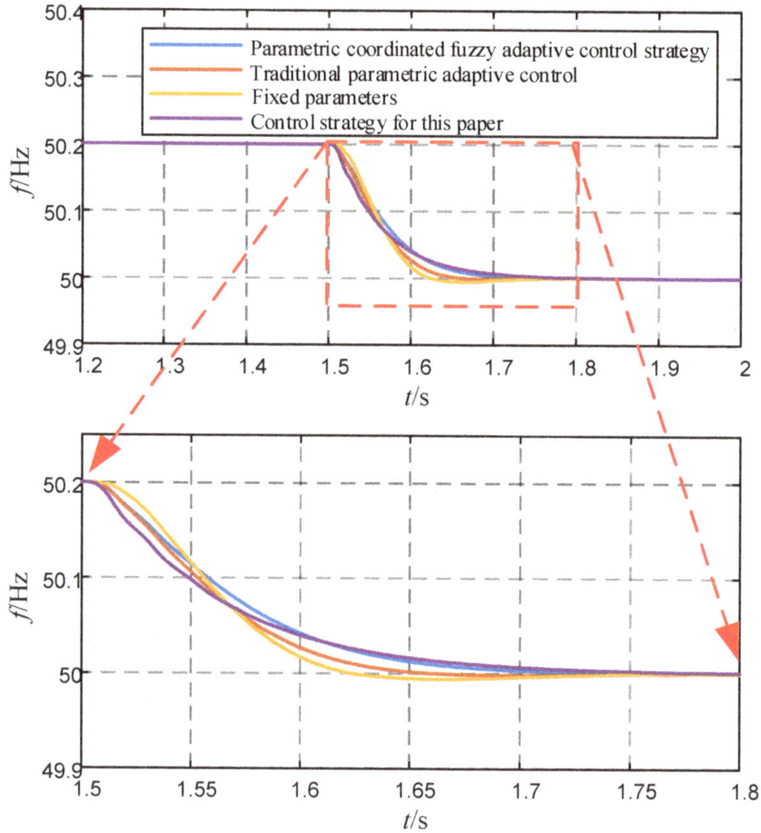

Figure 14. Frequency curves.

4.3.2. Variation of the Three Parameters

It can also be seen in Figure 15a that the VSG needs to adjust its output when the grid frequency rises by 0.2 Hz throughout; therefore, it needs to adjust the droop coefficient, which varies dynamically between 5000 and 5600. The virtual inertia also varies at this point between around 1.1 and 3.9, as shown in Figure 15b. The damping coefficient is adjusted in real time according to the mathematical logic between the three and varies between 36 and 82, as shown in Figure 15c. Unlike the load variation, the sag coefficient, virtual inertia, and the damping coefficient keep interacting with each other throughout the process, without stopping to vary when the frequency is stable.

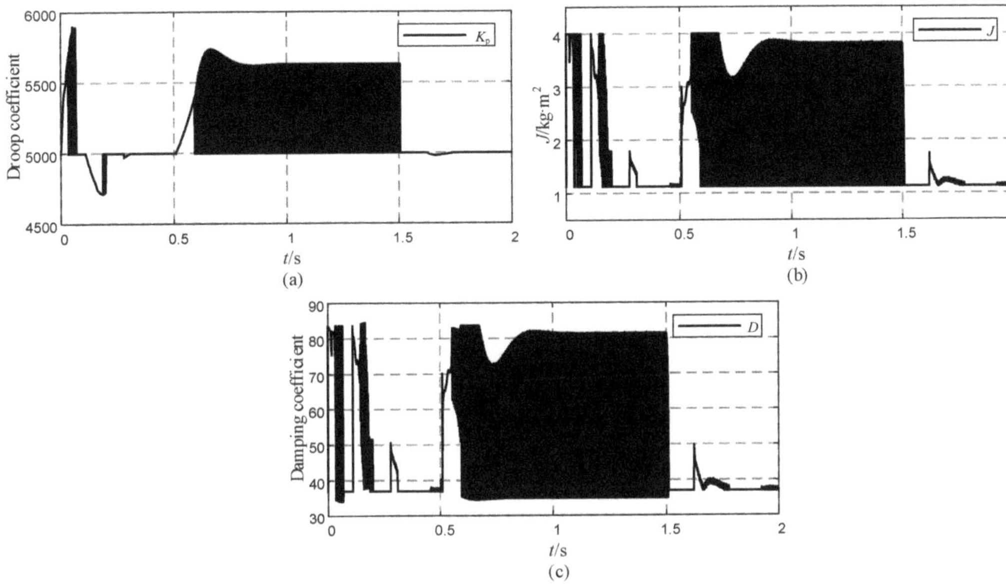

Figure 15. (**a**) Droop coefficient change curve. (**b**) Virtual inertia change curve. (**c**) Damping coefficient change curve.

5. Conclusions

Although the traditional virtual synchronous generator technology can improve the stability of the system by simulating the operating characteristics of the synchronous generator, it has problems, such as the inability to take into account multiple performance indicators for parameter design, long adjustment time, and large transient overshoot. To address these problems, this paper uses an improved VSG parameter optimization cooperative control strategy, and the effectiveness of this control strategy is verified by simulation through simulated load variation and frequency variation. The main work and contributions are as follows:

(1) Analyze the effect of droop factor, virtual inertia, and damping factor on the VSG system and determine the range of values for the system parameters.

(2) The existing J and D coordinated adaptive control is optimized. The droop coefficient and virtual inertia can be adjusted in real time according to the system frequency state, and the damping coefficient can be changed cooperatively according to the corresponding relationship. The three are always coordinated and adjusted during the change process, which effectively improves the dynamic performance of the system frequency. The effectiveness and reliability of the proposed multiparameter cooperative adaptive control strategy have been verified by simulation.

Author Contributions: Conceptualization, B.R. and Z.F.; methodology, Q.L.; software, Y.S.; validation, B.R., Z.F. and Y.S.; formal analysis, Q.L.; investigation, B.R.; resources, Q.L.; data curation, Q.L.; writing—original draft preparation, B.R.; writing—review and editing, Y.S.; visualization, B.R.; supervision, Y.S.; project administration, Y.S.; funding acquisition, Y.S. All authors have read and agreed to the published version of the manuscript.

Funding: This work was supported by the Science and Technology Project of State Grid Jiangsu Electric Power Company Limited (J2021177) (Research on the influence mechanism of high aggregation access of offshore wind farms on the inertia of Jiangsu Power grid and key technologies to improve the accommodation capacity).

Data Availability Statement: Not applicable.

Conflicts of Interest: The authors declare no conflict of interest.

References

1. Yang, X.Z.; Su, J.H.; Ding, M. Frequency control strategy for microgrid island operation. *Power Syst. Technol.* **2010**, *34*, 164–168.
2. Sun, H.D.; Xu, T.; Guo, Q. Analysis on blackout in Great Britain power grid on August 9th, 2019 and its enlightenment to power grid in China. *Proc. CSEE* **2019**, *39*, 6183–6191.
3. Wen, Y.F.; Yang, W.F.; Lin, X.H. Review and prospect of frequency stability analysis and control of low-inertia power systems. *Electr. Power Autom. Equip.* **2020**, *40*, 211–222.
4. Zhong, Q.C.; Nguyen, P.L.; Ma, Z.Y. Selfsynchronized synchronverters: Inverters without a dedicated synchronization unit. *IEEE Trans. Power Electron.* **2014**, *29*, 617–630. [CrossRef]
5. Du, W.; Jiang, Q.R.; Chen, J.R. Design and application of reactive power control system for wind farm. *Autom. Electr. Power Syst.* **2011**, *35*, 26–31.
6. Ren, H.P.; Chen, Q.; Zhang, L.L. Parameter adaptive strategy for virtual synchronous generator control. *Control Theory Appl.* **2020**, *37*, 2571–2580.
7. Wang, F.; Zhang, L.J.; Feng, X.Y. An adaptive control strategy for virtual synchronous generator. *IEEE Trans. Ind. Appl.* **2018**, *54*, 5124–5133. [CrossRef]
8. Alipoor, J.; Miura, Y.; Ise, T. Power system stabilization using virtual synchronous generator with alternating moment of inertia. *IEEE J. Emerg. Sel. Top. Power Electron.* **2015**, *3*, 451–458. [CrossRef]
9. Zhu, Z.B.; Zhang, C.Y.; Zeng, X.B. Photovoltaic energy storage microgrid system based on adaptive rotating inertia VSG control strategy. *Proc. CSU-EPSA* **2021**, *33*, 67–72.
10. Zhou, P.G.; Meng, J.H.; Wang, Y. Influence analysis of the main control parameters in FVSG on the frequency stability of the system. *High Volt. Eng.* **2018**, *44*, 1335–1342.
11. Yang, Y.; Mei, F.; Zhang, C.Y. Coordinated adaptive control strategy of rotational inertia and damping coefficient for virtual synchronous generator. *Electr. Power Autom. Equip.* **2019**, *39*, 125–131.
12. Li, D.D.; Zhu, Q.W.; Lin, S.F. A self-adaptive inertia and damping combination control of VSG to support frequency stability. *IEEE Trans. Energy Convers.* **2017**, *32*, 397–398. [CrossRef]
13. Li, Z.J.; Yang, M.W.; Zhang, J.A.; Liu, H.J. Research on the cooperative adaptive control of VSG inertia and damping. *J. Electr. Power Syst. Autom.* **2023**, *35*, 36–43.
14. Zhang, X.W.; Tan, L.P.; Chen, W.H.; Liu, Y.; Liu, M.S. Inertia lifting method of energy storage converter based on secondary system optimization. *Power Syst. Prot. Control* **2021**, *49*, 128–135. (In Chinese)
15. Fan, Y.N.; Liu, T.Y.; Jiang, X.C.; Sheng, G.H.; Zhang, X.S. Optimal control strategy based on optimal damping ratio dummy synchronous generator. *Electr. Meas. Instrum.* **2020**, *57*, 60–67. (In Chinese)
16. He, P.; Li, Z.; Jin, H.R.; Zhao, C.; Fan, J.L.; Wu, X.P. An adaptive VSG control strategy of battery energy storage system for power system frequency stability enhancement. *Int. J. Electr. Power Energy Syst.* **2023**, *149*, 142–615. [CrossRef]
17. Zhang, Y.N.; Sun, X.L.; Fu, W.L.; Hu, S.L. Parametric coordinated fuzzy adaptive VSG control strategy. *Electron. Meas. Technol.* **2022**, *45*, 1–7. (In Chinese)
18. Peng, C.; Hu, L.R.; Zhang, L.M.; Wang, B.H. Adaptive damping control of microgrid based on distributed VSG. *Power Electron. Technol.* **2023**, *57*, 85–88. (In Chinese)
19. Shao, Y.Y. *Research on Energy Storage Inverter Control Based on Virtual Synchronous Generator*; Zhejiang University: Hangzhou, China, 2021.
20. Zhang, G.F.; Yang, J.Y.; Wang, H.X.; Xie, C.J.; Fu, Y. Coordinated Frequency Modulation Control Strategy of Wind Storage System Based on Virtual Synchronous Machine Technology. *J. Electrotech. Technol.* **2022**, *37*, 83–92.
21. Lü, Z.P.; Sheng, W.X.; Zhong, Q.C. Virtual synchronous generator and its applications in micro-grid. *Proc. CSEE* **2014**, *34*, 2591–2603. (In Chinese)
22. Du, Y.; Su, J.H.; Zhang, L.C. A mode adaptive frequency controller for microgrid. *Proc. CSEE* **2013**, *33*, 67–75.
23. Xu, H.Z.; Zhang, X.; Liu, F. Virtual Synchronous Generator Control Strategy Based on Lead-lag Link Virtual Inertia. *Proc. CSEE* **2017**, *37*, 1918–1927. (In Chinese)
24. Xu, H.Z.; Zhang, X.; Liu, F. Control strategy of virtual synchronous generator based on differential compensation virtual inertia. *Autom. Electr. Power Syst.* **2017**, *41*, 96–102. (In Chinese)
25. Xu, H.; Yu, C.; Liu, C.; Wang, Q.; Zhang, X. An Improved Virtual Inertia Algorithm of Virtual Synchronous Generator. *J. Mod. Power Syst. Clean Energy* **2020**, *8*, 377–386. [CrossRef]

energies

MDPI

Article

Joint Detection and State Estimate with GSAs in PMU-Based Smart Grids

Feng Hua [1,2], Wengen Gao [1,2,*], Yunfei Li [1,2], Pengfei Hu [1,2] and Lina Qiao [1,2]

1 School of Electrical Engineering, Anhui Polytechnic University, Wuhu 241000, China
2 Key Laboratory of Advanced Perception and Intelligent Control of High-End Equipment, Chinese Ministry of Education, Wuhu 241000, China
* Correspondence: ahpuchina@ahpu.edu.cn

Abstract: The Phasor Measurement Unit (PMU) with a GPS signal receiver is a synchronized sensor widely used for power system state estimation. While the GPS receiver ensures time accuracy, it is vulnerable to network attacks. GPS spoofing attacks can alter the phase angle of PMU measurement signals and manipulate system states. This paper derives a power system state model based on PMUs under GPS spoofing attacks, according to the characteristics of changes in bus voltages and branch currents after GSA. Based on the characteristics of this model, a detection and correction algorithm for attacked data is proposed to detect GSA and correct attacked measurements. The corrected measurements can be used for power system state estimation. Simulation results on the IEEE 14-bus system show that the proposed algorithm improves the accuracy of state estimation under one or multiple GSAs, especially when multiple GSAs are present, compared to classical Weighted Least Squares Estimation (WLSE) and Alternating Minimization (AM) algorithms. Further research indicates that this algorithm is also applicable to large-scale networks.

Keywords: phasor measurements unit (PMU); GPS spoofing attacks (GSAs); detection; correction; state estimate

Citation: Hua, F.; Gao, W.; Li, Y.; Hu, P.; Qiao, L. Joint Detection and State Estimate with GSAs in PMU-Based Smart Grids. *Energies* **2023**, *16*, 5731. https://doi.org/10.3390/en16155731

Academic Editors: Gianfranco Chicco and Abu-Siada Ahmed

Received: 7 June 2023
Revised: 18 July 2023
Accepted: 28 July 2023
Published: 31 July 2023

1. Introduction

Phasor Measurement Units (PMUs) based on global positioning system have been widely used in wide area measurement system [1,2]. PMU is an advanced digital instrument used for real-time monitoring of grid operation in smart grids. By installing a PMU on one bus, it can obtain the bus voltage phasor and all current phasors of the branches connected to that bus [3]. PMU data is an important data source for situation awareness, dispatching control, and early warning of modern power systems. Many advanced applications have been developed based on PMU data, such as state estimation, wide area damping control, generator state monitoring, static voltage stability evaluation, etc. Accurate PMU data is a prerequisite for advanced applications.

The PMU uses a common time source for synchronization and can measure the electrical waves on the grid. The PMU does this by converting the analog signals of voltage and current to digital signals and applying anti-alias filters and discrete Fourier transforms to isolate the fundamental frequency components and compute their phase representation [4,5]. This allows obtaining the fundamental frequency amplitude and phase information of the voltage and current signals at each bus in the power system, instead of the actual measurements in conventional Supervisory Control and Data Acquisition (SCADA) systems. When the number of PMUs deployed in the system is sufficient to make the system globally observable, state estimation is only required based on the PMU data [6,7]. In this case, the state of the power system can be expressed in the form of a linear equation that can be solved in a single iteration, which improves the processing speed [8,9]. Replacing nonlinear state estimation with linear state estimation using PMU measurements allows direct manipulation of the Jacobi matrix. The reliability of the simplified model and

PMU state variables was explained in [10]. The solution of the linear model is direct and non-iterative.

In recent years, with the intelligent development of power grid and the importance of the power grid security, PMUs have become increasingly important in power system state estimation [10–15]. The low reporting rate and complex nonlinear state estimation of traditional SCADA systems for power grids make it difficult to meet high-precision state analysis and real-time safety monitoring. However, PMU can provide synchronized phasor measurements, generating a linear model for state estimation. Their sampling rate is much higher than that of the SCADA system, and it can estimate the state of power system in real-time and quickly respond to abnormal conditions.

Each PMU is equipped with a GPS receiver, which is synchronized using GPS civilian signals, which are not encrypted like the military signals [11,16–18]. GPS provides sub-microsecond precision timing [16], which plays a crucial role in the time synchronization of PMU measurements. Wireless communication between civilian receivers and satellites is then vulnerable to cyber attacks. GPS spoofing is caused by transmitters mimicking GPS signals with the intent of altering the GPS time estimated by the PMU's GPS receiver [19]. These attacks maliciously introduce incorrect timestamps, which leads to incorrect phase angles in PMU measurements [20], making the system state estimation problem nonlinear, counteracting the original motivation for the introduction of PMUs as well as posing a serious threat to the correct state estimation of the power system. The issue of time delay in PMU measurements and its compensation methods have been discussed in [21], but these methods have limited effectiveness compared to GSAs. As a result, GSAs detection and countermeasures are widely studied and methods are proposed to prevent serious destructions to the power system.

2. Prior Work

In recent years, there have been much research on GSA, and the research directions were mostly divided into three categories. They were respectively the feasibility of GSA [22–25], the impact of GSA on power grid state [19,26], and some solutions for GSA [21,27–30].

The feasibility details of GSAs were shown in [22–25]. Shepard et al. [24] have performed field tests of GPS spoofing on PMUs, exposing the vulnerability of PMUs to this malicious attack; by spoofing the GPS signal receiver inside the PMU, the attacker can introduce timing errors, which result in a consequent shift in the phase angle of the PMU, which posed a serious problem for real-time monitoring of the smart grid. Humphreys et al. [22] first implemented GPS spoofing attack in the laboratory and gave a preliminary mitigation method for non-encrypted GSA. Jiang et al. [25] manipulated the acquired GPS signal data, and they found that this operation could change the phase angle of the PMU-measured signal.

However, more research were on the countermeasures of GSA [19,26–30]. Liang et al. [26] comprehensively discussed the theoretical basis of false data injection attacks (FDIA) and gave the most basic defense strategy for FDIA. After measuring and collecting the data provided by the PMUs, a cross-layer detection method was proposed in [19]. Its principle is to use the angle of arrival of GPS as the initial guess, and then detect whether GSA has occurred through the state estimation of the system. Mahapatra et al. [27] proposed a method to detect bad data in PMU measurements during interference. Which was based on the principal component analysis (PCA) method to distinguish the safety data caused by bad data from the manipulation data caused by interference. In particular, Zhang et al. [28] proposed a novel distribution system identification and correction algorithm for simultaneous GSA with multiple PMU positions. The algorithm first analyzed the sensitivity of the residual of the phase angle state estimation under a single GSA, and then used the proposed detection technology identification algorithm to locate the attacked PMU and the shift range of the phase angle after the attack. Finally, the phase shift was determined by minimizing the offset of measurement and system state estimation. PMU is vulnerable to attack because it receives unencrypted civil signals. To solve this problem, Mina et al. [29] proposed a wide-area spoofing detection algorithm for PMU, which used the hybrid communication architecture of NAPSInet. They created conditional signal fragments containing military P (Y) signals, whose precise code sequences are not available to civilian users, thus protecting PMUs from attack from the source. Unlike adding military encrypted signal fragments, Bhamidipati et al. [30] proposed a new algorithm for jointly spoofer location and GPS time using multiple receivers direct time estimation (MRDTE). The experimental results show that it can locate the attacked PMU within a certain range and estimate the time change. For the error synchronization in PMU measurement data, Zhang et al. [21] gave the source of this error, and proposed the method of using the Kalman filter to compensate for this error, which effectively alleviates the error caused by different.

The impact of a GPS spoofing attack on the secure operation of the power grid system was shown in [25]. We can find that once the PMU is subject to GSA, it is impossible to conduct real-time monitoring and correct state estimation of the system, which will bring huge hidden dangers.

For system state estimation, the static estimation method was used to detect attacks of PMU in the power grid [31]. More PMUs were installed in the power grid to make the measurements of the bus redundant, and then various PMU signals are compared mathematically to obtain the GPS attack detection formula [32]. Most of the above literature treats the system as static. The dynamic changes of the system were not considered. Phase shift caused by PMU delay under system dynamic characteristics was considered in [33]. However, this phase shift is not caused by spoofing attacks, but the deployed PMUs come from different standards, protocols, and designs. Siamak et al. [34] also described the attacked power system under the dynamic framework, which was an extension of the dynamic model of the methods in [13,26], and then proposed a method to detect multiple non-constant attacks on the system. By using Kalman filter, this method can estimate the data and measurements of spoofing attack with higher accuracy, and determine the phase shift of spoofing attack. In addition, the traditional measurement model assumed that the incomplete synchronization of PMU was modeled as additive noise [27,35–37]. However, according to the numerical example given in [33], if the synchronization error and/or the time between successive synchronizations increase, the traditional estimation method will deteriorate significantly.

In this paper, we consider the introduced multiplicative noise by GSAs in the proposed model, which results in the attacked residuals being greater than the residuals of the nominal system. We propose a spoofing algorithm for attacked data that classifies the attacked data and secure data, and then inputs the attacked data into the proposed data correction algorithm to obtain measurement data under secure residuals. Finally, a static estimation method is employed to estimate the system state using all the data. The main contributions of this paper can be summarized as follows:

- The PMU's measurement model under GSA attack is derived by comparing changes in bus voltage and branch current before and after the attack.
- According to the characteristic that the measurement residual of PMU data will change after being attacked, an attack detection method is proposed, which can effectively detect one or more GSAs.
- Through the particularity of the matrix in the measurement model, a bad data correction algorithm is proposed to restore the corrected attacked to the security measurements residual range. This method effectively avoids the challenge of coupling two unknown parameters in the model, which is difficult to estimate.

The rest of this paper is organized as follows. Section 3 introduces the security measurement model of PMU and derives the measurement model of PMU after being attacked by GPS spoofing. Section 4 describes the proposed algorithm and explains the simulation environment and implementation. Section 5 presents the simulation results, and our conclusions are peserented in Section 6.

3. System Model

This section summarizes the measurement model for the presence of GPS spoofing attacks in the network. The PMU measurements are correlated by derivation to the network state and the phase angle shift caused by GPS spoofing attacks.

3.1. Measurement Model

We consider a power network with N buses connected via l transmission lines, e.g., the IEEE 14-bus model, that is observed by M PMUs installed on several buses. This collection of measured quantities (in rectangular corrdinate) at bus k is concatenated in a vector $\boldsymbol{z}_k \in \mathbb{R}^{(2+2l)\times 1}$. The PMU measurements at bus k, which is connected to l different buses, are given by

$$\boldsymbol{z}_k = [U_k^r, U_k^i, I_{k1}^r, I_{k1}^i, \ldots, I_{kl}^r, I_{kl}^i]^T \tag{1}$$

where U_k^r, U_k^i denote the real and imaginary parts of the complex voltage at bus k, respectively. I_{kl}^r, I_{kl}^i are the real and imaginary parts of the complex current injected into line (k, l), T is the transpose operator. The system state $\mathbf{x} \in \mathbb{R}^{2N\times 1}$ can be written as

$$\mathbf{x} = [U_1^r, U_1^i, \ldots, U_k^r, U_k^i, \ldots, U_N^r, U_N^i]^T \tag{2}$$

Thus, using the bus admittance matrix of the network, $\boldsymbol{z}_k$ can be written as a linear function of the system state $\mathbf{x}$:

$$\boldsymbol{z}_k = \boldsymbol{H}_k\mathbf{x} + \boldsymbol{e}_k \tag{3}$$

where $\boldsymbol{z}_k$ denotes the PMU measurements at the bus k, $\boldsymbol{H}_k$ is the admittance matrix associated with a bus at bus k, and $\boldsymbol{e}_k$ denotes gauss measurement noise. Thus, the overall PMUs' measurements are given by

$$\boldsymbol{z} = \boldsymbol{H}\mathrm{x} + \boldsymbol{e} \tag{4}$$

where $\boldsymbol{z} = [\boldsymbol{z}_1, \ldots, \boldsymbol{z}_M]^T$, $\boldsymbol{H} = [\boldsymbol{H}_1, \ldots, \boldsymbol{H}_M]^T$, $\boldsymbol{e} = [\boldsymbol{e}_1, \ldots, \boldsymbol{e}_M]^T$. In this case, the estimation of the state variable $\mathbf{x}$ can be obtained from the least square estimation (LSE), which is expressed as

$$\hat{\mathbf{x}} = (\boldsymbol{H}^T\boldsymbol{H})^{-1}\boldsymbol{H}^T\boldsymbol{z} \tag{5}$$

In the network, the continuous voltage signal on the bus k when the time t is defined as

$$\bar{U}_k(t) = U_k(t)\cos(2\pi f_c t + \varphi_k(t)) \tag{6}$$

where f_c is the frequency, $\bar{U}_k$ can be expressed in the form of phasor $U_k(t)e^{j\varphi_k(t)}$, where $U_k(t)$ denotes magnitude and $\varphi_k(t)$ denotes the phase at time t. Since the subsequent analysis of the GPS spoofing attack is based on the data collected at a certain time point, we omit the symbol t in the subsequent formula for simplicity of notation. Therefore, according to its phasor form, it can be obtained that the real and imaginary parts of the complex voltage are

$$U_k^r = U_k\cos\varphi_k \tag{7}$$

$$U_k^i = U_k\sin\varphi_k \tag{8}$$

Here, a branch line is approximated using a π model, as illustrated in Figure 1. The admittance matrix relates the complex current flowing in a line with the complex voltages at the buses of the π model.

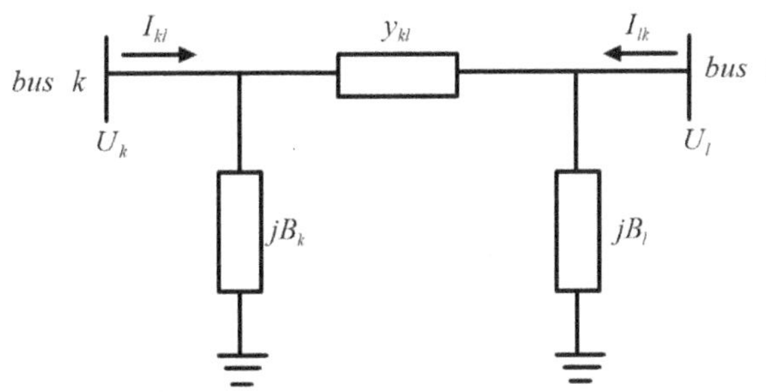

Figure 1. Bus branch model.

In Figure 1, we denote the susceptance at bus k as B_k and the admittance at branch (k,l) as y_{kl} with

$$y_{kl} = g_{kl} + jb_{kl} \tag{9}$$

where g_{kl} is the conductance and b_{kl} is the susceptance. In this paper, we assume these parameters are known and constant. Therefore, we can calculate the real and imaginary parts of the branch current as follows:

$$I^r_{kl} = (U^r_k - U^r_l)g_{kl} - (U^i_k - U^i_l)b_{kl} - B_k U^i_k \tag{10}$$

$$I^i_{kl} = (U^r_k - U^r_l)b_{kl} + (U^i_k - U^i_l)g_{kl} + B_k U^r_k. \tag{11}$$

3.2. Spoofing Attack Model

Consider attackers can manipulate the synchronization of PMUs through GPS spoofing, such that the time reference of an attacked PMU is delayed or advanced. For each attacked measurement, we consider GPS spoofing attack will shift the phase angle of the phasor z_n by an angle α_n while the phasor magnitude is unchanged [25]. Thus the attacked measurement are defined as

$$z^{spf}_n = z_n e^{j\alpha_n} \tag{12}$$

where z^{spf}_n denotes the change of measurement under attack. The resulting measurement vector $\mathbf{z}^{spf}$ is given by

$$\mathbf{z}^{spf} = (z^{spf}_1, \ldots, z^{spf}_k, \ldots z^{spf}_M)^T = \mathbf{z}^{spf} \odot \boldsymbol{w} \tag{13}$$

where $\boldsymbol{w} = (e^{j\alpha_1}, \ldots, e^{j\alpha_k}, \ldots, e^{j\alpha_M})^T$ is the attack vector, $\odot$ is the Hadamard product and if a PMU is security, the shift angle $\alpha_n = 0$, i.e., $e^{j\alpha_n} = 1$ can be obtained. Thus, the measurments are still z_n. However, the PMU installed on bus k is attacked, and the voltage measurements are

$$\bar{U}^{spf}_k = \bar{U}_k \cdot e^{j\alpha_k} = U_k \cdot e^{j(\varphi_k + \alpha_k)} \tag{14}$$

where α_k is the phase shift angle caused by the attack, and

$$\alpha_k = 2\pi f_c t_k \tag{15}$$

where t_k is the time delay of the kth bus due to a spoofing attack. Thus, by transforming (14) in the rectangular coordinates, we can obtain the real and imaginary parts of attacked voltage can be expressed as

$$\bar{U}^{spf,r}_k = U_k \cos(\alpha_k + \varphi_k) = U^r_k \cos\alpha_k - U^i_k \sin\alpha_k \tag{16}$$

and

$$\bar{U}_k^{spf,i} = U_k \sin(\alpha_k + \varphi_k) = U_k^r \sin\alpha_k + U_k^i \cos\alpha_k \tag{17}$$

Note that trigonometric identities $\cos(x+y) = \cos(x)\cos(y) - \sin(x)\sin(y)$ and $\sin(x+y) = \sin(x)\cos(y) + \cos(x)\sin(y)$ are introduced in the operation of the above formula. In addition, the delay expression of the current after being attacked are given by

$$\begin{aligned} I_{kl}^{spf,r} &= (U_k\cos(\alpha_k+\varphi_k) - U_l\cos(\alpha_k+\varphi_k))g_{kl} - (U_k\sin(\alpha_k+\varphi_k) \\ &\quad - U_l\sin(\alpha_k+\varphi_k))b_{kl} - B_kU_k\sin(\alpha_k+\varphi_k) \\ &= U_k^r(g_{kl}\cos\alpha_k - b_{kl}\sin\alpha_k - B_k\sin\alpha_k) + U_l^r(-g_{kl}\cos\alpha_k + b_{kl}\sin\alpha_k) \\ &\quad + U_k^i(-g_{kl}\sin\alpha_k - b_{kl}\cos\alpha_k - B_k\cos\alpha_k) + U_l^i(g_{kl}\sin\alpha_k + b_{kl}\cos\alpha_k) \end{aligned} \tag{18}$$

$$\begin{aligned} I_{kl}^{spf,i} &= (U_k\cos(\alpha_k+\varphi_k) - U_l\cos(\alpha_k+\varphi_k))b_{kl} + (U_k\sin(\alpha_k+\varphi_k) \\ &\quad - U_l\sin(\alpha_k+\varphi_k))g_{kl} + B_kU_k\cos(\alpha_k+\varphi_k) \\ &= U_k^r(b_{kl}\cos\alpha_k + g_{kl}\sin\alpha_k + B_k\cos\alpha_k) + U_l^r(-b_{kl}\cos\alpha_k - g_{kl}\sin\alpha_k) \\ &\quad + U_k^i(-b_{kl}\sin\alpha_k + g_{kl}\cos\alpha_k - B_k\sin\alpha_k) + U_l^i(b_{kl}\sin\alpha_k - g_{kl}\cos\alpha_k) \end{aligned} \tag{19}$$

Note that since GSA has the same effect on all signals measured by PMU at any time, phase angles of voltage and currents shift are identical [38]. Thus, the measurement of PMU installed at bus k after the spoofing attack is given by

$$z_k^{spf} = \boldsymbol{\tau}_k \boldsymbol{H}_k \mathbf{x} + \boldsymbol{e}_k \tag{20}$$

where $\boldsymbol{\tau}_k$ is a block diagonal matrix with following submatrix

$$\boldsymbol{G} = \begin{bmatrix} \cos\alpha_k & -\sin\alpha_k \\ \sin\alpha_k & \cos\alpha_k \end{bmatrix} \tag{21}$$

and the matrix $\boldsymbol{H}_k$ can be written as

$$\boldsymbol{H}_k = \begin{bmatrix} \boldsymbol{I} & 0 & \cdots & 0 \\ \boldsymbol{\Psi}_{k1} & \tilde{\boldsymbol{\Psi}}_{k1} & \cdots & 0 \\ \vdots & \vdots & \ddots & \vdots \\ \boldsymbol{\Psi}_{kl} & 0 & \cdots & \tilde{\boldsymbol{\Psi}}_{kl} \end{bmatrix} \tag{22}$$

where

$$\boldsymbol{I} = \begin{bmatrix} 1 & 0 \\ 0 & 1 \end{bmatrix} \tag{23}$$

$$\boldsymbol{\Psi}_{kl} = \begin{bmatrix} g_{kl} & B_k + b_{kl} \\ -B_k - b_{kl} & g_{kl} \end{bmatrix} \tag{24}$$

$$\tilde{\boldsymbol{\Psi}}_{kl} = \begin{bmatrix} -g_{kl} & b_{kl} \\ -b_{kl} & -g_{kl} \end{bmatrix} \tag{25}$$

Stacking (20) for all into one large model yields the measurements of all PMUs under attack as

$$z^{spf} = \boldsymbol{\phi} \boldsymbol{H} \mathbf{x} + \boldsymbol{e} \tag{26}$$

where the measurement error vector $\boldsymbol{e}$ is assumed to be Gaussian $\boldsymbol{e} \sim N(0, \delta^2 \boldsymbol{I})$, and $\boldsymbol{\phi}$ is given by

$$\boldsymbol{\phi} = \begin{bmatrix} \boldsymbol{I}_1 & & & & 0 \\ & \ddots & & & \\ & & \boldsymbol{\tau}_k & & \\ & & & \ddots & \\ 0 & & & & \boldsymbol{I}_M \end{bmatrix} \quad (27)$$

where $\boldsymbol{I}$ denotes an identity matrix and the presence of $\boldsymbol{\tau}_k$ in the diagonal element of the matrix indicates that the kth PMU is attacked. Note that $\boldsymbol{\tau}_k$ depends on α_k, when $\alpha_k = 0$ (no spoofing attack), identity matrix $\boldsymbol{\tau}_k = \boldsymbol{I}_{2l+2}$ can be obtained. Without given $\boldsymbol{\phi}$, the solution of the least squares of the state variable $\mathbf{x}$ in (26) is

$$\hat{\mathbf{x}}^{spf} = (\boldsymbol{H}^T \boldsymbol{H})^{-1} \boldsymbol{H}^T \boldsymbol{z}^{spf} \quad (28)$$

4. Detection of Attacked Data and State Estimate of System

After collecting all the PMU measurement data, we propose a residual-based detection and correction method for attacked data. The specific steps of the entire algorithm are as follows:

(1) Initialization: Use the measurements of all PMUs to estimate the power system state $\hat{\mathbf{x}}^{spf}$ and calculate the measurement residuals $\boldsymbol{r}^{spf}$.

(2) Detection of Attacked Data: Compare the residuals $\boldsymbol{r}^{spf}$ calculated in step 1 with the predetermined threshold ε. If the norm of the measurement residuals $\boldsymbol{r}^{spf}$ is greater than the predetermined threshold, we consider the PMU's measurements to be generated under attack.

(3) Correction of Attacked Data: Correct the attacked data in step 2. We first estimate the attack angle and then use the special property of the PMU measurement model under attack to easily correct the attacked data.

(4) State Estimate: Use the corrected data and the non-attacked data to estimate the power system state $\mathbf{x}$. The detection and correction of attacked data are mainly composed of two sub-algorithms: the detection of attacked data and the correction of attacked data. We will provide specific details in the following sections.

4.1. Detection of Attacked Data

In Section 3.2, we theoretically derived the measurement model of PMU under GSAs. Equation (26) show that the attack model introduces multiplicative attacked data into the security measurement model. The more simplified relationship between the measurements before and after the attack is

$$\boldsymbol{z}^{spf} = \boldsymbol{\phi} \boldsymbol{z} \quad (29)$$

Given the security power grid, the measurement residual of PMU is expressed as

$$\boldsymbol{r} = \boldsymbol{z} - \boldsymbol{H}\hat{\mathbf{x}} = \boldsymbol{z} - \boldsymbol{H}(\boldsymbol{H}^T \boldsymbol{H})^{-1} \boldsymbol{H}^T \boldsymbol{z} = (\boldsymbol{I} - \boldsymbol{H}(\boldsymbol{H}^T \boldsymbol{H})^{-1} \boldsymbol{H}^T) \boldsymbol{z} \quad (30)$$

However, when PMU in the power grid is affected by GSAs, the residual becomes

$$\boldsymbol{r}^{spf} = \boldsymbol{z}^{spf} - \boldsymbol{H}\hat{\mathbf{x}}^{spf} = \boldsymbol{z}^{spf} - \boldsymbol{H}(\boldsymbol{H}^T \boldsymbol{H})^{-1} \boldsymbol{H}^T \boldsymbol{z}^{spf} = (\boldsymbol{I} - \boldsymbol{H}(\boldsymbol{H}^T \boldsymbol{H})^{-1} \boldsymbol{H}^T) \boldsymbol{z}^{spf} \quad (31)$$

Chauhan et al. [39] revealed that when GSAs occur in the power grid, the deviation of the residual error will be caused, which is embodied as $||\boldsymbol{r}^{spf}||^2 \geq ||\boldsymbol{r}||^2$. However, a necessary condition to be met is that the matrix $(\boldsymbol{I} - \boldsymbol{H}(\boldsymbol{H}^T \boldsymbol{H})^{-1} \boldsymbol{H}^T)$ must be Semi-positive definite.

Therefore, we can set a residual threshold ε for the measurement data to identify whether certain data is affected by GSA. When the residual error of a measurement is greater than a given threshold, the algorithm considers the PMU of the measured data as being attacked, otherwise the data is secure. For setting the predetermined threshold, as we mentioned in the previous section, the measurement residual increases under GSA attack. Based on this change, we perform Monte Carlo simulations under secure power system conditions and take the maximum measurement residual as the threshold to distinguish between secure and attacked scenarios. The specific simulation process is as follows: we use MATPOWER to simulate a power system, generate PMU measurements using Equation (4), and estimate the power system state using the least squares estimate. We perform multiple simulations and record the maximum residual norm as the threshold ε.

4.2. Attacked Data Correction and System State Estimate

In Section 4.1, we used the attack detection algorithm to determine the measurements. If the norm of the measurement residuals is greater than the selected threshold, we correct the measurements in the measurement correction algorithm; otherwise, we can directly use the measurements to estimate the power system state. In this section, specific algorithm is proposed to correct the attacked data so that measurements can also be used for state estimation normally.

Equation (29) shows the relationship between the measurements of a PMU under GSA attack and the secure condition. For simplicity, we only consider an attack on one measurements.

$$\boldsymbol{\phi} = \begin{bmatrix} \cos\alpha & -\sin\alpha \\ \sin\alpha & \cos\alpha \end{bmatrix} \tag{32}$$

where α is the shift angle generated by the attack. According to the property $\cos^2\theta + \sin^2\theta = 1$ of trigonometric function and the particularity of matrix $\boldsymbol{\phi}$, (33) can be derived.

$$\boldsymbol{\phi}^T\boldsymbol{\phi} = \begin{bmatrix} \cos\alpha & \sin\alpha \\ -\sin\alpha & \cos\alpha \end{bmatrix}\begin{bmatrix} \cos\alpha & -\sin\alpha \\ \sin\alpha & \cos\alpha \end{bmatrix} = \begin{bmatrix} 1 & 0 \\ 0 & 1 \end{bmatrix} \tag{33}$$

Therefore, for the attack angles matrix introduced by the attacked measurements, we can obtain

$$\boldsymbol{\phi}^T\boldsymbol{\phi} = \boldsymbol{I} \tag{34}$$

Inspired by (34), the correction method of the attacked data can be given by

$$\boldsymbol{z}_{new} = \boldsymbol{\phi}^T\boldsymbol{z}^{spf} \tag{35}$$

where $\boldsymbol{z}_{new}$ represents a new measurement $\boldsymbol{z}_{new}$ that can be used for normal state estimation obtained by correcting the measurements of the attacked PMU(s). With these data, we can use LSE to estimate the state of the system, that is

$$\hat{\mathbf{x}} = (\boldsymbol{H}^T\boldsymbol{H})^{-1}\boldsymbol{H}^T\boldsymbol{z}_{new} \tag{36}$$

When a PMU is detected to be under attack, i.e., the residual of its measurements exceeds the given threshold, we use an iterative method to minimize these residuals $||\boldsymbol{r}_i||$ and bring this set of data back to the level where safe state estimation can be performed. To this end, the goal of the question is

$$\underset{\alpha}{minimize}\, ||\boldsymbol{r}_{max}^{spf}|| \tag{37}$$

The specific steps of the correction and safe state estimation process are as follows:

(1) If the calculated residual norm is less than the predetermined threshold, i.e., the PMU is not under attack, the algorithm outputs the same measurements, which can be directly used for state estimation.

(2) When the measurements are under attack, we estimate the attack angle by solving the objective function (37). The specific method for minimizing the objective function is as follows:

- Firstly, we select the PMU with the largest residual norm, calculate its residual.
- Then give a priori state estimation $\mathbf{x}$, and minimize the objective function with respect to the angle to estimate the attack angle.
- Finally, use the estimated attack angle to correct the phase angle of the PMU's measurements using equation (29) and update the power system state. Repeat this process until the estimation converges.

(3) Repeat from step 1 until the residual satisfies $||r^{spf}|| < \varepsilon$.

In this way, the difficult problem that the objective function (37) under the coupling of two unknown parameters is a non-convex becomes relatively simple.

We have presented the principles and implementation of the entire algorithm process for attack detection, correction of attacked data, and safe state estimation. The process is summarized in the flowchart shown in Figure 2.

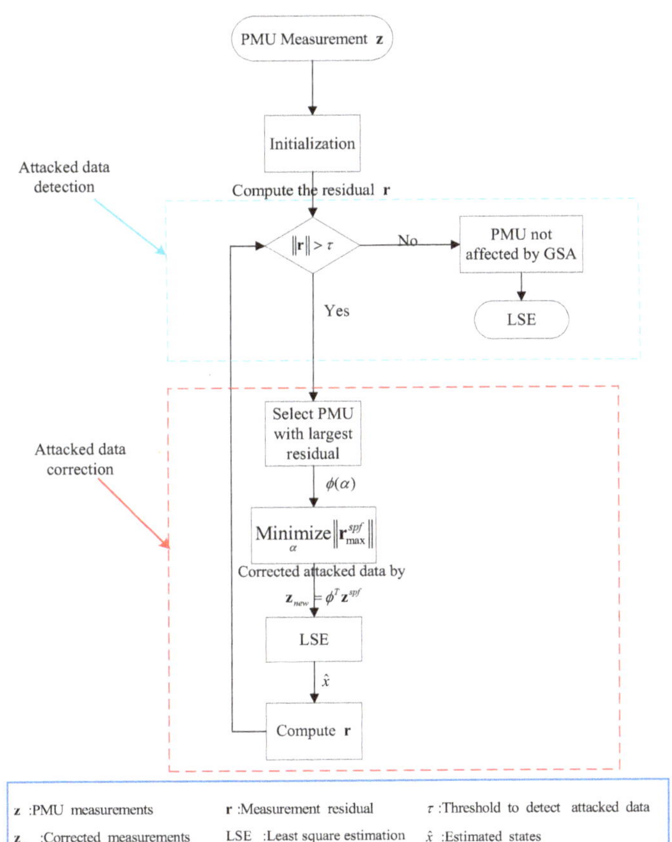

Figure 2. Flow chat of the combined attack detection, attacked data correction, and power system state estimation method.

5. Simulation Results

In this section, we conducted numerical experiments on the GSA detection, data correction, and system state estimation methods proposed in Section 3. The main simulation in this paper were conducted on the IEEE 14-bus system. The IEEE 14-bus test system (i.e., Figure 3) is a commonly used standard test system for evaluating the performance of power system load flow calculation and optimization algorithms. It is based on the IEEE standard 14-bus reference model and is widely used in research and applications in the field of power systems. The bus data, line data, and generator data of this test system all conform to the definitions and specifications of the IEEE standard, making it considered as a standard test system. It is extensively used for validating and comparing different load flow calculation algorithms, optimization algorithms, and other power system analysis methods. Using the IEEE standard test system as a benchmark ensures comparability of research and application results and provides a common platform for the academic and industrial communities. MATPOWER is a MATLAB toolbox used for power system load flow calculation and optimization. For the IEEE 14-bus test system, MATPOWER can provide the required buse data (bus number, bus types), line data (line resistance and reactance, etc.), and load flow calculation results (bus voltage magnitude and phase angle, etc.) as requested in this paper. With this data, the system model described in the paper has a standardized and realistic representation.

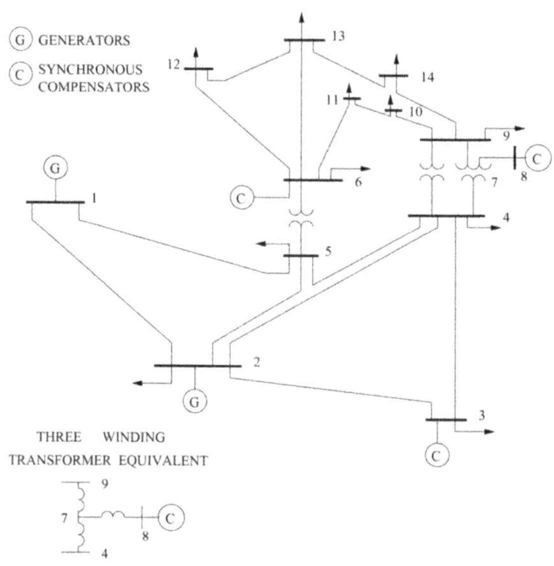

Figure 3. IEEE 14-bus system.

In addition, in the IEEE 14-bus system attack simulation, the threshold value ε that distinguishes the nominal scenario from the spoofing scenario is set to 0.0215. In all our simulations, we assume that the system bus network are observable. Table 1 provides the placement locations of PMUs for different test cases. And in this paper the measurement noise covariance of the PMUs is a diagonal matrix, with a standard deviation of 0.01 for each measurement.

Table 1. PMU buses for different IEEE bus test case.

Test Case	Number of PMUs	PMU Buses
IEEE 14	6	2,4,6,7,10,14
IEEE 30	12	1,3,5,7,9,10,12 18,24,25,27,28
IEEE 118	54	1,3,4,5,6,8,9 11,12,15,17,19,21 23,25,26,28,30,34 35,37,40,43,45,46 49,52,54,56,59,62 63,65,68,70,71,75 76,77,78,80,83,85 86,89,90,92,94,96 100,105,108,110,114

5.1. Attacked Data Detection

In this paper, attacked data detection algorithm uses the residual norm of the measurements as the threshold to distinguish whether data is attacked data generated by GSA. Therefore, the selection of this threshold is very important. We conducted 2000 Monte Carlo simulations for nominal and spoofing scenarios. In the nominal scenario, no modifications were made to the PMU measurements. The highest residual values obtained from static state estimation in these nominal scenarios were utilized as the threshold. Remarkably, during over 1000 simulations of the nominal scenarios, the occurrence rate of instances reaching the maximum residual value is estimated to be around 2 to 3 thousandths of a percent. This suggests that using this threshold to distinguish between secure and attacked data exhibits exceptional robustness. Figure 4 shows the residual norm distributions under different simulation scenarios. The results in Figure 4 clearly demonstrate that the presence of GSA causes an increase in the measurement residuals for PMUs, thus further validating the feasibility of using a set residual threshold to differentiate between attacked and non-attacked systems.

Through the Monte Carlo simulation results, we found that the maximum value of the measurement residual for PMUs under secure conditions is 0.025, which serves as the threshold for differentiating between attacked and non-attacked PMUs based on residual values. The specific method is to generate a series of PMU measurements according to the security measurement Equation (4), then use the traditional static state estimation (5) to estimate the state of the power grid, and finally calculate the measurement residual $\boldsymbol{r}$ under each simulation condition through (30), taking $r_{max} = \varepsilon = 0.0215$.

Since the GSA process requires special equipment and a long time to achieve, and the actual geographical span of the grid is huge, especially the distance between adjacent substations is long. Therefore, it is difficult to coordinate the generation of GSA in multiple locations or PMUs. Therefore, we first consider that only one PMU is subject to GSA on the system. For a single GSA, we consider that the measured phase of bus 2 has shifted by 12° due to the GSA. We record the measurements of all PMUs of the system under the security state and the attack of a single GSA in Figure 5. For multiple GSAs, we consider that the PMU measurement phase on bus 2 and bus 7 has a phase shift of 12° and 60°, respectively. Figure 6 shows a record of the measurements of all PMUs when the above attack occurs on PMUs on bus 2 and bus 7.

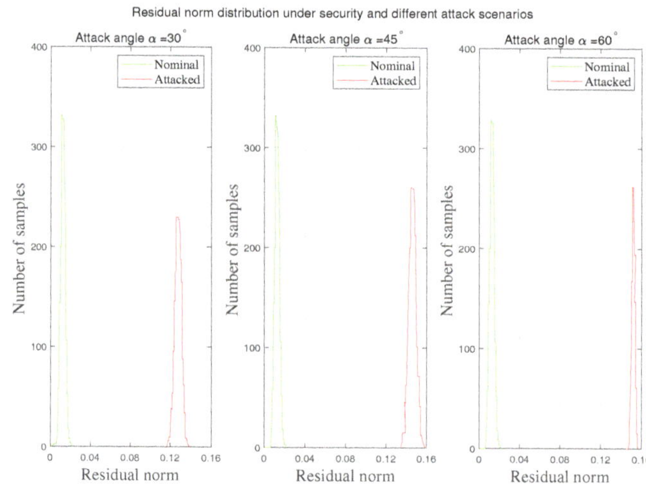

Figure 4. Residual norm distributions for both nominal and attacked scenarios.

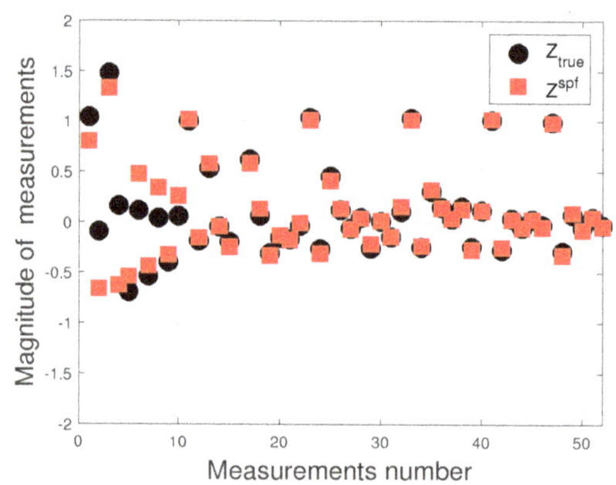

Figure 5. PMU measurements under security conditions and after the PMU on bus 2 has a 12° phase shift due to GSA.

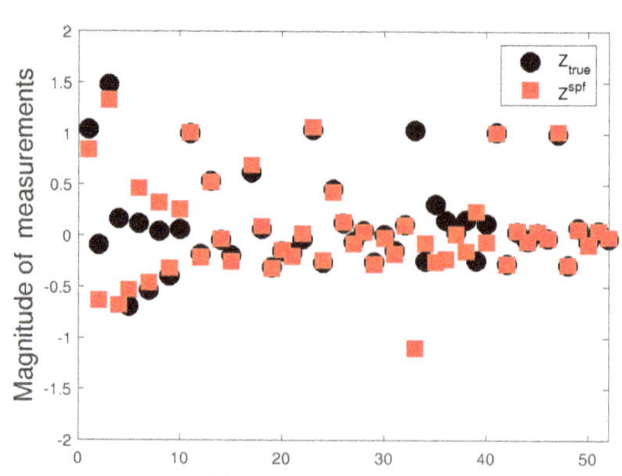

Figure 6. PMU measurements under security conditions and after the PMU on bus 2 and bus 7 have a phase shift of 12° and 60°, respectively.

Figures 5 and 6 show the measurements of the system PMU under a single GSA and multiple GSAs, and the measured results of each PMU under the security state. And the scatter diagram shows that when a PMU was attacked, the measurements will have a large deviation from the actual value, which is enough to significantly deteriorate the system state estimation. On the other hand, from the location of the data with obvious deviation, we can also accurately locate the location of the PMUs under attack. At the same time, this also verifies that it is reasonable for us to use the maximum deviation norm threshold to classify the security and attacked data within a certain range.

5.2. Correction of Attacked Data under Different Number of GSA

To observe the effects after correction, for a single GSA, we still take the PMU on bus 2 as an example, which is attacked to produce a 12° phase shift. In the measurement data detection algorithm, we can obtain a dataset of PMU measurements containing attacked data. After correcting these data with the data correction algorithm, a new set of effective measurements z_{new} can be obtained, satisfying $||r^{new}|| < \varepsilon$. The data correction process is the same for multiple GSAs. Therefore, the measurement data correction results of two different attack scenarios in Section 5.1 are shown in Figure 6 and Figure 7, respectively. The results in Figures 8 and 9 illustrated that the corrected measurements are almost identical to the true values with very small RMSE, indicating their suitability for power system state estimation for secure operation.

In addition, we apply 1–3 GSAs to the IEEE 14-bus system to study the performance of the algorithm. We take the root mean square error (RMSE) of corrected measurements as the performance metric, and RMSE is defined as $\sqrt{(\hat{\mathbf{x}} - \mathbf{x})^2/n}$, where n represents the number of data samples, $\hat{\mathbf{x}}$ represents estimated value and $\mathbf{x}$ is actual value.

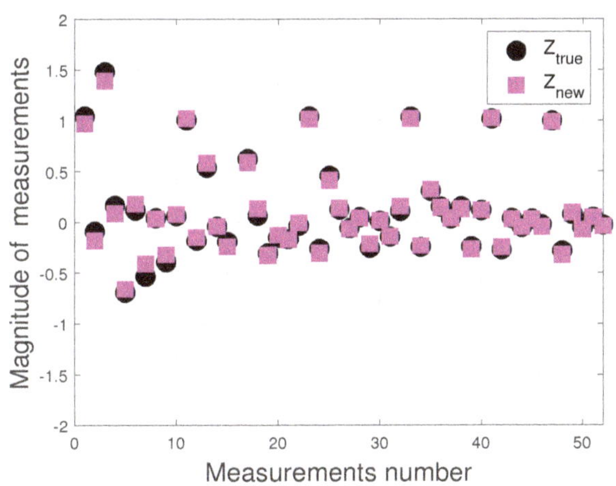

Figure 7. Scatter diagram of the measurements of 12° shift under a single GSA after data correction vs the security measurements.

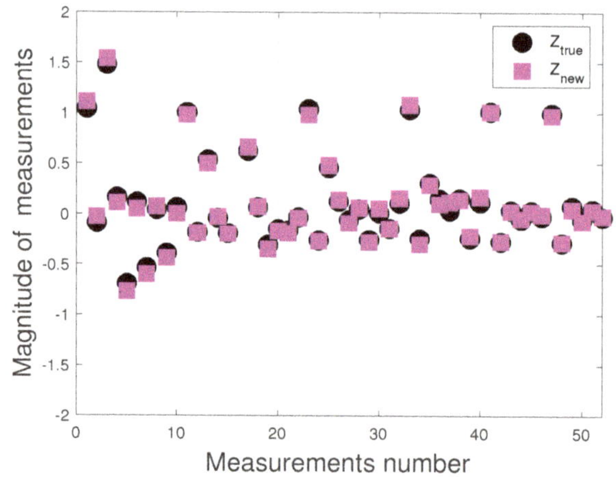

Figure 8. Scatter diagram of the measurements of 12° and 60° shift under multiple GSAs after data correction vs the security measurements.

Figure 9 shows that for the IEEE 14-bus system, the corrected measurements in this paper have a lower RMSE, and the algorithm also has good robustness with the increase of GSAs. Even when the system is subject to three GSAs at the same time, the RMSE of the corrected measurements remains below 0.02.

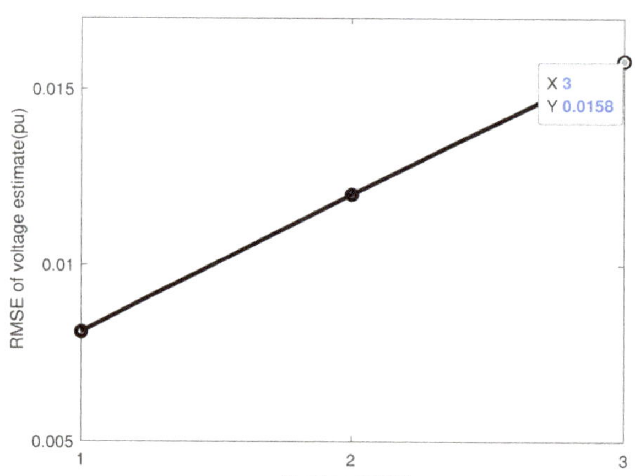

Figure 9. RMSE of corrected measurements at different GSAs.

On the other hand, in our simulation study, we examined the applicability of the algorithm to larger scale networks to showcase its performance on larger networks. We proposed algorithm on the IEEE 14, IEEE 57, and IEEE 118-bus test cases. We performed Monte Carlo simulations in which the number of GSAs are varied from 1 to 3. For each GSA, we perform 100 Monte Carlo simulations in which we randomly spoofed a given number of PMU buses with the attack angles. Table 2 presents the RMSE in the corrected measurements obtained from AM and proposed algorithms. The RMSE of proposed algorithm corrected measurements is smaller than AM for all test cases under multiple GSAs. In addition, Table 2 gave the computation time of all algorithms for 100 Monte Carlo simulations. The computation time of the proposed algorithm is less than AM in all the test cases. The computation time of the proposed algorithm increases with the increase of

GSA as it is an iterative estimator that mitigates one GSA at a time. For the scenario with 3 GSAs, the maximum computation time observed in the IEEE 118-bus test case is 0.0068 s. It is worth noting that all simulation tests were conducted in real-time calculations. This demonstrates that the proposed algorithm performs effectively when applied to large-scale networks and real-time monitoring.

Table 2. Comparison of the proposed algorithm, the computation time of the AM algorithm, and the RMSE of the measurements corrected for different GSAs under the IEEE 14, IEEE 30, and IEEE 118-bus test cases.

Test Case	Scenario	Corrected Measurements RMSE (pu) of Proposed Algorithm	Computation Time (s) of Proposed Algorithm	Corrected Measurements RMSE of AM Algorithm	Computation Time (s) of AM Algorithm
IEEE 14	1 GSAs	0.0093	0.0294	0.0413	8.4909
	2 GSAs	0.0120	0.1144	0.0429	14.6323
	3 GSAs	0.0158	0.2830	0.0522	21.3417
IEEE 30	1 GSAs	0.0026	0.1426	0.0193	87.9859
	2 GSAs	0.0219	0.2421	0.0396	94.3241
	3 GSAs	0.0413	0.3029	0.0641	107.4123
IEEE 118	1 GSAs	0.0033	0.1487	0.0137	91.3233
	2 GSAs	0.0057	0.4023	0.0366	122.4132
	3 GSAs	0.0068	0.7011	0.0539	129.5605

5.3. *State Estimate of System with Different Number of GSAs*

In the previous section, we used the attacked data correction algorithm to correct the data damaged by GSA into secure data within the specified threshold range. Then we can use these data to estimate the state of the system through (36). We compared the proposed algorithm in this paper with the Alternating minimization (AM) [20] algorithm and the traditional weighted least square (WLS) algorithm under the influence of 1–3 GSAs. The results are shown in Figures 10 and 11.

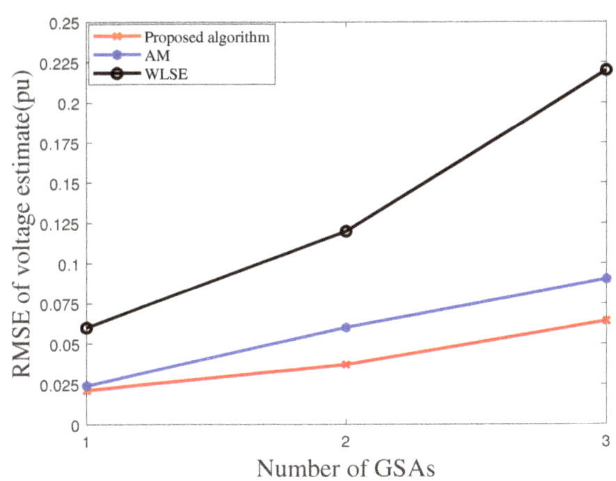

Figure 10. Proposed algorithm vs. AM and WLSE voltage estimate for different GSAs.

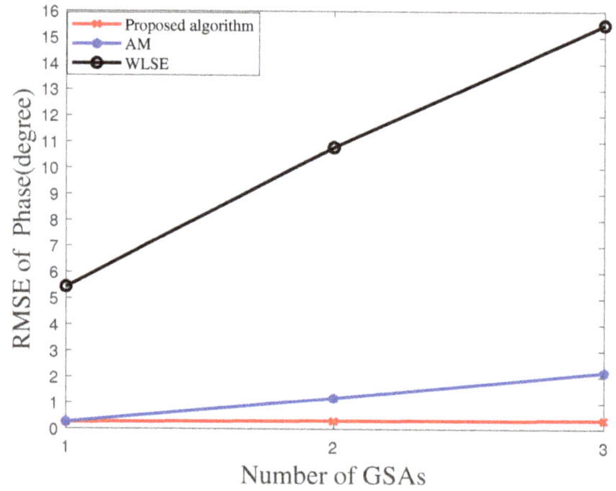

Figure 11. Proposed algorithm vs. AM and WLSE phase estimate for different GSAs.

The results in Figures 10 and 11 show that the proposed algorithm in this paper has a small RMSE for different GSAs. Even under three GSAs, the RMSE of voltage amplitude is below the same order of magnitude. The proposed algorithm can be well used in the process of system state estimation under GSAs.

6. Conclusions

This paper proposes a detection and data correction algorithm for the measurement data deviation caused by the GSAs of PMUs in the power grid, and uses the corrected data to estimate the system state. Through the simulation test on IEEE 14-bus system, we observed that the algorithm can detect the location of GSAs and the corrected data are very close to the original real measurements. And the root mean square error of the system state estimation for the corrected data is also very small, which greatly improves the estimation accuracy. In the generalized simulation, we also found that the algorithm is also applicable to larger scale networks.

Subsequent research will consider the study of the problem under different PMU placements, and develop a new joint estimation algorithm to estimate the two coupled unknown parameters in the model more accurately.

Author Contributions: Conceptualization, F.H.; methodology, Y.L. and W.G.; software, F.H. and P.H.; validation, W.G.; formal analysis, L.Q.; resources, W.G.; data curation, L.Q.; writing—original draft preparation, F.H.; writing—review and editing, W.G. and P.H.; visualization, F.H.; supervision, W.G.; project administration, Y.L.; funding acquisition, W.G. All authors have read and agreed to the published version of the manuscript.

Funding: This research was funded by the National Natural Science Foundation of China (U21A20146), Collaborative Innovation Project of Anhui Universities (GXXT-2020-070) and Open Research Fund of Anhui Province Key Laboratory of Electric Drive and Control (DQKJ202103).

Institutional Review Board Statement: Not applicable.

Informed Consent Statement: Not applicable.

Data Availability Statement: Not applicable.

Acknowledgments: We thank the anonymous reviewers for their valuable comments.

Conflicts of Interest: The authors declare no conflict of interest.

References

1. Singh, R.S.; vanden Brom, H.; Babaev, S.; Cobben, S.; Ćuk, V. Estimation of Impedance and Susceptance Parameters of a 3-Phase Cable System Using PMU Data. *Energies* **2019**, *12*, 4573. [CrossRef]
2. James, J.Q.; Hill, D.J.; Li, V.O.; Hou, Y. Synchrophasor recovery and prediction: A graph-based deep learning approach. *IEEE Internet Things J.* **2019**, *6*, 7348–7359. [CrossRef]
3. Gou, B. Generalized Integer Linear Programming Formulation for Optimal PMU Placement. *IEEE Trans. Power Syst.* **2008**, *23*, 1099–1104. [CrossRef]
4. Terzija, V.; Valverde, G.; Cai, D.; Regulski, P.; Madani, V.; Fitch, J.; Skok, S.; Begovic, M.M.; Phadke, A. Wide-Area Monitoring, Protection, and Control of Future Electric Power Networks. *Proc. IEEE* **2011**, *99*, 80–93. [CrossRef]
5. Matsumoto S.; Serizawa Y.; Fujikawa F. Wide-Area Situational Awareness (WASA) system based upon international standards. In Proceedings of the 11th IET International Conference on Developments in Power Systems Protection (DPSP 2012), Birmingham, UK, 23–26 April 2012; pp. 1–6. [CrossRef]
6. Xie, N.; Torelli, F.; Bompard, E.; Vaccaro, A. A graph theory based methodology for optimal PMUs placement and multiarea power system state estimation. *Electr. Power Syst. Res.* **2015**, *119*, 25–33. [CrossRef]
7. Xia, N.; Gooi, H.B.; Chen, S.X.; Wang, M.Q. Redundancy based PMU placement in state estimation. *Sustain. Energy Grids Netw.* **2015**, *2*, 23–31. [CrossRef]
8. Hongga, Z. A New State Estimation Model of Utilizing PMU Measurements. In Proceedings of the 2006 International Conference on Power System Technology, Chongqing, China, 22–26 October 2006; pp. 1–5. [CrossRef]
9. Srdjan, S.; Igor, I.; Zdeslav, C. Hybrid State Estimation Model Based on PMU and SCADA Measurements. *IFAC-PapersOnLine* **2016**, *49*, 390–394. [CrossRef]
10. Ghahremani, E.; Kamwa, I. Local and Wide-Area PMU-Based Decentralized Dynamic State Estimation in Multi-Machine Power Systems. *IEEE Trans. Power Syst.* **2016**, *31*, 547–562. [CrossRef]
11. Ioannides, R.T.; Pany, T.; Gibbons, G. Known Vulnerabilities of Global Navigation Satellite Systems, Status, and Potential Mitigation Techniques. *Proc. IEEE* **2016**, *104*, 1174–1194. [CrossRef]
12. Antončič, M.; Papič, I.; Blažič, B. Robust and Fast State Estimation for Poorly-Observable Low Voltage Distribution Networks Based on the Kalman Filter Algorithm. *Energies* **2019**, 12, 4457. [CrossRef]
13. Zhao, J.; Gómez-Expósito, A.; Netto, M.; Mili, L.; Abur, A.; Terzija, V.; Kamwa, I.; Pal, B.; Singh, A.K.; Qi, J.J.; et al. Power System Dynamic State Estimation: Motivations, Definitions, Methodologies, and Future Work. *IEEE Trans. Power Syst.* **2019**, *34*, 3188–3198. [CrossRef]
14. Zhao, J.B.; Netto, M.; Huang, Z.Y.; Yu, S.S.; Gómez-Expósito, A.; Wang, S.B.; Kamwa, I.; Akhlaghi, S.; Mili, L.; Terzija, V.; et al. Roles of Dynamic State Estimation in Power System Modeling, Monitoring and Operation. *IEEE Trans. Power Syst.* **2021**, *36*, 2462–2472. [CrossRef]
15. Lin, C.H.; Wu, W.C.; Guo, Y. Decentralized Robust State Estimation of Active Distribution Grids Incorporating Microgrids Based on PMU Measurements. *IEEE Trans. Smart Grid.* **2020**, *11*, 810–820. [CrossRef]
16. Enge, P.P. The Global Positioning System: Signals, measurements, and performance. *Int. J. Wirel. Inf. Netw.* **1994**, *1*, 83–105. [CrossRef]
17. Dempster, A.G.; Cetin, E. Interference Localization for Satellite Navigation Systems. *Proc. IEEE* **2016**, *104*, 1318–1326. [CrossRef]
18. Vyskocil, P.; Sebesta, J. Relative timing characteristics of GPS timing modules for time synchronization application. In Proceedings of the International Workshop on Satellite and Space Communications, Siena, Italy, 9–11 September 2009; pp. 230–234. [CrossRef]
19. Fan, Y.; Zhang, Z.; Trinkle, M.; Dimitrovski, A.D.; Song, J.B.; Li, H. A Cross-Layer Defense Mechanism Against GPS Spoofing Attacks on PMUs in Smart Grids. *IEEE Trans. Smart Grid.* **2015**, *6*, 2659–2688. [CrossRef]
20. Risbud, P.; Gatsis, N.; Taha, A. Vulnerability Analysis of Smart Grids to GPS Spoofing. *IEEE Trans. Smart Grid.* **2019**, *10*, 3535–3548. [CrossRef]
21. Zhang, Q.; Vittal, V.; Heydt, G.; Chakhchoukh, Y.; Logic, N.; Sturgill, S. The time skew problem in PMU measurements. In Proceedings of the 2012 IEEE Power and Energy Society General Meeting, PES 2012, San Diego, CA, USA, 22–26 July 2012. [CrossRef]
22. Navid, R.; Lotfollah, S. A Compact Microstrip Patch Antenna for Civilian GPS Interference Mitigation. *IEEE Antennas Wireless Propagat. Lett.* **2018**, *17*, 381–384. [CrossRef]
23. Bonebrake, C.; O'Neil, L.R. Attacks on GPS Time Reliability. *IEEE Secur. Priv.* **2014**, *12*, 82–84. [CrossRef]
24. Akkaya, I.; Lee, E.A.; Derler, P. Evaluation of the vulnerability of phasor measurement units to GPS spoofing attacks. In Proceedings of the 2013 Workshop on Modeling and Simulation of Cyber-Physical Energy Systems (MSCPES), Berkeley, CA, USA, 20 May 2013; pp. 1–6. [CrossRef]
25. Jiang, X.; Zhang, J.; Harding, B.J.; Makela, J.J.; Domínguez-García, A.D. Spoofing GPS Receiver Clock Offset of Phasor Measurement Units. *IEEE Trans. Power Syst.* **2013**, *28*, 3253–3262. [CrossRef]
26. Liang, G.; Zhao, J.; Luo, F.; Weller, S.; Dong, Z.Y. A Review of False Data Injection Attacks Against Modern Power Systems. *IEEE Trans. Smart Grid.* **2017**, *8*, 1630–1638. [CrossRef]
27. Mahapatra, K.; Chaudhuri, N.R.; Kavasseri, R. Bad data detection in PMU measurements using principal component analysis. In Proceedings of the 2016 North American Power Symposium (NAPS), Denver, CO, USA, 18–20 September 2016; pp. 1–6. [CrossRef]

28. Zhang, Y.; Wang, J.; Lui, J. Attack Identification and Correction for PMU GPS Spoofing in Unbalanced Distribution Systemss. *IEEE Trans. Smart Grid.* **2017**, *11*, 762–773. [CrossRef]
29. Mina, T.Y.; Bhamidipati, S.; Gao, G.X. Detecting GPS Spoofing via a Multi-Receiver Hybrid Communication Network for Power Grid Timing Verification. In Proceedings of the 31st International Technical Meeting of the Satellite Division of the Institute of Navigation (ION GNSS+ 2018), Miami, FL, USA, 24–28 September 2018. [CrossRef]
30. Bhamidipati, S.; Gao, G.X. GPS multireceiver joint direct time estimation and spoofer localization. *IEEE Trans. Aerosp. Electron. Syst.* **2019**, *55*, 1907–1919. [CrossRef]
31. Abbaspour, A.; Sargolzaei, A.; Forouzannezhad, P.; Kang, K.Y.; Sarwat, A.I. Resilient Control Design for Load Frequency Control System Under False Data Injection Attacks. *IEEE Trans. Ind. Electron.* **2020**, *67*, 7951–7962. [CrossRef]
32. Yasinzadeh, M. Detection of PMU Spoofing in Power Grid Based on Phasor Measurement Analysis. *IET Transm. Distrib.* **2018**, *12*, 1980–1987. [CrossRef]
33. Yang, P.; Tan, Z.; Wiesel, A. Power System State Estimation Using PMUs With Imperfect Synchronization. *IEEE Trans. Power Syst.* **2013**, *28*, 4162–4172. [CrossRef]
34. Siamak, S.; Dehghani, M.; Mohammadi, M. Dynamic GPS Spoofing Attack Detection, Localization, and Measurement Correction Exploiting PMU and SCADA. *IEEE Syst. J.* **2021**, *15*, 2531–2540. [CrossRef]
35. Jain, A.; Shivakumar, N.R. Impact of PMU in dynamic state estimation of power systems. In Proceedings of the 2008 40th North American Power Symposium, Calgary, AB, Canada, 28–30 September 2008; pp. 1–8. [CrossRef]
36. Faranttatos, E.; Stefopoulos, G.K.; Cokkinides, G.J.; Meliopoulos, A.P. PMU-based dynamic state estimation for electric power systems. In Proceedings of the 2009 IEEE Power and Energy Society General Meeting, Calgary, AB, Canada, 26–30 July 2009; pp. 1–8. [CrossRef]
37. Kekatos, V.; Giannakis, G.B. An Alternative for Including Phasor Measurements in State Estimators. *IEEE Trans. Power Syst.* **2013**, *28*, 1617–1626. [CrossRef]
38. Shereen, V.; Ramakrisshan, R.; Dan, G. Detection and Localization of PMU Time Synchronization Attacks via Graph Signal Processing. *IEEE Trans. Power Syst.* **2022**, *13*, 3241–3254. [CrossRef]
39. Chauhan, S.; Gao, G.X. Synchrophasor Data Under GPS Spoofing: Attack Detection and Mitigation Using Residuals. *IEEE Trans. Smart Grid.* **2021**, *12*, 3415–3424. [CrossRef]

MDPI
St. Alban-Anlage 66
4052 Basel
Switzerland
www.mdpi.com

Energies Editorial Office
E-mail: energies@mdpi.com
www.mdpi.com/journal/energies

www.ingramcontent.com/pod-product-compliance
Lightning Source LLC
Chambersburg PA
CBHW041940260926
51850CB00018B/319
9783036599137